Modern Nonlinear Equations

Thomas L. Saaty

University Professor, University of Pittsburgh

DOVER PUBLICATIONS, INC., NEW YORK

To ROZANN

NOTE TO THE 1981 REVISION

The first two chapters of this revision have been carefully read and corrected by my friend Dr. M. Z. Nashed of the University of Delaware. For this I owe him my deep gratitude.

T. L. S.

Published in Canada by General Publishing Company, Ltd., 30 Lesmill Road, Don Mills, Toronto, Ontario.
Published in the United Kingdom by Constable and Company, Ltd., 10 Orange Street, London WC2H 7EG.

This Dover edition, first published in 1981, is a revised republication of the work originally published in 1967 by the McGraw-Hill Book Company, N.Y.

International Standard Book Number: 0-486-64232-1
Library of Congress Catalog Card Number: 81-68481

Manufactured in the United States of America
Dover Publications, Inc.
180 Varick Street
New York, N.Y. 10014

PREFACE

Equations are the lifeblood of applied mathematics and science. This book and its companion volume† bring together the widest coverage of the field of equations at present available for the scientist, technologist, and mathematician. From the theoretical standpoint it provides a perspective of the field of equations, discusses new techniques from the literature, and outlines recent developments. From the applied standpoint it will help the reader acquire new models for formulating problems, acquaint him with the variety of existing types of equations, and give him some methods for their solution. Thus the book should prove to be of value not only to the teacher and student but to the research worker as well. Our approach of treating an entire subject in a single chapter is intended to stimulate interest in the subject and some of its techniques. Needless to say, it does not elaborate exhaustively on each technique; references are given for that purpose.

As before, our emphasis is on nonlinear equations because this is what is needed today. As we said in the preface to "Nonlinear Mathematics":

> Although we are impressed with the power and diversity of linear methods, we must respond to the challenge to find, in the widely scattered literature, effective ways of viewing nonlinear problems. One such method is the iterative procedure based on the fixed-point property of contraction mappings, which has provided us with a versatile technique made effective by modern high-speed computing devices that can rapidly perform step-by-step procedures. However, in special cases an abundance of other ingenious techniques can be found.

† Thomas L. Saaty and Joseph Bram, "Nonlinear Mathematics," McGraw-Hill Book Company, New York, 1964, for which we use the abbreviation "NM" in cross references.

Preface

To the connoisseur, progress in existence theory and in the construction of solutions of equations is just as exciting and impressive as any scientific discovery. Each new result little by little furthers the growth of the whole field of equations. In this book we continue our presentation of equations begun in the companion volume, bringing the total number of different types of equations considered to eight.

Algebraic and differential equations were covered in "Nonlinear Mathematics." The others are treated here, viz., difference, integral, and functional and such combinations of these as delay-differential and integrodifferential equations. Thus with the two volumes we have covered all the major types of classical equations except partial differential equations, which require a volume to themselves. With ideas from probability theory, our deterministic equations become stochastic, e.g., stochastic differential equations, which are covered in Chap. 8. The existing literature on many of these subjects is still fragmentary.

Particular emphasis has been placed on linear and nonlinear equations in function spaces and on general methods of solving different types of such equations. The first two chapters supplement Chap. 1 and the appendix of "Nonlinear Mathematics" to provide the student with a background of functional analysis as it applies to equations. The reader may skip the first two chapters and some sections of remaining chapters and still learn about various equations studied without the use of terminology from functional analysis. See for example Chaps. 3, 4, 7, and 8.

Functional analysis is the study of functions defined on abstract sets. The latter may vary in their structure and the nature of their elements, but it is possible to define on them the notion of convergence, limit, etc. That is, in addition to their algebraic properties of addition, scalar multiplication, and sometimes multiplication even of their elements, they have topological properties. Functional analysis derives its name from functionals, which are functions from an abstract space to the real (or complex) numbers. An ordinary function is a special case of a functional (see Chap. 1). In its simplicity this subject provides depth, unity, and a framework for the transfer of techniques in mathematical analysis. It has been one of the great incentives for generalization. The more of it given to the student, the better prepared he will be to face nonlinear problems.

In functional analysis, in addition to an integrated approach to the vast field of classical analysis, one studies problems and classifies solutions according to the function spaces in which these solutions are required. Even when classical theorems have been extended and generalized, functional analysis can offer proofs that are simple and elegant. Systems of equations or inequalities are treated as a single equation or inequality. From a practical standpoint, functional analysis serves as a pivot for transferring methods of constructing solutions, e.g., Newton's method, from one type of equation to another, such as from ordinary algebraic equations to integral equations. Another advantage of functional analysis is illustrated in the fact that existence and embedding theorems for differential equations are special cases of a general implicit-function theorem (see "Nonlinear Mathematics"). The reader is urged to examine the excellent expository papers by Cooper, Ficken, Graves, and Hewitt and other references at the end of Chap. 1. Many types of equations of classical analysis are still in an embryonic stage, since before a field can be subsumed by functional analysis, its own theory and methods must have reached an adequate stage of development.

In writing this book we have in mind particularly the student who may not have been exposed to equations of the types illustrated here. We therefore concentrate on exemplifying and giving a broad exposure to each subject in a balance of solution methods and existence theorems, paying particular attention to the former. Stability theory for some of these topics has not yet been developed coherently; however, we do illustrate some ideas and techniques from stability theory.

From this book and "Nonlinear Mathematics" it should be possible to conduct a full year course on nonlinear mathematics when supplementary material from the references is included.

Chap. 6, Integral Equations, was written by Professor Donald Hyers of the University of Southern California, who has lectured on this topic in my two-week courses on nonlinear analysis at the University of California, Los Angeles. Chap. 8, Stochastic Differential Equations, was written by my colleague and coinstructor in the above-mentioned courses, Professor Ryszard Syski, of the University of Maryland.

I am particularly grateful to Joseph Bram for penetrating suggestions and to J. Alderman, Harold Anderson, John Clark, Rodney Driver, James Enstrom, Jack Hale, Gordon Heffron, I. Heller, Einar Hille, Zuhair M. Nashed, Robert J. Oberg, Roger H. Prager, John Turtora, G. S. Wheeler, Robert Whitley, and David Wood for many helpful comments and suggestions.

Thomas L. Saaty

CONTENTS

Contents

Contents

ONE

BASIC CONCEPTS IN THE SOLUTION OF EQUATIONS

1-1 OPERATOR EQUATIONS†

A serious and indispensable pursuit of mathematics is concerned
with the existence and construction of solutions of equations and
inequalities.   Since applied mathematics is concerned mainly with
representations of the real world (or certain aspects of it), and
since these representations are in the form of equations which
essentially describe the laws governing a physical system, it is
clear that once the representation has been achieved, the main
objective of (applied) mathematics is to prove existence and find
methods of solution.   Equations, for example, can be given in
explicit or in implicit form, with or without auxiliary conditions.
This chapter and the next provide a broad view of equations from
the point of view of functional analysis, which is an abstract

† The author is grateful to several students, particularly G. S. Wheeler,
for help in preparing some of the material of this chapter.

theory of functions. This material will be better understood if it is studied in conjunction with some standard elementary text on functional analysis, e.g., that of Vulikh,[67] Simmons,[60] or Goffman and Pedrick.[19] In this section we shall not dwell on definitions as we take a brief look at the modern terrain of operators and equations. This is perhaps the most condensed chapter of the book, as compared with Chap. 3, for example, which is more classical in style.

An equation is an expression of a correspondence between elements of a space, e.g., points of euclidean space or continuous functions on a closed interval, and other elements from the same space or from another. Using the properties of the transformation that defines the correspondence and the properties of the spaces themselves, one then attempts to prescribe conditions for the solvability of the equation. If the transformation is linear, a solution can sometimes be obtained by finding the inverse transformation. If the transformation is nonlinear, a solution, if it exists, can occasionally be obtained by means of convergent recursive applications of the transformation. Conditions for existence of solutions and their construction by inversion (rare) and by iteration (more common) are important. Other techniques embed the problem in a more general one and obtain the solution of the given problem as a particular case. Additional methods are mentioned in the introduction to Chap. 2.

Some of the most powerful tools for the treatment of systems of equations of any kind are those found in the field of functional analysis. Many existence theorems are best phrased in the language of this subject. Many construction techniques used in this field, whether they are iterative, direct, variational, or approximations of other kinds, are themselves generalizations of ideas developed in the solution of specific types of equations. Generalizations have been made of Newton's method ordinarily used to solve algebraic equations, of Picard's method in differential equations, of the methods of Ritz in integral equations, and so on. Although the generalizations are extremely useful and can be transferred to other types of equations, much inventiveness in the analysis of equations arises in the study of specific types and even of specific examples of equations.

Functional analysis provides unity since it regards ordinary function spaces, e.g., the real-line, n-dimensional euclidean space and the space of Lebesgue-integrable functions, as special cases of a more abstract space. This permits the introduction of a metric (and hence a topology) and the analysis of convergence and limits. The existence and techniques of solution of particular types of equations are then examined in this general context, which also permits the transfer of specific knowledge obtained from the analysis of one kind of equation to another, e.g., from ordinary to integral equations. Let us now repeat some of the foregoing concepts in slightly more specific terms.

A transformation T is linear if $T(\alpha x_1 + \beta x_2) = \alpha T x_1 + \beta T x_2$; that is, it is both additive and homogeneous. Here α and β are real or complex scalars, and x_1 and x_2 belong to the space X on which T acts (see Sec. 1-2 for details). Many mathematicians in the Soviet Union define T to be linear if it fulfills the conditions of additivity, that is,

$$T(x_1 + x_2) = T x_1 + T x_2$$

and continuity, that is, $x_n \rightarrow x$ implies $T x_n \rightarrow T x$. Homogeneity, that is, $T(\alpha x) = \alpha T x$, $T(ix) = iT x$, $i = \sqrt{-1}$, for all x, where α is a scalar, follows from these two properties.

An operator that is not linear is called *nonlinear*. It is important to note that in all problems considered here the operator may be nonlinear but the space of elements on which it operates, i.e., its domain, is a linear space. This is also true of the space into which these elements are mapped by the operator, i.e., its range.

Examples of specific forms of classical equations, of which those studied here are abstractions, are ordinary algebraic equations, ordinary and partial differential equations, difference equations, integral equations, e.g., those of Volterra, Fredholm, and Hammerstein, functional equations, and mixtures of these types. With the exception of partial differential equations, later chapters of this book, like "Nonlinear Mathematics," are concerned with nonlinear classical equations of these types.

Functional analysis has several basic precursors in classical analysis, two of which we mention. (1) Functionals (first investigated by Volterra in 1887) are encountered in the analysis of variational problems, e.g., minimize $I(y) = \int_a^b \sqrt{1 + y'^2} \, dx$, where I gives the length of the curve $y(x)$ defined on the interval $[a,b]$. Analysis of the existence of solutions requires a theory of limits and convergence on spaces of functions and also requires a generalization of the idea that a continuous function on a compact set must attain a minimum. (2) Another precursor is found in operational calculus, e.g., the use of the derivative operator D in algebraic manipulations of a differential equation. By a generalization of this idea in differential equations and in integral equations (as demonstrated by Fredholm and Volterra) it is possible to investigate the solvability of linear differential and integral equations by analogy with an algebraic system of equations.

A simple way of motivating the concept of an operator is to examine matrix operators first. Here A is a transformation which maps n-dimensional vectors to other n-dimensional vectors. From a system of ordinary linear equations $\sum_{j=1}^{n} a_{ij}x_j = y_i$, $i = 1, \ldots, n$, one abstracts, by means of the notion of a square $n \times n$ matrix (see the discussion in Sec. 1-3 on pseudo inverses for a more general view of matrix equations) and of n vectors x and y, to equations having the operator form $Ax = \mathbf{0}$, $Ax = y$, $Ax = \lambda x$ [or simply $(A - \lambda I)x = \mathbf{0}$], and $(A - \lambda I)x = y$, where I is the identity matrix and λ is a complex number.

The first of these, $Ax = \mathbf{0}$, is called a *homogeneous equation* and is known to have a nontrivial solution, that is, $x \neq \mathbf{0}$, if and only if A is a singular matrix. The set of solutions of $Ax = 0$ is a subspace of the space X. This subspace is called the *null space* of A and is denoted by N_A. The second equation, $Ax = y$, is known as a *nonhomogeneous* equation. It has a solution if and only if y is in the range of A, that is, if A has the same rank before and after it is augmented with the additional column y. Clearly, if A is singular, then $N_A \neq \{0\}$.

The third equation, $(A - \lambda I)x = \mathbf{0}$, called a *homogeneous eigenvalue equation*, has a nontrivial solution if and only if λ satisfies the *characteristic equation* of A obtained uniquely by equating the determinant of the matrix $(A - \lambda I)$ to zero, thus ensuring that $(A - \lambda I)$ be singular. Such a λ is known as a *characteristic* or *eigenvalue* of A.

To solve a *nonhomogeneous eigenvalue equation* of the form

$$(A - \lambda I)x = y$$

where y is given, one must exclude the characteristic values of A, so that $(A - \lambda I)$ has an inverse, and thus obtain $x = (A - \lambda I)^{-1}y$. Consequently a unique solution exists if and only if λ is not an eigenvalue of A.

Similar ideas can be applied to the analysis of solutions of a linear differential operator on functions on an interval

$$I \equiv \{t \mid a \le t \le b\} \qquad x \to Tx = \sum_{k=0}^{n} a_k(t)x^{(k)}(t)$$

with $a_n(t) \neq 0$ on I, all $a_k(t)$ continuous, and $x^{(k)}(t)$ the kth derivative of $x(t)$. The domain of T may be the space C^n of functions with n continuous derivatives on I, and its range C^0, the space of continuous functions on I.

Exercise 1-1   Discuss the solvability of the equations $Tx = 0$ and $Tx = y$ using the well-known notions of a complementary function and of a particular integral.

Remark: Consider now two equations in two different linear operators, $Tu = y$ and $Su = \eta$, with y and η given, e.g., a differential equation and its boundary conditions. Suppose that one has u_0 such that $Su_0 = \eta$.

Let $u = u_0 + v$, thus obtaining for v

$$Tv = y - Tu_0$$
$$Sv = 0$$

Hence u is found in the affine subspace determined by translating the null space of S by u_0. This shows also that one can reduce, for example, a boundary value problem for a linear differential equation with non-homogeneous boundary conditions to an equivalent problem with homogeneous boundary conditions.

Now let us look briefly at another example. The equation $dx(t)/dt = 0$, with $x(t)$ a real differentiable function in t whose derivative is continuous, has a nontrivial solution if and only if $x(t)$ is constant. Here d/dt transforms the subspace of functions with a continuous derivative into the space of continuous functions. The equation $dx/dt - \lambda x = 0$ has, for example, continuous nontrivial solutions of the form $e^{\lambda t}$ belonging to the space of continuous bounded functions on $[0, \infty]$ including continuity at ∞ (denoted by $C[0, \infty]$) if and only if the real part of λ is negative or if $\lambda = 0$. For these λ, also called eigenvalues of d/dt, the nonhomogeneous equation $(d/dt - \lambda I)x = y$ need not have a unique solution. Here I is the identity operator in the space of functions whose first derivative is continuous, x is an element of this space, and y is an element of the space of continuous functions $C[0, \infty]$. But the exclusion of these λ does not by itself ensure that the inverse operator $(d/dt - \lambda I)^{-1}$ exists and is also bounded. Thus, for example, if $\lambda = i\alpha$, $\alpha \neq 0$ real, a solution of the homogeneous equation takes the form $e^{i\alpha t} = \cos \alpha t + i \sin \alpha t$ and does not belong to $C[0, \infty]$ because, as $t \to \infty$, it does not tend to a definite value. Thus it turns out that even for these values of λ the nonhomogeneous equation has no solution of the required form. Such values of λ for which the nonhomogeneous equation has no unique solutions are known as *spectral values* and are said to belong to the spectrum of the operator $T = d/dt$. The remaining values for which there is a bounded inverse $(T - \lambda I)^{-1}$ whose domain is dense are said to belong to the resolvent set of T. For these values $(T - \lambda I)^{-1}$, called the *resolvent of T*, exists. Thus the analogy with matrices does not give the full answer. The reader can attempt to find the answer by investigating the nonhomogeneous equation for solutions in $C[0, \infty]$, but later, when we discuss the question of inverse operators, the ideas will become clearer. What about solutions in $C[0,1]$, or in $L(0,1)$, the space of Lebesgue-integrable functions on $(0,1)$?

It is often important to know the type of solution desired; i.e., given $y \in Y$ find $x \in X$ with x having given properties such that $Tx = y$.

The study of the spectrum of an operator leads to far-reaching results. The reader may recall Sylvester's theorem for matrices (see "NM,"† page 194), which enables one to calculate a variety of functions of a matrix in a rather simple way, once its characteristic values are known. Thus, for example, one can compute e^{At} or A^m for any m. A similar theory is available for special types of linear bounded operators.

Exercise 1-2 Show that the matrix $A = \begin{pmatrix} 3 & 0 \\ 2 & 2 \end{pmatrix}$ has the eigenvectors $\begin{pmatrix} 0 \\ 1 \end{pmatrix}$ and $\begin{pmatrix} 1 \\ 2 \end{pmatrix}$ with corresponding eigenvalues 2 and 3. Raise this matrix to the power 10^{10}. Also compute e^{At}.

An important class of nonlinear problems, which we do not discuss here, is that of optimization or extremum problems, in which it is desired to find an element φ of an appropriate space which maximizes or minimizes a functional $F(\varphi)$ subject to constraints of the form $T\varphi = f$ and $T\varphi \leq g$, etc., where T is a known operator and f and g are given elements of the same space as φ or another space. Such problems are meaningful when studied in partially ordered spaces. The problems are, of course, generalizations of extremum problems in the calculus, in the calculus of variations, in the theories of linear and nonlinear programming, and in the theory of optimum control.

The analysis of equations and that of optimization problems are closely related. Some criteria of extremum problems lead to the solution of equations, e.g., Euler's equation in the calculus of variations. Conversely, there are problems involving equations, e.g., integral equations, which are more conveniently treated through variational techniques, a procedure successfully applied by Hammerstein to certain integral equations.

In chap. 1 of Ref. 58 there is a sketch of some ideas leading to nonlinear transformations operating on elements of a complete normed linear vector space, known more simply as a *Banach space*. Banach spaces are defined, and some fixed-point theorems are given as an aid in proving the existence of solutions of operator equations. Of particular importance is the use of linear functionals and the associated weak topology. By means of linear functionals and the Hahn-Banach theorem, it is possible to generalize properties which hold on linear submanifolds to an entire

† As noted in the preface, the abbreviation "NM" will be used to refer to the companion volume, Thomas L. Saaty and Joseph Bram, "Nonlinear Mathematics," McGraw-Hill Book Company, New York, 1964. (Dover, 1981)

Banach space. Also given is a brief account of the spectral decomposition of a self-adjoint bounded linear operator. Here we go even further into the existence and construction of solutions of operator equations.

In this and in the next chapters, we essentially have three major topics to discuss.

1. The question of inversion of operators.
2. Existence and uniqueness, through local and global theorems in the case of implicit functions and through algebraic and topological theorems in the case of equations in explicit form.
3. Solution by iterative and direct methods in Banach space and in Hilbert space. We shall be particularly concerned with questions of convergence, approximation, and error.

Both ordinary and proper-value equations having the following general explicit forms will also be studied:

1. $Tx = y$.
2. $Tx = 0$.
3. $Tx = x$ or $(T - I)x = 0$.
4. $x = Tx + y$.
5. $x = \epsilon Tx$ (ϵ small).
6. $Ax + Bx = y$.
7. $x + KFx = Tx$ (Hammerstein).
8. $x - ATx = y$ (potential).
9. $x = \lambda Tx + y$.

Note that an operator equation may involve several different operators and may take on a variety of forms, depending on the usefulness and generality for which the equation is studied.

1-2   REVIEW OF BASIC IDEAS

For furture reference, we assemble here those basic concepts for analysis which will find use in later chapters, particularly in dealing with the existence and construction of solutions of equations.

When we write an equation, we automatically presuppose that elements of a certain space are mapped into those of another. Thus we seek solutions in the appropriate space or subspace on which the transformation operates. For instance, to find all solutions to a system of the form $Ax = y$ amounts to finding the complete inverse image of y under the mapping A. (For example, if A is a linear transformation, this complete inverse is an affine subspace.) We shall more or less presuppose knowledge of the material and motivation included in chap. 1 of "NM." In addition to the algebraic properties of the spaces with which we shall be concerned, we shall also need topological properties which enable us to analyze convergence and the existence of limits. On the real line these properties are stated in five equivalent (and famous) theorems given in the next chapter. The idea of a covering (see later) is an abstraction of ideas occurring in one of them, the Heine-Borel theorem.

Definition: A pair (E,O) is called a *toplogical space* if O is a family of subsets of E such that:

1. Any union of elements of O is in O.
2. Any finite intersection of elements of O is in O.
3. $E \in O$, $\emptyset \in O$, where $\emptyset$ denotes the null set.

Members of O are called open sets. The complements of members of O relative to E are called closed sets.

Definition: A *metric space* is a pair (E,ρ) such that for any pair of elements x, y of the set E there corresponds a unique real number, denoted by $\rho(x,y)$, which satisfies the following metric axioms:

1. $\rho(x,y) = 0$ if and only if $x = y$.
2. $\rho(x,y) = \rho(y,x)$.
3. For $x, y, z \in E$, $\rho(x,y) \leq \rho(x,z) + \rho(z,y)$ (the triangular inequality).

Examples of metric spaces are the set of all real-valued continuous functions on the closed interval $[a,b]$, with $\rho(x,y) = \max\limits_{a \leq t \leq b} |x(t) - y(t)|$ (we denote this space by $C[a,b]$); the set of all bounded sequences $x = \{x_n\}$, x_n real numbers with $\rho(x,y) = \sup\limits_n |x_n - y_n|$ (we denote this space by m); the set of all infinite real sequences $\{x_n\}$ such that $\Sigma x_n{}^2 < \infty$ with $\rho(x,y) = \left[\sum\limits_{n=0}^{\infty} (x_n - y_n)^2 \right]^{1/2}$ (we denote this space by l^2). For $C[a,b]$, the notion of convergence means precisely that $x_n(t) \to x(t)$ uniformly in t as $n \to \infty$.

In a metric space E, a sphere S with center $x_0 \in E$ and radius $\epsilon > 0$ is given by $S(x_0,\epsilon) = \{x | \rho(x,x_0) < \epsilon\}$.

The metric topology of a metric space has the family O of subsets of E given by $A \in O$ (that is, A is an open set) if and only if for every point $x_0 \in A$ there exists an $\epsilon > 0$ such that $S(x_0,\epsilon) \subset A$.

Definition: A triple $(L,+,.)$ is called a *real (complex) linear vector space* if addition (x, $y \in L$ implies $x + y \in L$) and scalar multiplication by a real (complex) number [$x \in L$ implies $a \cdot x$ or simply $ax \in L$, where a is real (complex)] are defined in the set L such that:

1. $x + y = y + x$.
2. $x + (y + z) = (x + y) + z$.
3. Given $x, z \in L$, there exists an element y such that $x + y = z$.
4. $a(bx) = (ab)x$, $x \in L$, and a and b are real (complex) numbers.
5. $1 \cdot x = x$.
6. $(a + b)x = ax + bx$.
7. $a(x + y) = ax + ay$.

Some consequences of these axioms, using the definition $-x = (-1)x$, are:

1. $x + \mathbf{0} = x$.
2. $x + (-x) = \mathbf{0}$.
3. $x + (-y) = x - y$, where $-y$ satisfies $y + (-y) = \mathbf{0}$.

Definition: A *Hausdorff space* E is a topological space with the property that for any two distinct points x, $y \in E$ there exists an open set U with $x \in U$ and an open set V with $y \in V$ such that $U \cap V = \emptyset$.

Definition: A topological space E is *compact* if every open covering of E contains a finite subcovering. An *open covering* of E is a collection of open sets $\{O_\alpha\}$ of E such that $x \in E$ implies $x \in O_\alpha$ for some α. A finite *subcovering* by elements of $\{O_\alpha\}$ is a subcollection $\{O_{\alpha_i}\}$, $i = 1, \ldots, N$, such that $\bigcup_{i=1}^{N} O_{\alpha_i} \supset E$. A characterization of compactness of a subset A of a topological space in metric spaces is that every infinite sequence of elements in A contains a convergent subsequence with a limit point in A. It is easy to show that on the real line compactness is equivalent to the requirement that a set be closed and bounded.

Remark: Recall the existence of a maximum and a minimum of a continuous real-valued function when it is defined on a nonvoid compact set. It is important to note that without compactness one can always define an unbounded continuous real-valued function. One can also define a bounded real-valued function which assumes neither its least upper bound nor its greatest lower bound.[26] With compactness one can prove other useful theorems, such as the Stone-Weierstrass theorem, viz., every continuous function defined on a compact subset of the reals is the uniform limit of a sequence of appropriately chosen polynomials.

Definition: A *partially ordered* set is a set in which a partial ordering $\leq$ is defined; i.e.,

1. $x \leq y$, $y \leq x$ implies $x = y$.
2. $x \leq y$, $y \leq z$ implies $x \leq z$.

Note that for some elements x and y we may have neither $x \leq y$ nor $y \leq x$; that is, some elements may not be comparable.

Definition: A *partially ordered linear space* L is a set L which is partially ordered and is a linear space satisfying:

1. $x \leq y$ implies $x + z \leq y + z$.
2. $x \geq 0$ and λ real implies $\lambda x \geq 0$.

If we let $A \subset E$, $A \neq \emptyset$, for some set E, $b \in E$ is called an *upper bound* of A if $x \leq b$ for all $x \in A$. Note that $[0,1)$ is bounded by 1 on the right but has no maximum element in the set.

Definition: If A and B are subsets of a linear space L, we make the following designations:

1. $A + B \equiv \{x + y : x \in A, y \in B\}.$
2. $A - B \equiv \{x - y : x \in A, y \in B\}.$
3. If $x_0 \in L$, $x_0 + B \equiv \{x_0 + b : b \in B\}.$
4. If a is a real number, $aA = \{ax : x \in A\}.$

Definition: A nonempty subset S of a real (complex) linear space L is a *linear subspace* if whenever $x, y \in S$, α real (complex), we have $x + y \in S$ and $\alpha x \in S$. Thus S is a linear space with the operations $+$ and $\cdot$, inherited from L.

Definition: A *linear manifold*, or an *affine space*, is a translation of a linear subspace; i.e., it is of the form $x_0 + L$ where L is a subspace.

Definition: A (Hamel) *basis* of a linear space L is a set S of linearly independent elements such that every element of L can be expressed as a finite linear combination of elements of S.

It is useful to observe that many properties of interest in analysis, e.g., boundedness, continuity, smoothness, measurability, oddness and evenness, and integrability, are preserved under the formation of linear combinations; convexity, however, is not.

Definition: A set E in a linear space L is *convex* if $E \neq \emptyset$ and if $x, y \in E$ and $\alpha \geq 0$, $\beta \geq 0$, $\alpha + \beta = 1$ implies $\alpha x + \beta y \in E$; that is, for any two points in E, the line segment joining them is in E. (It is an affine space if the conditions $\alpha \geq 0$, $\beta \geq 0$ are omitted.)

A real linear topological space is a real vector space which is a topological space such that addition, that is, $(x,y) \to x + y$, and scalar multiplication, that is $(\alpha,x) \to \alpha x$, are continuous transformations (see below) for all x and y in the space and all real α. Such a space is called *locally convex* if every neighborhood of an arbitrary vector x contains a convex neighborhood of x.

Definition: In a linear space L a set $E \neq \emptyset$ is a *cone* if $x \in E$, $\alpha \geq 0$ implies $\alpha x \in E$, that is, $\alpha E \subset E$ for all $\alpha \geq 0$ and $x, y \in E$ implies $x + y \in E$. If $x \in E$, then for $\alpha = 0$, $0 \cdot x \in E$ and clearly $\mathbf{0} \in E$. Also, the whole space is a cone. A *convex cone* is a cone which is convex.

Definition: A *normed linear space* (a particular case of a metric space) is a linear space L such that for each $x \in L$ there exists a real number, called its *norm* and denoted by $\|x\|$, satisfying the following axioms:

1. $\|x\| \geq 0$ for all $x \in L$,; $\|x\| = 0$ only for $x = \mathbf{0}$.
2. $\|\alpha x\| = |\alpha| \, \|x\|$ for all $x \in L$ and any α.
3. $\|x + y\| \leq \|x\| + \|y\|$ for all $x, y \in L$.

Every normed space is a metric space with the metric ρ given by $\rho(x,y) = \|x - y\|$ but not conversely. A simple example of a metric space which cannot be made into a normed linear space so that the metric coincides with the metric ρ given by the norm no matter how addition and scalar multiplication are defined is a space of n points with the metric $\rho(x,y) = 0$ if $x = y$ and $\rho(x,y) = 1$ if $x \neq y$.

As an example of a normed space, we have the space E^n with elements of the form $x = (x_1, \ldots, x_n)$, x_i real or complex numbers, and with either the norm $\|x\| = \sqrt{\sum_{i=1}^{n} x_i^2}$ or $\|x\| = \max(|x_1|, \ldots, |x_n|)$. In the space s of all numerical sequences, we can define a metric by

$$\rho(x,y) = \sum_{n=1}^{\infty} \frac{1}{2^n} \frac{|x_n - y_n|}{1 + |x_n - y_n|}$$

Exercise 1-3 Show that the last metric does not satisfy the norm axiom of scalar multiplication, i.e., axiom 2.

Definition: We say that the sequence of points x_n of the metric space E converges to the limit point $x^* \epsilon E$, that is, $\lim_{n \to \infty} x_n = x^*$, if for each positive ϵ there is some integer N depending on ϵ such that $\rho(x_n, x^*) < \epsilon$ whenever $n \geq N$.

Definition: A sequence of points x_n of a metric space E is called a *Cauchy* (fundamental) *sequence* if $\rho(x_n, x_m) \to 0$ as $n, m \to \infty$.

Every convergent sequence is a Cauchy sequence, but the converse does not hold in general.

Definition: A metric space E is *complete* if every Cauchy sequence of elements of E is convergent to an element in the space.

In euclidean space of finite dimensions any Cauchy sequence converges to a point of the space. Most of the spaces we shall encounter are complete. Functional analysis makes profound use of completeness.

An example of a space which is not complete is the space of all polynomials with real coefficients and with the metric

$$\rho(f,g) = \max_{a \leq t \leq b} |f(t) - g(t)|$$

Thus we can construct a sequence of polynomials which converges uniformly to a continuous function which is not a polynomial (Weierstrass's theorem). The space of real sequences taken on $(0,1)$ is not complete; for example, $\{1/n\} \to 0$, which is not in $(0,1)$. The sequence .5, .505, .505005, .5050050005, . . . is a nonconvergent Cauchy sequence in the space of rational numbers, since its limit is not a rational number.

Definition: A *Banach* space X is a complete normed linear space. A *Hilbert* space H is a Banach space in which the norm is defined by $\|x\|^2 = (x,x)$, where (x,y) is a scalar product (often called inner product in functional analysis) satisfying the following axioms [we use $(\overline{x,y})$ to denote the complex conjugate of (x,y)]:

1. $(x,y) = (\overline{y,x})$.
2. $(x + y, z) = (x,z) + (y,z)$.

3. $(\alpha x, y) = \alpha(x, y)$.

4. $(x,x) \geq 0$ for any $x \in H$, and $(x,x) = 0$ if and only if $x = 0$.

Remark: (The reader may skip this remark or look for definitions below.) Not every norm can be generated by a scalar product. Every inner product induces a norm via $\|x\| := (x,x)^{1/2}$. However, not every norm can be generated by an inner product. A necessary and sufficient condition for a norm $\| \cdot \|$ to be generated by an inner product is that the norm satisfies the parallelogram equality $\|x + y\|^2 + \|x - y\|^2 = 2\|x\|^2 + 2\|y\|^2$ for all x and y. The Banach space C with norm $\|x\| = \sup |x(t)|$ does not satisfy this law.

We shall be interested in considering general Hilbert spaces H with real or complex scalars. If $x, y \in H$, then Re (x,y) will denote the real part of (x,y) if the scalars are complex. Of course, when the scalars are real, Re (x,y) and (x,y) are identical. We have the Cauchy-Schwarz inequality

$$\|x\| \cdot \|y\| \geq |(x,y)| \geq |\text{Re } (x,y)| \geq \text{Re } (x,y)$$

Two examples of a Hilbert space are the spaces l^2 of all sequences $x = \{x_n\}$ such that $\sum_{n-1}^{\infty} |x_n|^2 < \infty$, with $(x,y) = \sum_1^{\infty} x_n \bar{y}_n$, $x = \{x_n\}$, $y = \{y_n\}$, and the space $L^2(a,b)$, where (a,b) is an interval of the real line and an element x of L^2 is a class of functions such that any two members of the class are equal almost everywhere (see "NM," chap. 1) and if $x(t)$ is a member of x, then $\|x\|^2 = \int_a^b |x(t)|^2 \, dt < \infty$ and $(x,y) = \int_a^b x(t)\overline{y(t)} \, dt$. Another example is n-dimensional euclidean space E^n. (As a space without the euclidean norm it is denoted by R_n, the equivalent n-dimensional vector space.) Here $(x,y) = \sum_{i=1}^{n} x_i y_i$. Sometimes E_n is used instead of E^n.

Definition: A linear space with a scalar-product norm which is not complete is called a *pre-Hilbert* space.

Definition: The *closure* of a subset E_1 of a topological space E is the intersection of the members of the family of all closed sets containing E_1. An alternate definition may be given as follows. The *closure* of any set is the union of the set and the set of its accumulation points. A point x is an *accumulation point* of a subset E_1 of a topological space E if and only if every neighborhood of x contains points of E_1 other than x. A closed set is identical with its closure.

For a Hilbert space, the notion of separability plays an important role.

Definition: A topological space H is *separable* if it contains a countable dense subset E; that is, E is a countable set such that $\bar{E}$, the closure of E, is the entire space H.

The presence of a dense subset in a space sometimes enables one to extend properties which hold on the dense set to the entire space by means of limiting operations. It is possible to uniquely extend a bounded linear operator defined on a dense subset of a normed linear space to the entire space. It can be shown that a finite-dimensional Hilbert space is algebraically isomorphic, that is, algebraic operations are preserved, to a finite-dimensional euclidean space. Similarly, an infinite-dimensional (separable) Hilbert space H is algebraically isomorphic to the space l^2.

Remark: The unit sphere $\|x\| \leq 1$ in infinite-dimensional Banach spaces, is not compact, as can be seen by taking the vectors $(1,0,0,0, \ldots)$, $(0,1,0,0, \ldots)$, $(0,0,1,0, \ldots)$ etc., on the sphere. No subsequence converges even though the columns of the components converge to the zero vector. This leads to many topological difficulties.

Remark: The space of continuous functions is an example of a Banach algebra which is a Banach space with the additional property that the product of two elements of the space is also an element of the space. Thus, for example, we define $\|x\| = \sup |x(t)|$ and because

$$\sup |x(t)y(t)| \leq \sup |x(t)| \sup |y(t)|$$

we have

$$\|xy\| \leq \|x\| \, \|y\|$$

as an additional property of the norm in a Banach algebra.

Remark: A main result, due to Gelfand, for a commutative Banach algebra with no nilpotent elements and with a multiplicative unit element e, that is, $ex = x$ for every x, is that there is a compact Hausdorff space such that the algebra is in one-to-one correspondence with the continuous functions $f(t)$ on this space and such that x in the algebra corresponds to some $f_x(t)$ and xy to $f_x(t)f_y(t)$. This theorem was used to prove a theorem of Wiener's, viz., the inverse of an absolutely convergent nowhere-vanishing Fourier series is absolutely convergent. Another example of a commutative Banach algebra without a unit element is the space $L(-\infty, \infty)$ of functions summable over $(-\infty, \infty)$, in which multiplication, called *convolution*, is defined by $f * g = \int_{-\infty}^{\infty} f(x - t)g(t) \, dt$.

Exercise 1-4 Show that $f * g = g * f$ and $\|f * g\| \leq \|f\| \|g\|$.

Examples of simple operators are a matrix T in the algebraic system of equations $y = Tx$ (here T maps a real n-dimensional vector space R_n into itself); the integral operator $Tx = \int_a^b x(s) \, ds$ (mapping the space $C[a,b]$ into the real numbers); or an operator with a kernel $K(t,s)$, $Tx = \int_a^b K(\cdot \, ,s)ds$ (mapping $C[a,b]$ into itself for $K(\cdot \, ,s) \equiv (Tx)\,(t)$ continuous on $[a,b] \times [a,b]$); the differential operator $(Tx)(t) = (d/dt)x(t)$ (mapping the space $C'[a,b]$ of functions with a continuous derivative into, say, the space $C[a,b]$); and the difference operator $Tx_n = px_{n+1} + qx_{n-1}$ (mapping the space of sequences into itself).

As examples of more complex operators, we have:

1. The operators K and F in Hammerstein's equation (we use K in conformity with standard notation) $y + KFy = x$, where

$$(Fy)(t) = f(t,y(t))$$
$$(Ky)(t) = \int_a^b k(s,t)y(t) \, dt$$

so that
$$KFy(s) = \int_a^b k(s,t)f(t,y(t)) \, dt$$

To see how such an equation can arise from physical considerations consider the equation for forced oscillation of a simple pendulum

$$\ddot{x}(t) + \alpha^2 \sin x(t) = f(t)$$

with
$$f(-t) = -f(t)$$

Let $f(t) = \beta \sin (\pi t/T)$; then we seek a solution of the same form. If $\bar{x}$ is a solution on $0 \le t \le T$, with $\bar{x}(0) = \bar{x}(T) = 0$, then

$$x(t + nT) = \bar{x}(t) \qquad 0 \le t \le T, n = 0, 1, \ldots$$
$$x(-t) = -x(t)$$

is the desired odd-periodic solution. On the other hand, if we are given such a solution, it must satisfy the boundary conditions $x(0) = x(T) = 0$. For simplicity let $T = 1$, and after transposing the second term on the left to the right, integrating twice on $[0,t]$ [the double integral can be replaced with a single integral with kernel $(t - s)$], and determining the two constants of integration, our equation becomes

$$x(t) = \int_0^t s(1 - t)[f(s) - \alpha^2 \sin x(s)] \, ds$$
$$+ \int_t^1 t(1 - s)[f(s) - \alpha^2 \sin x(s)] \, ds$$
$$\equiv - \int_0^1 K(t,s)[f(s) - \alpha^2 \sin x(s)] \, ds$$

where $K(t,s)$ is Green's function defined in terms of the two kernels on $[0,t]$ and on $[t,1]$, respectively. This is a Hammerstein type of equation.

2. The operator K_v in the Volterra equation

$$y(x) + \int_0^x k(x,t;y(t)) \, dt = g(x) \qquad x \ge 0$$

where
$$K_v y = \int_0^x k(x,t;y(t)) \, dt$$

and a similar operator in Fredholm's equation

$$y(x) + \int_0^1 k(x,t;y(t))\, dt = g(x) \qquad 0 \le x \le 1$$

3. A difference operator

$$Tf(t) = f(t+1) \qquad \text{on } C[0, \infty]$$

4. A convolution operator on $L_2(-\infty, \infty)$

$$y(s) = Tx \equiv \int_{-\infty}^{\infty} K(s-t)x(t)\, dt$$

with
$$\int_{-\infty}^{\infty} |K(s)|\, ds < \infty$$

The best-known and most widely used operators are linear operators. The first definition repeats some ideas appearing in the previous section.

Definition: An operator T mapping a set X into a set Y is *linear* if it is both additive, that is, $T(x_1 + x_2) = Tx_1 + Tx_2$, x_1, $x_2 \, \epsilon \, X$, and homogeneous, that is, $T(\alpha x) = \alpha Tx$, $x \, \epsilon \, X$, and α a scalar (real or complex). T is continuous if $x_n \to x$ implies $Tx_n \to Tx$. This definition of continuity is due to Heine and is equivalent to Cauchy's familiar ϵ, δ definition. T is bounded if there is a constant C such that $\|Tx\| \le C\|x\|$ for all $x \, \epsilon \, X$. Note that additivity and continuity imply homogeneity, for real scalars.

For additive homogeneous operators in a normed space, continuity and boundedness are equivalent. Consequently, additive homogeneous unbounded operators generally require separate treatment. One often refers to such operators as *unbounded operators*, distinguishing them from unbounded nonlinear operators.

An example of an unbounded operator is obtained by considering the sequence in $L_2(0,1)$

$$x_n(t) = \sqrt{2} \sin \pi n s$$

so that $\|x_n\| = 1$ for all n, and defining $T \equiv d/dt$. Thus

$$Tx_n = \sqrt{2}\, n\pi \cos n\pi t$$

$\|Tx_n\| = n\pi \to \infty$ as $n \to \infty$. Now $\|x_n\| = 1$, and therefore $\|Tx_n\| \le C\|x_n\|$ cannot hold.

Definition: If T is an operator mapping pairs (x,y) of points of a Banach space X into a Banach space Y such that T is linear in x for all y and linear in y for all x, then T is called *bilinear*.

If, in addition,

$$T(x,y) = T(y,x)$$

T is said to be a *bilinear symmetric operator*.

The totality of linear operators which map a normed space X into a normed space Y form a space L. If $T, S \epsilon L$, then $T \equiv S$ means that $Tx = Sx$ for all $x \epsilon X$. For any $T, S \epsilon L$, we denote the sum by an operator W such that

$$Wx = Tx + Sx$$

and similarly the product αT by an operator W such that

$$W = \alpha(Tx) = (\alpha T)x$$

With these definitions, bounded linear operators form a linear system which is a normed space inheriting its norm from the range space. We define

$$\|T\| \equiv \sup_{\|x\| \leq 1} \|Tx\| = \sup_{\|x\| = 1} \|Tx\|$$

An operator whose range consists of real or complex numbers is called a *functional*. The space of continuous linear functionals on a normed linear space X is called the conjugate space of X and is denoted by X^*. For X, convergence in the norm, that is, $x_n \rightarrow x$ if and only if $\|x_n - x\| \rightarrow 0$, is called *strong convergence*. It is called *weak convergence* if for all functionals $f \epsilon X^*$ the sequence $f(x_n) \rightarrow f(x)$ as $n \rightarrow \infty$. In a finite-dimensional space weak and strong convergence coincide.

We define convergence in the space of operators $T:X \rightarrow Y$ defined on a linear space X to a linear space Y as follows.

Definition: If $\|T_n - T\| \rightarrow 0$, where T_n maps a normed linear space X into a normed linear space Y, then the sequence of operators $\{T_n\}$ converges to T, and we speak of *convergence in the norm* or *uniform convergence*.

Definition: In the space of linear operators mapping X into Y the sequence of operators $\{T_n\}$ is *strongly or pointwise convergent* to the linear operator T if for every $x \epsilon X$, the sequence $\{T_n x\}$ converges to Tx.

Note that uniform convergence of a sequence T_n implies strong convergence of this sequence but not conversely. Weak convergence of operators is defined by $fT_n x \rightarrow fTx$ for all $f \epsilon Y^*$, $x \epsilon X$.

An example of a sequence of operators on $L(-\infty, \infty)$ is

$$T_n x = \frac{n}{2} \int_{t-1/n}^{t+1/n} x(s)\, ds \equiv x_n(t)$$

Thus $$\|T_n x\| = \int_{-\infty}^{\infty} |x_n(t)|\, dt \leq \int_{-\infty}^{\infty} dt \left(\frac{n}{2} \int_{t-1/n}^{t+1/n} |x(s)|\, ds \right)$$

and one can show that $\|T_n\| = 1$ and $f_n(t) \to f(t)$ in the mean for $f \,\epsilon$ $L(-\infty, \infty) \cap C[-\infty, \infty]$, that is, everywhere in $L(-\infty, \infty)$.

A linear operator need not be continuous. Since continuity is a desirable property, e.g., a compact set is mapped into a compact set by such an operator, we give an important condition for an operator to be continuous. A linear operator T is *closed* if whenever $x_n \,\epsilon\, D[T]$ (the domain of T), $x_n \to x_0$, and $y_n = Tx_n \to y_0$, then $x_0 \,\epsilon\, D[T]$, and $Tx_0 = y_0$. Every bounded transformation is closed, since $D[T]$ is the entire space; i.e., T can be extended to the whole space if it has a dense domain and Tx_0 is defined as the limit point at the transformed value. However, a closed transformation need not be bounded. The closed-graph theorem asserts that if a linear operator is closed and maps a whole Banach space X into a Banach space Y, then it is continuous.

We show that $T = t(d/dt)$, with domain $D[T]$ of all functions $f \,\epsilon\, C^1[0, \infty]$, for which $tf'(t)$ is in C, is closed. Let $f_n \,\epsilon\, D[T]$, $g_n = t(df_n/dt)$, $f_n \to f_0$, and $g_n \to g_0$. We show that $f_0 \,\epsilon\, D[T]$ and $g_0 = t(df_0/dt)$. Now our sequences converge in the norm of $C[0, \infty]$ and hence uniformly in t on $[0, \infty]$ to f_0 and g_0, respectively. First consider the derivatives h_n of f_n. th_n are in C by hypothesis. Now if a sequence of continuous functions $\{f_n\}$ converges uniformly on $[a,b]$ to f_0 and the sequence of continuous derivatives $\{h_n\}$ converges uniformly to a function h_0, then f_0 is differentiable, h_0 is continuous, and $df_0/dt = h_0$ for all $t \,\epsilon\, [a,b]$.

Applying this theorem, we have the fact that f_0 is differentiable, with derivatives in $C[0, \infty]$. We also have

$$g_0(t) = \lim_{n \to \infty} t \frac{df_n(t)}{dt} = t \lim_{n \to \infty} \frac{df_n(t)}{dt} = t \frac{df_0(t)}{dt} \qquad (1\text{-}1)$$

by this theorem. Since, by hypothesis, $g_0(t) \,\epsilon\, C[0, \infty]$, hence also $t[df_0(t)/dt] \,\epsilon\, C[0, \infty]$ with $f_0(t) \,\epsilon\, D[T]$. Note that the domain $D[T]$ is dense in $C[0, \infty]$. Thus consider the set $\{e^{-nt}\}$, $n = 0, 1, 2, \ldots$, which is fundamental in $C[0, \infty]$. It is transformed by T into $\{-nte^{-nt}\}$, $n = 0, 1, 2, \ldots$, which is fundamental in $C[0, \infty]$. A set is *fundamental* if and only if the smallest linear space containing it is dense in the space. Hence the range $R[T]$ is dense in $C[0, \infty]$.

We now consider an unbounded operator $T[f] = f''(t)$ with $f \,\epsilon\, C^2[0,1]$ which takes the space $L_p[0,1]$ into itself (note that $C^2 \subseteq L_p$). This operator need not be closed, but it has a closure together with other interesting properties. Note that an operator can be identified with its graph. The closure of the operator is the closure of its graph [contained in the cartesian product $D \times R$ of its domain D and range R—the elements of such a product are of the form (f, Tf) with $f \,\epsilon\, D$ and $Tf \,\epsilon\, R$—] when considered as a point set. The reader should have no problem in restating the definition of closure using sequences of the form $\{(f_n, T[f_n])\}$.

Let $D[T] = C^2[0,1]$, $T[f] = f''(t)$ for all $f \in D[T]$; then T has a closure $\bar{T}$. We have $D[\bar{T}] = \{f \in L_p[0,1] : f$ equal a.e. to some absolutely continuous function, f' equal a.e. to some absolutely continuous function, and $f'' \in L_p[0,1]\}$, and for all $f \in D[T]$ we have $T[f] = f''$. Here a.e. means "almost everywhere."

Let $f_n \in C^2$, $f_n \xrightarrow{\text{mean}} f$, $f_n'' \xrightarrow{\text{mean}} g$. For all φ and ψ in C^2 such that

$$\psi(0) = \psi(1) = \psi'(0) = \psi'(1) = 0 \qquad (*)$$

we have

$$\int_0^1 (f_n\varphi + f_n''\varphi)\, dt \to \int_0^1 (f\varphi + g\psi)\, dt$$

and hence

$$\int_0^1 f_n(\varphi + \psi'')\, dt \to \int_0^1 (f\varphi + g\psi)\, dt$$

If also $\varphi + \psi'' \equiv 0$ we have

$$\int_0^1 (f\varphi + g\psi)\, dt = 0 \qquad (**)$$

Let $\varphi \in C^2$ with $\varphi \perp 1$ and $\varphi \perp t$ (we say φ is orthogonal to 1 and φ is orthogonal to t), i.e.,

$$\int_0^1 \varphi(t)\, dt = \int_0^1 t\varphi(t)\, dt = 0$$

Take $\psi(t) = \int_0^1 (s - t)\varphi(s)\, ds$; then $\psi(t) \in C^1$ and $\psi'(t) = -\int_0^t \varphi(s)\, ds$. Therefore $\psi \in C^2$ and $\psi''(t) = -\varphi(t)$. So $\varphi + \psi'' \equiv 0$ and (*) holds. By (**) we have, for all $\varphi \in C^2[0,1]$, $\varphi \perp 1$ and $\varphi \perp t$.

$$\int_0^1 f(t)\varphi(t)\, dt + \int_0^1 g(t)\left[\int_0^t (s - t)\varphi(s)\, ds\right] dt = 0$$

Thus

$$\int_0^1 \varphi(s)\left[f(s)t \int_0^1 g(t)(s - t)\, dt\right] ds = 0$$

We conclude that the quantity in brackets which is orthogonal to every $\varphi(s)$ which is orthogonal to 1 and t is a linear combination $c_0 + c_1 t$ of 1 and t. Thus

$$f(s) = -\int_s^t g(t)(s - t)\, dt + c_0 + c_1 s$$

and hence $f(s)$ is absolutely continuous and

$$f'(s) - \int_s^1 g(t)\, dt + c_1$$

Almost everywhere, therefore, $f'(s)$ is absolutely continuous and thus $f''(s) = g(s)$.

Definition: If an operator T is defined on a domain D in a normed linear space, and if there is a constant $C > 0$ such that

$$\|Tx_1 - Tx_2\| \leq C\|x_1 - x_2\| \qquad \text{for all } x_1,\, x_2 \in D$$

then T is said to satisfy a *Lipschitz condition*, and T is called *lipschitzian*. T is a *contraction operator* if $C < 1$. For a metric space, this property reads $\rho(Tx_1, Tx_2) \leq C\rho(x_1, x_2)$. Later we give a well-known fixed-point theorem for contraction operators appearing in equations of the form $Tx = x$.

A generalization of the notion of a functional is a functional transformation whose values are functions. An illustration of a functional transformation is given by

$$z(x) = \int_a^b K(x,t,y,(t))\, dt$$

The derivative operation is another example of a functional transformation. As we shall need the notion of derivative later, we shall motivate and introduce it below. In the calculus of variations one computes the variation of a functional $I(y)$ in the direction of an increment η as follows:

$$D_\eta I(y) = \lim_{\epsilon \to 0} \frac{I(y + \epsilon\eta) - I(y)}{\epsilon}$$

From this analysis one obtains, for example, Euler's differential equation as a necessary condition for a minimum when it is desired to determine the curve $y(x)$ which has the shortest length on $[a,b]$.

Abstract differentiation is a generalization of this notion.[66]

Definition: Let T be an operator from a real normed space X to another normed space Y. If for some $x \in X$ and all $h \in X$ we have

$$\lim_{t \to 0} \left\| \frac{T(x + th) - Tx}{t} - VT(x,h) \right\| = 0$$

then the operator $VT(x,h)$ is called a *Gateaux differential* (or weak differential) of the operator T at the point x in the direction h. If $VT(x,h)$ is a bounded linear operator, then it is denoted by $DT(x,h)$.

From the above equation it is easily seen that $VT(x,h)$ (which need not be linear) is homogeneous in h. That is, for all real α, $VT(x,\alpha h) = \alpha VT(x,h)$.

If the Gateaux differential of T exists at each point of some convex set $E \subset X$, then for all x, $x + h \in E$ the following mean-value theorem holds: For *some* t, $0 \leq t \leq 1$

(1) $$T(x + h) - T(x) = VT(x + th, h).$$

Thus,

(2) $$\|T(x + h) - Tx\| \leq \sup_{0 \leq t \leq 1} \|VT(x + th, h)\|$$

If the differential is linear and bounded in h for all $x \in E$, then this last inequality becomes

$$\|T(x + h) - Tx\| \leq \sup_{0 \leq t \leq 1} \|DT(x + th)\|\, \|h\|$$

The Gateaux differential is linear and bounded in h for $x = x_0$ if:

1. $VT(x,h)$ exists in some neighborhood of x_0 and is continuous in x at x_0 and

2. $VT(x,h)$ is continuous in h at $h = \mathbf{0}$.

Theorem 1-1   The Gateaux differential is linear and bounded in h if and only if:

1. T satisfies a weak Lipschitz condition at x_0, that is,

$$\|T(x_0 + th) - Tx_0\| \le C\|th\|$$

where $C > 0$ does not depend on h, $\|h\| = 1$, and for each h there exists a $\delta(h)$ such that this inequality holds when $|t| < \delta(h)$ and

2. $T(x_0 + th_1 + th_2) - T(x_0 + th_2) + Tx_0 = \theta(t)$ where $\theta(t) \to \mathbf{0}$ as $t \to 0$.

Definition: If at the point $x \,\epsilon\, X$

$$T(x + h) - Tx = u(x,h) + w(x,h)$$

where $u(x,h)$ is a linear bounded operator in $h \,\epsilon\, X$ at x, and

$$\lim_{\|h\| \to 0} \frac{\|w(x,h)\|}{\|h\|} = 0$$

then $u(x,h)$ is the *Fréchet* (strong) *differential* of T at x and is denoted by $dT(x,h)$.

We now illustrate a Fréchet differential.   Before doing so we give an ϵ,δ definition of a continuous functional.[47]   Consider $C[a,b]$.   A functional $f(x)$ on a neighborhood of $x_0(t)$ of $C[a,b]$ to $C[a,b]$ is continuous at $x(t) = x_0(t)$ of $C[a,b]$ if, given $\epsilon > 0$, there exists a $\delta(\epsilon) > 0$ such that $\|f(x_0 + y) - f(x_0)\| < \epsilon$ for all $y(t) \,\epsilon\, C(a,b)$ which satisfy $\|y(t)\| < \delta(\epsilon)$. Now consider a linear continuous operator on $C[a,b]$ defined by

$$f(x) = x(t) \int_a^b K(t,s)x(s)\, ds$$

We have

$$f(x_0 + y) - f(x_0) = y(t) \int_a^b K(t,s)x_0(s)\, ds$$
$$+ x_0(t) \int_a^b K(t,s)y(s)\, ds + y(t) \int_a^b K(t,s)y(s)\, ds$$

and the Fréchet differential $df(x_0,y)$ exists and is given by

$$y(t) \int_a^b K(t,s)x_0(s)\, ds + x_0(t) \int_a^b K(t,s)y(s)\, ds$$

since $f(x_0 + y) - f(x_0) - df(x_0,y)$

is given by

$$y(t) \int_a^b K(t,s)y(s)\, ds$$

with the condition

$$\max_{a \leq t \leq b} |y(t) \int_a^b K(t,s)y(s) \, ds \,| \leq \epsilon \max_{a \leq t \leq b} |y(t)|$$

for all $y(t)$ such that $\max_{a \leq t \leq b} |y(t)| < \delta(\epsilon)$

If $dT(x,h)$ exists, then so does $DT(x,h)$, but not conversely.

If the operator T from a Banach space X into a Banach space Y has a bounded linear Gateaux differential $DT(x, h)$ at the point x, then $DT(x, \cdot)$ is an element of the space of bounded linear operators from X to Y. This operator is called the *Gateaux derivative* of T at x and is denoted by $T'x$.   Therefore $DT(x,h) = T'(x)h$.

Similar remarks hold for the *Fréchet derivative*

$$dT(x,h) = T'(x)h$$

Theorem 1-2   If the Gateaux derivative exists in a neighborhood of x_0 and is continuous at x_0, then

$$dT(x_0,h) = DT(x_0,h)$$

Proof: Let

$$w(x_p,h) = T(x_0 + h) - Tx_p - T'(x_0)h$$

then by the mean-value theorem

$$(w(x_0,h),e) = ((T'(x_0 + th) - T'(x_0))h,e),$$

where e is an arbitrary element of the dual space Y^*, and (x,e) means $e(x)$ for some t, $a < t < b$.

By the Hahn-Banach Theorem (see "NM," p. 37), e can be chosen so that

$$|(w(x_p,h),e)| = \|w(x_0,h)\| \qquad \text{and} \qquad \|e\| = 1$$

Therefore $\|w(x_0,h)\| \leq \|T'(x_0 + th) - T'x_0\| \, \|h\|$
Since T' is continuous at x_0,

$$\frac{\|w(x_0,h)\|}{\|h\|} \to 0 \qquad \text{as } \|h\| \to 0$$

Therefore $DT(x_0,h) = dT(x_0,h)$

If the Gateaux differential of a functional $f(x)$ is linear and bounded in h for all x on some $E \subset X$, X a Banach space, then $Df(x,.)$ is a linear bounded functional mapping E into the conjugate space X^*.   We denote this linear functional by Tx.

Definition: The operator Tx defined by

$$\lim_{t \to 0} \frac{1}{t} (f(x + th) - f(x)) = (Tx,h)$$

is called the *gradient* of f

$$T = \text{grad } f$$

If f has a Fréchet derivative, then T is called a *strong gradient*.

Example: Let $f(x) = \|x\|$, where $x \, \epsilon \, H$, H is a Hilbert space, and $x \neq 0$.   Then

$$\|x + th\|^2 - \|x\|^2 = 2(x,th) + \|th\|^2 = (\|x + th\| + \|x\|)(\|x + th\| - \|x\|)$$

$$\lim_{t \to 0} \frac{\|x + th\| - \|x\|}{t} = \frac{(x,h)}{\|x\|} = \left(\frac{x}{\|x\|}, h \right)$$

Hence

$$\text{grad } \|x\| = \frac{x}{\|x\|}$$

Example: If $f(x) = \|x\|^2$, where $x \, \epsilon \, H$ and H is a Hilbert space,

then

$$\|x + th\|^2 - \|x\|^2 = 2t(x,h) + t^2\|h\|^2$$

Thus

$$\text{grad } \|x\|^2 = 2x$$

$T'(x) \, (\, \cdot \,)$ is continuous as operator on X into Y by definition of Fréchet differentiability. Fréchet differentiability of T at x_0 implies continuity of T at x_0.   However, if we start by assuming that T is continuous at x_0, then the requirement of continuity of $DT(x,h)$ in h is redundant, since

$$\|T'(x)h\| \leq \|T(x+h) - Tx - T'(x)h\| + \|T(x+h) - Tx\| \to 0 \text{ as } \|h\| \to 0$$

If $T'(x)h$ has a Fréchet differential with respect to x, that is,

$$T'(x + h_2)h_1 - T'(x)h_1 - T''(x)h_1h_2 \to 0$$

the operator $T''x$ is called the second Fréchet derivative of T and is a bilinear operator operating on h_1 and h_2.

Using an argument similar to that given by Kantorovich (see "NM," chap. 2, with the ordinary derivative replaced by the Fréchet derivative), one can show that if the operator T admits Fréchet derivatives up to order $n + 1$ in a convex domain, and if the norm of the $(n + 1)$st derivative is majorized by a positive number C in this domain, the following generalization of Taylor's formula is valid:

$$T(x) = T(x_0) + T'(x_0)(x - x_0) + \frac{1}{2!} T''(x_0)(x - x_0)^2 + \cdots$$

$$+ \frac{1}{n!} T^{(n)}(x_0)(x - x_0)^n + R_{n+1}(T,x_0,x)$$

with

$$\|R_{n+1}(T,x_0,x)\| \leq \frac{1}{(n + 1)!} C\|x - x_0\|^{n+1}$$

Here C can be replaced by

$$\sup_{\substack{\bar{x} = x_0 + \theta(x - x_0) \\ 0 < \theta < 1}} \|T^{(n+1)}(\bar{x})\|$$

If T is a polynomial of degree n, its $(n + 1)$st Fréchet derivative is zero, and we have the classical version of Taylor's formula for scalar polynomials.

Adjoint self-adjoint, and symmetric operators can be defined in a Hilbert space H.

Definition: Let T be a linear operator on a Hilbert space H with dense domain. We define a related operator T^*, called the *adjoint* of T, as follows. The domain of T^* is the set $D(T^*) = \{y \,|\, (Tx,y)$ is continuous as a function of x on $D(T)$, the domain of $T\}$. For all $y \in D(T^*)$, T^*y is the unique element defined by the equation $(Tx,y) = (x,T^*y)$. T is symmetric if $(Tx,y) = (x,Ty)$ for all x in $D(T)$. T is self-adjoint if it is symmetric and $T = T^*$ [which implies that $D(T) = D(T^*)$].

Exercise 1-5 Show that T^* is closed.

For example, in R_n we have for a self-adjoint operator

$$(Tx,y) = \sum_{i=1}^{n} \sum_{k=1}^{n} t_{ik}x_k \bar{y}_i$$

$$(x,Ty) = \sum_{i=1}^{n} \sum_{k=1}^{n} x_i \bar{t}_{ik} \bar{y}_k = \sum_{i=1}^{n} \sum_{k=1}^{n} x_i t_{ki} \bar{y}_k$$

That is, for the self-adjoint property to hold, $\bar{t}_{ik}$ must equal t_{ki}. Self-adjoint operators are also called *hermitian* operators.

The operator T^* adjoint to the linear operator T is also linear, and $\|T\| = \|T^*\|$ if T is bounded.

Remark: If T^* is the adjoint of T then $Tx = y$ has a nontrivial solution only if y is orthogonal to all u satisfying $T^*u = 0$. Indeed if x is a solution, then $0 = (T^*u,x) = (u,Tx) = (u,y)$ and hence y is orthogonal to u. The converse is true if, for example, the range of T is closed.

An operator T is called *normal* if $TT^* = T^*T$. This is equivalent to $\|Tx\| = \|T^*x\|$ for all $x \in H$. A self-adjoint operator is also a normal operator. A special case of a normal operator is a unitary operator T which satisfies

$$TT^* = T^*T = I$$

Note that a linear operator is unitary if it does not change the norm of an element, that is $\|Tx\| = \|x\|$, and transforms H onto all of H.

Exercise 1-6 Show that the convolution operator defined previously is a normal operator.

Definition: An operator T defined on a normed linear space E whose range is in the normed linear space E_1 is called *completely continuous* if it maps every bounded set of E into a set in E_1 whose closure is compact.

Example: If $K(s,t)$ is a continuous kernel on $0 \leq s,\ t \leq 1$, then the operator

$$(Tx)\ (s)\ =\ y(s)\ \equiv\ \int_0^1 K(s,t)x(t)dt$$

is completely continuous on $L_2\ (0,1)$.   This is shown by proving that all $y(s)$ are uniformly bounded and equicontinuous and hence form a set with compact closure.   This gives uniform convergence and hence also convergence in the mean, as required in L_2.   This is also the case if $K(s,t)$ is an arbitrary square-integrable kernel.

If T is completely continuous and S is a bounded linear operator, TS and ST are completely continuous.   This follows from the fact that S maps a bounded set to a bounded set and T maps a bounded set to a set whose closure is compact.   Again the map by T is compact, and since S is continuous, it maps a compact set into a compact set.

If T is completely continuous and linear, it cannot have a bounded inverse in an infinite-dimensional space, since $TT^{-1} = I$ would be completely continuous.   But I cannot be completely continuous.   In finite-dimensional space every linear operator is completely continuous, since it maps a bounded set into a bounded set, and such a set always has compact closure.

Definition: An operator T is *positive* if it is self-adjoint and $(Tx,x) \geq 0$ for all x; it is positive definite if for some $m > 0$

$$(Tx,x)\ \geq\ m\ \|x\|^2$$

for all x.

We now introduce the concept of a projection operator.   Let L be a closed subspace of a Hilbert space H so that any $x \epsilon H$ is uniquely representable in the form $x = y + z$, where $y \epsilon L$ and $z \perp L$ [$x \perp y$ means $(x,y) = 0$, and $x \perp S$, where S is a set, means that $x \perp y$ for all $y \epsilon S$]. Then $(x - y) \perp L$, and the operator which determines y is called a *projection operator*, and y is called the *projection* of x onto L.

We give the following without proof.

Theorem 1-3   A linear operator P on H is a projection operator if and only if:

1. $P(Px) = Px$ (idempotence).
2. $Px \perp (y - Py)$ for all $x,\ y \epsilon H$.
3. $\|Px\| \leq \|x\|$.

A projection operator is self-adjoint (hermitian) and positive.   To show this let $x,\ y \epsilon H$.   Then $(Px,\ y - Py) = 0$ from condition 2, and thus $(Px,y) = (Px,Py) = (x,Py)$, and P is therefore both positive and hermitian.

A widespread application of projection operators is the approximation of functions $x \, \epsilon \, L_2$ by an infinite sequence of orthonormal functions (see "NM"), e.g., Legendre polynomials and trigonometric systems, to which the following theorem of Riesz and Fischer[67] applies.

Theorem 1-4   If $\{\varphi_n\}$ is an arbitrary orthonormal system of functions in L_2, given real numbers λ_n such that $\sum_{n=1}^{\infty} \lambda_n{}^2 < \infty$, then:

1.   The series $\sum_{n=1}^{\infty} \lambda_n \varphi_n$ converges in the mean to some function $x \, \epsilon \, L_2$.

2. λ_n are the Fourier coefficients of this function, determined by the projection operator

$$\lambda_n = \int_a^b \varphi_n(t) x(t) \, dt$$

the integral being taken in the Lebesgue sense.

3.   For every x

$$\|x\|^2 = \sum_{n=1}^{\infty} \lambda_n{}^2$$

An important property of a self-adjoint positive operator T is that it has a "square root," i.e., there exists a unique positive self-adjoint operator S such that $SS = T = S^2$.

In some cases the hypothesis of a completely continuous or of a contractive operator is too restrictive. With this in mind, Minty, Dolph, and Browder[4,7,48,49] have worked toward the development of an intermediate theory that will be applicable to integral equations of the Hammerstein type when the associated operators do not have the above properties. Below we shall describe Minty's concept of a monotone operator, defined on a Hilbert space. It is developed in the study of existence of solutions of Hammerstein equations given earlier, where a solution $y \, \epsilon \, H$ is such that, for a given $u \, \epsilon \, H$, the integral exists, and the equation is satisfied as an identity in s.

Recall that if f is a real-valued function defined on R_1, then it is said to be *monotone nondecreasing* if, whenever $x_1 < x_2$, we have $f(x_1) \leq f(x_2)$; or, equivalently,

$$(x_1 - x_2)(f(x_1) - f(x_2)) \geq 0 \tag{1-2}$$

Now if R_1 is considered as a Hilbert space with norm defined as the usual absolute value $|x|$ and scalar product by the usual product $(x,y) = xy$, then the function f is monotone nondecreasing if it satisfies

$$(x_1 - x_2, f(x_1) - f(x_2)) \geq 0$$

This concept is used below in the definition of a monotone operator.

An operator T (not necessarily linear) which maps a Hilbert space H into itself is called *monotonic* if, for x_1, $x_2 \epsilon H$,

$$\text{Re } (x_1 - x_2, Tx_1 - Tx_2) \geq 0$$

It is strictly monotonic if $\geq$ can be replaced by $>$ for $x_1 \neq x_2$, strongly monotonic if 0 on the right can be replaced by $C\|x_1 - x_2\|^2$ with $C > 0$, and linear if $(x,Tx) \geq 0$. Thus a linear operator T on a real Hilbert space is monotone if $(x,Tx) \geq 0$; note that T is not required to be self-adjoint.

Some of the theorems to be introduced later become intuitively clear when the operator is thought of as a real-valued continuous monotone nondecreasing function on R_1.

As an example, suppose f is such a function, and suppose further that there exists a positive M such that

$$|x| > M \text{ implies } (x,f(x)) \geq 0$$

Then the equation $f(x) = 0$ has a solution.

Geometrically this means that if x_1 is sufficiently large and negative, with $|x_1| > M$, then $f(x_1)$ must be nonpositive to ensure that the scalar product is nonnegative, and, conversely, when x_2 is sufficiently positive, then $f(x_2)$ must be nonnegative. Then, since f is continuous, it must take on all values between $f(x_1)$ and $f(x_2)$ and thus must cross the x axis at least once.

Theorem 1-5   If $E \subset H$ is convex and T is defined and has a directional derivative, then T is monotone on E provided that, for any $x_1 \epsilon E$, the *directional* derivative is ≥ 0, that is,

$$\frac{d}{dt} \text{Re } (h,T(x + th))_{t=0} \geq 0$$

where $h = x - x_1$ and t is real.

Proof: For any x_1, $x_2 \epsilon E$, $0 \leq \alpha \leq 1$, consider the real function

$$f(\alpha) = \text{Re } (x_1 - x_2, T(\alpha x_1 + (1 - \alpha)x_2))$$

By the mean-value theorem there exists a ξ such that $0 < \xi < 1$ and

$$
\begin{aligned}
\text{Re}\,(x_1 - x_2,\, Tx_1 - Tx_2) &= f(1) - f(0) \\
&= \left[\frac{d}{dt}\,\text{Re}\,(x_1 - x_2,\, T(\alpha x_1 + (1 - \alpha)x_2))\right]_{\alpha=\xi} \\
&= \lim_{\Delta\alpha\to 0} \frac{\text{Re}\,(x_1 - x,\, T(x + (\Delta\alpha/\xi)(x - x_2)) - Tx)}{\Delta\alpha/\xi} \\
&= \left[\frac{d}{dt}\,\text{Re}\,(h, T(x + th))\right]_{t=0}
\end{aligned}
$$

where $x = \xi x_1 + (1 - \xi)x_2$.

That is, $f(1) - f(0)$ coincides with the length of the interval $\alpha = 0$ to $\alpha = 1$ times the derivative evaluated at some point h in the interval; if the product is positive, then T is monotonic. Q.E.D.

Here and in a later section we give a review of some results contained in the book[66] by Vainberg concerned with the use of potential operators in variational methods applicable to the study of existence of solutions of nonlinear operator equations. The principal results in Vainberg's work have been summarized by Rall in Ref. 4. Variational methods are also discussed in the last chapter of Ref. 40.

Vainberg's method is used to prove existence of solutions of nonlinear operator equations. The proofs are nonconstructive in nature, the actual solution of an equation requiring other techniques.

The usefulness of variational methods rests on the fact that it is simpler to produce theorems concerning the existence of maximum and minimum points of functionals than those concerning the existence of solutions of nonlinear operators. Therefore, if a functional can be found which is the derivative, in some sense, of the operator whose equation is to be solved, then the points at which the functional achieves its extreme values are the solutions of the operator equation.

Two differential operators are used, the Gateaux (weak) and the Fréchet (strong).

An operator which is the derivative of a functional on a subset of a Banach space is said to be potential on the subset or to be the gradient of the functional.[66]

Definition: If there exists a functional f such that $Tx = \text{grad } f(x)$ for all $x \in E \subset X$, where X is a Banach space, then T is said to be a *potential operator* on E.

Theorem 1-6 Let:

1. T be an operator from X to X^*, where X^* denotes the conjugate space of X.

2. T have a linear bounded Gateaux differential in the ball

$$
S = \{x : \|x - x_0\| < r\}
$$

3. The functional $(DT(x,h_1),h_2)$ be continuous in x at every point of S.

Then T is potential in S if and only if $(DT(x, h_1), h_2)$ is a bilinear symmetric operator for every $x \, \epsilon \, S$; that is,

$$(DT(x,h_1),h_2) = (DT(x,h_2),h_1)$$

Proof: Necessity is shown by assuming the existence of a functional $f(x)$ such that $Tx = \mathrm{grad}\, f(x)$ and forming

$$\Delta = f(x + ah_1 + bh_2) - f(x + ah_1) - f(x + bh_2) + f(x)$$

where $x \, \epsilon \, S$, and $h_1, \, h_2 \, \epsilon \, X$, $\|h_1\| = \|h_2\| = 1$, and a and b are such that $x + uh_1 + vh_2 \, \epsilon \, S$, for $0 \leq u \leq a$, $0 \leq v \leq b$.

Δ is now transformed by applying successfully the mean-value theorem:

$$f(x + h) - f(x) = Vf(x + th,h), \text{ for some } t \, \epsilon \, (0,1)$$

Grouping the four terms of Δ in two ways gives

$$\Delta = ab(DT(x + t_1bh_2 + t_2ah_1, h_1),h_2)$$
$$\Delta = ab(DT(x + t_3bh_2 + t_4ah_1, h_2),h_1)$$

Then
$$\lim_{a,b \to 0} \frac{\Delta}{ab} = (DT(x,h_1),h_2) = (DT(x,h_2),h_1)$$

To prove sufficiency, consider the functional

$$f(x) = f(x_0) + \int_0^1 (T(x_0 + t(x - x_0)), x - x_0)\, dt \qquad (1\text{-}3)$$

Then

$$f(x + h) - f(x)$$
$$= \int_0^1 (T(x_0 + t(x - x_0) + th) - T(x_0 + t(x - x_0), x - x_0))\, dt$$

Representing this integral by

$$I = \int_0^1 dt \int_0^t \frac{\partial}{\partial s} (T(x_0 + t(x - x_0) + sh), x - x_0)\, ds$$
$$= \int_0^1 dt \int_0^t (DT(x_0 + t(x - x_0) + sh, h), x - x_0)\, ds$$

(this integral exists since DT is continuous in X), we note that since

$$(DT(x,h_1),h_2) = (DT(x,h_2),h_1)$$

we have
$$I = \int_0^1 dt \int_0^t (DT(x_0 + t(x - x_0)$$
$$+ sh, x - x_0),h)\, ds$$

Interchanging the order of integration, we have

$$I = \int_0^1 ds \int_s^1 (DT(x_0 + t(x - x_0) + sh, x - x_0), h) \, dt$$

$$= \int_0^1 (T(x_0 + (x - x_0) + sh) - T(x_0 + s(x - x_0) + sh), h) \, ds$$

When this last expression is substituted back, the last integral cancels the first, and we have

$$f(x + h) - f(x) = \int_0^1 (T(x + sh), h) \, ds$$

By condition 3 of the theorem, $(T(x + sh), h)$ is a continuous function of s on $[0,1]$. By the mean-value theorem,

$$f(x + h) - f(x) = (T(x + \tau h), h) \text{ for some } \tau, \, 0 < \tau < 1$$

so that $\quad \dfrac{1}{t} (f(x + th) - f(x)) = (T(x + \tau th), h)$

As was shown earlier, if T has a Gateaux differential, then T is continuous. Thus by condition 2,

$$\lim_{t \to 0} \frac{1}{t} (f(x + th) - f(x)) = (Tx, h)$$

and therefore, grad $f(x) = Tx$. Q.E.D.

Theorem 1-7 If A is a linear bounded operator on a real Hilbert space H and T is a potential operator in H with a linear bounded Gateaux differential $DT(x,h)$ continuous in X, then AT is potential if and only if

$$(ADT(x,h_1), h_2) = (DT(x, A^*h_1), h_2)$$

for all $h_1, h_2 \epsilon H$. Therefore

$$ADT(x,h_1) = DT(x, A^*h_1) \qquad \text{for all } h_1 \epsilon H$$

Proof: To show sufficiency note that

$$(ADT(x,h_1), h_2) = (DT(x, A^*h_1), h_2)$$

since $\quad (ADT(x,h_2), h_1) = (DT(x,h_2), A^*h_1) = (DT(x, A^*h_1), h_2)$

because T is potential and by Theorem 1-6. Then

$$(ADT(x,h_1), h_2) = (DT(x, A^*h_1), h_2)$$

and since $ADT = DAT$, because of the linearity of A, the operator $DATx$ is potential by the statement of the theorem.

Necessity is easy to prove.

Example: If A is linear in H and I is the identity operator in H, since $DI(x,h) = h$, I is a potential operator in H which satisfies the conditions of the above theorem.   Therefore, $AI = A$ is a potential operator if and only if

$$ADI(x,h) = Ah = DI(x,A^*h) = A^*h$$

that is, only if $A^* = A$.

From Eq. (1-3) it turns out that the self-adjoint linear operator A is the gradient of the quadratic functional $f(x) = \frac{1}{2}(Ax,x)$.

Theorem 1-8   If:
1. T is an operator from X to the conjugate space X^*,
2. T is continuous in an open simply connected region of X, $W \subset X$, then in order that T be a potential operator in W it is necessary and sufficient that the curvilinear integral

$$\int_L (Tx,dx)$$

be independent of the path.[48]

A potential operator can always be used to solve equations of the type

$$x = KTx$$

in real Hilbert spaces, where K is positive, self-conjugate, and linear on H and Tx is the gradient of a functional $f(x)$.   To do so, we introduce an operator A such that $A^2 = AA = K$, where A is also self-conjugate and linear on H, and the functional

$$g(x) = \frac{1}{2}(x,x) - f(Ax)$$

We form $G(x)$, the gradient of $g(x)$, using

$$\operatorname{grad} f(Ax)A = AT(Ax)$$

which is obtained from the self-conjugate property.   Then

$$G(x) = x - AT(Ax)$$

Let

$$G(x) = 0 \text{ at } x = z_0$$

that is, $z_0 = AT(Az_0)$

then $Az_0 = A^2T(Az_0) = KT(Az_0)$

and hence $x_0 = Az_0$

is a solution of $x = KTx$.

As we shall see later, potential operators can also be used in proving existence of solutions of the Hammerstein equation.

1-3 INVERSE OPERATORS AND THE SOLVABILITY OF EQUATIONS[43,61]

The concept of an inverse operator arises in connection with questions about the existence and uniqueness of solutions of the equation $Ax = y$ for given y, where $x \in X$, $y \in Y$, and X and Y are linear spaces. Familiarity with many equations of this form serves to emphasize the importance of the inverse operator. In this section we direct our attention to two types of problems:

1. Those concerned with the existence of an inverse of a linear (additive and homogeneous) operator.

2. Those of implicit-function theory, which provides conditions for the existence of local, and in some cases global, inverses of more general operators.

These will be discussed in order after the following introductory remarks. Suppose that T is a linear mapping from a linear space X to a linear space Y such that the equation

$$Tx = y$$

has a unique solution for *every* $y \in Y$. We then write $x = T^{-1}y$, where the inverse T^{-1} of T is *defined* by the one-one correspondence between $y \in Y$ and $x \in X$. One can easily show that $(T^{-1})^{-1} = T$ and hence that $TT^{-1} = T^{-1}T = I$, where I is the identity operator. We see that in such a case the operator T^{-1} is both a left and a right inverse of the operator T and (as above) is simply called the *inverse of T*.

However, an operator T may have only a left inverse S such that $ST = I$ or only a right inverse U such that $TU = I$. If T has only a right inverse, $Tx = y$ has a solution for all $y \in Y$, but this solution need not be unique. If T has only a left inverse, the equation need not have a solution, but if it does, the solution is unique. In general neither of these inverses S, U need be unique, and there may be an infinite number of either. However, it can be shown that if a *linear bounded* operator has a left inverse S and a right inverse U, then they are unique, and $S = U = T^{-1}$. Also if S (or U) exists and is unique, then also U (or S) exists, and again $S = U = T^{-1}$. (An important theorem, due to Banach, asserts: Let X and Y be Banach spaces and let T be a *one-to-one* bounded linear map from X onto Y. Then T^{-1} is a bounded linear operator.)

The problem of defining an inverse to a linear operator T (bounded or unbounded) involves three basic considerations:

1. An inverse exists if the operator T is one-one, for then the inverse correspondence is automatically given. It is easy to show that a necessary and sufficient condition that $Tx = y$, T additive, defines a one-one correspondence between $x \in X$ and $y \in R(T)$ (the range of T) is that $Tx = \mathbf{0}$

have only the $x = 0$ solution; also that if the inverse exists, it is automatically additive and homogeneous.

2. Suppose now that T is one-one, and hence has an inverse. The range of T need not be the entire space; that is, T may not be *onto*. In that case T^{-1} is not defined for all $y \in Y$. However, we can come close to requiring that T be onto by settling for the fact that its range, which is also the domain of T^{-1}, may be dense in the space Y.

3. Finally it is desired that the inverse operator exist and be continuous, i.e., bounded. This requirement is equivalent to the existence of a positive number C such that $\|Tx\| \geq C\|x\|$ for every $x \in X$. To prove this let $Tx = y$ and let T^{-1} exist and be bounded; then $\|x\| \leq \|T^{-1}\| \, \|y\|$, and if $\|T^{-1}\| = 1/C$, we have necessity. Conversely, if $\|Tx\| \geq C\|x\|$, then $T(0) = 0$ implies $x = 0$, and thus T^{-1} exists. Also $x = T^{-1}y$ implies that $\|T^{-1}y\| \leq (1/C)\|y\|$, and that T^{-1} is bounded.

Definition: A linear bounded operator mapping a normed space X to a normed space Y is said to be reversible (or invertible) if it is one-one and onto and if its inverse is bounded, i.e., considerations 1 to 3 hold. From the above considerations we see, for example, that the operator T may be linear, bounded, one-one, and defined for all of X, but T^{-1} may exist and be defined only on a subspace $R[T]$ of Y. In that case T^{-1} maps $R[T]$ onto X according to $y = Tx$. On the other hand, T may be a linear unbounded operator that is one-one but defined only on an everywhere-dense subspace of X, and T^{-1} may be linear, bounded, and defined on the entire space. For example $Tx = \int_0^t x(s) \, ds$ is bounded on $X = Y = C[0,1]$, but its inverse $T^{-1}y = (d/dt)y(t)$ is unbounded on the dense subset of continuously differentiable functions. The Sturm-Liouville operator $\left[\dfrac{d}{dt}\left(f(t)\,\dfrac{d}{dt}\right) + g(t)\right]$ is unbounded on the subspace of twice continuously differentiable functions, but its inverse $\int_0^1 K(t,s)y(s) \, ds$ $[K(t,s)$ being the Green's function] is linear, bounded, and defined on all of X. We have given in consideration 3 a condition which ensures the boundedness of the inverse.

If we now consider the problem of finding an inverse to the operator $T - \lambda I$, where T is linear and λ is an arbitrary complex number, the same set of conditions applies. However here we have the flexibility of choosing λ (usually possible if T is bounded) for which properties 1 to 3 are satisfied and excluding the remaining values in the complex λ plane. Such values of λ as are excluded are said to belong to the spectrum of T. The complement of this set is called the *resolvent set*. On this set $T - \lambda I$ has a bounded inverse with dense domain called the *resolvent of* T and given by $R_\lambda \equiv (T - \lambda I)^{-1}$. Those λ which belong to the resolvent set are called *regular*. One method of applying the resolvent to both sides of $(T - \lambda I)x = y$ in order to solve for x is to use the expansion

$$(T - \lambda I)^{-1} = - \sum_{n=0}^{\infty} \frac{T^n}{\lambda^{n+1}}$$

which converges if the series $- \sum_{n=0}^{\infty} \frac{\|T^n\|}{|\lambda^{n+1}|}$ converges. The latter numerical series converges whenever $\|T\| < |\lambda|$, which, as we shall see later in this section, is true only if λ is regular. Thus if λ is in the spectrum, then $\|T\| > |\lambda|$. There might be regular points inside the circle with radius $|\lambda|$

Self-adjoint operators have wide applicability in the theory of matrices and of linear differential and integral equations. They enjoy many properties not satisfied by operators in general. For example if T is self-adjoint, the range of $T - \lambda I$ is dense for λ regular. If, *in addition*, T is completely continuous, then $T - \lambda I$ is an onto mapping.

Remarks: The adjoint T^* of a linear operator T is one-one if and only if the range of T is dense. A unitary operator transforms Hilbert space H in a one-one manner onto itself and has a bounded inverse defined by

$$U^{-1} = U^*$$

where U^{-1} is also a unitary operator and so does not change a scalar product.

If a closed linear transformation has an inverse T^{-1}, then T^{-1} is also closed.

We need the following definition before we give another important inversion theorem. A set in a topological space is said to be of the first category if it is the union of a countable number of sets none of which is dense in any open subset of the space. Otherwise it is said to be of the second category.

If $y = Tx$ is a closed linear transformation with domain $D[T] \subset X$, and if its range $R[T]$ is of the second category in Y, then $R[T] = Y$, and there is a constant $C > 0$ such that $y \epsilon Y$ implies the existence of $x \epsilon X$, with $y = Tx$ and $\|x\| \leq C\|y\|$. If T^{-1} exists, it is bounded.

If $y = Tx$ is a closed linear transformation between Banach spaces whose domain is the whole space X, then T is bounded, and if T^{-1} exists and T is onto, then T^{-1} is also bounded.

If $y = Tx$ is a linear transformation mapping Hilbert space H into itself, and if $(Tx,y) = (x,Ty)$ holds for $x\ y \epsilon H$, then T is bounded.

After the foregoing preliminaries we shall present a somewhat more elaborate discussion of an inverse.

***Theorem* 1-9.** If T is a linear bounded operator mapping the Banach space X into itself, and if:

1. $\|T\| \leq \alpha < 1$, for some α, and
2. I is the identity operator in X,

then $I - T$ is reversible, and $\|(I - T)^{-1}\| \leq 1/(1 - \alpha)$.

Proof: For all $x \in X$

$$\|(I - T)x\| = \|Ix - Tx\| = \|x - Tx\| \geq \|x\| - \|Tx\|$$
$$\geq \|x\| - \|T\|\,\|x\| \geq \|x\| - \alpha\|x\|$$
$$= (1 - \alpha)\|x\|$$

Thus $(I - T)^{-1}$ exists and is bounded. If we assume $y = (I - T)x$, then $x = (I - T)^{-1}y$, and $\|(I - T)^{-1}y\| \leq 1/(1 - \alpha)\|y\|$, and therefore

$$\|(I - T)^{-1}\| \leq \frac{1}{1 - \alpha}$$

It remains only to show that $I - T$ is an onto mapping. Take any $y \in X$ and construct the operator $\tilde{T}x = Tx + y$ (note that if $y \neq \mathbf{0}$, $\tilde{T}$ is nonlinear). Choosing $x_1, x_2 \in X$ we have

$$\|\tilde{T}x_1 - \tilde{T}x_2\| = \|Tx_1 - Tx_2\| = \|T(x_1 - x_2)\| \leq \alpha\|x_1 - x_2\|$$

Under the conditions of the Banach contraction-mapping theorem (see "NM" or the next section), $\tilde{T}$ is a contraction operator with a single fixed point; i.e., there exists an $x \in X$ such that $\tilde{T}x = x = Tx + y$ or

$$(I - T)x = y \tag{1-4}$$

Therefore, any arbitrary $y \in X$ is a member of the set of values of the operator $(I - T)$, and hence this set of values coincides with X. This proves the theorem. For one application, see Picard's method in Chap. 2.

Remark: The contraction-mapping theorem also ensures that the solution of $(I - T)x = y$ is unique and can be found by the method of successive approximations beginning with an arbitrary x_0. Thus the series $\Gamma(y) = y + Ty + T^2y + \cdots + T^ny + \cdots$ converges, since $\|T^ny\| \leq \alpha^n\|y\|$, and we obtain $\|\Gamma(y)\| \leq \|y\|/(1 - \alpha)$. The successive approximations to the solution x^* of (1-4) will be

$$x_1 = Tx_0 + y$$
$$x_2 = Tx_1 + y = T^2x_0 + Ty + y$$
$$\cdots \cdots \cdots \cdots \cdots \cdots \cdots$$
$$x_n = Tx_{n-1} + y = T^nx_0 + T^{n-1}y + \cdots + Ty + y$$
$$\cdots \cdots \cdots \cdots \cdots \cdots \cdots$$

These approximations differ from the partial sums of the series $\Gamma(y)$ by T^nx_0, which approaches zero. Therefore, as $\Gamma(y)$ converges, so do the successive approximations to x^*.

Example: As an application of Theorem 1-9, consider the system of equations

$$\sum_{k=1}^{\infty} a_{ik}x_k = b_i \qquad i = 1, 2, \ldots$$

which can be simply transformed into

$$x_i - \sum_{k=1}^{\infty} c_{ik}x_k = b_i \qquad i = 1, 2, \ldots$$

where $c_{ik} = \delta_{ik} - a_{ik}$. Suppose that for all i

$$\sum_{k=1}^{\infty} |c_{ik}| \leq \alpha < 1 \qquad \alpha < 1$$

Define the matrix operator $Tx = y$ in the space m of bounded sequences as

$$y_i = \sum_{k=1}^{\infty} c_{ik}x_k \qquad i = 1, 2, \ldots$$

Obviously, T is linear, and the sequence $\{y_i\}$ is bounded. Let

$$\|y\| = \sup_i |y_i| \leq \alpha \|x\|$$

Then $\|T\| \leq \alpha < 1$. Assume further that $\{b_i\}$ is bounded. Then the second system can be written in the form $(I - T)x = b$. And since $\|T\| < 1$, Theorem 1-9 guarantees that $I - T$ is reversible, and by the previous remark the solution can be obtained by the method of successive approximations.

Theorem 1-10 Let T be a reversible bounded linear operator mapping a Banach space X onto a Banach space Y. If $\tilde{T}$ is linear and bounded, maps X into Y, and satisfies:

$$\|\tilde{T} - T\| \, \|T^{-1}\| \leq \alpha < 1,$$

then $\tilde{T}$ is reversible, and $\|\tilde{T}\| \leq [1/(1 - \alpha)]\|T^{-1}\|$.

Proof: Let $S = (T - \tilde{T})$. If α is such that $\|\tilde{T} - T\| \, \|T^{-1}\| \leq \alpha < 1$, then $\|S\| \leq \alpha/\|T^{-1}\|$. Therefore,

$$T^{-1}S = I - T^{-1}\tilde{T}$$

From the fact that $\|T^{-1}S\| \leq \|T^{-1}\| \, \|S\|$ there follows $\|T^{-1}S\| \leq \alpha$. Applying Theorem 1-9 to the operator $T^{-1}\tilde{T} = I - T^{-1}S$, we obtain the existence of an operator $U = (T^{-1}\tilde{T})^{-1}$ on the entire space X such that $\|U\| \leq 1/(1 - \alpha)$.

Since T^{-1} establishes a one-one and onto correspondence between Y and X and the set of values of $T^{-1}Y$ is the whole space X, the set of values of $\tilde{T}$ coincides with Y. Let A be a linear operator with values in

X such that $A = UT^{-1}$. Then

$$A\tilde{T} = UT^{-1}\tilde{T} = (T^{-1}\tilde{T})^{-1}(T^{-1}\tilde{T}) = I$$

Therefore, $A = \tilde{T}^{-1}$, and $\tilde{T}$ is reversible. In addition,

$$\|\tilde{T}^{-1}\| = \|A\| = \|UT^{-1}\| \leq \|U\|\,\|T^{-1}\| \leq \frac{\|T^{-1}\|}{1 - \alpha}$$

This completes the proof.

The next theorem and corollary are concerned with approximating the solution of a linear equation of the form $x - Tx = y$ by a solution of the linear equation $\tilde{x} - \tilde{T}\tilde{x} = y$, where y in both equations is the same and T and $\tilde{T}$ are linear and bounded and related in a prescribed manner.[67]

Theorem 1-11   If T and $\tilde{T}$ are bounded linear operators mapping a Banach space X into itself and:
  1. $I - T$ is reversible,
  2. $\|(I - T)^{-1}\| < 1/\alpha$, where α is a real constant, and
  3. $\|\tilde{T} - T\| \leq \alpha$,
then the operator $(I - \tilde{T})$ is also reversible.

The following corollary gives the nearness of the approximation $\tilde{x}$ to x.

Corollary to Theorem 1-11: If $x = (I - T)^{-1}y$ and $\tilde{x} = (I - \tilde{T})^{-1}y$ for the same y, then

$$\|\tilde{x} - x\| \leq \alpha\|(I - \tilde{T})^{-1}\|\,\|x\| \tag{1-5}$$

  Proof:

$$\begin{aligned}
\tilde{x} = (I - \tilde{T})^{-1}y &= (I - \tilde{T})^{-1}(I - T)x \\
&= (I - \tilde{T})^{-1}[(I - \tilde{T}) + (\tilde{T} - T)]x \\
&= (I - \tilde{T})^{-1}(I - \tilde{T})x \\
&\qquad\qquad + (I - \tilde{T})^{-1}(\tilde{T} - T)x \\
&= x + (I - \tilde{T})^{-1}(\tilde{T} - T)x
\end{aligned}$$

Therefore $\tilde{x} - x = (I - \tilde{T})^{-1}(\tilde{T} - T)x$

thus $\|\tilde{x} - x\| \leq \|(I - \tilde{T})^{-1}\|\,\|\tilde{T} - T\|\,\|x\| \leq \alpha\|(I - \tilde{T})^{-1}\|\,\|x\|$

Thus Theorem 1-11 and the above corollary guarantee the existence of a solution to the approximate equation if T is sufficiently near to $\tilde{T}$. As (1-5) makes use of $\|x\|$, the norm of the unknown solution to the original equation, we as yet have no estimate of the error. Such an estimate is given by the following.

Theorem 1-12   If $\alpha\|(I - T)^{-1}\| \leq \beta < 1$, and if x and $\tilde{x}$ are defined as above, then

$$\|\tilde{x} - x\| \leq \frac{\beta}{1 - \beta}\,\|\tilde{x}\|$$

Proof: Using (1-5), we have

$$\|\bar{x} - x\| \leq \beta\|x\| \leq \beta(\|x - \bar{x}\| + \|\bar{x}\|)$$
$$= \beta\|\bar{x} - x\| + \beta\|\bar{x}\|$$

Therefore $\quad (1 - \beta)\|\bar{x} - x\| \leq \beta\|\bar{x}\|.$

The following variational principle for linear operator equations in Hilbert space provides an extremal characterization of a solution and is useful for the application of the Ritz method, studied in a later section.

Theorem 1-13 Let T be a linear bounded positive self-adjoint operator. Any element x_0 in the real Hilbert space H is a solution of $y = Tx$ for any fixed $y \epsilon H$ if and only if the Ritz functional

$$Fx \equiv (Tx,x) - 2(x,y)$$

has its least value at $x = x_0$.[67]

Proof: For any $x_0, z \epsilon H$

$$F(x_0 + z) = (T(x_0 + z), x_0 + z) - 2(x_0 + z, y)$$
$$= (Tx_0,x_0) + (Tz,x_0) + (Tx_0,z) + (Tz,z) - 2(x_0,y) - 2(y,z)$$

Since T is self-adjoint, $(Tz,x_0) = (Tx_0,z)$, and

$$F(x_0 + z) = Fx_0 + 2(Tx_0 - y, z) + (Tz,z)$$

If x_0 is a solution, $Tx_0 - y = 0$; and because T is positive-definite, we have

$$(Tz,z) \geq 0$$

Thus $\quad F(x_0 + z) = Fx_0 + (Tz,z) \geq Fx_0$

or $\quad\quad Fx \geq Fx_0 \quad\quad x, x_0 \epsilon H, Tx_0 = y$

We shall omit the other half of the proof.

PROPER - VALUE PROBLEMS; THE SPECTRUM AND RESOLVENT OF AN OPERATOR

We now discuss equations of the form

$$(T - \lambda I)x = y \tag{1-6a}$$

which is often investigated in the form

$$(\lambda I - T)x \equiv T_\lambda x = y \tag{1-6b}$$

The linear operator T_λ now depends on the complex parameter λ. We assume that T itself is linear *and* bounded. The operator T_λ maps the Banach space X onto $X_\lambda \subset X$. Many of the previous ideas on inverses occurring in the study of $Tx = y$ can now be transferred to this equation by choosing the appropriate values of λ for which a bounded linear inverse exists. Thus we exclude those λ for which the linear operator T_λ is not one-one and also those for which in addition it does not have a bounded inverse. A characterization of both these conditions has been given before. Also those λ are excluded for which $\bar{X}_\lambda \neq X$. For a self-adjoint operator T, the range of $(T - \lambda I)$ is dense in X for all λ regular.

Let T be a linear bounded operator and consider the inverse operator $T_\lambda^{-1} \equiv (\lambda I - T)^{-1}$ if it exists. The operator T_λ^{-1} is also linear, and if it is bounded and has domain dense in X, then it is known as the *resolvent operator* of T_λ and is denoted by $R(\lambda, T)$.

Those values of λ for which $(\lambda I - T)x = y$ has a bounded inverse with dense domain for all y are called *regular values* of T. The totality of non-regular values of T is known as the *spectrum* of T. The spectrum of T may be divided into the following categories:

1. Those values of λ for which T_λ is not one-one are said to belong to $\sigma_p(T)$, the point spectrum of the operator T. Thus for these λ, $(\lambda I - T)x = \mathbf{0}$ has nontrivial solutions. They are known as eigenvalues and the corresponding solutions are known as eigenelements. To say that T_λ is not one-one for λ_0 is equivalent to the requirement that $(T - \lambda_0 I)x = \mathbf{0}$ has a nonzero solution. For if x_1 and $x_2 \, \epsilon \, X$, $x_1 \neq x_2$, and $T_{\lambda_0}x_1 = y_0$, $T_{\lambda_0}x_2 = y_0$, then $T_{\lambda_0}(x_1 - x_2) = \mathbf{0}$; that is, $T_{\lambda_0}x = \mathbf{0}$, $x = x_1 - x_2$. The converse is obvious, as T_{λ_0} maps x and $\mathbf{0}$ to zero. Also note from $Tx = \lambda_0 x$ that $\|T\| \cdot \|x\| \geq |\lambda_0| \cdot \|x\|$ and hence that $\|T\| \geq |\lambda_0|$. Thus those λ for which T_λ is not one-one satisfy $|\lambda| \leq \|T\|$.

2. Those values of λ for which $\lambda \, \notin \, \sigma_p(T)$ and X_λ is not dense in X are said to belong to $\sigma_r(T)$, the residual spectrum of the operator T. The condition of a dense range of T_λ corresponds to a dense domain for a linear bounded inverse T_λ^{-1}.

3. Those values of λ for which $\lambda \, \notin \, \sigma_p(T) \cup \sigma_r(T)$ and for which the map of the unit sphere in X is not bounded away from the origin are said to belong to $\sigma_c(T)$, the continuous spectrum of the operator T. For these λ the inverse exists but is unbounded, since it violates the boundedness theorem given before and consequently cannot be a continuous operator. Thus note that if in fact we write $\inf \|T_\lambda x\| = M > 0$, then because T_λ is also one-one, from $y = T_\lambda x$ we have $x = T_\lambda^{-1}y$, $y \, \epsilon \, X_\lambda$, $x \, \epsilon \, X$. Thus $\|T_\lambda T_\lambda^{-1}y\| \geq M\|T_\lambda^{-1}y\|$, or simply $\|y\| \geq M\|T_\lambda^{-1}y\|$, and hence

$$\|x\| = \|T_\lambda^{-1}y\| \leq (1/M)\|y\| = (1/M)\|T_\lambda x\|$$

ensuring boundedness.

Thus those λ for which there is an operator T_λ^{-1} such that

$$T_\lambda T_\lambda^{-1} = T_\lambda^{-1} T_\lambda = I$$

are regular and belong to $\rho(T)$, the complement of $\sigma(T)$ called the *resolvent set* of T.

Theorem 1-14   If T is a continuous linear operator, then:

1. $\rho(T)$ and $\sigma(T)$ are disjoint, and their union is the whole complex plane.

2. $\rho(T)$ is open, and $\sigma(T)$ is a bounded closed set.

3. $T_\lambda - T_\mu = (\mu - \lambda)T_\lambda T_\mu$ if λ, $\mu \, \epsilon \, \rho(T)$.

4. T_λ in the neighborhood of $\lambda_0 \, \epsilon \, \rho(T)$ is given by

$$T_\lambda = \sum_0^\infty (\lambda_0 - \lambda)^n T_{\lambda_0}{}^{n+1} \qquad |\lambda - \lambda_0| < \|T_{\lambda_0}\|^{-1}$$

5. σ is not void (unless the space X consists of a single element).

6.   If $|\sigma(T)| \equiv \inf_{\lambda \, \epsilon \, \sigma(T)} |\lambda|$, then

$$|\sigma(T)| \le \|T\| \qquad \text{and} \qquad -T_\lambda^{-1} = \sum_0^\infty \frac{T^n}{\lambda^{n+1}}$$

if $|\lambda| > |\sigma(T)|$; the series does not converge for $|\lambda| < |\sigma(T)|$.

Remark: Let T be a bounded *nonlinear* operator for which $T\mathbf{0} = \mathbf{0}$, and consider the equation (in a Banach space X)

$$y = \lambda T y + f \tag{1-7}$$

whose solution is y_0 for a given λ and f.   The *resolvent*, $R(\lambda, f)$, is the name given to the operator which associates a particular solution y_0 with each λ and f.

If T satisfies a Lipschitz condition

$$\|Tx_1 - Tx_2\| \le C\|x_1 - x_2\| \qquad C > 0$$

where x_1, $x_2 \, \epsilon \, S$ and S is a sphere of radius r about $\mathbf{0}$, then for $|\lambda| < 1/C$ and $f \, \epsilon \, X$, $\|f\| < (1 - |\lambda C|)\, r$, (1-7) has a unique solution in S.   From this and the fact that $T\mathbf{0} = \mathbf{0}$ we have $R(\lambda, f)\mathbf{0} = \mathbf{0}$ and

$$R(\lambda, f) = \lambda T R(\lambda, f) + f$$

In addition, it can be shown that $R(\lambda, f)$ is a continuous operator.   If $|\lambda| > \|T\|$, then $\|T/\lambda\| < \|I\| = 1$, and hence $\lambda I - T = \lambda(I - T/\lambda)$ is invertible by Theorem 1-9, and λ is regular.

A linear operator T has a resolvent $T_\lambda^{-1} = R(\lambda, T)$ if and only if there is at least one λ for which $R(\lambda, T)$ has an inverse.   This is always the case for bounded operators but need not be the case if an operator is unbounded.

The resolvent can be shown to be analytic in each component of $\rho(T)$ by 4 of Theorem 1-14; there may be several components. The component with $|\lambda| > \|T\|$ is called the *principal component* and in it

$$(\lambda I - T)^{-1} = \sum_{n=0}^{\infty} \frac{T^n}{\lambda^{n+1}}$$

The radius of convergence is given by $\lim_{n \to \infty} \|T^n\|^{1/n} \equiv r$, and the spectrum lies within this value; that is, $|\lambda| \leq r$.

One can introduce and generalize notions from function theory and show that a holomorphic bounded function in a domain containing the spectrum of a linear bounded operator T can be represented by

$$f(T) = \frac{1}{2\pi i} \int_{\Gamma} f(\lambda) R(\lambda, T) \, d\lambda \tag{1-8}$$

where Γ is a contour in the complex λ plane surrounding the spectrum of T.

The spectrum of an unbounded operator may occupy the whole complex plane. The resolvent of such an operator can be defined if and only if the resolvent set is nonempty.

For an unbounded closed linear operator we have

$$f(T) = f(\infty)I + \frac{1}{2\pi i} \int_{\Gamma} f(\lambda) R(\lambda, T) \, d\lambda$$

whenever f is analytic in an open set containing the spectrum, where the integration is taken over an appropriate contour Γ containing the spectrum of T and a neighborhood of ∞. For this representation to hold the spectrum must not cover the whole plane.

In a Hilbert space it is possible to obtain an integral representation of $f(T)$ by means of a spectral theorem for a wide class of f, for example, bounded and measurable, and for T linear and bounded and normal or unbounded and self-adjoint. This representation has the abstract form

$$f(T) = \int f(\lambda) \, dE_\lambda \tag{1-9}$$

which is a generalization of a similar result[43] for matrices discussed in "Nonlinear Mathematics."

Example: In $L(0,1)$ let $(Tx)(t) = \int_0^t x(s) \, ds$, $0 \leq t \leq 1$. The spectral radius of this operator is zero, and $\lambda = 0$ is in the continuous spectrum and is the only point in the spectrum. To see this we obtain the resolvent. We can compute powers of T, for example, T^n, by integrating n times from zero to t and using the formula (page 329) for reduction to a single integral with kernel $(t - s)^{n-1}$. We get

$$T^n x = \frac{1}{(n-1)!} \int_0^t (t - s)^{n-1} x(s)\, ds$$

Multiplying by $1/\lambda^{n+1}$ and summing over n to obtain the resolvent $R(\lambda, T)$, we have

$$R(\lambda, T) = \frac{x(t)}{\lambda} + \frac{1}{\lambda^2} \int_0^t x(s) e^{(t-s)/\lambda}\, ds$$

which converges for all λ except $\lambda = 0$.

Example: For linear unbounded operators the spectral approach is applicable provided that the spectrum does not occupy the entire λ plane. Here is an example of an unbounded operator whose spectrum occupies the whole plane. Consider the unbounded operator $T = d^2/dt^2$ acting on $f \in C^2[0, \infty]$, which is dense in the range of T, $C[0, \infty]$ (T is unbounded since it maps $\{e^{int}\}$ to $\{-n^2 e^{int}\}$ which $\rightarrow \infty$ as $n \rightarrow \infty$). Now $\sigma_p[T]$ is the entire plane except for those λ which are real and negative. This is easily seen by examining all λ for which solutions of $d^2f/dt^2 - \lambda f = 0$ are in $C[0, \infty]$. If $\lambda = 0$, then $f_1 = 0$, $f_2 = 1$ are such solutions. For any $\lambda \neq -\alpha^2$, α real, we have the two general independent solutions $f_1 = \exp(\sqrt{\lambda}t)$, $f_2 = \exp(-\sqrt{\lambda}t)$, and hence it follows that f_2 is always in $C[0, \infty]$. Let $\lambda = -\alpha^2$, α real. Then the functions $\exp(-1/n + i\alpha)t$, $n = 1, 2, \ldots$ (which clearly have unit norms), are transformed to the same set multiplied by $(1/n^2 - 2i\alpha/n)$, which is not bounded away from the origin as $n \rightarrow \infty$. Thus

$$\lambda = -\alpha^2 \in \sigma_c[T]$$

To show that $\lambda = -\alpha^2$ is not in $\sigma_r[T]$ (we know it is not in $\sigma_p[T]$) we note that the fundamental set $\{e^{-nt}\}$ is transformed to the fundamental set $(n^2 + \alpha^2)e^{-nt}$, and hence the mapping is one-one for $\lambda = -\alpha^2$.

Example: Now let us consider the transformation $[\lambda I - t(d/dt)]f = g$, f and $g \in C[0, \infty]$. This has the solutions in $C[0, \infty]$

$$f(t) = R(\lambda, T)g = \begin{cases} t^\lambda \int_t^\infty \dfrac{g(s)}{s^{n+1}}\, ds & \text{Re } (\lambda) > 0 \\[2ex] -t^\lambda \int_0^t \dfrac{g(s)}{s^{\lambda+1}}\, ds & \text{Re } (\lambda) < 0 \end{cases}$$

Note that one must show that $\|f(t)\|$ is bounded and that $f(t)$ is continuous at both end points. [For boundedness use sup and estimate $g(s)$ by its sup, which, of course, is $\leq \|g(s)\|$. For continuity use the mean-value theorem.] Thus the imaginary axis is in the spectrum, of which $\lambda = 0$ is the only value in the point spectrum [t^λ satisfies $\lambda f - t(df/dt) = 0$ and is in $C[0, \infty]$ only for $\lambda = 0$ as $\cos(\alpha \log t) + i \sin(\alpha \log t)$ does not tend to a limit for $\alpha \neq 0$]. The fundamental set $\{e^{-nt}\}$, $n = 0, 1, \ldots$, is transformed to a fundamental set by $i\alpha - t(d/dt)$. Thus the remainder of the imaginary axis is in the continuous spectrum.

Exercise 1-7 Show that the spectrum of the difference operator $Tf(t) = f(t + 1)$ on $C[0, \infty]$ is the unit circle. Show that $\lambda = 1$ is in $\sigma_p[T]$ and that $\lambda = e^{i\theta}$, $\theta \not\equiv 0$ mod 2π, is in $\sigma_c[T]$.

Let T be a linear self-adjoint operator. The range of $(T - \lambda I)$ is dense for regular λ; hence the residual spectrum is empty. The rest of the spectrum is real and lies between the upper and lower (norm) bounds of the operator, and these two values also belong to the spectrum. The entire λ space can be divided into two disjoint subsets, on each of which the operator has only one type of spectral values.

If T is both self-adjoint and completely continuous, then all nonzero values of the spectrum consist only of eigenvalues and are isolated points of the real λ axis. In the case of some operators the origin may be a limit point of such values. To each $\lambda \neq 0$ there corresponds a finite-dimensional eigensubspace. From each subspace we select an orthonormal basis and use the entire collection to obtain a Fourier expansion for any element in the range of T. This set of elements is complete if $\lambda = 0$ is not an eigenvalue of T.

If we consider an equation of the form $(T - \lambda I)x = y$, where T is self-adjoint and completely continuous, and where $\lambda \neq 0$ is an eigenvalue, we can in this case assert that this equation has a solution (without uniqueness) if and only if y is orthogonal to the eigensubspace corresponding to this eigenvalue λ, that is, if the scalar product of y with any fundamental solution of the homogeneous equation is zero. Note that the scalar product in a Hilbert space has definite meaning; e.g., it defines a norm, and orthogonality can be tested, for example, in L_2 by showing that $\int_a^b x(s)\bar{y}(s) \, ds$ vanishes for each solution of the homogeneous equation. There is a unique solution of $(T - \lambda I)x = y$ for any y if and only if $\lambda \neq 0$ is regular. For a singular matrix A, for $Ax = y$, we can write $A^T A x = A^T y$, and $A^T A$ is self-adjoint. We can assert that this equation has solutions if and only if $A^T y$ is orthogonal to every solution of $A^T A x = 0$.

Theorem 1-15 Let T be self-adjoint and completely continuous. The equation $(T - \lambda I)x = y$, where λ is nonzero, has a unique solution for all $y \in H$ if and only if the equation $(T - \lambda I)x = 0$ has $x = 0$ as its only solution.

Proof: The proof is based on the relationship between an eigensubspace M relative to a fixed λ and the orthogonal complement N of that subspace. In the eigensubspace the homogeneous equation has no solution, as λ is an eigenvalue. However, in the orthogonal complement λ is a regular value, and hence $T - \lambda I$ is reversible on that subspace. For any $x \in H$, x can be written $x = x_1 + x_2$, where $x_1 \in M$ and $x_2 \in N$. Therefore $(T - \lambda I)x = (T - \lambda I)(x_1 + x_2) = Tx_1 - \lambda x_1 + Tx_2 - \lambda x_2$. However, $Tx_1 - \lambda x_1 = 0$, and it is easily seen that the nonhomogeneous equation has a solution only for $y \in N$. Uniqueness follows only if $M = \emptyset$ (the empty set) since if x is a solution for some y, $x + \Delta$, where $\Delta \in M$, is also a solution for all $\Delta \in M$. Q.E.D.

In a Hilbert space, for example, the eigenelements, which we denote by φ_i, form an orthonormal system, and equations of the form $y = Tx$ can be expressed as

$$y = \sum_n c_n \varphi_n$$

where $c_n = (y, \varphi_n) = (Tx, \varphi_n) = (x, T\varphi_n) = \lambda_n(x, \varphi_n)$. If the operator is self-adjoint and completely continuous, we can find a complete set of φ_i which span the domain of T.

Self-adjoint operators can be used to develop a solution of the Fredholm equation

$$f(t) = \int_a^b K(s,t)g(s) \, ds$$

in a pre-Hilbert space G of continuous complex-valued functions in $I = [a,b]$, with $(f,g) = \int_a^b f(t)\overline{g(t)} \, dt$. If K is a hermitian kernel, in the Fredholm equation we define the operator T by

$$T(f(t)) = \int_a^b K(s,t)f(s) \, ds$$

which has eigenvalues λ_n and eigenfunctions φ_n such that

$$T\varphi_n = \lambda_n \varphi_n$$

Theorem 1-16   If K is a hermitian kernel, the series $\sum_n \lambda_n^2$ is convergent and $\sum_n \lambda_n^2 \leq \int_a^b dt \int_a^b |K(s,t)|^2 \, ds$.

This is shown using Bessel's inequality and $(\varphi_n, \varphi_m) = \delta_{nm}$ (see "NM").

Theorem 1-17   If K is a hermitian kernel and $y = Tx$ for $x \in X$, then the series $\sum_n c_n \varphi_n(t)$, where $c_n = (y, \varphi_n)$, converges absolutely to $y(t)$ in I.

Proof: The function x is decomposed into $x = \sum_n d_n \varphi_n$ with coefficients d_n.

Since T is a continuous linear mapping of X into Y (the space of all continuous complex-valued functions on I), we have $y = Tx = \sum_n \lambda_n d_n \varphi_n$

which converges. Let $c_n \equiv (y,\varphi_n) = (x,T\varphi_n) = \lambda_n(x,\varphi_n) = \lambda_n d_n$. The Cauchy-Schwarz inequality for finite-dimensional spaces gives

$$\Big(\sum_{n=1}^{N} |c_n \varphi_n(t)| \Big)^2 \leq \Big(\sum_{n=1}^{N} |d_n|^2 \Big) \Big(\sum_{n=1}^{N} \lambda_n{}^2 |\varphi_n(t)|^2 \Big)$$

where, by Bessel's inequality, the right side is bounded by a number independent of N and t. Thus the convergence is absolute.

Let T be a self-adjoint operator in Hilbert space H.[46] Corresponding to an eigenvalue λ_k is a linear manifold of eigensolutions which satisfy $T_{\lambda_k} x = 0$. (Why?) Let M_k denote the subspace of these solutions and let M be the subspace spanned by the entire set of eigensolution manifolds M_k. Every element of M can be developed into a sum of pairwise-orthogonal elements, each belonging to a different M_k for all k. If N is the orthogonal complement of M, that is, N is the subspace of elements in H orthogonal to M (prove that it is a subspace), we have, for $x \, \epsilon \, H$, the unique decomposition $x = y + z$, where $y \, \epsilon \, M$ and $z \, \epsilon \, N$. Now T, with domain D, can be decomposed into T_1 and T_2 with domains M and N, respectively. Now if $y \, \epsilon \, M$, then $y = \Sigma y_k$, and

$$T_1 y = T_1 \Sigma y_k = \Sigma T_1 y_k = \Sigma \lambda_k y_k$$

Thus the range of T_1 is also in M, and we have a spectral representation of T_1. If $\lambda \neq \lambda_k$ for any k, then $T_{1\lambda}^{-1}$ exists, and $T_{1\lambda} y = p$; that is, $T_1 y - \lambda y = p$ has a unique solution $y = T_{1\lambda}^{-1} p$ for p in $R[T_{1\lambda}]$; and because $y = \Sigma y_k$, $p = \Sigma p_k$, and $T_1 y - \lambda y = p$, which holds on account of $\Sigma \lambda_k p_k - \lambda \Sigma y_k = \Sigma p_k$; we equate coefficients (since the decomposition into M_k is unique), and we have $(\lambda_k - \lambda) y_k = p_k$, which, if solved for y_k and summed, yields

$$y = T_{1\lambda}^{-1} p = - \sum \frac{p_k}{(\lambda - \lambda_k)}$$

Here $y = y(\lambda)$ is analytic in λ except at the poles λ_k with respective residues $-p_k$, and $p = \Sigma p_k$ is given as a sum involving the negatives of the residues at λ_k. The last formula shows that the continuous spectrum of T_1 is empty.

Now T_2 operates in N and has no eigenvalues, since they are all related to M and there is no easy way of expanding an element $q \, \epsilon \, N$. However, if the above results are restated in a Stieltjes-integral form, it is possible to obtain an expansion for q, and an element of the space has an expansion which is the sum of those of p and q. The reader is urged to consult Ref. 45 on this subject.

Exercise 1-8[46]   Consider the operator $T \equiv -d^2/dt^2$, $0 \leq t \leq 1$, whose domain is restricted by the boundary conditions $x(0) = x(1) = 0$. Show that T is self-adjoint in $L_2[0,1]$. The equation $(\lambda I - T)x = f(x)$ is equivalent to $\ddot{x} + \lambda x = f(x)$, $x(0) = x(1) = 0$. Show that its solution is

$$x(t,\lambda) = \frac{\sin \sqrt{\lambda}\,(1-t)}{\sqrt{\lambda}\,\sin \sqrt{\lambda}} \int_0^x f(s)\,\sin \sqrt{\lambda}\,s\,ds$$

$$+ \frac{\sin \sqrt{\lambda}\,t}{\sqrt{\lambda}\,\sin \sqrt{\lambda}} \int_x^1 f(s)\,\cos \sqrt{\lambda}\,(1-s)\,ds$$

The only singularities of $x(t,\lambda)$ are the poles $\lambda_k = k^2\pi^2$, $k = 1, 2, \ldots$, and hence the continuous spectrum is empty, and the poles are the eigenvalues. Show that the residue of the pole $\lambda_k = k^2\pi^2$ is $-2 \sin k\pi t \int_0^1 f(s)\,\sin k\pi s\,ds$, from which obtain the eigensolution expansion

$$f(t) = 2 \sum_{k=1}^{\infty} \sin k\pi t \int_0^1 f(s)\,\sin k\pi s\,ds$$

which in fact is the Fourier sine expansion of $f(t)$ on $(0,1)$.

PSEUDO INVERSES [21,22,53]

The idea of a generalized, or pseudo, inverse has been introduced for operators in a manner similar to its definition for matrices. We shall discuss briefly the ideas for matrices and leave it to the student to pursue the ideas in the literature. If a matrix of real elements A is a rectangular matrix and we wish to solve $Ax = y$, we can multiply both sides by the transpose of A, obtaining $A^T A x = A^T y$. The matrix $A^T A$ is a square matrix, and if it is nonsingular, then the solution $x = (A^T A)^{-1} A^T y$ is unique. This is a least-squares solution of $Ax = y$ (a notion introduced for the sake of defining a solution when A is rectangular) in that it minimizes $(Ax - y)^T(Ax - y)$ and hence satisfies $A^T A x = A^T y$. If $A^T A$ is singular, then a number of solutions are possible. If the elements of A are complex, the conjugate transpose A^* is used.

Now we wish to introduce a concept of an inverse which subsumes the cases of A being a square (singular or not) and a rectangular matrix. In general one can show that for a given matrix A the four equations $AXA = A$, $XAX = X$, $(XA)^* = XA$, and $(AX)^* = AX$ have a unique solution $X = A^+$, where A^+ is a matrix known as the *generalized*, or pseudo, *inverse* of A. It has the following properties

$$A^+ = A^+AA^+ = (A^+A)^*A^+ = A^*(A^+)^*A^+$$
$$A^+ = A^+AA^+ = A^+(AA^+)^* = A^+(A^+)^*A^*$$

and hence $A^+(A^+)^*A^* = A^+ = A^*(A^+)^*A^+$

Exercise 1-9   Show that

$$A^+AA^* = A^* = A^*AA^+$$
$$(A^+)^+ = A$$
$$(A^*)^+ = (A^+)^*$$
$$A^+ = A^{-1} \qquad \text{if } A \text{ is nonsingular}$$
$$(\lambda A)^+ = \frac{1}{\lambda}A^+ \qquad \lambda \neq 0$$
$$(A^*A)^+ = A^+(A^*)^*$$
$$(AB)^+ \neq B^+A^+ \qquad \text{in general}$$

Exercise 1-10   If $A = \begin{pmatrix} 1 & 2 \\ 1 & 2 \end{pmatrix}$, then $A^+ = \begin{pmatrix} \frac{1}{10} & \frac{1}{10} \\ \frac{1}{5} & \frac{1}{5} \end{pmatrix}$.   Verify the above properties using this choice of A and A^+.

Theorem 1-18   A necessary and sufficient condition for the matrix equation $AXB = C$ to have a solution for given A, B, and C is $AA^+CB^+B = C$.   The general solution is given by

$$X = A^+CB^+ + Y - A^+AYBB^+$$

where Y is an arbitrary (dimensionally consistent) matrix.

Proof.   Note that if X is a solution, then

$$C = AXB = AA^+AXBB^+B = AA^+CB^+B$$

On the other hand, if $C = AA^+CB^+B$, then A^+CB^+ is a particular solution.   In addition, $X = Y - A^+AYBB^+$ gives all solutions of $AXB = \mathbf{0}$. The sum of these two solutions is the desired general solution.

Corollary: If the vector equation $Ax = b$ has a solution, the general solution is given by $x = A^+b + (I - A^+A)y$ for arbitrary y.

IMPLICIT - FUNCTION THEOREMS—LOCAL AND GLOBAL INVERSES

The concept of a general implicit function (see "NM," theorem 1-7) can be obtained by starting with the usual definition of implicit function [for example, $T(x,u) \equiv e^{xu} - x = 0$], in which we seek an expression for $u(x)$ from

$$T(x)u \equiv T(x,u) = \mathbf{0}$$

and by generalizing u, $T(x)u$, and x to real or complex numbers, vectors, elements of a Banach space, etc.   (As an exercise, solve for u in $u^2x^3 + 2u \sin x + 1 = 0$.)   Note that most of the equations listed in Sec. 1-1 are special cases of this equation.   In 1927, Hildebrandt and Graves[27] gave a general-implicit-function theorem for such an equation. Ehrmann[16] studies the equation $Tu = \mathbf{0}$, when u and Tu belong in a Banach space but no assumption is made that T itself belongs to any particular space. The solution u depends only on T; that is, $u = u(T)$. The continuity of $u(T)$ at a point $T = T_0$ can be established without assuming that T is continuous. The Hildebrandt-Graves theorem and other theorems on implicit functions, neighborhoods, i.e., existence of a neighborhood of a point in which a solution exists, perturbation, and continuous dependence on a parameter theorems are special cases of Ehrmann's theorems, which we illustrate here.

Remark: Even though we shall not be directly occupied with the following concept until Chap. 6, it is appropriate that the reader be aware of it now.[54]

Definition: A regular point of $T(x,y) = 0$, $x \in X$, $y \in Y$, $T(x,y) \in Z$, is any point which satisfies the equation, and $(f(y),y)$ is its unique solution, where $x = f(y)$ is a continuous transformation from Y into X. A branch point of the equation is a point which satisfies the equation but is not regular. A branch point $(0,y_0)$ is a bifurcation point of $T(0,y) = 0$ if for some $\alpha, \beta > 0$, $T(x,y) = 0$ for x and y satisfying

$$\|x\| < \alpha \qquad \|y - y_0\| < \beta$$

In conformity with our previous notation we have replaced u by y. Let ΔT denote $T - T_0$, and let D_T and D_{T_0} be the domains of T and T_0, respectively.

Definition: $T \in \Omega \equiv (x_0,r,a,b)$ neighborhood of an operator T_0 if and only if:

1. $\|(T - T_0)x_0\| = \|\Delta Tx_0\| < a$.
2. $\|\Delta Tx - \Delta Ty\| \leq b\|x - y\|$ for all x, y in the sphere $S(x_0,r)$ and in $D_T \cap D_{T_0}$.

Theorem 1-19 (*Implicit-function theorem*) Let X and Y be Banach spaces. If:

1. T_1 is an operator mapping the set $E \subset X$ into Y,
2. $S^* \equiv S(x^*,r^*) \cap E$ and $T_1x^* = 0$,
3. There exists a linear operator A on S^* to Y such that:
   a. A^{-1} exists, is defined on Y, and is bounded,
   b. There exists a constant $m < \|A^{-1}\|^{-1}$ such that $\|T_1x - T_1y - A(x - y)\| \leq m\|x - y\|$ for all $x, y \in S^*$,

then there is some $\Omega = (x^*,r,a,b)$ neighborhood of T_1 such that for all $T \in \Omega$

$$Tx = 0$$

has a unique solution $x = x(T)$ in $S(x^*,r)$. This solution is continuous in T at $T = T_1$; that is,

$$\|x(T) - x^*\| \to 0 \qquad \text{as } \|Tx^*\| \to 0$$

Note that T and A need not be continuous.

Proof: If $T \epsilon \Omega = (x^*, r, a, b)$, with $r \leq r^*$, then by condition $3b$ of the theorem and the definition of Ω neighborhood

$$\|Ty - Tx - A(y - x)\| \leq \|\Delta Ty - \Delta Tx\|$$
$$+ \|T_1 y - T_1 x - A(y - x)\| \leq (b + m)\|y - x\| \quad (1\text{-}10)$$

for $x, y \epsilon S(x^*, r) \cap E \subset S^*$.

Hence the equation $Tx = \mathbf{0}$ is equivalent to

$$x = T_2 x \equiv A^{-1}(A - T)x \qquad x \epsilon S(x^*, r) \cap E$$

For all positive b, with $\alpha = (b + m)\|A^{-1}\| < 1$, (1-10) yields

$$\|T_2 x - T_2 y\| = \|A^{-1}[A(x - y) - Tx + Ty]\| \leq C\|x - y\|$$

$C < 1$ for $x, y \epsilon S(x^*, r)$. It is easy to verify that if

$$\|T_2 x^* - x^*\| = \|A^{-1}Tx^*\| < (1 - C)r \qquad (1\text{-}11)$$

the hypotheses of the contraction-mapping theorem are satisfied.

Thus the unique solution $x = x(T)$ exists and satisfies

$$\|x - x^*\| \leq (1 - C)^{-1}\|A^{-1}Tx^*\|$$
$$\leq (1 - C)^{-1}\|A^{-1}\| \, \|Tx^*\| \qquad C < 1 \qquad (1\text{-}12)$$

Equation (1-12) implies the continuity of the solution T at $T - T_1$, and Eq. (1-11) is satisfied if $T \epsilon \Omega$ with

$$a = [\|A^{-1}\|^{-1} - (b + m)]r$$

This completes the proof.

1-4 EXISTENCE THEOREMS

In this section we present, for reference purposes, a number of useful existence theorems, some with proofs and some without. The purpose is to make the reader aware of the types of theorem available for the treatment of questions pertaining to equations in general. Some proofs and applications are also found in "NM," chap. 1.

General existence theorems for the solution of various explicit operator equations fall into two classes: (1) topological fixed-point theorems and (2) algebraic, or constructive, theorems. In the first class, the existence of a fixed point is guaranteed, but no method of finding this point is given. In the second, some form of a recursion definition is given by which the fixed point can be obtained through repeated application of the definition. If the space satisfies the Hausdorff separation axiom, then the limit of these iterations, if it exists, must be unique. Later we also give existence theorems, both local and global, for equations in implicit form.

TOPOLOGICAL FIXED-POINT THEOREMS

Here, we shall be interested in sets K which have the fixed-point property: for every continuous mapping T of K into K there exists an $x \in K$ such that $Tx = x$. We begin our discussion of topological fixed-point theorems with a basic theorem, due to Brouwer, which has served as the foundation for several other well-known theorems.

Theorem 1-20 Let S be a closed unit sphere in R_n and let T be a continuous mapping of S into S. Then there exists a point $x \in S$ such that $Tx = x$.

Proof: See "NM," page 42.

We shall first extend the Brouwer fixed-point theorem to Banach spaces, obtaining thus the Birkhoff-Kellogg-Schauder theorem, which appears in the following weak form, due to Schauder.[6]

Theorem 1-21 If:

1. X is a Banach space,
2. E is a convex compact subset of X,
3. T is a continuous mapping of E into E,

then T has a fixed point.

Proof:

1. First we show that for all $\epsilon > 0$ there exists a continuous mapping h of E into a finite-dimensional subset E_1 of E such that

$$\|h(x) - x\| < \epsilon$$

Since E is compact, we can find a finite number of points $x_1, \ldots, x_r$ in E such that every point in E will be at a distance $< \epsilon$ from at least one of the x_i.

Let $\omega(t)$ be any continuous function such that

$$\omega(t) \equiv \begin{cases} 0 & t \geq 1 \\ >0 & 0 \leq t < 1 \end{cases}$$

Let $\psi_i(x) = \omega(\rho(x,x_i)/\epsilon)$. Then, on E, $\sum_i \psi_i(x) > 0$.

Form

$$\varphi_i(x) = \frac{\psi_i(x)}{\sum_j \psi_j(x)}$$

φ_i is continuous on E, $\varphi_i(x) = 0$ if $\rho(x,x_i) \geq \epsilon$, and $\sum_i \varphi_i(x) = 1$.

Let F be the intersection of all convex sets containing $x_1, \ldots, x_r$, that is, the convex hull of $\{x_1, \ldots x_r\}$. As E is convex, $F \subset E$. Define

$$h(x) = \sum_{i=1}^{r} \varphi_i(x)x_i$$

Then $h(x) \in F$, and

$$\rho(h(x),x) = \|h(x) - x\| = \left\| \sum_{i=1}^{r} \varphi_i(x)x_i - \sum_{i=1}^{r} \varphi_i(x)x \right\|$$

$$\leq \sum_{\|x - x_i\| < \epsilon} \varphi_i(x)\|x_i - x\| < \epsilon$$

$h(x)$ is the sum of continuous functions and hence continuous.

2. By part 1, there exist finite-dimensional subspaces F_n of X and continuous mappings h_n such that

$$h_n : E \to F_n \cap E \qquad n = 1, \ldots, k$$

and

$$\| h_n(x) - x \| \leq \frac{1}{n}$$

Consider $T_n x = h_n[Tx]$. Then T_n maps E continuously into F_n. Also consider

$$T_n : F_n \cap E \to F_n \cap E$$

We can now use Brouwer's fixed-point theorem to show that each T_n has a fixed point, say x_n; that is, $T_n x_n = x_n$. Now

$$\|Tx - T_n x\| \leq \frac{1}{n}$$

and as $n \to \infty$

$$\|Tx - T_n x\| \to 0$$

This is true in particular for x_n; that is,

$$\|Tx_n - x\| \to 0$$

Since E is compact, the sequence $\{x_n\}$ has a convergent subsequence $\{x_{n_i}\}$. Denote $\lim x_{n_i} \equiv x^*$. Now

$$\|Tx_{n_i} - x_{n_i}\| \to 0$$

and therefore, in particular,

$$\|Tx^* - x^*\| = 0$$

and hence $x^* = Tx^*$ is a fixed point.   Q.E.D.

The following generalization of Theorem 1-21 is a stronger form of the Brouwer theorem, due to Leray.

Theorem 1-22 (*Schauder-Leray fixed-point theorem*)

1. Let E be a closed convex subset of a Banach space X.

2. Let T be a continuous operator mapping E into itself such that the image of E is compact.

Then there exists at least one $x^* \epsilon E$ such that $Tx^* = x^*$.

Proof: See "NM," page 45.

An important corollary of Theorem 1-22, due to Rothe,[57] follows.

Corollary: If, in addition to the hypotheses of the foregoing theorem, there exist α and β, $0 < \alpha < 1$ and $\beta \geq 0$, such that

$$\|Tx\| \leq \alpha\|x\| + \beta$$

then there exists a fixed point x^* satisfying $Tx^* = x^*$ such that

$$\|x^*\| \leq \frac{\beta}{1 - \alpha}$$

The following application of this corollary, given by Wouk,[69] deals with periodic continuous functions.

Let $f(x,t)$ be continuous on $0 \leq x, t < \infty$, be uniformly bounded by M, and have unit period in x. For each fixed t define the mapping Ty as follows

$$Ty = f(x,t) + \int_{x-\frac{1}{2}t}^{x+\frac{1}{2}t} y(s) \, ds \tag{1-13}$$

Note that Ty maps the space of continuous periodic functions of period one into itself, and we have that if

$$\|y\| = \max_x |y(x)| \leq M$$

then
$$\|Ty\| \leq t\|y\| + \max_{|x| \leq 1} |f(x,t)| \tag{1-14}$$

and one can show that the hypotheses of Theorem 1-22 are satisfied, and assures that (1-13) has a fixed point among the continuous functions of period one. The Rothe corollary gives a bound for the norm of the fixed point in case $t < 1$.

Another fixed-point theorem[65] follows.

Theorem 1-23 (*Schauder-Tychonoff theorem*). A compact convex subset E of a locally convex linear topological space X has the fixed-point property; i.e., for every continuous mapping T of E into itself there is an $x^* \epsilon E$ such that $Tx^* = x^*$.

Proof: By Zorn's lemma (if in a partially ordered set E_1 every totally ordered subset has a lower bound, then E_1 has a minimum element) there exists a minimal convex subset E_1 of E with the property that $TE_1 \subseteq E_1$. This minimal subset contains only one point by the following.

Lemma: Let $T:E \to E$ be continuous. If E contains at least two points, then there exists a proper closed convex subset $E_1 \subset E$ such that $T(E_1) \subseteq E_1$.

Altman[3] has developed a fixed-point theorem in Banach space for completely continuous operators based on the notion of a Schauder-Leray degree of a mapping. Some necessary results involving this notion follow.

Let the equation

$$T_1x \equiv x + T_2x$$

be a mapping on the closure of an open convex set E in a Banach space X into X. Let T_2x be completely continuous, and let F be the boundary of E. Then it can be shown[3] that to every point $y \notin T_1(F)$ there corresponds an integer $i = i(T_1,E,y)$, called the *degree* at the point y of $T_1(E)$, with the property that if $i \neq 0$, then $y \in T_1(E)$; that is, there exists an $x \in E$ such that $T_1(x) = y$.

Let X be a Banach space. Let S be a sphere with radius r in X with center at the zero element $\mathbf{0}$ of X; that is, $S(\mathbf{0},r) \equiv \{x \in X : \|x\| \leq r\}$. Denote the boundary B of S by $B \equiv \{x \in X : \|x\| = r\}$.

Consider the operator $T_1 = I - T_2$, where T_2 is completely continuous on the bounded closed convex set $E \subset X$, and let I be the identity operator of X. Suppose

$$T_1x \neq \mathbf{0} \qquad \text{for all } x \in E \tag{1-15}$$

Then the Schauder degree $i[T_1,S,\mathbf{0}]$ of T_1 is given by the following.

Theorem 1-24   If in a Banach space X, $T_2:X \to X$:

1. T_2 is completely continuous,
2. $T_1x = x - T_2x$ is continuous, defined on E, and satisfies (1-15),

and

3. $\|x - T_2x\|^2 \geq \|T_2x\|^2 - \|x\|^2$ for all $x \in E$,

then the degree of T_1 is equal to unity.

Theorem 1-24 can be used to prove[3] the following theorem.

Theorem 1-25   If in a Banach space X:

1. T is defined on the sphere $S(\mathbf{0},r)$ and is completely continuous,
2.   Any of the following conditions are satisfied:
    a. $\|x - Tx\|^2 \geq \|Tx\|^2 - \|x\|^2$ for all $x \in E$,
    b. $\|x - Tx\| \geq \|Tx\|$ for all $x \in E$,
    c. $\|Tx\| \leq \|x\|$ for all $x \in E$,

then there exists an element x^* in S such that $Tx^* = x^*$.

Now consider the real Hilbert space H. Condition $2a$ of the above theorem is equivalent to

$$(Tx,x) \leq (x,x) \tag{1-16}$$

for all $x \in E$. Therefore we can reformulate Theorem 1-24 for a Hilbert space as follows:

Theorem 1-26   If:

1. T is completely continuous and defined on the sphere S of H, and
2. T satisfies conditions (1-15) and (1-16),

then the degree of $U = I - T$ is unity.

Restating Theorem 1-25 for a real Hilbert space gives the following theorem, due to Krasnosel'skii.[40]

Theorem 1-27   If:
1. T is completely continuous and defined on a sphere S of H, and
2. T satisfies (1-16),

then there exists an element x^* in S such that $Tx^* = x^*$.

Altman[2] gives the following fixed-point theorem in a Hilbert space for a weakly closed operator T; that is, T is such that if x_n converges weakly to x and Tx_n converges weakly to y, then $y = Tx$.

Theorem 1-28   If:
1. H is a real separable Hilbert space,
2. $S \equiv \{x : \|x\| < r\}$, and the boundary of S, $S' \equiv \{x : \|x\| = r\}$, $x \in H$, $r > 0$,
3. T is a weakly closed mapping of S into H, and
4. $(Tx, x) \leq (x, x)$ for all $x \in S'$ implies T maps S into a bounded subset of H,

then there exists a point $x^* \in S$ such that $Tx^* = x^*$.

It is now convenient to give a topological theorem due to Krasnosel'skii,[41] which combines topological and algebraic fixed-point theorems, as both Schauder's theorem and the Banach theorem, given below, are special cases of it.

Theorem 1-29   Let E be a closed bounded convex subset of a Banach space X. Let T_1 and T_2 be operators on X to X such that the following conditions hold:
1. $T_1 x + T_2 y \in E$, when $x, y \in E$.
2. T_1 is a contraction operator with constant $0 < \alpha < 1$.
3. T_2 is completely continuous.

Then there exists at least one point x^* in E such that $T_1 x^* + T_2 x^* = x^*$.

Remark: An interesting recent development has a general relation to the idea of a fixed point and is due to Bernstein and Robinson.[5]   It is as follows: Aronszajn and Smith, generalizing a result of Aronszajn and von Neumann for Hilbert space, have shown that every compact, i.e., completely continuous, operator acting on an infinite-dimensional complex Banach space has a proper invariant subspace, i.e., a subspace which is mapped into itself by the operator.   Bernstein and Robinson have shown that if T is a linear bounded operator on an infinite-dimensional Hilbert space H over the complex numbers, and if $p(z) \neq 0$ is a polynomial with complex coefficients such that $p(T)$ is a compact operator, then T leaves invariant at least one closed subspace of H other than H or $\{0\}$.

ALGEBRAIC FIXED - POINT THEOREMS

These theorems give the existence of a solution through a successive approximations procedure.   The best-known theorem of this type is:

Theorem 1-30 (*Banach fixed-point theorem*) If T is a contraction operator with $\|T\| \leq \alpha < 1$ mapping a complete metric space X into itself, then the successive approximations $x_{n+1} = Tx_n$ for any initial point $x_0 \, \epsilon \, X$ converge to a unique limit x^*.

The error of approximation is bounded by

$$\rho(x^*,x_n) \leq \frac{\alpha^n}{1 - \alpha} \, \rho(x_1,x_0)$$

Proof: See "NM," chap. 1.

Next, we give the contraction-mapping theorem for a Banach space, i.e., a complete normed linear space, in the following form:

Theorem 1-31 If T is a contraction operator with $\|T\| \leq \alpha < 1$ mapping a closed region $E \subset X$ of the Banach space X into itself, then in E, T has exactly one fixed point x^*, which is the limit of the sequence $\{x_n\}$ given by $x_{n+1} = Tx_n$. The estimate of error is

$$\begin{aligned}
\|x^* - x_{n+1}\| &\leq \alpha(1 - \alpha)^{-1}\|x_{n+1} - x_n\| \\
&\leq \alpha^{n+1}(1 - \alpha)^{-1}\|x_1 - x_0\| \qquad x_0 \, \epsilon \, E
\end{aligned} \qquad (1\text{-}17)$$

Example: Consider the system of linear equations

$$\sum_{j=1}^{n} a_{ij}x_j = b_i \qquad i = 1, \ldots, n$$

This system can be simply transformed to the system

$$x_i = \sum_{j=1}^{n} c_{ij}x_j + b_i \qquad i = 1, \ldots, n$$

The solution of this system is a fixed point of

$$y_i = \sum_{j=1}^{n} c_{ij}x_j + b_i \qquad i = 1, \ldots, n$$

which for arbitrary $x = (x_1, \ldots, x_n)$ defines the mapping $y = Tx$ of the space R_n to itself. Thus $Tx = Cx + b$, where C is the matrix of the c_{ij} and $b = (b_1, \ldots, b_n)$. To obtain a sufficient condition that T be a contraction we must satisfy for any two points x^1 and x^2

$$\begin{aligned}
\rho(Tx^1,Tx^2) = \|Tx^1 - Tx^2\| &= \|Cx^1 - Cx^2\| \\
&= \|C(x^1 - x^2)\| \leq \|C\|\rho(x^1,x^2)
\end{aligned}$$

Hence $\|C\| < 1$ must hold. One can use, for example,

$$\|C\|^2 = \sum_{i,j=1}^{n} c_{ij}^2$$

Exercise 1-11 Consider the system $\dot{x} = f(t,x)$, $x(t_0) = x_0$, and let $|t - t_0| \leq a$, $|x - x_0| \leq b$; let f be continuous

$$|f(t,x)| \leq B$$
$$|f(t,x_1) - f(t,x_2)| \leq A|x_1 - x_2|$$

for any $t \epsilon [t_0 - a, t_0 + a] \equiv I$ and any $x_1, x_2 \epsilon [x_0 - b, x_0 + b] \equiv J$. To prove the local existence of a solution on an interval $[t_0 - h, t_0 + h]$ for $h > 0$ small, one notes that the differential equation is equivalent to

$$x(t) = x_0 + \int_{t_0}^{t} f(t,x)\, dt \qquad t \epsilon I$$

As in the case of the linear systems of algebraic equations, denote the right side of this equation by Tx. Here T maps the subset S of continuous functions with $x(t) \epsilon J$ (whenever $t \epsilon I$) into itself. Thus S is a closed sphere with radius b and with center at x_0. By writing $y = Tx$ show that $|y - x_0| \leq b$ and hence conclude that $y \epsilon S$. Using the definition $\rho(x_1,x_2) = \max_{t \epsilon I} |x_1(t) - x_2(t)|$, show that $\rho(y_1,y_2) \leq \alpha\rho(x_1,x_2)$, $\alpha = Bh$, and hence that T is a contraction if $\alpha < 1$.

Many algebraic fixed-point theorems are attempts to weaken the hypotheses of Banach's theorem. A weaker version is due to Caccioppoli.[8]

Theorem 1-32 (**Caccioppoli's corollary**) Let $T^n x \equiv T(T^{n-1}x)$. Let T map a complete metric space X into itself and satisfy:

1. $\rho(Tx, Tx^*) \leq \|T\|\rho(x,x^*)$, $x \epsilon X$.

2. $\sum_{n=1}^{\infty} \|T^n\| < \infty$.

Then the sequence $x_n = T^{n-1}x_0$ converges to $x^* = Tx^*$ and $\rho(x^*,x_n) \leq \sum_{j=n}^{\infty} \|T^j\|\rho(x^*,x_0)$.

Note that if $\|T\| < 1$, we have Banach's theorem.

This theorem has been applied to the Hammerstein equation, $x = KFx + x$, by Caccioppoli[8] and Kolodner.[36]

Nemytskii[52] has proved:

Theorem 1-33 Let the operator T satisfy a Lipschitz condition with some constant α in a sphere $\|x\| \leq \rho$ of a Banach space X. Then the equation

$$x = \epsilon Tx$$

has a solution for sufficiently small values of ϵ.

The above is a direct result of Banach's theorem, the problem being essentially to find an ϵ compatible with the requirement that T be a contraction.

Nemytskii's result has an application[31] to Uryson's equation $x = \mu Tx$, where

$$Tx(r) = \int_B K[r,s,x(s)]\, ds$$

and B is a closed bounded set in R_n. Assume that the kernel $K[r,s,t]$ is continuous for $r, s \epsilon X, |t| \leq \epsilon, \epsilon > 0,$ and that K has a bounded partial derivative with respect to t; that is,

$$\left| \frac{\partial K[r,s,t]}{\partial t} \right| \leq M \qquad r, s \epsilon x, |t| < \epsilon$$

The Uryson operator T satisfies the following Lipschitz condition

$$\|Tx - T\xi\| = \max_{r \epsilon B} \left| \int_B \{K(r,s,x(s)) - K(r,s,\xi(s))\} \, dt \right|$$
$$\leq M \int_B |x(s) - \xi(s)| \, dt$$
$$\leq M \text{ meas } B\|x - \xi\|$$

If we assume that for some number $N > 0$

$$\|Tx\| \leq N \qquad \|x\| \leq \epsilon$$

then the equation $x = \mu Tx$ has a unique solution in the sphere $\|x\| \leq \epsilon$.

The following theorem is due to Collatz; its proof closely parallels that of the contraction-mapping theorem. The requirement here is that the contraction operator map a certain sphere whose radius is related to both the contraction constant and the initial point into the complete subspace (in which case the fixed point is also in the sphere).

Theorem 1-34   Let:

1. T map the metric space E into itself.
2. E_1 be a complete subspace of E.
3. T be contractive with constant $\alpha < 1$ on E_1.
4. $x_1 = Tx_0$ where $x_0 \epsilon E_1$ and also let the sphere S defined by

$$S \equiv \left\{ x : \rho(x,x_1) \leq \frac{\alpha}{1 - \alpha} \rho(x_1,x_0) \right\} \text{ be contained in } E_1$$

Then $x = Tx$ has a unique solution $x^* \epsilon S$; the iterations $x_{n+1} = Tx_n$ converge to x^*, that is, $\lim_{n \to \infty} \rho(x_n,x^*) = 0$; and the error of the approximation is bounded by

$$\rho(x^*,x_1) \leq \frac{\alpha}{1 - \alpha} \rho(x_1,x_0)$$

Proof: Apply Theorem 1-30 to T restricted to S.

The ideas of Kantorovich and Schröder on operators in partially ordered spaces give an interesting example of abstract existence, unique-ness, and convergence of successive approximations.

Let R be a partially ordered normed real vector space containing a set K closed under addition and scalar multiplication by positive real scalars. We shall order R in the following fashion. If x, $y \in R$ and $y - x \in K$, then x and y are ordered (written $x:y$).

Assume the following axioms of order on K:

1. For any set x_1, . . . , x_k in R, with k finite, there is an element y of R called $\sup_k x_k$ such that $x_i:y$, while $x_i:z$, $i = 1$, . . . , k, implies $y:z$.

2. $x:y$ whenever, for all n, $x_n:y_n$ and $x_n \to x$, $y_n \to y$ in the norm as $n \to \infty$.

3. If $\mathbf{0}:x:y$, then $\|x\| \leq \|y\|$.

4. There exists an element x^* to which x_n converges in the norm if x_n is an infinite sequence such that $x_n:x_{n+1}$ . . . while $\|x_n\| \leq M$ for all n.

Note that if the above axioms hold on K, obviously they hold on R.

We define an arbitrary Banach space X to be R-metrized by an R metric if the following three axioms hold for a continuous function ρ on $X \times X$ (i.e., the set of ordered pairs of elements of X) into the positive cone K of R:

1. $\rho(x,y) = 0$ if and only if $x = y$.
2. $\rho(x,y) = \rho(y,x)$.
3. $\rho(x,y):\rho(x,z) + \rho(z,y)$.

***Theorem* 1-35 (*Uniqueness theorem for partially ordered spaces*)** Let φ be a continuous mapping of K into K such that:

1. $\|\varphi(p)\| \leq M$ for all $p \in K$.
2. $p_1:p_2$ implies $\varphi(p_1):\varphi(p_2)$.
3. $\varphi(p) = p$ if and only if $p = \mathbf{0}$.

Let T map a Banach space X into itself such that

$$\rho(Tx,Ty):\varphi(\rho(x,y)) \qquad \text{for all } x, y \in X$$

Then T has at most one fixed point.

Proof: Assume the existence of two fixed points x_1 and x_2, with $x_1 \neq x_2$. Then $\rho(x_1,x_2) \neq 0$ and

$$\mathbf{0}:\rho(x_1,x_2) = \rho(Tx_1,Tx_2):\varphi(\rho(x_1,x_2))$$

Consider the sequence defined as follows: $\rho_1 = \rho(x_1,x_2) \neq 0$, $\rho_n = \varphi(\rho_{n-1})$. From assumption 2 and the fact that x_1 and x_2 are both fixed points, so that $\rho_1:\varphi(\rho_1)$, it is easy to see that $\rho_n:\rho_{n+1}$ for all n. Also $\|\rho_n\| = \|\varphi(\rho_{n-1})\| \leq M$. By axiom 4 on K, this implies that ρ_n converges in the norm. Since φ is continuous, the limit ρ satisfies $\rho = \varphi(\rho) \neq 0$. This violates assumption 3 of the theorem; therefore $\rho(x_1,x_2) = 0$, and hence $x_1 = x_2$. Q.E.D.

Theorem 1-36 (*Existence theorem for partially ordered spaces*)
Under the hypotheses of Theorem 1-35 the sequence

$$x_n = Tx_{n-1}$$

converges for an arbitrary initial point x_0 to a fixed point of T.

EXISTENCE THEORY FOR MONOTONE OPERATORS

Now we prove an existence theorem for a solution to the Hammerstein integral equation $y + KTy = x$ in the Hilbert space H.

We shall use the following theorem to complete one of the proofs below.

Theorem 1-37 (*Minty*[49])   If $T:H \to H$ is continuous and monotonic, and if x', $y' \in H$ satisfy

$$\text{Re } (x - x', Tx - y') \geq 0 \qquad (1\text{-}18)$$

for all $x \in H$, then $y' = Tx'$.

Geometrically this means that if T satisfies the above hypotheses, and if $(x',y') \in H \times H$ satisfies (1-18) for all points (x,Tx) in the graph of T, then (x',y') is a point in the graph of T.

Proof: Suppose x', $y' \in H$ satisfy (1-18) for all $x \in H$ and that $\|y' - Tx'\| > 0$; that is, $y' \neq Tx'$.

By the continuity of T we can find a $\delta > 0$ small enough so that if $\|x - x'\| < \delta$, we have

$$\|Tx - Tx'\| < \|y' - Tx'\| \qquad (1\text{-}19)$$

Now, choose a real number $t > 0$ such that $\|t(y' - Tx')\| < \delta$. Then for $x = t(y' - Tx') + x'$, (1-19) holds. By assumption, (1-18) holds, and hence

$$\text{Re } (t(y' - Tx'), Tx - y') \geq 0$$

Since t is real and positive, we can factor it out and divide by it, and by adding $0 = -Tx' + Tx'$, we have

$$\text{Re } (y' - Tx', Tx - Tx') - \text{Re } (y' - Tx', y' - Tx') \geq 0$$

Using the properties stated in Sec. 1-2, we then have

$$\|y' - Tx'\| \, \|Tx - Tx'\| \geq \|y' - Tx'\|^2$$

Since, by assumption, $\|y' - Tx'\| > 0$, division yields

$$\|Tx - Tx'\| \geq \|y' - Tx'\|$$

which contradicts (1-19). Thus $y' = Tx'$.

The following theorem requires a rather lengthy proof, which is omitted, and depends in essence on a theorem by Kirzbaum dealing with the extension of an operator.

Theorem 1-38 (*Browder*[7])   If $T:H \to H$ is continuous and strongly monotonic, then the equation $Tx = y$ has, for any $y \in H$, a unique solution for x which depends continuously on y.

Corollary (*Minty*,[48] *Browder*[7]):   If $T:H \to H$ is continuous and monotonic, then, for any $y \in H$, the equation $x + Tx = y$ has a unique solution for x which depends continuously on y.

Proof: The required observation is that the operator $\bar{T}$ defined by $\bar{T}x = x + Tx$ is strongly monotonic if T is monotonic.

Choose any $x_1, x_2 \in H$.   Then

$$\begin{aligned}
\mathrm{Re}\,(x_1 - x_2, \bar{T}x_1 - \bar{T}x_2) &= \mathrm{Re}\,(x_1 - x_2, x_1 + Tx_1 - x_2 - Tx_2) \\
&= \mathrm{Re}\,(x_1 - x_2, Tx_1 - Tx_2) \\
&\qquad + \mathrm{Re}\,(x_1 - x_2, x_1 - x_2)
\end{aligned}$$

Since T is monotonic, the leftmost term is nonnegative, and thus

$$\mathrm{Re}\,(x_1 - x_2, \bar{T}x_1 - \bar{T}x_2) \geq \|x_1 - x_2\|^2$$

That is, $\bar{T}$ is strongly monotonic with constant $C = 1$.

Since T is continuous, $\bar{T}$ is continuous; and thus by Theorem 1-38, with $\bar{T}x = x + Tx$, the conclusion follows.

Following is the main theorem on the solution of a general Hammerstein equation using the concept of a monotone operator. Its proof is long and will not be given here. Note that no requirement of self-adjointness or complete continuity is placed on the linear operator K. Indeed, the purpose of introducing the idea of monotone operators was to remove these particular restrictions.

Theorem 1-39 (*Minty, Dolph*[4])   Suppose $K:H \to H$ is linear, continuous, and monotonic and $T:H \to H$ is continuous, bounded, and monotonic.   In addition, suppose there is an M such that

$$\|x\| > M \text{ implies } \mathrm{Re}\,(x, Tx) \geq 0$$

for some positive constant M, then the equation $y + KTy = \mathbf{0}$ has a solution for y.

It should be noted that a nonlinear operator is called *bounded* if the image of each bounded set is bounded; the range of the operator need not be bounded.

We now give a uniqueness theorem.

Theorem 1-40 (*Minty, Dolph*[4])   If K and T are monotonic and either K or T is strictly monotonic, then a solution of the equation $v + KTv = y$ is unique.

Proof: Assume that y' and y'' are such that

$$y' + KTy' = y \quad \text{and} \quad y'' + KTy'' = y \quad\quad (1\text{-}20)$$

Subtracting, we have $(y' - y'') + (KTy' - KTy'') = \mathbf{0}$, and forming the scalar product with $(Ty' - Ty'')$ gives

$$\text{Re } (Ty' - Ty'', y' - y'')$$
$$+ \text{ Re } (Ty' - Ty'', KTy' - KTy'') = 0 \quad (1\text{-}21)$$

Each term in (1-21) must be zero, since K and T are both monotonic. So, if T is strictly monotonic, then $y' = y''$, and if K is strictly monotonic, then $Ty' = Ty''$. But if $Ty' = Ty''$, then $y' = y''$ by (1-20). So in either case, y' and y'' are identical.

The following theorem is an interesting result because it generalizes almost word for word the rather obvious analogue for real-valued functions given in Sec. 1-2.

Theorem 1-41   If $T:H \to H$ is continuous and monotonic and there exists a positive number M such that

$$\|x\| \geq M \text{ implies Re } (x,Tx) \geq 0$$

then the equation $Tx = \mathbf{0}$ has a solution.

Minty and Dolph give sufficient conditions for the hypotheses of the main theorem in this section to be satisfied for the Hammerstein equation

$$y(s) + \int_\Gamma k(s,t)f(t,y(t)) \, dt = 0$$

using a real Hilbert space.

Browder generalizes this main result and applies it to the existence of solutions for the Dirichlet problem for a class of nonlinear elliptic partial differential equations.

EXISTENCE THEORY FOR SOLUTIONS OF POTENTIAL - OPERATOR EQUATIONS[54,66]

For motivation of potential operators, see Sec. 1-2.

Definition: A point $x_0 \, \epsilon \, X$ is called an *extreme point* of a functional $f(x)$ on X if there exists a neighborhood of x_0 such that for all points x in this neighborhood either $f(x) \leq f(x_0)$ or $f(x) \geq f(x_0)$.

Definition: A point x_0 is called a *critical point* of the functional $f(x)$ if grad $f(x_0) = \mathbf{0}$.

Theorem 1-42   Let the functional f be defined on a region E of a Banach space X, and let x_0 be an interior point of E at which f has a linear bounded Gateaux differential. Then, in order for the point x_0 to be extreme it is necessary that it be critical.

For if h is an arbitrary vector in X, then $f(x + th)$ is a real function of t defined in some neighborhood of $t = 0$ and having a derivative at $t = 0$. If x_0 is extreme, then

$$\frac{d}{dt} f(x_0 + th) \Big|_{t=0} = 0$$

Therefore, $(d/dt)f(x_0) = Df(x_0,h) = (\text{grad } f(x_0),h) = 0$. Since h is an arbitrary vector, grad $f(x_0) = \mathbf{0}$.

Definition: A set is weakly closed if it contains its weak limit points in the sense of weak convergence. *Weak compactness* in E^n implies that every bounded infinite sequence has a weakly convergent subsequence.

Definition: A functional is weakly lower semicontinuous at x_0 if for any sequence $\{x_n\}$ weakly convergent to x_0

$$f(x_0) \leq \lim_{n \to \infty} \inf f(x_n)$$

Theorem 1-43   If a weakly lower semicontinuous functional f is defined on a bounded weakly closed set σ in a reflexive Banach space (one having weakly compact spheres), then f has a lower bound, which it achieves on σ.

Definition: A functional f is said to have the t property if there is some weakly closed set $E^- \subset X$ such that f is weakly lower semicontinuous on E^- and if E' is the boundary of E^- and $E \cup E' = E^-$ (E is the largest weakly open set in E^-), then there is a point $x_0 \, \epsilon \, E$ such that $f(x) > f(x_0)$ for all $x \, \epsilon \, E'$.

Theorem 1-44   If a weakly lower semicontinuous functional f defined on a space X having weakly compact spheres has the t property, then f has at least one critical point.

By Theorem 1-43, f achieves its lower bound on E^-, and since $f(x) > f(x_0)$ for all $x \, \epsilon \, E'$, the lower bound cannot be an element of E'. Therefore f has an extreme point in the open set E, and by Theorem 1-42 it will be a critical point.

Consider the equation

$$x = BTx$$

defined on a Hilbert space H, where T is a potential operator in H and B is a self-adjoint operator defined on all of H.

Lemma: If, in H, grad $g = T$, and if B is a self-adjoint operator, then

$$\text{grad } g(Bx) = BT(Bx)$$

since $\displaystyle \lim_{t \to 0} \frac{g(Bx + tBh) - g(Bx)}{t} = (T(Bx),Bh) = (BT(Bx),h)$

Theorem 1-45   Let B be a positive self-adjoint operator with domain H and $T = \text{grad } g$, where

$$2g(x) \leq a_1(x,x) + a_2(x,x)^r + a_3$$
$$a_2, a_3 > 0 \qquad 0 < r < 1 \qquad a_1\|B\| < 1 \qquad \text{if } a_1 > 0$$

and let one of the following two conditions be satisfied:
1. g is a continuous functional, and B is completely continuous.
2. g is weakly continuous (or weakly upper semicontinuous).

Then the operator equation $x = BTx$ has a solution.

Let A be the positive square root of B, and consider the functional

$$f(x) = (x,x) - 2g(Ax)$$

If the first condition is fulfilled, then the completely continuous linear operator A carries every weakly convergent sequence into a strongly convergent sequence, and therefore $g(Ax)$ is weakly continuous. If condition 2 is satisfied, then the bounded self-adjoint operator A carries every weakly convergent sequence into a weakly convergent sequence and, since $g(x)$ is weakly upper semicontinuous, $g(Ax)$ is also. (x,x) is weakly lower semicontinuous, and so is $-2g(Ax)$. Therefore $f(x)$ is weakly lower semicontinuous.

From the inequality for $g(x)$ in the theorem, we have

$$f(x) \geq \begin{cases} (x,x)^r[(x,x)^{1-r} - a_2\|B\|^r] - a_3 & \text{if } a_1 \leq 0 \\ (x,x)^r[(x,x)^{1-r}(1 - a_1\|B\|)] - a_2\|B\|^r - a_3 & \text{if } a_1 > 0 \end{cases}$$

Therefore, $f(x)$ is unbounded and increases with $\|x\|$. Consequently, there exists a sphere $\|x\| = r$ on which $f(x) > f(0)$, and so f has the t property.

By Theorem 1-44, f has a critical point, $\text{grad } f(x_0) = 0$.

But $\text{grad } (x,x) = 2x$, and $\text{grad } g(Ax) = AT(Ax)$ by the lemma to Theorem 1-44, so that

$$\text{grad } f(x_0) = 0 = 2x - 2AT(Ax)\Big|_{x_0}$$

or
$$Ax_0 - A^2T(Ax_0) = 0$$

Setting $Ax_0 = z_0$, and remembering that $A^2 = B$, we get

$$z_0 - BT(z_0) = 0$$

Therefore z_0 is a solution of the equation.

If the potential operator T satisfies the Lipschitz condition

$$\|Tx_2 - Tx_1\| \leq a_1\|x_2 - x_1\| \qquad 0 < a_1\|B\| < 1$$

then if x_1 and x_2 are two solutions of $x = BTx$,

$$\|x_2 - x_1\| = \|BTx_2 - BTx_1\| \leq \|B\|a_1\|x_2 - x_1\| \leq \|x_2 - x_1\|$$

and so $\|x_2 - x_1\| = 0$, $x_2 = x_1$, and the equation has a unique solution.

Theorem 1-46 Let B be a positive self-adjoint operator defined on all of a real Hilbert space H, and let the potential operator T satisfy the condition

$$(DT(x,h),h) \leq a_1(h,h)$$

where $a_1\|B\| < 1$ for $a_1 > 0$. Then $x = BTx$ has at least one solution.

Proof: Let $T = \text{grad } g$. Consider

$$f(x) = \tfrac{1}{2}(x,x) - g(Ax)$$

where A is the positive square root of B. We have

$$Df(x,h) = \lim_{t \to 0} \frac{f(x + th) - f(x)}{t} = (x,h) - (T(Ax),Ah)$$

$$D^2f(x,h,k) = \lim_{t \to 0} \frac{Df(x + tk, h) - Df(x,h)}{t} = (k,h) - (DT(Ax,Ak),Ah)$$

Because of the inequality in the statement of the theorem,

$$D^2f(x,h,h) \geq \begin{cases} (h,h) & \text{for } a_1 \leq 0 \\ (h,h)(1 - a_1\|B\|) & \text{for } a_1 > 0 \end{cases}$$

From this inequality the existence of a relative minimum for the functional $f(x)$ can be shown. Since f is a differentiable functional, the point x_0 is a critical point of f:

$$Df(x,h) = (x,h) - (T(Ax),Ah) = (x - AT(Ax),h)$$

so that $\quad \text{grad } f(x_0) = x_0 - AT(Ax_0) = \mathbf{0}$

Therefore, $Ax_0 = A^2T(Ax_0)$. For $Ax_0 = z_0$, $z_0 = BT(z_0)$ is a solution of the equation.

Definition: A self-adjoint operator B is called *quasi-positive* if its negative spectrum is nonempty and contains a finite number of proper values of finite multiplicity and its positive spectrum is nonempty and arbitrary.

Quasi-negative is defined similarly.

It can be shown that the equation $x = BTx$ has solution if:

1. The self-adjoint operator B is quasi-negative, and its positive spectrum belongs to some interval $[m,\beta]$, $m > 0$.

2. $T = \text{grad } g$, $g(x) \geq (1/m)(x,x) + a_2(x,x)^r + a_3$; $a_2, a_3 < 0$, $r \in (0,1)$.

3. Either g is a continuous functional and B is a completely continuous operator or g is weakly lower semicontinuous.

A similar result holds if B is quasi-positive.

The equation also has a solution if condition 1 above is satisfied and the potential operator T satisfies

$$(DT(x,h),h) \geq \frac{2}{m}(h,h)$$

Consider now the equation $Tx = y$, where T is some operator, y is an element of the Hilbert space H, and x is the unknown element.

Theorem 1-47   If the potential operator T given in the Hilbert space H satisfies

$$(DT(x,h),h) \geq c(h,h) \qquad \text{for all } h \text{ and } x$$

where c is some positive number, then the equation

$$Tx = y$$

where y is a fixed element in H, has at least one solution.

This result is shown directly by considering the functional

$$f(x) = g(x) - (x,y)$$

for which
$$Df(x,h) = (Dg(x,h) - D(x,y))$$
$$= (Tx,h) - (h,y)$$
and
$$D^2f(x,h,h) = (DT(x,h),h) \geq c(h,h) = c\|h\|^2$$

Then, as in Theorem 1-46, there exists an x_0 such that

$$\text{grad } f(x_0) = \mathbf{0}$$

Since grad $f(x) = Tx - y$, then $Tx = y$ is the solution.

If A is a linear bounded operator with an inverse, then $Tx = A^{-1}y$ has a solution, and so does $ATx = y$.

EXISTENCE THEORY FOR PROPER - VALUE PROBLEMS[66]

When eigenvalues of linear bounded operators are considered, the existence of one eigenvalue implies the existence of eigenvectors of any norm, since the multiplication of an eigenvector of norm unity by any scalar is also an eigenvector.   For nonlinear operators the corresponding situation is not true in general.   The existence of one eigenvector need not imply the existence of others.

In fact a distinction is made between nonlinear operators which map the zero element into the zero element, that is, $T\mathbf{0} = \mathbf{0}$, and those which do not.

Definition: If $T\mathbf{0} = \mathbf{0}$, then an element $x_0 \in X$ such that Tx_0 is a scalar multiple of x_0 is called a *proper element* of T.   A proper element is also called a *proper vector* or *proper function* (if X is a space of functions).

The number μ_0 corresponding to the proper element x_0 is called the *proper value* or *proper number* of T. The reciprocal of a nonzero μ is called a *characteristic value* of T. (If $\mu_0 = 0$ for a proper element, it is not of fixed direction.)

Variational methods are concerned with the proper elements of a product AT of operators, where A is a self-adjoint operator defined on all of a real Hilbert space H and T is a strongly potential operator defined on H.

Vainberg states that the results given in his work[66] include all the existence theorems for proper elements and fixed directions proved by variational methods.

Definition: Consider two continuous real-valued functionals f and g on a Banach space X. Denote by M_0 the manifold generated by $g(x) = c$, c a constant. Then a point $x_0 \epsilon M_0$ is called a *conditionally relative minimum point* of the functional f with respect to the manifold M_0 if there exists a δ neighborhood $V \equiv \{x : \|x - x_0\| < \delta\}$ of x_0 such that for every $x \epsilon V \cap M_0$ we have $f(x) \geq f(x_0)$.

A similar definition holds for conditional relative maximum.

Definition: A conditional relative maximum or minimum of f with respect to M_0 is called a *conditionally relative extreme point* of f with respect to M_0.

Definition: If $\|\text{grad } g(x_0)\| > 0$, $x_0 \epsilon M_0$ is called an *ordinary point* of the manifold M_0 if f and g are strongly differentiable on X.

Definition: A point $x_0 \epsilon M_0$ is called a *conditionally critical point* of the functional f with respect to M_0 if grad $f(x_0) = \mu$ grad $g(x_0)$, where μ is some number.

Theorem 1-48 For an ordinary point of the manifold M_0 to be a conditionally extreme point of the functional f it is necessary that this point be a conditionally critical point of f (f has a Fréchet derivative).

This theorem is needed only for the case when M_0 is a sphere generated by $g = (x,x) = a^2$. It is proved for this case by showing:

1. Every point of the sphere is an ordinary point since

$$\|\text{grad } (x,x)\| = \|2x\| > 0 \text{ for } \|x\| = a$$

2. If $x_0 \epsilon S_a = \{x : \|x\| = a\}$ is a conditionally extreme point and $h \epsilon S_a$ is an arbitrary vector such that $(x_0,h) = 0$, then $(\text{grad } f(x_0),h) = 0$, and therefore grad $f(x_0) = \mu(x_0)$.

Theorem 1-49 Let T be a positive and strongly potential operator defined on all of a real Hilbert space H. Let A be a positive self-adjoint operator also defined on all of H. Suppose that at least one of the following conditions is satisfied:

1. A is completely continuous.
2. T is the gradient of a weakly continuous functional.

Then on any sphere $\|x\| = r > 0$ in H there exists at least one vector x_r such that the element $z_r = A^{1/2}x_r$, where $A^{1/2}$ is the positive square root of A, is a fixed direction of AT.

Proof: Let $T = \text{grad } f$, $\varphi(x) = f(A^{1/2}x)$. It follows from the strong differentiability, and hence continuity of f, and from either of conditions 1 or 2 that $\varphi(x)$ is weakly continuous.

From Theorem 1-43, and the corresponding theorem for a weakly upper semicontinuous functional, and upper bounds one obtains the generalized Weierstrass theorem, viz., a weakly continuous functional on a bounded weakly closed set in a Banach space with weakly compact spheres achieves its infimum and supremum.

Therefore, since the ball $S \equiv \{x:\|x\| \leq r\}$ in H is weakly compact, $\varphi(x)$ assumes its infimum and supremum on S. Since $(Tx,x) > 0$ if $\|x\| \neq \mathbf{0}$, then for $t > 0$,

$$\frac{d}{dt}f(tx) = (T(tx),x) = \frac{1}{t}(T(tx),x) > 0$$

Therefore $f(x)$ is monotonically increasing along every ray which starts from $\mathbf{0}$, and similarly for

$$\varphi(x) = f(A^{1/2}x) \qquad \text{if } \|A^{1/2}x\| > 0$$

Therefore φ has a minimum m at $\mathbf{0}$ and a maximum M on the sphere $\|x\| = r$ at some point x_r.

$\|A^{1/2}x_r\| > 0$, since otherwise $m = M$, and $f(tx)$ is an increasing function of x for $x \neq \mathbf{0}$, implying that $A^{1/2}$ is the zero operator.

The point x_r is a conditionally extreme point of the functional $\varphi(x)$ with respect to $S \equiv \{x:\|x\| = r\}$. Therefore, by Theorem 1-48 x_r is a conditionally critical point, so that

$$\text{grad } \varphi(x_r) = \mu_1 \text{ grad } (x_r,x_r) = \mu_1 x_r$$

by definition.

By the lemma to Theorem 1-44,

$$\text{grad } \varphi(x_r) = A^{1/2}T(A^{1/2}x_r) = \mu x_r$$
so that $\qquad AT(A^{1/2}x_r) = \mu A^{1/2}x_r$

Setting $z_r = A^{1/2}x_r$ gives $AT(z_r) = \mu z_r$.

Finally,

$$(\mu x_r,x_r) = \mu(x_r,x_r) = \mu r^2 = (A^{1/2}T(A^{1/2}x_r),x_r)$$
$$= (T(A^{1/2}x_r),A^{1/2}x_r) > 0$$

since $\|A^{1/2}x_r\| > 0$. Therefore $\mu > 0$, and z_r is a fixed direction.

Remark: z_r is a proper element of the operator AT, since $f(\mathbf{0}) = m$ is an absolute minimum of $f(x)$, so that, by Theorem 1-42,

$$T\mathbf{0} = \text{grad } f(\mathbf{0}) = \mathbf{0}$$

Remark: Under the conditions of the theorem, AT has a continuum of proper elements with proper values.   To prove this, it is shown that two proper vectors z_{r_1}, z_{r_2}, $r_1 \neq r_2$, are distinct.   In fact if

$$z_{r_1} - z_{r_2} = A^{1/2}(x_{r_1} - x_{r_2}) = \mathbf{0}$$

then, since $x_{r_1} - x_{r_2} \neq \mathbf{0}$, $x_{r_1} - x_{r_2}$ must be a nonzero element of H_0, the zero subspace of the operator $A^{1/2}$.

But if $x \in H_0$, then $A^{1/2}(A^{1/2}x) = A^{1/2}(\mathbf{0}) = \mathbf{0}$, so that H_0 is a subset of the zero subspace of A.   Conversely, if x belongs to the zero subspace of A, then $(Ax,x) = 0 = (A^{1/2}x, A^{1/2}x)$, so that $A^{1/2}x = \mathbf{0}$, and the zero subspace of A is a subset of H_0.   Therefore they coincide.

Consider now the orthogonal complement H_1 of H_0.   The gradient

$$\text{grad } \varphi(x) = A^{1/2}T(A^{1/2}x)$$

carries H_1 into H_1, since if $h \in H_0$, then

$$(A^{1/2}T(A^{1/2}x),h) = (T(A^{1/2}x), A^{1/2}h) = (T(A^{1/2}x),\mathbf{0}) = 0$$

Thus, if $x \in H_1$, then $A^{1/2}T(A^{1/2}x) \in H_1$.

If, in the proof of Theorem 1-49, we consider the ball S in the subspace H_1, rather than in H, every step remains valid, and the vector x_r on the sphere $\|x\| = r$ will be an element of H_1.   Therefore, the difference of two vectors $x_{r_1} - x_{r_2}$ is also an element of H_1, and so

$$A^{1/2}(x_{r_1} - x_{r_2}) \neq \mathbf{0}$$

and the second remark is proved.

If the requirement that T be a positive operator is dropped from the hypothesis of Theorem 1-49 but $T\mathbf{0} = \mathbf{0}$, then the existence of proper elements can also be shown.   In fact, under these conditions, one obtains:

Theorem 1-50   For any positive number a there exists a continuum of proper elements of AT whose norms do not exceed a.

Consider the functional $\varphi(x) = f(A^{1/2}x)$ where $T = \text{grad } f$ on the orthogonal complement H_1 of the null subspace of A.   If φ has a critical point on each sphere $(x,x) = r^2 > 0$, where $\|A^{1/2}\|r \leq a$ in H_1, then from the definition of a conditionally critical point

$$A^{1/2}T(A^{1/2}x) = \text{grad } \varphi(x) = \tfrac{1}{2}\mu \text{ grad }(x,x) = \mu x$$

and $\|A^{1/2}x\| > 0$, since $x \neq \mathbf{0}$.   Thus, $ATz = \mu z$, where $z = A^{1/2}x$, and z is a proper element of AT.

For two distinct spheres, the proper elements are distinct, as in the second remark.

Suppose that on some sphere the function does not have critical points. Then it can be shown that the equation

$$\text{grad } \varphi(x) = \mu(x)$$

has a continuum of nonnull solutions in H_1 whose norms are less than r, with nonzero μ.

The fact that $T0 = 0$ is not used in the proof, but it allows one to speak of proper elements rather than merely nonnull solutions.

If A is the identity operator, it satisfies the conditions of the theorem, and an obvious corollary follows.

By using Theorem 1-48, the following theorem can also be proved.

Theorem 1-51   If the strong gradient of a weakly continuous functional equals the zero element only at zero of the Hilbert space H, then it has proper vectors of all norms.

Definition: If $T(-x) = -Tx$, the operator T is said to be odd.

The following three theorems deal with the existence of proper elements of odd operators.

Theorem 1-52   Let T be potential, strongly continuous, odd, and positive in the ball $\|x\| \leq R$, of a real separable Hilbert space H. Then, on every sphere $S \equiv \{x : \|x\| = r \leq R\}$, there exists a sequence of proper elements of T which converges weakly to $\mathbf{0}$, and the corresponding proper values are positive and converge to zero.

Theorem 1-53   If:

1. T is a continuous potential operator defined in a real Hilbert space H,

2. T is odd and positive, and

3. A is a positive self-adjoint completely continuous operator defined on all of H, which has a countable set of positive proper values,

then for any $c > 0$ the operator AT has at least countably many pairwise linearly independent proper elements, representable in the form

$$\varphi_\gamma = \sum_{k=1}^{\infty} \frac{\eta_k^{(\gamma)}}{\sqrt{\lambda_k}} x_k$$

$$(\eta_1^{(\gamma)}, \eta_2^{(\gamma)}, \ldots) = \eta^{(\gamma)} \in l^2 \qquad \gamma = 1, 2, \ldots$$

where

$$\|\eta^{(\gamma)}\| = \left[\sum_{k=1}^{\infty} (\eta_k^{(\gamma)})^2 \right]^{1/2} = c$$

and x_k and λ_k are corresponding (normalized) proper vectors and characteristic numbers of A.   To these proper elements of AT correspond the proper numbers

$$\mu_\gamma = c^{-2}(\varphi_\gamma, T(\varphi))$$

They form a decreasing sequence which converges to zero.

Theorem 1-54 If:

1. T is potential, strongly continuous, odd, and positive in the ball $\|x\| \leq R$ in a real separable Hilbert space H, and

2. A is a positive self-adjoint operator defined on all of H and the orthogonal complement $H_1 = H - H_0$ of its null space H_0 is infinite-dimensional,

then for any number a satisfying $0 < a < R\|A^{\frac{1}{2}}\|^{-1}$ there is a sequence of elements $\{h_n\}$, $\|h_n\| = a$, $n = 1, 2, \ldots$, such that the elements $\varphi_n = A^{\frac{1}{2}}h_n$ are proper elements of AT, the sequence $\{\varphi_n\}$ converges weakly to 0, and the proper values φ_n corresponding to the φ_n, which are given by

$$\mu_n = a^{-2}(\varphi_n, T(\varphi_n))$$

are positive and converge to zero as $n \to \infty$.

Several theorems are given for quasi-definite operators by Vainberg, of which the following is representative.

Theorem 1-55 Let T be a strongly potential operator in a real Hilbert space H which satisfies the condition

$$\lim_{\|x\| \to \infty} \int_0^1 (T(tx), x) \, dt = +\infty$$

for $x \in AV_0$; A is the principal square root of the quasi-negative operator B, and V_0 is the conical region generated by B; and at least one of the following conditions is satisfied:

1. B is completely continuous.

2. T is the gradient of the weakly lower semicontinuous function f.

Then the operator equation

$$x = BTx$$

has a continuum of solutions whose norms exceed any given number.

REFERENCES

1. Akhiezer, N. I., and I. M. Glazman: "Theory of Linear Operators in Hilbert Space," Frederick Ungar Publishing Co., New York, 1961.

2. Altman, M.: A Fixed Point Theorem in Hilbert Space, *Bull. Acad. Polon. Sci. Sér. Sci. Math. Astronom. Phys.*, cl. 3, vol. 5, no. 1, pp. 19–22, 1957.

3. ———: A Fixed Point Theorem in Banach Space, *Bull. Acad. Polon. Sci. Sér. Sci. Math. Astronom. Phys.*, cl. 3, vol. 5, no. 2, pp. 89–92, 1957.

4. Anselone, P. M.: "Nonlinear Integral Equations," The University of Wisconsin Press, Madison, Wis., 1964.

5. Bernstein, A. R., and A. Robinson: Solution of an Invariant Subspace Problem of K. T. Smith and P. R. Halmos, to appear.

6. Bers, L.: Topology, notes produced at New York University, 1956–1957.

7. Browder, F.: The Solvability of Non-linear Functional Equations, *Duke Math. J.*, vol. 30, pp. 557–566, 1962.

8.   Caccioppoli, R.: Un teorema generale sull'esistenza di elementi uniti una traetor-maxione funzionale, *Atti. Accad. Naz. Lincei Rend. Cl. Sci. Fis. Mat. Natur.*, ser. 6, vol. 11, pp. 794–799, 1930.

9.   Collatz, L.: Einige Anwendungen funktional-analytischer Methoden in der praktischen Analysis, *Z. Angew. Math. Phys.*, vol. 4, pp. 327–357, 1953.

10.   Cooke, R.: "Linear Operators," The Macmillan Company, New York, 1953.

11.   Cooper, J. L. B.: Functional Analysis, *Math. Gaz.*, pp. 102–109, 1959.

12.   Desoer, C. A., and B. H. Whalen: A Note on Pseudo-inverses, *J. Soc. Indust. Appl. Math.*, vol. 2, no. 2, pp. 442–447, June, 1963.

13.   Dieudonné, J.: "Foundations of Modern Analysis," Academic Press Inc., New York, 1960.

14.   Dolph, C., and G. Minty: On Nonlinear Integral Equations of the Hammerstein Type, in Ref. 4, pp. 99–154.

15.   Dunford, N., and J. T. Schwartz: "Linear Operators," part I, "General Theory," 1958; part II, "Spectral Theory, Self Adjoint Operators in Hilbert Space," 1963, Interscience Publishers, Inc., New York.

16.   Ehrmann, H.: On Implicit Function Theorems and the Existence of Solutions of Nonlinear Equations, *Enseignement Math.*, vol. 9, no. 3, pp. 129–176, 1963.

17.   Ficken, F. A.: Some Uses of Linear Spaces in Analysis, *Amer. Math. Monthly*, vol. 66, pp. 259–275, April, 1959.

18.   Friedricks, K. O.: "Functional Analysis and Applications," Courant Institute of Mathematical Sciences, New York University Press, New York, 1953.

19.   Goffman, C., and George Pedrick: "First Course in Functional Analysis," Prentice-Hall, Inc., Englewood Cliffs, N.J., 1965.

19a.   Goldberg, S.: "Unbounded Linear Operators," McGraw-Hill Book Company, New York, 1966.

20.   Graves, L. M.: What is a Functional? *Amer. Math. Monthly*, vol. 55, pp. 467–472, October, 1948.

21.   Greville, T. N. E.: The Pseudoinverse of a Rectangular or Singular Matrix and Its Application to the Solution of Systems of Linear Equations, *SIAM Rev.*, vol. 1, pp. 15–22, 1959.

22.   ———: Some Applications of the Pseudoinverse of a Matrix, *SIAM Rev.*, vol. 2, no. 1, pp. 15–22, 1960.

23.   Halmos, P. R.: "Introduction to Hilbert Space and the Theory of Spectral Multiplicity," Chelsea Publishing Company, New York, 1951.

24.   ———: "Introduction to Hilbert Space," pp. 41–43, Chelsea Publishing Company, New York, 1957.

25.   Hamburger, H. L., and M. E. Grimshaw: "Linear Transformations," Cambridge University Press, New York, 1956.

26.   Hewitt, Edwin: The Role of Compactness in Analysis, *Amer. Math. Monthly*, vol. 67, pp. 499–516, June–July, 1960.

27.   Hildebrandt, T. H., and L. M. Graves: Implicit Functions and Their Differentials in General Analysis, *Trans. Amer. Math. Soc.*, vol. 29, pp. 127–153, 1927.

28.   Hille, E., and R. S. Phillips: "Functional Analysis and Semi-groups," Colloquium Publications, American Mathematical Society, Providence, R.I., 1958.

29.   Inselberg, Alfred: On Classification and Superposition for Nonlinear Operators, Ph.D. dissertation (Department of Electrical Engineering), University of Illinois, Urbana, Ill., 1965.

30.   James, Robert C.: Linear Functionals as Differentials of a Norm, *Math. Mag.*, vol. 24, pp. 237–244, May–June, 1951.

31.   Kantorovich, L. V., and G. P. Akilov, "Functional Analysis in Normed Spaces," Pergamon Press, New York, 1964.

32. ————: "Functional Analysis and Applied Mathematics," translated by Curtis D. Benster, edited by G. E. Forsythe, University of California Press, Berkeley, Calif., 1963.

33. Karlin, S.: Positive Operators, *J. Math. Mech.*, vol. 8, pp. 907–937, 1959.

34. Kelley, John L.: "General Topology," D. Van Nostrand Company, Inc., Princeton, N.J., 1955.

35. Kolmogorov, A. N., and S. V. Fomin: "Elements of the Theory of Functions and Functional Analysis," Graylock Press, Albany, N.Y., vol. 1, 1957, vol. 2, 1961.

36. Kolodner, I. I.: Equations of the Hammerstein Type in Hilbert Space, *Univ. New Mexico Dept. Math. Tech. Rep.* 47, November, 1963.

37. Kolomy, Josef: Remark to the Solution of Non-linear Functional Equations in Banach Spaces, *Comment. Math. Univ. Carolinae*, vol. 5, no. 2, pp. 97–116, 1964.

38. ————: Contributions to the Solution of Non-linear Equations, *Comment. Math. Univ. Carolinae*, vol. 4, no. 4, pp. 165–171, 1963.

39. Krasnosel'skii, M. A.: "Positive Solutions of Operator Equations," Erven P. Noordhoff, NV, Groningen, Netherlands, 1964.

40. ————: "Topological Methods in the Theory of Nonlinear Integral Equations," Pergamon Press and The Macmillan Company, New York, 1964.

41. ————: Two Remarks on the Method of Succession Approximations, *Uspehi Mat. Nauk.*, n.s., vol. 10, no. 1, pp. 123–127, 1955.

42. Krein, M. G.: "Functional Analysis," in Russian, Moscow, 1964.

43. Liusternik, L., and V. Sobolev: "Elements of Functional Analysis," translated by A. E. LaBarre, Jr., H. Izbicki, and H. W. Crowley, Frederick Ungar Publishing Co., New York, 1961.

44. Loomis, L. H.: "An Introduction to Abstract Harmonic Analysis," D. Van Nostrand Company, Inc., Princeton, N.J., 1953.

45. Lorch, E. R.: The Spectral Theorem and Classical Analysis, in "Proceedings of the Symposium on Spectral Theory and Differential Problems," pp. 259–265, Oklahoma Agricultural and Mechanical College, Stillwater, Okla., 1951.

46. McEwen, W. H.: Spectral Theory and Its Application to Differential Eigenvalue Problems, *Amer. Math. Monthly*, vol. 60, pp. 223–233, April, 1953.

47. Michal, Aristotle D.: Integral Equations and Functionals, *Math. Mag.*, vol. 24, pp. 83–95, November–December, 1950.

48. Minty, G.: Two Theorems on Nonlinear Functional Equations in Hilbert Space, *Bull. Amer. Math. Soc.*, vol. 69, pp. 691–692, September, 1963.

49. ————: Monotone (Non-linear) Operators in Hilbert Space, *Duke Math. J.*, vol. 29, pp. 341–346, 1962.

50. Nahano, K.: On Characteristics of Projection Operators, *Proc. Japan Acad.*, vol. 36, pp. 196–199, 1960.

51. Nemytskii, V. V.: A Method for Finding All Solutions of Non-linear Operator Equations, translated by R. A. Demarr.

52. ————: Fixed Point Methods in Analysis, *Uspehi Mat. Nauk.*, vol. 1, 1936.

53. Penrose, R.: A Generalized Inverse for Matrices, *Proc. Cambridge Philos. Soc.*, vol. 51, pp. 406–413, 1955.

54. Rall, L. B.: Variational Methods for Integral Equations, in Ref. 4, pp. 155–189.

55. Rickart, C. E.: "General Theory of Banach Algebras," D. Van Nostrand Company, Inc., Princeton, N.J., 1960.

56. Riesz, F., and B. Sz.-Nagy: "Leçons d'analyse fonctionelle," Academiai Kiado, Budapest, 1962.

57. Rothe, E.: Zur Theorie der topologischen Ordnung und der Vektorfelder in Banachschen Räumen, *Compositio Math.*, vol. 5, pp. 177–196, 1937–1938.

57a. Rota, Gian-Carlo: An "Alternierende Verfahren" for General Positive Operators, *Bull. Am. Math. Soc.*, vol. 68, No. 2, pp. 95–102, 1962.

58. Saaty, Thomas L., and Joseph Bram: "Nonlinear Mathematics," McGraw-Hill Book Company, New York, 1964.

59. ———: An Application of Semigroup Theory to the Solution of a Mixed type Partial Differential Equation, *Univ. e Politec. Torino Rend. Sem. Mat.*, vol. 18, pp. 53–76, 1958–1959.

60. Simmons, George F.: "Introduction to Topology and Modern Analysis," McGraw-Hill Book Company, New York, 1963.

61. Smirnov, V. I.: "Integration and Functional Analysis," Addison-Wesley Publishing Company, Inc., Reading, Mass., 1964.

62. Stone, M. H.: "Linear Transformations in Hilbert Space and Their Applications to Analysis," Colloquium Publications, American Mathematical Society, Providence, R.I., 1932.

63. Taylor, A. E.: "Introduction to Functional Analysis," John Wiley & Sons, Inc., New York, 1958.

64. Thielman, H. P.: Applications of the Fixed Point Theorem by Russian Mathematicians, in Ref. 4, pp. 35–48.

65. Tychonoff, A.: Ein Fixpunktsatz, *Math. Ann.*, vol. 111, pp. 767–776, 1935.

66. Vainberg, M. M.: "Variational Methods for the Study of Nonlinear Operators," Holden-Day, Inc., San Francisco, Calif., 1964.

67. Vulikh, B. Z.: "Introduction to Functional Analysis for Scientists and Technologists," Pergamon Press and The Macmillan Company, New York, 1964.

68. Wilansky, Albert: "Functional Analysis," The Blaisdell Publishing Company, New York, 1964.

69. Wouk, Arthur: Direct Iteration, Existence and Uniqueness, in Ref. 4, pp. 3–33.

70. Zaanen, A. C.: "Linear Analysis," North-Holland Publishing Company, Amsterdam, 1960.

71. Zarantonello, E. H.: Solving Functional Equations by Contractive Averaging, *U.S. Army Math. Res. Center Tech. Sum. Rep.* 160, June 1960.

72. Nashed, M. Z.: Some Remarks on Variations and Differentials, *Amer. Math. Monthly*, vol. 73, no. 4, part II, pp. 63–76, April 1966.

TWO

SOME ITERATIVE
AND DIRECT
TECHNIQUES FOR
NONLINEAR
OPERATOR
EQUATIONS

2-1 INTRODUCTION—REMARKS ON THE THEORY OF CONVERGENCE

Since solution of nonlinear equations is usually accomplished by approximation techniques, among which iterative, or successive, approximations occupy a prominent place, a thorough understanding of the concept of convergence is essential. In the next few paragraphs we shall discuss these ideas in broad outline. We urge the reader to increase his familiarity with these concepts through the literature cited for Chaps. 1 and 2.

Solution of an equation by iterative methods usually leads to the consideration of an infinite sequence of approximations. We hope we can obtain a solution by means of limiting operations on this sequence or on a new sequence constructed from the original one. A major concern then is the convergence of the sequence or a subsequence and whether the limit is the possibly unique solution.

Let X be a topological space, and let A be a subset of X. An element $x \in X$ is a limit point of A if every neighborhood of x contains a point of A distinct from x. Recall that a neighborhood of a point is a set which contains an open set containing the point. A sequence $\{x_n\}$ is a mapping of the natural numbers into a set. The sequence $\{x_n\}$ is said to converge to x if x is a limit point of $\{x_n\}$. A sequence is called *convergent* if it has a limit point.

A topological space in which every sequence contains a convergent subsequence is called *sequentially compact*. A topological space X is said to be *countably compact* if every countable covering of X by open sets contains a finite subcovering. It is said to be *compact* if every open covering contains a finite subcovering. In a metric space all these properties are equivalent. Thus, in a metric space, when speaking of sequential compactness, we simply refer to it as *compactness*. In general it is ambitious to require that the entire space be compact, and hence one confines the study to subsets which are compact. Our problem then is this: Under what conditions is a subset of a particular metric space compact?

Exercise 2-1 Guided by the last remarks and specializing to the real line, prove the equivalence of compactness and sequential compactness. Thus, show that the Bolzano-Weierstrass theorem implies the Heine-Borel theorem and conversely. (For statements of these theorems, see below.)

The essential point of the foregoing remarks is that if we wish to ensure the existence of a limit point of a sequence (which may consist of iterative approximations to the solution of an equation), it is enough to know that this sequence lies in a compact set (which may itself be a sequence).

Whether one has compactness or not, it must be shown that the sequence of approximations or a specific subsequence of it converges. This is usually accomplished by means of the Cauchy criterion [viz., $\{x_n\}$ is a convergent sequence if $\{x_n\}$ is a fundamental, or Cauchy, sequence; that is, if $\rho(x_n, x_m) \to 0$ in the metric ρ as $m, n \to \infty$] as a sufficient condition for convergence. If the space is complete, the Cauchy criterion is both necessary and sufficient for the convergence of a sequence.

However, the Cauchy criterion may fail for the original sequence but hold for a subsequence. If the sequence belongs to a compact set, then there is reason to apply the criterion to various subsequences of the given sequence in the hope of finding one which converges.

How does one prove the compactness of a subset of a Banach space? This is a difficult task in a Banach space. A general theorem states that if a Banach space has a Schauder basis, i.e., every element has a unique convergent expansion in terms of a basis as an infinite sum, then a set is compact if and only if the representing expansions converge uniformly on the set. The problem here is to find such a basis and obtain the representation. Alternatively, we have theorems characterizing the compactness of a subset of a specific Banach space, e.g., closed and bounded in R_n or the equicontinuity and uniform boundedness of the elements of a subset in $C[a,b]$ (see "NM").

It is important to observe that compactness is a relative notion depending on the topology. A topology is defined by open sets, and in a metric space the open sets can be defined in terms of the spheres, which, in turn, are defined by the metric. Thus a sequence may converge for one choice of metric but not for another. For a space with several norms it may be that convergence is obtained in one norm but not in another. This calls for the concept of a topology's being as strong or as fine as a second one; i.e., every neighborhood of a point in the second topology is also a neighborhood in the first topology. Thus if a sequence converges in the stronger topology, it automatically converges in the weaker one, but the converse need not hold.

Exercise 2-2   Applying the foregoing concepts, investigate the convergence of a sequence of real numbers $\{x_n\}$ by means of two different metrics. How are the spheres defined? Suggest a third metric which gives a weaker topology and for which the sequence need not converge. Consider now the same problem for sequences of the form $\{(x_n,y_n)\}$, where x_n and y_n are real. Such sequences may occur, for example, in applying Newton's method to a pair of algebraic equations. Finally consider the same questions for a sequence of continuous functions.

Having obtained the limit of a convergent sequence of approximations, one proceeds to show that it is the desired solution and perhaps also that it is unique. Here again there are problems to be considered.

First note that a metric space has the Hausdorff property, which ensures that a convergent sequence has exactly one limit point. However, two subsequences of our original sequence may converge to two distinct limit points. In addition, there may be other solutions to which none of the subsequences of our sequence converges. Here then one must start with an idea of the properties of the solution desired. Again recall, for example, that a polynomial equation has several solutions; to obtain a specific one an initial point near this solution must be chosen. Such may also be the case for a general operator equation.

A final concern in the theory of convergence is the rate at which a given sequence converges. For example, one iterative scheme may be preferred to another because it requires fewer steps to approach the limit point within a prescribed error.

The reader will find it helpful to keep the foregoing remarks in mind while reading about the various iterative methods below.

Remark: The concept of compactness in topological spaces, discussed above, has gone far beyond the ideas developed for the real line, with which the reader must be familiar if he has already dealt with convergence on the real line. He will recall that there the following five properties, which are equivalent, play a fundamental role in characterizing the real line:

1. *The least-upper-bound theorem*   Any set of real numbers which is bounded above has a least upper bound.

2. *The Dedekind cut theorem*   Let the real numbers be divided into two nonempty classes A and B such that (1) $a \epsilon A$, $b \epsilon B \rightarrow a < b$ and (2) $A \cup B$ is the set of real numbers. Then either there is a greatest number in A or a least number in B.

3. *The Bolzano-Weierstrass theorem*   Any infinite bounded set of real numbers has a limit point.

4. *The Heine-Borel theorem*   If $\{O_\alpha\}$ is a collection of open sets covering a closed bounded set M of the real line, then there exists a finite subcollection of $\{O_\alpha\}$ covering M.

5. *The Cantor-product, or nested-sets, theorem*   If $M_1 \supset M_2 \supset \cdots$ is a sequence of closed sets on the real line such that $\lim\limits_{n \to \infty} \text{diam } M_n = 0$ (where diam $M_n = \sup\limits_{x,y \epsilon M_n} |x - y|$), then $\bigcap\limits_{n=1}^{\infty} M_n \neq \emptyset$.

Let T designate a nonlinear transformation from a Banach space X to a Banach space Y. Consider first the operator equation $T(x) = \mathbf{0}$. Let us construct a sequence, $x_1 = f(x_0)$, $x_2 = f(x_1)$, $\ldots$, $x_n = f(x_{n-1})$, where f is an endomorphism of X, that is, a mapping of X into X. Suppose that $x^* = \lim\limits_{n \to \infty} x_n$ exists and satisfies $T(x) = \mathbf{0}$; then f constitutes an iteration or iterative process for the equation $T(x) = \mathbf{0}$. An iteration f is said to be of order p if the sequence $\{x_n\}$ satisfies

$$\|x_{n+1} - x_n\| = \|f(x_n) - f(x_{n-1})\| \leq q\|x_n - x_{n-1}\|^p$$

where q is a positive number independent of n.

A transformation f satisfies a Lipschitz condition of order p if the second inequality in the foregoing expression holds. Thus an iteration of order p satisfies a Lipschitz condition of order p.

Exercise 2-3   Let $p = 1$, and show that if $q < 1$, then for all m

$$\|x_{n+m} - x_n\| \leq \frac{q^n}{1 - q} \|x_1 - x_0\|$$

Note that since a Banach space is complete, the sequence tends to a limit x^*, and

$$\|x^* - x_n\| \leq \frac{q^n}{1 - q} \|x_1 - x_0\|$$

Exercise 2-4 If $p > 1$ and $q > 0$ is arbitrary, then if $q\|x_1 - x_0\|^{p-1} = \alpha < 1$, show that

$$\|x_{n+1} - x_n\| \leq q^{(p^n-1)/(p-1)}\|x_1 - x_0\|^{p^n}$$

and that

$$\|x_{n+m} - x_n\| \leq \frac{1}{1 - \alpha}\|x_{n+1} - x_n\|$$

and that passing to the limit gives

$$\|x^* - x_n\| = \frac{1}{1 - \alpha}q^{(p^n-1)/(p-1)}\|x_1 - x_0\|^{p^n} = C\lambda^{p^n}$$

with
$$C = \frac{1}{1 - \alpha}q^{-1/(p-1)} \qquad \lambda = \alpha^{1/(p-1)} < 1$$

Remark: The above definition of an iteration f of order p is equivalent to the requirement that if $x_n \to x^*$, then

$$\|x_n - x^*\| \leq r\|x_{n-1} - x^*\|^p$$

is satisfied, where r satisfies

$$r\|x_0 - x\|^{p-1} < 1$$

Remark: Conditions very similar to those used by Kantorovich to show the convergence of the Newton-Raphson method (see "NM," sec. 2-2) can be applied to an iteration of the first order, $f(x) \equiv x + \lambda T(x)$, where λ is a bounded linear transformation from Y to X, $Y \subset X$. Thus under suitable hypotheses the sequence $\{x_n\}$, defined by

$$x_1 = f(x_0)$$
$$x_n = f(x_{n-1}) = x_{n-1} + \lambda T(x_{n-1})$$

converges to a root x^* of the equation $T(x) = 0$. If λ satisfies a Lipschitz condition of order one in a sphere $S(x_0, r)$, that is, if for all x and x' in $S(x_0, r)$ we have

$$\|\lambda(x) - \lambda(x')\| \leq L\|x - x'\|$$

and if λ^{-1} exists for all $x \in S(x_0, r)$, then a similar theorem also applies to an iteration of the form $f(x) \equiv x + \lambda(x)T(x)$.

Exercise 2-5 Show that the requirement that $T(x)$ be twice Fréchet-differentiable in a sphere $S(x_0, r)$ enables one to write

$$x_2 - x_1 = \lambda f(x_1) = \lambda[f(x_0) + f'(x_0)(x_1 - x_0) + R_2(f, x_1, x_0)]$$
$$= [I + \lambda f'(x_0)](x_1 - x_0) + \lambda R_2(f, x_1, x_0)$$

Before we begin our discussion of various iterative methods a word of caution is in order. Although we illustrate several iterative schemes for operators in Banach and Hilbert spaces, we shall not dwell on finding particular applications for each one, as our intent is to write a chapter and not a book on the subject.   Indeed, we do give some illustrations, such as the application of Newton's method to the solution of an integral equation due to Rall.   It is obvious, however, that in order to apply any of these methods to the solution of an equation, the equation must be represented in appropriate operator notation, as our illustration of Newton's method shows.

Exercise 2-6   Consider the system of equations

$$f_i(x_1, \ldots, x_n) = 0, \ i = 1, \ldots, n$$

Show that it can be represented by a single equation $f(x) = \mathbf{0}$.   Also show that a Taylor-series expansion of all f_i around a point $x^0 \equiv (x_1{}^0, \ldots, x_n{}^0)$ can be represented by the single equation

$$f(x) = f(x^0) + [(x - x^0)\nabla]f + \frac{1}{2!}[(x - x^0)\nabla]^2 f + \cdots$$

where on the right f is evaluated at x^0 after operating on it by each power of ∇. Show how Newton's method relies on this expansion and indicate the method for obtaining the $(k+1)$st iteration x^{k+1} from the kth iteration x^k.

Exercise 2-7   Most iterative techniques relate the $(k+1)$st iteration and the kth iteration approximately as

$$x^{k+1} = x^k + \epsilon_k y^k$$

This form is convenient for applying the Cauchy criterion by forming $\|x^{k+p} - x^k\|$. To show that this difference $\to 0$ as $k \to \infty$, with $p > 0$ arbitrary, one must apply boundedness conditions on y_0 obtained at the first estimate x^0 and then prove that such conditions continue to hold for subsequent iterations and provide the desired bound, which $\to 0$ as $k \to \infty$.   Form by repetitive application the difference $x^{k+p} - x^k$, and then conjecture some boundedness conditions for the term corresponding to $\epsilon_0 y_0$ in Newton's method which may be needed to prove convergence of the iterations.

In the remainder of the chapter we shall discuss iterative methods (see also "NM") and direct methods useful in the solution of equations.

2-2   ITERATIVE METHODS (USEFUL FOR BOUNDED OPERATORS)

NEWTON'S METHOD

Here we give a few ideas which supplement the development of Newton's method, used to solve $T(x) = \mathbf{0}$, given in "NM," chap. 2.   First we give Bartle's theorem, which is concerned with the convergence of Newton's iterative process without imposing conditions on the second derivative. We then follow Altman in a theorem and give his example of what to do if $[T'(x_0)]^{-1}$ does not exist at the initial point x_0.   Finally we illustrate an application of Newton's method (developed by Rall) to the solution of an integral equation.

Bartle's Theorem[4]

If G is an open set in the Banach space X, then f is said to be in the class $C'(G)$ if $f'(x)$ exists for $x \in G$ and if the mapping $x \to f'(x)$ is a continuous map of G into the space $B(X,Y)$ of bounded linear operators from X to Y equipped with the uniform operator topology. Convergence in the norm in the space of bounded linear operators is known as *uniform convergence*, and the topology generated by this norm is the uniform topology.

For f in the class $C'(G)$, the *modulous of continuity* is a number $\delta > 0$ depending on x_0 and ϵ such that whenever

$$\|x - x_0\| \leq \delta(x_0, \epsilon)$$

then
$$\|f'(x) - f'(x_0)\| \leq \epsilon$$

We state without proof two lemmas used in proving Bartle's theorem.

Lemma 1: If f is in the class $C'(S(x_0, \alpha))$ and x_1 and x_2 are such that $\|x_i - x_0\| \leq \delta(x_0, \epsilon)$, $i = 1, 2$, then

$$\|f(x_1) - f(x_2) - f'(x_0)(x_1 - x_2)\| \leq \epsilon\|x_1 - x_2\| \qquad (2\text{-}1)$$

Lemma 2: Let f be in the class $C'(S(x_0, \alpha))$, and suppose that $f'(x_0)$ has a bounded inverse. For any $\lambda > \|[f'(x_0)]^{-1}\|$ there is a number $0 < \beta \leq \min\{1, \alpha\}$ such that:

1. If $\|x - x_0\| \leq \beta$, then the operator $f'(x)$ has an inverse, and $\|[f'(x)]^{-1}\| < \lambda$.

2. If $\|x_i - x_0\| \leq \beta$, $i = 1, 2, 3$, then

$$\|f(x_1) - f(x_2) - f'(x_3)(x_1 - x_2)\| \leq \frac{1}{2\lambda}\|x_1 - x_2\| \qquad (2\text{-}2)$$

Theorem 2-1 Let $f : S(x_0, \alpha) \to Y$ be in class $C'(S(x_0, \alpha))$, and suppose that $f'(x_0)$ has an inverse with $\|[f'(x_0)]^{-1}\| < \lambda < \infty$. Let $\|f(x_0)\| < \beta/2\lambda$, where $\beta = \min\{1, \alpha_1, \delta(x_0, 1/4\lambda)\}$, and let z_n, $n = 0, 1, 2, \ldots$, be arbitrary points with $\|z_n - x_0\| \leq \beta$. Then the sequence $\{x_n\}$ obtained by the iterative process

$$x_{n+1} = x_n - [f'(z_n)]^{-1}f(x_n) \qquad n = 0, 1, 2, \ldots$$

converges to a solution $\bar{x}$ of the equation $f(x) = 0$. Further, $\|\bar{x} - x_0\| \leq \beta$ and is the only solution of the equation in this neighborhood of x_0. The rapidity of convergence is given by $\|x_n - \bar{x}\| < 2^{-n}\beta$, $n = 0, 1, 2, \ldots$.

Note: S_1 is chosen such that if $\|x - x_0\| < \delta_1$ then $\|f'(x) - f(x_0)\| \leq 1/4\lambda$. Let α_1 be such that $\alpha_1 \leq \min\{\alpha, \delta_1\}$ and such that $\|x - x_0\| \leq \alpha_1$ implies $\|[f'(x)]^{-1}\| < \lambda$. This choice is possible by Lemma 2, part 1.

Proof: By definition $x_1 = x_0 - [f'(z_0)]^{-1}f(x_0)$, and so

$$\|x_1 - x_0\| = \|[f'(z_0)]^{-1}f(x_0)\| \le \|[f'(z_0)]^{-1}\| \, \|f(x_0)\|$$
$$< \lambda\|f(x_0)\| < \lambda\frac{\beta}{2\lambda} = \frac{\beta}{2}$$

since $\|[f'(z_0)]^{-1}\| < \lambda$.   Also, from

$$(x_1 - x_0) = -[f'(z_0)]^{-1}f(x_0)$$

we have

$$f'(z_0)(x_1 - x_0) = -[f'(z_0)][f'(z_0)]^{-1}f(x_0) = -f(x_0)$$

so that $$f(x_1) = f(x_1) - f(x_0) - f'(z_0)(x_1 - x_0)$$

and thus $\|f(x_1)\| = \|f(x_1) - f(x_0) - f'(z_0)(x_1 - x_0)\| \le \dfrac{1}{2\lambda}\|x_1 - x_0\|$

by replacing x_2 by x_0 and x_3 by z_0 in part 2 of Lemma 2.

By induction, suppose that for $i = 1, 2, \ldots, n$ we have:

1. $\|x_i - x_0\| < \beta$.
2. $\|x_i - x_{i-1}\| \le \lambda\|f(x_{i-1})\|$.
3. $\|f(x_i)\| \le \dfrac{1}{2\lambda}\|x_i - x_{i-1}\|$.

Then we wish to show that conditions 1 to 3 hold for $i = n + 1$.

Now $x_{n+1} - x_n = -[f'(z_n)]^{-1}f(x_n)$, and so, by part 1 of Lemma 2,

$$\|x_{n+1} - x_n\| \le \|[f'(z_n)]^{-1}\| \, \|f(x_n)\| \le \lambda\|f(x_n)\|$$

which proves condition 2.   Combining this with condition 3 in which we put $i = n$, we have

$$\|x_{n+1} - x_n\| \le \lambda\|f(x_n)\| \le \lambda\frac{1}{2\lambda}\|x_n - x_{n-1}\| = \tfrac{1}{2}\|x_n - x_{n-1}\|$$

Taking this result and iterating conditions 2 and 3, we find that

$$\|x_{n+1} - x_n\| \le \left\{\sum_{i=0}^{n} 2^{-i}\right\}\|x_1 - x_0\| < (1 - 2^{-(n+1)})\beta$$

which proves condition 1.   In the same manner as we had for x_1 we have

$$f(x_{n+1}) = f(x_{n+1}) - f(x_n) - f'(z_n)(x_{n+1} - x_n)$$

and because condition 1 has been proved, we can replace x_1 by x_{n+1}, x_2 by x_n, and x_3 by z_n in part 2 of Lemma 2 to obtain

$$\|f(x_{n+1})\| = \frac{1}{2\lambda}\|x_{n+1} - x_n\|$$

which is condition 3. For any positive integers n and p,

$$\|x_{n+p} - x_n\| \le \sum_{i=1}^{p} \|x_{n+i} - x_{n+i-1}\| \le \|x_{n+1} - x_n\| \sum_{i=0}^{p-1} 2^{-i}$$

$$\le \lambda\|f(x_n)\| \sum_{i=0}^{p-1} 2^{-i} \le \lambda\|f(x_0)\|2^{-n} \sum_{i=0}^{p-1} 2^{-i} < 2^{-n}\beta \quad (2\text{-}3)$$

and so $\{x_n\}$ is a Cauchy sequence, hence converges to an element $\bar{x}$ of X. From condition 1, $\|\bar{x} - x_0\| \le \beta$. Let $p = 1$ in (2-3); then from this and condition 3 we have $f(\bar{x}) = 0$.

Let $\hat{x} \ne \bar{x}$ be another solution of $f(x) = 0$ such that $\|\hat{x} - x_0\| \le \beta$. Then,

$$\|\bar{x} - \hat{x}\| = \|[f'(x_0)]^{-1}f'(x_0)(\bar{x} - \hat{x})\| \le \lambda\|f'(x_0)(\bar{x} - \hat{x})\|$$

But by part 1 of Lemma 1 we have

$$\|f'(x_0)(\bar{x} - \hat{x})\| \le \frac{1}{2\lambda}\|\bar{x} - \hat{x}\|$$

Since λ can be taken ≥ 1, $\|\bar{x} - \hat{x}\| \le \frac{1}{2}\|\bar{x} - \hat{x}\|$, which is impossible. The rapidity of convergence is immediately obtained from (2-3) as $p \to \infty$.

Altman's Theorem

An example where $[T'(x_0)]^{-1}$ may not exist is given by

$$f_1(x_1,x_2) \equiv x_1{}^2 + x_1 + x_2 + \tfrac{3}{2} = 0$$

$$f_2(x_1,x_2) \equiv x_1{}^2 + x_1 - x_2 - 1 = 0$$

Define $x = (x_1,x_2)$ and $T(x) = (f_1(x),f_2(x))$ and suppose that

$$x_0 = (x_1{}^{(0)},x_2{}^{(0)}) = (-\tfrac{1}{2},-1)$$

Now

$$T'(x) = \begin{pmatrix} 2x_1 + 1 & 1 \\ 2x_1 + 1 & -1 \end{pmatrix}$$

and has no inverse at $(-\tfrac{1}{2},-1)$.

The following theorem concerning the solution of $T(x) = 0$ in Hilbert space, which Altman[3] has proved, remedies the above difficulty. First suppose that the nonlinear continuous operator $T(x)$ with values in Hilbert space H is Fréchet-differentiable in a sphere $S(x_0,r)$. Consider the modified Newton iterations

$$x_{n+1} = x_n - \frac{\|T(x_n)\|}{2\|Q(x_n)\|} Q(x_n) \qquad n \ge 1 \quad (2\text{-}4)$$

where $Q(x) = T'^*(x) T(x)$ and the linear operator T'^* is the adjoint of $T'(x)$. Suppose that $Q'(x)$ also exists in $S(x_0,r)$ and is bounded; that is, $\|Q'(x)\| \leq K$, $x \, \epsilon \, S(x_0,r)$. Then we have:

Theorem 2-2 If:

1. $Q(x_0) = T'^*(x_0) T(x) \neq 0$,
2. $\|Q'(x)\| \leq K$, $x \, \epsilon \, S(x_0,r)$, and for some constant K.
3. $\dfrac{\|T(x_0)\|^2}{\|Q(x_0)\|^2} K \leq 2h_0 \leq 1$,

where r is defined by

$$r = \frac{1 - \sqrt{1 - 2h_0}}{h_0} \eta_0$$

in which $h_0 = B_0\eta_0 K \leq \tfrac{1}{2}$

where $\qquad \dfrac{1}{\|T(x_0)\|} \leq B_0 \qquad \|x_1 - x_0\| \leq \dfrac{\|T(x_0)\|^2}{2\|Q(x_0)\|^2} \leq \eta_0$

for some B_0 and η_0, then $T(x) = 0$ has a solution x^* in $S(x_0,r)$ to which the sequence of iterations defined by (2-4) converges, and the error estimate is given by

$$\|x_n - x^*\| \leq \frac{1}{2^{n-1}} (2h_0)^{n-1}\eta_0$$

When Theorem 2-2 is applied to the example, we have

$$Q(x) = T'(x)T(x) = (4x_1{}^3 + 6x_1{}^2 + 3x_1 + \tfrac{1}{2}, 2x_2 + \tfrac{5}{2})$$
$$Q'(x) = \begin{pmatrix} 12x_1{}^2 + 12x_1 + 3 & 0 \\ 0 & 2 \end{pmatrix}$$

Now $\|T(x_0)\|^2 = \tfrac{1}{8}$, $\|Q(x_0)\| = \tfrac{1}{2}$, and hence $2\eta_0 = \tfrac{1}{8}/\tfrac{1}{4} = \tfrac{1}{2}$. Also $\|Q'(x)\| \leq 2$ in the sphere $S(x_0,2\eta_0)$. Thus $K = 2$. Condition 3 is also satisfied since $(\tfrac{1}{8} \times 2)/\tfrac{1}{4} = 2h_0 = 1$. We have used the square root of the sum of the squares of the coefficients for the norms.

An Application of Newton's Method[40] (Rall's Example)

In this section we introduce ideas from Banach spaces which in effect apply Newton's method to a nonlinear integral equation.

A bounded bilinear transformation B from a Banach space X to a Banach space Y associates with every ordered pair of elements of X, that is, (x_1,x_2), a unique y in Y [we write $B(x_1,x_2) = y$] such that:

1. $\|B\| \equiv \sup\limits_{\substack{x_1 \neq 0 \\ x_2 \neq 0}} \dfrac{\|B(x_1,x_2)\|}{\|x_1\| \, \|x_2\|}$ exists and is finite.

2. B is bilinear; that is,

$$B((x_1 + x_1'),x_2) = B(x_1,x_2) + B(x_1',x_2)$$
$$B(x_1,(x_2 + x_2')) = B(x_1,x_2) + B(x_1,x_2')$$
$$B(\alpha x_1,x_2) = \alpha B(x_1,x_2) = B(x_1,\alpha x_2)$$

An example of a bilinear operator in the space of n-dimensional real vectors is a three-dimensional matrix $B = (b_{ijk})$, $i, j, k = 1, \ldots, n$.

The components of the vector $y = (\eta_1, \ldots, \eta_n)$, where $y = B(x_1,x_2)$, with $x_1 = (\zeta_1^{(1)}, \ldots, \zeta_n^{(1)})$, $x_2 = (\zeta_1^{(2)}, \ldots, \zeta_n^{(2)})$, are obtained from

$$\eta_i = \sum_{j=1}^{n} \sum_{k=1}^{n} b_{ijk} \zeta_k^{(1)} \zeta_j^{(2)} \qquad i = 1, \ldots, n$$

Thus
$$\|B\| \leq \max_i \sum_{j=1}^{n} \sum_{k=1}^{n} |b_{ijk}|$$

A continuous problem related to the foregoing is given by the transformation

$$y(s) = \int_0^1 K(s,t)x_1(s)x_2(t)\, dt \qquad 0 \leq s \leq 1$$

which defines a bounded bilinear operator B for all x_1 and x_2 in $C[0,1]$. Note that this is a special case of the generalization $\int\int K(s,t,u)x_1(s) x_2(t)\, ds\, dt$. Here K is also in $C[0,1]$, and

$$\|B\| \leq \max_{0 \leq s \leq 1} \int_0^1 |K(s,t)|\, dt$$

The operator B^* such that

$$B^*(x_1,x_2) = B(x_2,x_1)$$

is called the permutation of B, and clearly $\|B^*\| = \|B\|$. For the first of the foregoing examples

$$B^* = (b_{ikj}) \qquad i, j, k = 1, \ldots, n$$

and for the second

$$B^*(x_1,x_2) = \int_0^1 K(s,t)x_1(t)x_2(s)\, dt$$

If B is bilinear, then its Fréchet derivative at x is given by

$$B'(x) = Bx + B^*x$$

that is, $B'(x)$ is a linear operator whose value at y is given by

$$[B'(x)](y) = (B + B^*)(x,y) = B(x,y) + B^*(x,y)$$

for if we write $T(x) = B(x,x)$, then

$$T(x + \Delta x) - T(x) = B[(x + \Delta x),(x + \Delta x)] - B(x,x)$$
$$= B(x,x) + B(x,\Delta x) + B(\Delta x,x) + B(\Delta x,\Delta x)$$
$$- B(x,x)$$
$$= B(x,\Delta x) + B^*(x,\Delta x) + B(\Delta x,\Delta x)$$

and hence

$$B[(x + \Delta x),(x + \Delta x)] - B(x,x) - (B + B^*)(x,\Delta x) = B(\Delta x,\Delta x)$$

and the norm of the left side does not exceed $\|B\| \, \|\Delta x\|^2$. Note that the second Fréchet derivative of a linear operator T is zero, but for a bilinear operator B we have $B'' = B + B^*$ for all $x \, \epsilon \, X$. To interpret this, note that B' is a linear operator mapping a Banach space X into the Banach space of all linear operators on X. Thus for all $x, y \, \epsilon \, X$,

$$B'(x + \Delta x)(y) - B'(x)(y) = (B + B^*)(\Delta x,y)$$

We begin by writing Newton's process as

$$x_{n+1} = [I - T'^{-1}T]x_n$$

and consider the nonlinear integral equation

$$f(s) = 1 + \alpha s f(s) \int_0^1 \frac{f(t)}{s + t} \, dt$$

which can be written in $C(0,1)$, using x for $f(s)$, as

$$T(x) \equiv Ix - 1 - \alpha B(x,x) = 0$$

where $K(s,t) = s/(s + t)$ (and hence $\|B\| \leq \log 2$), and where I is the identity operator. Now if $x_0 \, \epsilon \, C(0,1)$, for all $x \, \epsilon \, C(0,1)$ we have

$$T'(x_0) = I - \alpha(Bx_0 + B^*x_0)$$
$$T''(x_0) = -\alpha(B + B^*)$$
$$\|T''(x)\| \leq 2\alpha \log 2 \equiv A$$

Thus the application of Newton's method requires the solution of a sequence of Fredholm integral equations of the third kind given by

$$[I - \alpha(Bx_{n-1} + B^*x_{n-1})]x_n = 1 - \alpha Bx_{n-1}x_{n-1}$$

that is,

$$\left[1 - \alpha \int_0^1 \frac{s}{s + t} x_{n-1}(t) \, dt\right] x_n(s) - \alpha \int_0^1 \frac{s}{s + t} x_{n-1}(s)x_n(t) \, dt$$
$$= 1 - \alpha \int_0^1 \frac{s}{s + t} x_{n-1}(s)x_{n-1}(t) \, dt$$

On the other hand, if the quantity in brackets, which we denote by $f_{n-1}(s)$, does not vanish for $0 \leq s \leq 1$, then we have a Fredholm equation of the second kind

$$(I - \alpha f_{n-1}^{-1}(s)Bx_{n-1})x_n = f_{n-1}^{-1}(s)y_{n-1} \tag{2-5}$$

where
$$y_{n-1} = 1 - \alpha B(x_{n-1},x_{n-1})$$

If
$$\|\alpha Bx_{n-1}\| < 1$$

we obtain the solution x_n of (2-5) by iteration. Note that

$$\|[T'(x_{n-1})]^{-1}\| \le \frac{\|f_{n-1}^{-1}\|}{1 - \|\alpha f_{n-1}^{-1}Bx_{n-1}\|}$$

Exercise 2-8 Starting with $x_0(s) = 1$ for $0 \le s \le 1$ in the original equation, obtain Tx_0 and show that

$$\frac{b_0}{a_0} = \alpha \log 2$$

Determine $f_0(s)$ and show that

$$\|f_0{}^{-1}\| = \frac{1}{1 - \alpha \log 2}$$

and that
$$a_0 = \frac{1}{1 - 2\alpha \log 2}$$

Remark (Bypassing a choice of an initial value): The choice of an initial approximation can be crucial for the convergence of Newton's method. How to bypass this difficulty of starting with an initial value close to the root will now be described briefly. Later it is discussed in more general terms in Sec. 2-3. The procedure is to embed the problem (for simplicity assume that it is a system of algebraic equations) in a larger problem in the presence of continuously varying parameters or parameter, e.g., the coefficients, such that a solution is known for a specific set of values of the parameters. Then by fixing a value of each parameter near the specific value, one can apply Newton's method to solve this problem with the known solution as an initial approximation. When this solution has been obtained, it is used as the initial value for a new set of parameter values (no specific rule is given for their choice). The procedure is thus continued, each time varying the coefficients (or parameters) in the direction of the known coefficients (or parameter values) of the original problem and hence gradually creeping up on its solution in steps (possibly a finite number). In Ref. 11 sufficiency conditions are mentioned for the convergence of this process, which essentially require that the parameterized functions be continuous for the range of values of the parameters under consideration, i.e., for any choice to be made for the parameter values, and that the solution as a function of a continuously varying number of steps also be continuous so that it will converge to the solution of the original problem. For example, to solve the two equations

$$-y^3 + 5y^2 - 2y + x - 13 = 0$$
$$y^3 + y^2 - 14y + x - 29 = 0$$

one starts with

$$-y^3 - 13y^2 - 50y + x - 71 = 0$$
$$y^3 + 19y^2 + 106y + x + 129 = 0$$

whose solution is at $(15, -2)$. This is used as the initial point for an application of Newton's method to the new system

$$-y^3 - 9.4y^2 - 40.4y + x - 59.4 = 0$$
$$y^3 + 15.4y^2 + 82y + x + 97.4 = 0$$

whose solution is at $(9.9, -1.91)$.

Exercise 2-9 Attempt in several steps to obtain the solution of the original problem at $(5,4)$.

Remark: Consider the real function of a real variable $y = f(x)$. An iterative procedure for picking a solution of $f(x)$ has been briefly explored by Khamis.[21] If f is continuous and monotone in the neighborhood of a solution point x^*, one can attempt a Taylor-series expansion of the inverse function $x = g(y)$ about a point y_0 near $y = 0$; that is, since $f(x^*) = 0$, $x^* = g(0)$. Thus,

$$x = g(y) = g(y_0) + g'(y_0)(y - y_0) + g(\eta) \frac{(y - y_0)^2}{2!} \qquad y_0 < \eta < y$$

By writing

$$x_0 = g(y_0)$$

and using the well-known relation

$$f'(x)g'(y) = 1$$

and hence

$$g''(y) = -\frac{f''(x)}{[f'(x)]^3}$$

we have

$$x = x_0 + \frac{f(x) - f(x_0)}{f'(x_0)} - \frac{f''(\xi)}{[f'(\xi)]^2} \frac{[f(x) - f(x_0)]^2}{2!} \qquad \xi = g(\eta)$$

In general one can use any y_n instead of y_0 and its corresponding x_n and then replace x by x_{n+1}, set $f(x_{n+1}) = 0$, and use the result as an iterative procedure.   The Newton-Raphson process is obtained by using the first two terms of this expansion.

ALTMAN'S[1] *METHOD OF TANGENT HYPERBOLAS*

Consider the equation

$$T(x) = 0 \qquad (2\text{-}6)$$

where T is a nonlinear operator from a Banach space X to a Banach space Y. Suppose that $T(x)$ has first-, second-, and third-order continuous derivatives in the sense of Fréchet.

The method of tangent hyperbolas utilizes the iterations

$$x_{n+1} = x_n - Q_n \gamma_n T(x_n) \qquad n = 0, 1, 2, \ldots \qquad (2\text{-}7)$$

where x_0 is the initial approximate solution of Eq. (2-6), $\gamma_n = [T'(x_n)]^{-1}$ denotes the inverse of $T'(x_n)$, and

$$Q_n = [I - \tfrac{1}{2}\gamma_n T''(x_n)\gamma_n T(x_n)]^{-1}$$

To apply the majorant principle to (2-7) consider the real equation

$$f(z) = 0 \qquad (2\text{-}8)$$

where $f(z)$ is a real function of the real variable z, which is three times continously differentiable in the interval (z_0, z'). Equation (2-6) possesses a real majorant if the following conditions are satisfied:

a. $\|T(x_0)\| \le f(z_0)$.
b. $\|\gamma_0\| \le B_0$ where $B_0 = -1/f'(z_0) > 0$.
c. $\|T''(x)\| \le f''(z)$ if $\|x - x_0\| \le z - z_0 \le z' - z_0$.
d. $\|T'''(x)\| \le f'''(z)$ if $\|x - x_0\| \le z - z_0 \le z' - z_0$.

We now proceed to construct the sequence of the approximate solutions z_n of Eq. (2-8) by the method of tangent hyperbolas. For this purpose we make use of the extended mean-value theorem to develop iterative methods of higher order. We first give a plausibility argument to set up the iteration scheme, and then we prove its convergence. In applying the mean-value theorem note that if the end points of the interval are close, the mean value of the function can be reasonably approximated by its value at either end point. Thus instead of

$$f(z_1) - f(z_0) = f'(\xi)(z_1 - z_0) \qquad z_0 < \xi < z$$

we write $\quad f(z_1) - f(z_0) \approx f'(z_0)(z_1 - z_0)$

where z_1 is close to z_0. Thus we have for an approximate solution of $f(z) = 0$

$$z_1 \approx z_0 - \frac{f(z_0)}{f'(z_0)}$$

Using Taylor's theorem, we can write

$$f(z_1) - f(z_0) = f'(z_0)(z_1 - z_0) + \tfrac{1}{2}f''(\xi)(z_1 - z_0)^2$$

where $z_0 < \xi < z_1$ and $f(z_1) = 0$. Thus

$$-f(z_0) \approx f'(z_0)(z_1 - z_0) + \tfrac{1}{2}f''(z_0)(z_1 - z_0)^2$$

which is also a good approximation if z_1 is close to z_0. This can be simplified to

$$z_1 - z_0 \approx \frac{-f(z_0)}{f'(z_0) + \tfrac{1}{2}f''(z_0)(z_1 - z_0)}$$

which by the first approximation, $z_1 - z_0 \approx -f(z_0)/f'(z_0)$, becomes the second approximation

$$z_1 - z_0 \approx - \frac{f(z_0)}{f'(z_0) - \tfrac{1}{2}[f''(z_0)f(z_0)/f'(z_0)]}$$

and

$$z_2 - z_1 \approx - \frac{f(z_0)}{(1 - \tfrac{1}{2}\{f''(z_0)f(z_0)/[f'(z_0)]^2\})f'(z_0)}$$

Generally we have

$$z_{n+1} = z_n - \frac{1}{1 - \tfrac{1}{2}\{f''(z_n)f(z_n)/[f'(z_n)]^2\}} \frac{f(z_n)}{f'(z_n)} \tag{2-9}$$

using the equality sign to indicate the sequence of iterations with whose convergence we shall be concerned.

Similarly, iterative methods of higher order can be developed. The first approximation given for roots of equations is essentially Newton's method, and the second is the method of tangent hyperbolas.

Lemma: If

$$f(z) \geq 0 \qquad \text{for } z_0 \leq z \leq z^*$$

where z^* satisfies $f(z^*) = 0$, and if $f''(z)$ is nondecreasing for $z_0 \leq z \leq z^*$, then

$$\frac{f''(z)f(z)}{[f'(z)]^2} \leq \frac{1}{2}$$

Note that in the neighborhood of the root z^*, the discriminant of the quadratic in z obtained from Taylor's expansion up to second-order terms must be nonnegative.

Theorem 2-3 If (2-6) has a majorant given in (2-7), and if:

1. Eq. (2-8) has a positive root, and
2. $f'(z) < 0$ for $z_0 \leq z \leq \tfrac{4}{3}z^*$, where z^* is the smallest positive root of (2-8), then the iterations (2-9) converge to the solution z^* of (2-8), and (2-6) has a solution x^*. Furthermore, the sequence of approximate values x_n defined by (2-7) converges to x^*. The error estimate is given by

$$\|x^* - x_n\| \leq z_2 - z_n \tag{2-10}$$

Proof: First we prove that the sequence z_n is convergent. It follows from the hypotheses of the theorem that this sequence is increasing.

Thus it is sufficient to show that it is bounded. Now using a simple argument and applying the lemma, we have $z_n \leq \frac{4}{3} z^*$, $n = 1, 2, \ldots$. From (2-9)

$$\frac{f(z_n)}{f'(z_n)} \to 0 \qquad \text{as } n \to \infty$$

Thus $f(z_n) \to 0$ as $n \to \infty$. Again from the lemma it follows, from the assumptions of the theorem, that

$$\|x_1 - x_0\| \leq (z_1 - z_0) \tag{2-11}$$

Next we show that the inverse γ_1 of $T'(x_1)$ exists and that condition b is satisfied for x_1. Let us estimate the expression

$$\|I - \gamma_0 T'(x_1)\| = \left\| \int_{x_0}^{x_1} \gamma_0 T''(x) \, dx \right\| \leq B \int_{z_0}^{z_1} Q''(z) \, dz$$

$$= 1 - \frac{Q'(z_1)}{Q'(z_0)} = q < 1 \tag{2-12}$$

Note that γ_0 is constant, and $\gamma_0 T'(x_0) = I$. Similarly B_0 is constant, and $B_0 Q'(x_0) = 1$. We have

$$\gamma_0 T'(x_1) = I - [I - \gamma_0 T'(x_1)]$$

Since $q < 1$, $T'(x_1)$ exists. We also have

$$T'(x_1)^{-1} = \gamma_1 = [T'(x_0)(I - [I - T'(x_1)])]^{-1}$$
$$= [T'(x_0)(I - \gamma_0[T'(x_0) - T'(x_1)])]^{-1}$$

and using (2-12), we obtain

$$\|\gamma_1\| \leq \|\gamma_0\| \frac{1}{1 - \|I - \gamma_0 T'(x_1)\|} \leq -\frac{1}{Q'(z_1)} \tag{2-13}$$

since $\quad \|I - \gamma_0 T'(x_1)\| \leq 1 - \dfrac{Q'(z_1)}{Q'(z_0)}$

The existence of the operator Q_1 follows from the lemma. It remains to be proved that condition a is fulfilled for x_1.

By the analogue of Taylor's formula in integral form and using (2-7), we have

$$T(x_1) = T(x_0) - T'(x_0)Q_0\gamma_0 T(x_0) + \frac{1}{2}T''(x_0)(Q_0\gamma_0 T(x_0))^2$$
$$+ \frac{1}{6} \int_{x_0}^{x_1} T'''(\bar{x})(x_1 - \bar{x})^3 \, d\bar{x}$$

Let $A = \frac{1}{2}\gamma_0 T''(x_0)\gamma_0 T(x_0)$; then we have $Q_0 = (I - A)^{-1} = I + AQ_0$.

Note that $Q_0 = I + AQ_0$ implies $Q_0(I - A) = I$, $Q_0 = (I - A)^{-1}$. Thus

$$T'(x_0)Q_0\gamma_0 T(x_0) = T(x_0) + \frac{1}{2}T''(x_0)\gamma_0 T'(x_0)Q_0\gamma_0 T(x_0) \tag{2-14}$$

and consequently

$$Q_0\gamma_0 T(x_0) = \gamma_0 T(x_0) + y_0 \tag{2-15}$$

where

$$y_0 = \tfrac{1}{2}\gamma_0 T''(x_0)\gamma_0 T(x_0)Q_0\gamma_0 T(x_0)$$

Further we have, by (2-15),

$$T''(x_0)[Q_0\gamma_0 T(x_0)]^2 = T''(x_0)(\gamma_0 T(x_0) + y_0)Q_0\gamma_0 T(x_0)$$
$$= T''(x_0)\gamma_0 T(x_0)Q_0\gamma_0 T(x_0) + T''(x_0)y_0 Q_0\gamma_0 T(x_0)$$

and using (2-14), we have

$$T(x_1) = \tfrac{1}{2}T''(x_0)y_0 Q_0\gamma_0 T(x_0) + \frac{1}{6}\int_{x_0}^{x_1} T'''(\bar{x})(x_1 - \bar{x})^3\, dx \tag{2-16}$$

On the other hand, we have

$$f(z_1) = f(z_0) - \frac{f(z_0)}{1 - a/2} + \frac{1}{2}\frac{f^2(z_0)}{[f'(z_0)]^2}\frac{f''(z_0)}{(1 - a/2)^2} + \frac{1}{6}\int_{z_0}^{z_1} f'''(z)(z_1 - z)^3\, dz$$

where

$$a = \frac{f(z_0)f''(z_0)}{[f'(z_0)]^2} \tag{2-17}$$

Thus

$$f(z_1) = \frac{1}{4}\left[\frac{a^2}{(1 - a/2)^2}\right]f(z_0) + \frac{1}{6}\int_{z_0}^{z_1} f'''(z)(z_1 - z)^3\, dz \tag{2-18}$$

It follows from (2-11), (2-16), and (2-18), using conditions a to d, that

$$\|T(x_1)\| \le f(z_1)$$

Hence finally

$$\|x_2 - x_1\| \le z_2 - z_1$$

One can show inductively that conditions a to d are satisfied for $x = x_n$, $n = 1, 2, \ldots$. As a consequence we get

$$\cdot\|x_{n+p} - x_n\| \le z_{n+p} - z_n$$

for any positive integers p and n. Hence, it follows that (2-10) is satisfied, and the sequence of approximations x_n defined by (2-7) converges toward an element x^* of the space X. Since

$$\|T(x_n)\| \le f(x_n)$$

it follows that

$$T(x_n) \to 0 \qquad \text{as } n \to \infty$$

and from the continuity of $T(x)$ we conclude that x^* is a solution of (2-6).

PICARD'S METHOD[38]

The reader will recall the classical application of Picard's method to solve $dy/dx = f(x,y)$ with given initial values [i.e., to find a differentiable function $y(x)$ such that $y'(x) = f(x,y(x)), y(x_0) = y^0$], by writing it in the integral-equation form

$$y(x) = y^0 + \int_{x_0}^{x} f(x,y)\, dx$$

and using the iterative technique obtained by replacing $y(x)$ on the left by $y_{n+1}(x)$ and in the integrand by y_n, starting with a given y_0, which often is taken as $y_0 = y^0$. This technique is sometimes also applicable to operator equations of the form $T(y) = y$, where the iterations are given by $x_{n+1} = T(x_n)$. Picard's method locates a fixed point by an iterative process. Convergence of the method for the above differential equation is shown by considering the series

$$y = y_0 + \sum_{n=0}^{\infty} (y_{n+1} - y_n)$$

of which the functions $y_n(x)$ are the partial sums, under the conditions:
1. $|x - x_0| \le a, |y - y_0| \le b$.
2. $f(x,y)$ is continuous in the rectangular region described in condition 1.
3. $|f(x,y)| < M < b/a$.
4. $|f(x,y') - f(x,y'')| < K|y' - y''|$ for $|y' - y_0| \le b$ and $|y'' - y_0| \le b$.

A slightly more general application of Picard's method is to the first-order system $dx_i/dt = f_i(x_1, \ldots ,x_n,t)$, $i = 1, \ldots , n$, with given initial condition $x_1^0, \ldots , x_n^0$ at t_0.

Now we illustrate a different use of Picard's method to show that under certain conditions (1) a Fredholm equation of the form $y = x - Tx$ is invertible and (2) the Neumann expansion of the kernel of a Volterra integral is convergent, and hence the integral equation is solvable.

Consider the Fredholm equation

$$y(s) = x(s) - \int_{a}^{b} k(s,t)x(t)\, dt \qquad (2\text{-}19)$$

subject to the following conditions:
1. $k(s,t)$ is continuous in s and t for $a \le s \le b, a \le t \le b$.
2. $|k(s,t)| \le M$.
3. $M(b - a) < 1$.

Under assumption 1, T maps the Banach space $C[a,b]$ of continuous functions on $[a,b]$ into itself where the norm on $C[a,b]$ is $\|x\| = \max_{a \le t \le b} |x(t)|$.

Consider the equation

$$y = (I - T)x \qquad (2\text{-}20)$$

where

$$Tx = \int_{a}^{b} k(s,t)x(t)\, dt \qquad (2\text{-}21)$$

Here we have a correspondence defined on C into C carrying x onto Tx. Clearly

$$\|Tx\| \leq \sup_{a \leq s \leq b} \int_a^b |k(s,t)| \, \|x(t)\| \, dt \leq M(b - a)\|x\| \qquad (2\text{-}22)$$

and hence from conditions 2 and 3

$$\|T\| < 1 \qquad (2\text{-}23)$$

a contraction. Therefore by Theorem 1-9, $(I - T)^{-1}$ exists and $(I - T)^{-1} = \sum_{k=0}^{\infty} T^k$. Thus, it follows that Eq. (2-19), which has the form

$$y = (I - T)x \qquad (2\text{-}24)$$

is solvable in the form

$$x = (I - T)^{-1}y \qquad (2\text{-}25)$$
$$x = (I + T + T^2 + \cdots)y \qquad (2\text{-}26)$$

To evaluate the inverse, we make the following designations:

$$k_1(s,t) \equiv k(s,t) \qquad (2\text{-}27)$$
$$k_n(s,t) \equiv \int_a^b k_1(s,u)k_{n-1}(u,t) \, du \qquad (2\text{-}28)$$

Define $\qquad\qquad T^2y = T(Ty) \qquad (2\text{-}29)$

Thus $\qquad\qquad T^2y = \int_a^b k(s,t) \int_a^b k(t,u)y(u) \, du \, dt$

$$= \int_a^b y(u) \int_a^b k(s,t)k(t,u) \, dt \, du$$

$$= \int_a^b y(u)k_2(s,u) \, du \qquad (2\text{-}30)$$

and inductively,

$$T^ny(s) = \int_a^b k_n(s,t)y(t) \, dt \qquad (2\text{-}31)$$

From (2-26) we obtain

$$x = y + \sum_{n=1}^{\infty} T^ny$$

$$= y + \sum_{n=1}^{\infty} \int_a^b k_n(s,t)y(t) \, dt \qquad (2\text{-}32)$$

From (2-28) and condition 2

$$|k_n(s,t)| \leq M \int_a^b |k_{n-1}(u,t)| \, |du|$$
and $\qquad\qquad |k_{n-1}(s,t)| \leq M \int_a^b |k_{n-2}(u,t)| \, |du| \qquad (2\text{-}33)$

Hence,
$$|k_n(s,t)| \leq M^2 \int_a^b \int_a^b |k_{n-2}(u,t)| \, |du|^2$$
$$\leq M^{n-1} \int_a^b \cdots \int_a^b |k_1(u,t)| \, |du|^{n-1}$$
$$\leq M^n(b-a)^{n-1} \tag{2-34}$$

which gives

$$|k_n(s,t)| \leq M^n(b-a)^{n-1}$$

Therefore, by condition 3, the series defined by

$$h(s,t) = \sum_{n=1}^{\infty} k_n(s,t)$$

converges uniformly. Now we can write series (2-32) as

$$x = y + \int_a^b h(s,t)y(t) \, dt \tag{2-35}$$

Let
$$Hy = \int_a^b h(s,t)y(t) \, dt$$

We have an equation of the form

$$x = y + Hy$$
$$= (I + H)y$$

From (2-25), we have

$$I + H = (I - T)^{-1} \tag{2-36}$$

As a second illustration let us examine the initial-value Volterra operator given by

$$y(s) = \int_a^s k(s,t)x(t) \, dt \tag{2-37}$$

Following the above pattern, we have

$$|k_n(s,t)| \leq M \int_a^s |k_{n-1}(u,t)| \, |du| \tag{2-38}$$

and
$$|k_{n-1}(s,t)| \leq M \int_a^s |k_{n-2}(u,t)| \, |du|$$

Thus we see that

$$|k_n(s,t)| \leq M^2 \int_a^s \int_a^u |k_{n-2}(u,t)| \, |du|^2$$
$$\leq M^{n-1} \int_a^s \int_a^u \cdots \int_a^u |k_1| \, |du| \cdots |du|$$
$$\leq M^n \int_a^s \int_a^u \cdots \int_a^u |du| \cdots |du|$$
$$\leq \frac{M^n(s-a)^{n-1}}{(n-1)!}$$

and the series
$$\sum_{n=1}^{\infty} \frac{M^n(s-a)^{n-1}}{(n-1)!} \tag{2-39}$$

is always convergent.   Hence, the Volterra integral equation is uniquely solvable.

REMARKS ON THE METHOD OF STEEPEST DESCENT

A recent development in the method of steepest descent is due to Nashed.[35] Variational methods used to prove existence and uniqueness of solutions to a nonlinear operator equation require the construction of a functional whose critical points are also solutions of the equation.   This approach is possible according to:

Theorem 2-4 (*Kerner's symmetry condition*)   Let $F(x)$ be a mapping of a Hilbert space H (or part of it) into itself.   Let $F(x)$ have a Gateaux differential $DF(x;h)$ at every point $x \, \epsilon \, S = \{x : \|x - x_0\| < r\}$, where $r > 0$.   Assume that at every point $x \, \epsilon \, S$, the functional $(DF(x;h),k)$ for h, $k \, \epsilon \, H$ is continuous in x.   Then in order that the mapping $F(x)$ be the gradient of some functional $f(x)$ defined on S, it is necessary and sufficient that the bilinear functional $(DF(x;h),k)$ be symmetric for every $x \, \epsilon \, S$; that is,

$$(DF(x;h),k) = (h,DF(x;k)) \qquad \text{for } h, \, k \, \epsilon \, H$$

Under these conditions

$$f(x) - f(x_0) = \int_0^1 (F(x_0 + t(x - x_0)), \, x - x_0) \, dt \qquad (2\text{-}40)$$

Consider now the nonlinear equation $Tx = 0$, where T satisfies the conditions of the foregoing theorem in a sphere $S \, \epsilon \, H$ and can be identified with the gradient of a real-valued functional $t(x)$ defined on H.   Using the iterations $x_{n+1} = x_n + \alpha_n z_n$ with initial x_0, we have for the direction of steepest descent from x_0:

Theorem 2-5   Let U_0 be a self-adjoint positive-definite bounded linear operator defined on the Hilbert space H with the inner product $(.,.)$, and let

$$(U_0 h,h) \geq m(h,h) \qquad m > 0, \text{ for all } h \, \epsilon \, H$$

For any two elements x, $y \, \epsilon \, H$, define the distance between x and y by the metric

$$\rho(x,y) = |x - y| = (U_0(x - y), \, x - y)^{1/2} \qquad (2\text{-}41)$$

Let x_0 be as before.   Then the direction of steepest descent from x_0 is given by

$$z_0 = -U_0^{-1} T(x_0)$$

From this, one has in general

$$x_{n+1} = x_n - \alpha_n U_n^{-1} T(x_n) \qquad (2\text{-}42)$$

where U_n is a positive-definite self-adjoint operator defined on H. To determine the α_n several procedures are possible.

1. The exact optimum-gradient method, in which α_n is chosen to minimize the functional $t(x_n - \alpha_n U_n^{-1} T_n)$; that is, it is the smallest positive root of $\partial t/\partial \alpha = 0$.

2. The approximate optimum-gradient method, in which α_n is chosen to minimize the second-order Taylor's-series approximation (i.e., ignoring the remainder) of $t(x_{n+1})$ expanded around x_n.

3. The minimization of the norm of the residue of the Taylor's-series expansion up to the second-order-derivative term of $T(x_{n+1})$ around x_n using $\bar{x}_n = x_n + \tau_n(x_{n+1} - x_n)$, $0 < \tau_n < 1$, in the last term. Thus α_n is chosen to minimize

$$\|T_n - \alpha_n DT(x_n; U_n^{-1} T_n)\| \tag{2-43}$$

for fixed x_n.

4. The sequential-descent method, in which a range is determined for α_n and particular choices of U_n^{-1} are made to expedite convergence. With the assumption that the path joining the successive iterations remains in S, Nashed proves the following theorem.

Theorem 2-6 Let $T(x)$ satisfy Kerner's symmetry condition. Let $T(x_0) \neq \mathbf{0}$, and let there exist a convex neighborhood Ω of x_0 such that the linear operator $dT(x; .)$ is continuous in x on Ω. If $x_n, x_{n+1} \in \Omega$, then for α_n sufficiently small and positive

$$t(x_{n+1}) < t(x_n) \qquad n = 0, 1, 2, \ldots$$

THE METHOD OF MINIMUM ERROR[10,12]

Consider the equation

$$T(x) = \mathbf{0} \tag{2-44}$$

where T maps the real Hilbert space H into itself and has a derivative $T'(x)$ in the sense of Fréchet. To approximate a solution to (2-44), we first define the gradient of the functional $t(x)$, where $t(x) = (Tx, Tx)$. Note that $t(x_n)$, for a given x_n, gives a measure of the "closeness" of x_n to a solution. We have

$$
\begin{aligned}
\frac{d}{d\epsilon} t(x + \epsilon z)\Big|_{\epsilon=0} &= \frac{d}{d\epsilon}(T(x + \epsilon z), T(x + \epsilon z))\Big|_{\epsilon=0} \\
&= \frac{d}{d\epsilon}[(T(x), T(x)) + 2(T(\epsilon z), T(x)) + (T(\epsilon z), T(\epsilon z))]\Big|_{\epsilon=0} \\
&= 0 + 2\,\mathrm{Re}\,(A(z), T(x)) + (T(\epsilon z), T(\epsilon z)) \\
&= 2\,\mathrm{Re}\,(z, A^*T(x))
\end{aligned}
$$

where A is a bounded linear operator defined by $A(x) = T'(x)$. We define grad $t(x)$ to be $A^*T(x)$.

Let

$$z_n = \text{grad } t(x_n) = A_n^* T(x_n) \qquad \text{where } A_n = T'(x_n) \qquad (2\text{-}45)$$

Then we have the following theorem.

Theorem 2-7 Let x_0 be an initial approximation to Eq. (2-44), and let $\|T(x_0)\| = a$. If in the ball S, defined by

$$\|x - x_0\| \leq 2\frac{a}{m} \qquad (2\text{-}46)$$

the following hold:

$$\frac{\|T'(x)z\|}{\|z\|} \geq m \qquad \text{or} \qquad \frac{\|z\|}{\|T'(x)z\|} \leq \frac{1}{m} \qquad (2\text{-}47)$$

$$\|T''(x)\| \leq K \qquad (2\text{-}48)$$

$$\frac{aK}{m^2} = r < 1 \qquad (2\text{-}49)$$

then (2-44) has (in the ball S) a unique solution x^*, and the sequence $\{x_n\}$, defined by

$$x_{n+1} = x_n - \frac{\|T(x_n)\|^2}{\|A_n^* T(x_n)\|^2} A_n^* T(x_n) \qquad (2\text{-}50)$$

converges monotonically and strongly to x^*. Furthermore the estimate of the error at the nth stage is

$$\|x_n - x^*\| \leq \|x_0 - x^*\| \left[1 - \frac{m^2}{M^2}(1-r) \right]^{n/2}$$
$$\leq \frac{a}{m} \left[1 - \frac{m^2}{M^2}(1-r) \right]^{n/2} \qquad (2\text{-}51)$$

where $M = \max \|T'(x)\|$.

Proof: From (2-50) and the assumption of a Hilbert space we have

$$\|x_{n+1} - x^*\|^2 = \|x_n - x^* - \frac{\|T(x_n)\|^2}{\|A_n^* T(x_n)\|^2} A_n^* T(x_n)\|^2$$
$$= \|x_n - x^*\|^2 - 2\frac{\|T(x_n)\|^2}{\|A_n^* T(x_n)\|^2}(x_n - x^*, A_n^* T(x_n))$$
$$+ \left\| \frac{\|T(x_n)\|^2}{\|A_n^* T(x_n)\|^2} A_n^* T(x_n) \right\|^2$$
$$= \|x_n - x^*\|^2 - \frac{2(T(x_n), T(x_n))}{(A_n^* T(x_n), A_n^* T(x_n))}(A_n(x_n - x^*), T(x_n))$$
$$+ \frac{(T(x_n), T(x_n))^2}{(A_n^* T(x_n), A_n^* T(x_n))}$$

If we use the Taylor's-series expansion

$$T(x^*) = 0 = T(x_n) + T'(x_n)(x^* - x_n) + \tfrac{1}{2}T''(\bar{x}_n)(x^* - x_n)^2$$

where $\qquad \bar{x}_n = \theta(x^* - x_n) \qquad$ and $\qquad 0 \le \theta \le 1$

we have

$$\frac{2(A_n(x_n - x^*), T(x_n))}{(T(x_n), T(x_n))} = \frac{2(T'(x_n - x^*), T(x_n))}{(T(x_n), T(x_n))}$$

$$= \frac{2 + (T''(\bar{x}_n)(x_n - x^*)^2, T(x_n))}{(T(x_n), T(x_n))} \equiv 2 + \alpha_n$$

Consequently

$$\|x_{n+1} - x^*\|^2 = \|x_n - x^*\|^2 - \frac{(T(x_n), T(x_n))^2}{(A^*T(x_n), A^*T(x_n))}(1 - \alpha_n)$$

For α_0 we have the estimate

$$|\alpha_0| \le \frac{\|T''(\bar{x}_0)(x_0 - x^*)^2\| \, \|T(x_0)\|}{\|T(x_0)\|^2} = \frac{\|T''(\bar{x}_0)(x_0 - x^*)^2\|}{\|T(x_0)\|}$$

$$= \frac{\|T''(\bar{x}_0)(x_0 - x^*)^2\|}{\|T(\bar{x}_0)(x_0 - x^*)\|}$$

where $\qquad \bar{x}_0 = \theta_1(x_0 - x^*) \qquad$ and $\qquad 0 \le \theta_1 \le 1$

This estimate is obtained by applying Schwarz's inequality, that is, $|(x,y)| \le \|x\| \, \|y\|$, to the numerator of α_n for $n = 0$. The denominator of the last expression on the right is obtained by replacing $T(x_0)$ by its equivalent in the Taylor's-series expansion, terminating with the first-order derivative.

Now inequalities (2-47) and (2-48), which are true in the ball S, are also true in the ball N defined by

$$\|x - x^*\| \le \|x_0 - x^*\| \le \frac{a}{m} \le \frac{K}{m} \qquad (2\text{-}52)$$

since $\qquad \|x - x_0\| \le \|x - x^*\| + \|x^* - x_0\| \le \dfrac{a}{m} + \dfrac{a}{m}$

It is possible that estimate (2-47) is known only at the point x_0, in which case we have the following modification of Theorem 2-7.

***Theorem* 2-8** Suppose that

$$\|T(x_0)\| = a \qquad \frac{\|T'(x_0)z\|}{\|z\|} \ge m \qquad (2\text{-}53)$$

and suppose also that in the ball S' defined by

$$\|x - x_0\| \le \frac{a}{m} \qquad (2\text{-}54)$$

we have $\qquad \|T''(x)\| \le K \qquad$ where $\dfrac{aK}{m^2} = \dfrac{r}{2} < \dfrac{1}{2} \qquad (2\text{-}55)$

then Eq. (2-44) has a unique solution x^* in the ball S', and the sequence $\{x_n\}$, defined by (2-50), converges monotonically and strongly to x^*, with the estimate of the error at the nth stage given by

$$\|x_n - x^*\| \leq \|x_0 - x^*\| \left[1 - \frac{m^2}{4M^2}(1 - r) \right]^{n/2}$$

$$\leq \frac{a}{m} \left[1 - \frac{m^2}{4M^2}(1 - r) \right]^{n/2} \quad (2\text{-}56)$$

Proof: This theorem is proved exactly as Theorem 2-7 was except that here, under condition (2-53), we have

$$\frac{\|T'(x)z\|}{\|z\|} \geq m - K\|x_0 - x^*\| \geq m - \frac{aK}{m} \geq m\left(1 - \frac{r}{2}\right) \geq \frac{m}{2}$$

Then $\|x - x_0\| \leq 2a/m$ holds. Clearly $\bar{x}_0 \in N$ and $\bar{x}_0 \in N$. Consequently, using (2-47) and (2-48), we obtain

$$|\alpha_0| \leq \frac{K\|x_0 - x^*\|}{m} = \frac{Ka}{m^2} = r < 1$$

Thus for the error of the first approximation the following estimate is valid

$$\|x_1 - x^*\|^2 \leq \|x_0 - x^*\|^2 - \|T(x_0)\|^2 \frac{(1 - r)}{M^2} = \|x_0 - x^*\|$$

$$- \|T'(\bar{x}_0)(x_0 - x^*)\|^2 \frac{(1 - r)}{M^2}$$

$$\leq \|x_0 - x^*\|^2 \left[1 - \frac{m^2}{M^2}(1 - r) \right] < \|x_0 - x^*\|^2$$

Now since x_1 lies in the interior of N, the above arguments can be repeated for the second approximation. We obtain

$$|\alpha_1| < r \qquad \|x_2 - x^*\|^2 \leq \|x_1 - x^*\|^2 \left[1 - \frac{m^2}{M^2}(1 - r) \right]$$

$$\leq \|x_0 - x^*\| \left[1 - \frac{m^2}{M^2}(1 - r) \right]^2$$

In this manner we can inductively estimate (2-51) for the nth approximation, and it follows that $\lim_{n \to \infty} x_n = x^*$, proving the theorem.

THE METHOD OF MINIMUM RESIDUALS[10,22,23]

Let $T(x)$ mapping Hilbert space H into H be first and second differentiable in the Fréchet sense. We consider the solution of

$$T(x) = 0 \quad (2\text{-}57)$$

by the method of minimum residuals. The iterations of this method are defined by the sequence

$$x_{n+1} = x_n + \epsilon_n y_n \qquad (2\text{-}58)$$

where $x_0 \,\epsilon\, H$ is a given initial approximation to the solution of (2-57), ϵ_n are real numbers (to be determined below), and $\{y_n\}$ is a certain sequence of elements in H. Since $T(x)$ is twice differentiable, we have

$$\|T(x_{n+1})\| \le \|T(x_n) + T'(x_n)(x_{n+1} - x_n)\| \\ + \tfrac{1}{2}\|T''(\bar{x}_n)\| \, \|x_{n+1} - x_n\|^2 \qquad (2\text{-}59)$$

where $\bar{x}_n = x_n + r_n(x_{n+1} - x_n)$, $0 < r_n < 1$. We choose ϵ_n so that, for fixed y_n, the expression

$$\|T(x_n) + T'(x_n)(x_{n+1} - x_n)\|^2 = \|T(x_n) + \epsilon_n T'(x_n)y_n\|^2$$

is a minimum. It is easy to verify that this minimum value is attained for

$$\epsilon_n = -\frac{(T(x_n), T'(x_n)y_n)}{\|T'(x_n)y_n\|^2} \qquad (2\text{-}60)$$

and we have

$$\|T(x_n) + T'(x_n)(x_{n+1} - x_n)^2\| = (T(x_n) + \epsilon_n T'(x_n)y_n, \, T(x_n) \\ + \epsilon_n T'(x_n)y_n) \qquad (2\text{-}61)$$
$$= \|T(x_n)\|^2 - \frac{(T(x_n), T'(x_n)y_n)}{\|T'(x_n)y_n\|^2}$$

Let $y_n = T(x_n)$, and our iterative sequence of approximation satisfies

$$x_{n+1} = x_n - \frac{(T(x_n), T'(x_n)T(x_n))}{\|T'(x_n)T(x_n)\|^2} \qquad (2\text{-}62)$$

We now turn to the question of the convergence of this sequence to a solution of (2-57).

Theorem 2-9 Let the following conditions be satisfied:
1. $\|T(x_0)\| \le \delta_0$.
2. For all $x \,\epsilon\, S(x_0, r) \equiv \{x \,\epsilon\, H \mid \|x - x_0\| \le r\}$, where

$$r = M\delta_0/(1 - q)$$

the following inequalities hold:
   a. $\|T'(x)\| \le A$.
   b. $\|T''(x)\| \le B$.
   c. $|(T'(x)h, h)| \ge M^{-1}\|h^2\|$, for all $h \,\epsilon\, H$ $(M > 0)$.
3. $q = \sqrt{1 - b^{-1}} + 1/2a_0 < 1$, where $b = M^2 A^2$ and $a_0 = M^2 B \delta_0$.

Then in the ball $S(x_0, r)$ Eq. (2-57) has a unique solution x^* to which the sequence defined by (2-62) converges and for which the following error estimate holds

$$\|x_n - x_0\| \le M\|T(x_n)\| \le M\delta_0 q^n \tag{2-63}$$

Proof: On putting $x = x_1$ in a Taylor's-series expansion up to second-order derivatives, that is,

$$T(x) \simeq T(x_0) + T'(x_0)(x - x_0) + \tfrac{1}{2}T''(x_0)(x - x_0)^2$$

taking the norm on both sides, then using the property of the norm of a sum and (2-61), we have, on estimating $\|x_1 - x_0\|$ from (2-62),

$$
\begin{aligned}
\|T(x_1)\| &\le \left[\|T(x_0)\|^2 - \frac{(T(x_0), T'(x_0)T(x_0))^2}{\|T'(x_0)T(x_0)\|^2}\right]^{\frac{1}{2}} + \tfrac{1}{2}\|T''(x_0)\|\,\|x_1 - x_0\|^2 \\
&\le \left[\|T(x_0)\|^2 - \frac{M^{-2}\|T(x_0)\|^4}{\|T'(x_0)\|^2\|T(x_0)\|^2}\right]^{\frac{1}{2}} \\
&\qquad + \frac{B}{2}\frac{(T(x_0), T'(x_0)T(x_0))^2}{\|T'(x_0)T(x_0)\|^4}\|T(x_0)\|^2 \\
&\le [1 - M^{-2}A^{-2}]^{\frac{1}{2}}\|T(x_0)\| + \frac{B}{2}\frac{\|T'(x_0)T(x_0)\|^2\|T(x_0)\|^4}{\|T'(x_0)T(x_0)\|^4} \\
&\le (1 - b^{-1})^{\frac{1}{2}}\|T(x_0)\| + \frac{BM}{2}\|T(x_0)\|^2 \\
&\le \left[(1 - b^{-1})^{\frac{1}{2}} + \frac{a_0}{2}\right]\|T(x_0)\| < q\|T(x_0)\|
\end{aligned}
$$

This means that there exists a constant δ_1 satisfying

$$\|T(x_1)\| \le \delta_1 \le q\|T(x_0)\| \le q\delta_0 \le \delta_0$$

Hence $S(x_1, r_1) \subset S(x_0, r)$, where $r_1 = M\delta_1/(1 - q_1)$,

$$q_1 = \sqrt{1 - b^{-1}} + \tfrac{1}{2}a_1 < q$$

and $a_1 = M^2 B\delta_1$. All the conditions are therefore satisfied for x_1, and we can continue the calculation of successive approximations. By induction we obtain for $n = 0, 1, 2, \ldots$ the relations $\|T(x_{n+1})\| \le \delta_{n+1} \le q\|T(x_n)\|$; $\|x_{n+1} - x_n\| \le M\|T(x_n)\|$. Using these two inequalities, we have for all n and p

$$\|x_{n+p} - x_n\| \le M(\|T(x_{n+p-1})\| + \cdots + \|T(x_n)\|) \le \frac{M\delta_0}{1 - q}q^n$$

This proves the existence of $\lim_{n \to \infty} x_n = x^* \epsilon\, S(x_0, r)$. Now since the operator $T(x)$ is continuous, we have

$$\|T(x^*)\| = \lim_{n \to \infty} \|T(x_n)\| \le \delta_0 \lim_{n \to \infty} q^n = 0$$

that is, x^* is a solution of (2-57). From condition 2c we see that x^* is unique in the sphere $S(x_0, r)$. Also from 2c we have

$$\|T(x_n)\| \, \|x_n - x^*\| \geq |(T(x_n) - T(x^*), x_n - x^*)|$$
$$= |T'(\bar{x}_n)(x_n - x^*)| \geq M^{-1}\|x_n - x^*\|$$

where $\bar{x}_n = x^* + \theta_n(x_n - x^*)$, $0 < \theta_n < 1$. This gives relation (2-63), and the theorem is proved.

In Theorem 2-9 we set $y_n = T(x_n)$. We now consider the case $y_n = T'^*(x_n)T(x_n)$, where $T'^*(x)$ is the operator adjoint to the linear operator $T'(x)$. In this case (2-62) is replaced by

$$x_{n+1} = x_n - \frac{\|T'^*(x_n)T(x_n)\|^2}{\|T'(x_n)T'^*(x_n)T(x_n)\|^2} T'^*(x_n)T(x_n) \qquad (2\text{-}64)$$

The question concerning us now is: Under what conditions does (2-64) converge to a solution of (2-57)? We have the following theorem as a partial answer.

***Theorem* 2-10**   Let the following conditions be satisfied:
1. $\|T(x_0)\| \leq \delta_0$.
2. For all $x \, \epsilon \, S(x_0, r)$, where $r = M\delta_0/(1 - q)$, the following inequalities hold:
   a. $\|T'(x)\| \leq A$.
   b. $\|T''(x)\| \leq B$.
   c. $\|T'(x)h\| \geq M^{-1}\|h\|$, and $\|T'^*(x)h\| \geq M^{-1}\|h\|$, for all $h \, \epsilon \, H$ $(M > 0)$.
3. $q = (b - 1)/(b + 1) + a_0/2 < 1$, where $b = M^2A^2$, $a_0 = M^2B\delta_0$.

Then in the ball $S(x_0, r)$ Eq. (2-57) has a solution x^* to which the sequence $\{x_n\}$ obtained from (2-64) converges, and for the error estimate at the nth stage we have the inequalities

$$\|x^* - x_n\| \leq \frac{M}{1 - q} \|T(x_n)\| \leq \frac{M\delta_0}{1 - q} q_n$$

We note that if we set $y_n = [T(x_n)]^{-1}T(x_n)$, then (2-61) gives rise to Newton's method.

A slightly more complex iterative process, which has minimum residuals, is defined by the recurrence relation

$$x_{n+1} = x_n - \frac{(T(x_n), T'(x_n)y_n)}{\|T'(x_n)y_n\|^2} y_n \qquad (2\text{-}65)$$

where $y_n \, \epsilon \, H$ for all n. We now define y_n by $y_n = U_n T(x_n)$, where for all n, U_n is a linear operator mapping H to H, so that (2-65) becomes

$$x_{n+1} = x_n - \frac{(T(x_n), T'(x_n)U_n T(x_n))}{\|T'(x_n)U_n T(x_n)\|^2} U_n T(x_n) \qquad (2\text{-}66)$$

from which we have

$$\|x_{n+1} - x_n\| \leq \frac{\|T_n\| \, \|U_n T_n\|}{\|T'_n U_n T_n\|} \tag{2-67}$$

and

$$\|T_n + T'_n(x_{n+1} - x_n)\|^2 = \|T_n\|^2 - \frac{(T'_n U_n T_n, T_n)^2}{\|T'_n U_n T_n\|^2} \tag{2-68}$$

where $T_n = T(x_n)$ and $T'_n = T'(x_n)$.

With regard to the convergence of the iterations defined by (2-66) we have:

Theorem 2-11   Let the following be true:

1. $\|T(x_0)\| = \delta_0 \leq \bar{\delta}_0$, $\|U_n\| \leq C$ for all n.
2.   For all $h \, \epsilon \, H$ and for $n = 0, 1, 2, \ldots$

$$\|T'(x_0) U_n h, h)| \geq B_0^{-1}\|h\|^2 \qquad (B_0 > 0)$$

3.   For all $x \, \epsilon \, S(x_0, r)$ and for $n = 0, 1, 2, \ldots$

$$\|T'(x) U_n\| \leq A, \quad \|T''(x)\| \leq K \tag{2-69}$$

4. $a_0 = B_0^2 C^2 K \delta_0$; $b_0 = A B_0$, and the sequence $\{a_n\}$, defined by $a_n = B_n^2 C^2 K \delta_n$, is convergent, where

$$B_{k+1} = \frac{B_k}{1 - B_k^2 C^2 K \bar{\delta}_k} \tag{2-70}$$

$$\delta_{k+1} = \delta_k[\sqrt{1 - (AB_k)^{-2}} + \tfrac{1}{2} B_k^2 C^2 K \bar{\delta}_k] \tag{2-71}$$

with $a_n < 1$ for all n.

5. $r \geq r_0 = \dfrac{1}{CK}\left(\dfrac{1}{B_0} - \dfrac{1}{B^*}\right)\dfrac{\delta_0}{\bar{\delta}_0}$, where $B^* = \lim\limits_{n \to \infty} B_n$.

Then Eq. (2-57) has a solution x^* in $S(x_0, r)$ to which the sequence $\{x_n\}$ defined by (2-66) converges, with the following error estimate at the nth stage

$$\|x^* - x_n\| \leq \frac{1}{CK}\left(\frac{1}{B_n} - \frac{1}{B^*}\right)\frac{\delta_n}{\bar{\delta}_n} \tag{2-72}$$

where $\delta_n = \|T(x_n)\|$ and B_n and $\bar{\delta}_n$ are defined by the recurrent relations (2-70) and (2-71), respectively.

Proof: Since $a_n < 1$ for all n, then $\{B_n\}$ is an increasing sequence, and since $\lim\limits_{n \to \infty} B_n = B^*$, we have $\lim\limits_{n \to \infty} a_n = 0$ and $\lim\limits_{n \to \infty} \delta_n = 0$. Now the first approximation x_1 can be evaluated. We show that if we consider x_1 instead of x_0, conditions 1 to 4 are all satisfied and that therefore the approximation x_2 can also be computed. By conditions 1 and 2 and relations (2-67) and (2-70) we have

$$\|x_1 - x_0\| \leq B_0 C \delta_0 = \frac{1}{CK}\left(\frac{1}{B_0} - \frac{1}{B_1}\right)\frac{\delta_0}{\bar{\delta}_0} < r_0$$

So that $x_1 \epsilon S(x_0, r_0)$. If we use (2-68), we have

$$\|T(x_1)\| \leq \|T(x_0) + T'(x_0)(x_1 - x_0)\| + \tfrac{1}{2}\|T''(\bar{x}_0)(x_1 - x_0)^2\|$$

$$\leq \left(\sqrt{1 - b_0{}^2} + \frac{a_0}{2}\right)\|T(x_0)\|$$

where $\bar{x}_0 \epsilon [x_0, x_1]$, and therefore

$$\|T(x_1)\| = \delta_1 \leq \left(\sqrt{1 - b_0{}^{-2}} + \frac{a_0}{2}\right)\delta_0 \leq \left(\sqrt{1 - b_0{}^{-2}} + \frac{a_0}{2}\right)\bar{\delta}_0 = \bar{\delta}_1 \tag{2-73}$$

So condition 1 for x_1 is satisfied. Also condition 2 is satisfied since

$$|(T'(x_1)U_nh,h)| \geq |(T'(x_0)U_nh,h)| - |(T''(\bar{x}_0)(x_1 - x_0)U_nh,h)|$$
$$\geq B_0{}^{-1}\|h\|^2 - B_0C^2K\bar{\delta}_0\|h\|^2 \equiv B_0{}^{-1}(1 - B_0{}^2C^2K\bar{\delta}_0)\|h\|^2 = B_1{}^{-1}\|h\|^2$$

where $\bar{x}_0 \epsilon [x_0, x_1]$. Since (2-73) implies that $\delta_1/\bar{\delta}_1 \leq \delta_0/\bar{\delta}_0$, it follows that

$$\|x - x_0\| \leq \|x - x_1\| + \|x_1 - x_0\| \leq \frac{1}{CK}\left(\frac{1}{B_1} - \frac{1}{B^*}\right)\frac{\delta_1}{\bar{\delta}_1} + B_0C\delta_0 \leq r_0$$

Hence $x \epsilon S(x_0, r_0)$. Now since condition 4 for $a_1 = b_1{}^2C^2K\delta_1$ and $b_1 = AB_1$ is satisfied, we can repeat the above argument starting with x_1. Using induction, we can show that conditions 2 to 4 remain true for all n for the elements x_n and that the process of evaluation of the elements by (2-66) is thus true for all n. In this way we get the inequalities

$$\|x_{n+1} - x_n\| \leq B_nC\delta_n = \frac{1}{CK}\left(\frac{1}{B_n} - \frac{1}{B_{n+1}}\right)\frac{\delta_n}{\bar{\delta}_n} \tag{2-74}$$

and

$$\delta_{n+1} \leq \left(\sqrt{1 - b_n{}^{-2}} + \frac{a_n}{2}\right)\delta_n \leq \left(\sqrt{1 - b_n{}^{-2}} + \frac{a_n}{2}\right)\bar{\delta}_n = \bar{\delta}_{n+1}$$

where $b_n = AB_n$. It now follows that $|(T'(x_n)U_kh,h)| \geq B_n{}^{-1}\|h\|^2$ and $x_{n+1} \epsilon S(x_n, r_n) \subset S(x_{n-1}, r_{n-1})$

where

$$r_k = \frac{1}{CK}\left(\frac{1}{B_k} - \frac{1}{B^*}\right)\frac{\delta_k}{\bar{\delta}_k}$$

So it remains to show that the sequence $\{x_n\}$ converges to a solution of (2-57) which is in the sphere $S(x_0, r_0)$. Using relation (2-74) and the inequalities

$$\frac{\delta_{n+k}}{\bar{\delta}_{n+k}} \leq \frac{\delta_n}{\bar{\delta}_n} \qquad k = 1, 2, \ldots$$

we have for $n = 0, 1, 2, \ldots$ and $p = 0, 1, 2, \ldots$ that

$$\|x_{n+p} - x_n\| \leq \|x_{n+p} - x_{n+p-1}\| + \cdots + \|x_{n+1} - x_n\|$$

$$\leq \frac{1}{CK}\left(\frac{1}{B_n} - \frac{1}{B_{n+p}}\right)\frac{\delta_n}{\bar{\delta}_n}$$

Note that for all p, $B_{n+p} \geq B_n \geq 0$ and hence $1/B_n - 1/B_{n+p} \geq 0$ and we have a monotone decreasing sequence in p for fixed n. Thus

$$0 \leq \lim_{p \to \infty} \frac{1}{B_n} - \frac{1}{B_{n+p}} = \frac{1}{B_n} - \frac{1}{B^*}$$

and

$$\frac{1}{B_n} - \frac{1}{B_{n+p}} \leq \frac{1}{B_n} - \frac{1}{B^*}$$

Thus the sequence $\{x_n\}$ is a Cauchy sequence, and since the space H is complete, there exists $\lim_{n \to \infty} x_n = x^* \epsilon H$. Passing to the limit as $p \to \infty$, we get relation (2-72). Obviously $x^* \epsilon S(x_0, r_0)$, and x^* is the solution of (2-57) since $\|T(x^*)\| = \lim_{n \to \infty} \|T(x_n)\| \leq \lim_{n \to \infty} \delta_n = 0$, and the theorem is proved.

2-3   DIRECT METHODS

These are useful for unbounded operators, e.g., partial differential equations. The idea in several of these methods is to embed the problem in a series of problems, solving each one, ultimately obtaining the solution of the original problem.

THE RITZ AND RITZ - GALERKIN METHODS†

The solution of boundary-value problems for a number of types of ordinary and partial differential equations can be reduced to the question of minimizing an integral I, that is, to a variational problem. Often such problems are more conveniently solved by an approximation technique, known as the *Ritz* or *Rayleigh-Ritz method*, in which the unknown solution is expressed as a linear combination of n arbitrary, linearly independent functions (which satisfy the initial or boundary conditions). These functions are suitably chosen to satisfy, for example, the physical requirement of the problem. Thus for periodic solutions one can use trigonometric functions. The integral will then depend on the arbitrary coefficients a_k, $k = 1, \ldots, n$, in the linear combination, and they must be chosen so as to yield a minimum value for I. This leads to a set of conditions of the form $\partial I / \partial a_k = 0$, $k = 1, \ldots, n$, from which the coefficients of the linear combination are determined and hence also the approximate expansion of the function.

Note that one can attempt successive improvement of the approximation by increasing the size of the set of approximating functions, and the problem is to give conditions under which, as $n \to \infty$, the value of I for

† See "NM," p. 264.

these approximations tends to the actual value. It is necessary to choose the infinite set of approximating functions (from which only the first n are used in the first approximation, the first $n + 1$ in the second, etc.) such that the unknown functions can actually be approximated by some linear combination of this set to any desired degree of accuracy. Often polynomials, sine or cosine functions, etc., are used. Specific examples are $\sin k\pi x$ and $(1 - x)x^k$, which are complete orthonormal systems in the space $L_2(0,1)$.

The method of Galerkin is concerned with ordinary rather than variational equations but utilizes the basic ideas of the Ritz approach. We broadly sketch the ideas leading to Galerkin's method; they are discussed in fine detail in Ref. 26. Many topological notions revolving around the idea of a vector field and the index of a mapping, etc., occur in that development, and no particular purpose would be served by repeating those ideas here.

Let X be a Banach space and $L_1 \subset L_2 \subset \cdots$ be a sequence of finite-dimensional subspaces of increasing dimension such that $\bigcup_{n=1}^{\infty} L_n$ is dense in X. Let $P_1, P_2, \ldots$ be a sequence of linear projection operators with uniformly bounded norms, that is, $\|P_n\| \leq K$, $n = 1, 2, \ldots$. Let $P_n X = L_n$, and let T be completely continuous on X to X. Consider the equation

$$x = Tx \tag{2-75}$$

assumed to have an isolated solution x^*. [For example, we note that $f(x) = e^{-1/x^2} \sin (1/x) = 0$ has $x = 0$ and $x = \pm 1/n\pi$, $n = 1, 2, \ldots$, as roots which converge to zero, and hence $x = 0$ is not isolated.] Also consider the solution x_n of the equation

$$x = P_n Tx \tag{2-76}$$

It can be shown that x_n exists for $n \geq N$ for some integer N and that as $n \to \infty$, $x_n \to x^*$. If we assume that T has a Fréchet derivative for which unity is not an eigenvalue, then the speed of convergence is governed by

$$\|x_n - x^*\| \leq (1 + \epsilon_n)\|P_n x^* - x^*\|$$

where $\epsilon_n \to 0$ as $n \to \infty$.

If, in addition, T is continuously differentiable, then for large n (2-76) has a unique solution in a neighborhood of the isolated solution x^*.

Galerkin's method uses the foregoing ideas in the following framework. Let the Banach space X have Schauder basis $\{y_i\}$. Then each element $x \in X$ can be represented uniquely as a convergent series

$$x = \sum_{1}^{\infty} f_i(x)y_i$$

where the f_i are continuous functionals. It can be shown that the projection operators P_n defined by

$$P_n x = \sum_1^n f_i(x) y_i$$

have uniformly bounded norms. We shall denote by L_n the linear manifold of the first n elements of the basis $\{y_i\}$.

Then the solutions x_n of the equations $x = P_n T x$ are called the *Galerkin approximate solutions* of $x = Tx$.

The foregoing statements regarding convergence are applicable. For the equation $Tx = y$, where T is positive-definite and self-adjoint, it can be shown that the residual in the Ritz-method approximation converges to zero if T has a pure point spectrum and if the coordinate elements chosen are the system of eigenelements of a certain suitable operator.

The ideas of Galerkin's method can be generalized in a separable Hilbert space H to equations of the form $Lx = Tx + Kx = y$, where T is self-adjoint in a domain dense in H and contained in the domain of K.

If T is completely continuous and linear and does not have unity as an eigenvalue, the method is applicable to the equation

$$x = Tx + y$$

If $\{y_i\}$ and $\{z_i\}$ are two bases in a Banach space X, then each $x \in X$ can be represented by convergent series

$$x = \sum_1^\infty f_i(x) y_i$$

$$x = \sum_1^\infty g_i(x) z_i$$

with functionals f_i and g_i. Let L_n be the linear manifold of $y_1, \ldots, y_n$ and M_n that of $z_1, \ldots, z_n$. Let M_n be the closed linear manifold of $z_{n+1}, z_{n+2}, \ldots$, and let Q_n be the projection operator defined by

$$Q_n x = \sum_1^n g_i(x) z_i \qquad x \in X$$

The norms of the operators Q_n are uniformly bounded.

A generalization due to Petrov seeks approximate solutions of the form

$$x_n = \sum_1^n a_i y_i$$

where the coefficients a_i are determined from the equations

$$g_j(x_n - Tx_n - y) = 0 \qquad j = 1, \ldots, n$$

The general proof of the convergence of this method is due to Pol'skii.[39] The requirement is that there exist a number $r > 0$ such that for all $n \geq N$ for some N

$$\|x_n\| \leq r \|\sum_1^n g_i(x)z_i\| \qquad x \in L_n \qquad (2\text{-}77)$$

If S_n is the operator defined on L_n by

$$S_n x = Q_n x \qquad x \in L_n$$

then S_n^{-1} is defined on M_n such that $\|S_n^{-1}\| \leq r$.

If we assume that this condition holds and consider

$$x = Tx$$

where T is completely continuous and nonlinear, the system of equations to determine a_i now is

$$g_j(x - Tx) = 0$$

The requirement that $x_n \in L_n$ be a solution of this is the requirement that $x_n - Tx_n$ belong to M_n for $x_n \in L_n$. By an appropriate convergence condition, e.g., there exists a number $R > 0$ such that for sufficiently large n, $\|x\| \leq R \| \sum_{i=1}^n f_i(x)y_i\|$. each $x \in X$ can be expressed in terms of $y_1, \ldots, y_n, z_{n+1}, z_{n+2}, \ldots$ (which for large n form a basis) as

$$x = \sum_1^n X_i^{(n)}(x)y_i + \sum_{n+1}^\infty X_i^{(n)}(x)z_i \qquad x \in X$$

Define the projection operators P_n by

$$P_n x = \sum_1^n X_i^{(n)}(x)y_i \qquad x \in X$$

and the problem of finding $x_n \in L_n$ is equivalent to finding in L_n the solution of $x = P_n Tx$. In fact if $x_n = P_n Tx_n$, then $P_n x_n = x_n$ implies $P_n(x_n - Tx_n) = 0$, that is, that $x_n - Tx_n \in M_n$. Conversely, if $x_n - Tx_n \in M_n$, then $x_n - P_n Tx_n = P_n(x_n - Tx_n) = 0$. To draw conclusions from the conditions on the convergence of Galerkin's method we must have $\|P_n\| \leq K$ for all n. But $P_n = S_n^{-1}Q_n$ implies $\|P_n\| \leq rK$, and hence if condition (2-77) is satisfied, the convergence statements given earlier apply to Petrov's generalization.

Remark: Note that it is sometimes possible to formulate a variational problem whose extremal function satisfies a given differential equation as a necessary condition. However, since there are several other necessary conditions for the variational problem, such as those of Legendre, Weierstrass, Clebsch, etc., it is possible to obtain more information about the character of the solution of the variational problem and thus also about that of the differential equation.

FRIEDRICHS CONTINUITY METHOD[14]

The continuity method studied by Friedrichs is not unlike the parameter-perturbation procedure in connection with Newton's method. We consider the problem of solving the equation $Tx = y$, where T is a linear transformation of a Banach space X into itself and y is a given element of X. The principle of the method lies in finding a linear transformation $\bar{T}$ such that $\bar{T}x = y$ has a known solution and such that $\bar{T}$ can be associated with T through a family of linear transformations $T_t(0 \leq t \leq 1)$ with $T_0 \equiv \bar{T}, T_1 \equiv T$. We wish to extend the solution x_0 of $T_0x \equiv \bar{T}x = y$ to x_t for all $0 \leq t \leq 1$, where x_t is a solution of $T_tx = y$, and hence also obtain the solution x_1 of $T_1x \equiv Tx = y$.

Friedrichs gives the following theorem for the existence of an inverse of T provided the inverse of $\bar{T}$ is known and $\bar{T}$ and T are connected as above.

Theorem 2-12   Let the following hold:

1.   The inverse of $\bar{T}$ exists.

2.   The linear transformation $Tx = y$ is connected with $\bar{T}x = y$ through the family $\{T_t : 0 \leq t \leq 1\}$, with $T_0 = \bar{T}$ and $T_1 = T$.

3.   Given $\epsilon > 0$, there is a $\delta > 0$ such that whenever $0 \leq t, t' \leq 1$, and $|t - t'| < \delta$, then $\|T_t - T_{t'}\| < \epsilon$; that is, T_t is uniformly continuous in t.

4.   If $t \epsilon [0,1]$ and x and y are arbitrary elements of the Banach space X which satisfy $T_tx = y$, then there exists a constant $C > 0$ such that $\|x\| \leq C\|y\|$.

Then T has a bounded inverse.

The two important assumptions are 3 and 4. Assumption 3 guarantees the continuity of T_t in both t and x. Assumption 4 guarantees that if T_t^{-1} exists for some t, it is bounded. Therefore the proof rests with showing that, for all $t \epsilon [0,1]$, T_t^{-1} exists. Friedrichs also extends the continuity method developed above for linear operators so as to be able to solve the nonlinear equation $Tx = 0$. Following the same principle, we need to find a transformation $\bar{T}$ and a point $\bar{x}$ such that $\bar{x}$ is the unique solution of $\bar{T}x = 0$ in a sphere $\|x - \bar{x}\| < r$.

Before giving the main theorem, the following definition and preliminary results are essential.

Definition: An operator T mapping a normed linear space X onto itself has a first variation $\delta T(x,y)$ at x if there exists a transformation $\delta T(x,y)$ such that:

1. $\delta T(x,y)$ is linear in y.
2. $R(x,y) \to 0$ as $y \to 0$, where

$$R(x,y) = \begin{cases} \dfrac{T(x+y) - Tx - \delta T(x,y)}{\|y\|} & y \neq 0 \\ 0 & y = 0 \end{cases}$$

In other words, for each $\epsilon > 0$ there exists an $\eta(\epsilon,x) > 0$ such that $\|y\| \leq \eta$ implies that $\|R(x,y)\| \leq \epsilon$.

It is sometimes possible to obtain a solution to $Tx = 0$ in the following manner. Suppose an x_0 can be found for which $\|T(x_0)\|$ is small. Set $x = x_0 + y$, and assume that a first variation $\delta T(x_0,y)$ exists. Then

$$Tx = Tx_0 + \delta T(x_0,y) + \|y\|R(x_0,y)$$

where $R(x_0,y)$ has the properties described above.

$Tx = 0$ can now be written in the form

$$\delta T(x_0,y) = -Tx_0 - \|y\|R(x_0,y)$$

The left side is linear in y, and the right side is small for small y. Therefore it is reasonable to expect that a solution y exists whose norm is small.

Certain assumptions are required to assure success of the procedure. These assumptions are stated prior to giving Theorem 2-13.

1. Tx has a first variation $\delta T(x,\delta x)$ in a sphere $\|x - x_0\| \leq r$. The remainder $\|\delta x\|R(x_0,\delta x)$ is bounded in δx locally uniformly in x, and $\delta T(x,\delta x)$ is continuous in x at x_0. That is, for all $\epsilon > 0$, there exist positive numbers α and β depending on ϵ and x_0, such that

   a. $\|x - x_0\| \leq \alpha$ and $\|\delta x\| \leq \beta$ imply

   $$\|R(x,\delta x)\| \leq \epsilon$$

   b. $\|x - x_0\| \leq \beta$ implies

   $$\|\delta T(x,\delta x) - \delta T(x_0,\delta x)\| \leq \epsilon\|\delta x\|$$

2. $\delta T(x_0,y)$ exists with a bounded inverse M. That is, there is a transformation $M(x_0,z)$, linear in z, for which

$$\delta T(x_0,M(x_0,z)) = z$$
$$M(x_0,\delta T(x_0,y)) = y$$

and for some finite $\mu(x_0) > 0$ and for all z

$$\|M(x_0,z)\| \leq \mu(x_0)\|z\|$$

3.
$$\|Tx_0\| \leq \frac{1 - \mu\epsilon}{\mu} \min\left(\alpha, \frac{\beta}{2}\right)$$
$$= \frac{2 - \theta}{2\mu} \min\left(\alpha, \frac{\beta}{2}\right) \qquad 0 < \theta < 1$$

Theorem 2-13   If:

1. x_0 is chosen such that assumption 2 holds,
2. θ is chosen so that $0 < \theta < 1$, with $\epsilon \equiv \theta/2\mu(x_0)$,
3.   In addition, x_0 is so chosen that $\alpha(\epsilon,x_0)$ and $\beta(\epsilon,x_0)$ can be determined to satisfy assumptions 1 and 3, then the equation $T(x) = \mathbf{0}$ has a unique solution in the sphere

$$\|x - x_0\| \; \min \; (\alpha(\epsilon,x_0), \; \tfrac{1}{2}\beta(\epsilon,x_0))$$

For general T, in choosing the family $T_t x = T(t,x)$, defined on $0 \le t \le 1$, it is important that the following five properties be satisfied:

a. $T(t,x)$ is continuous in t uniformly in t and x; that is, there exists a number $\sigma(\eta) > 0$ for each $\eta > 0$ such that $\|x\| \le K$ and $|t' - t| \le \sigma(\eta)$ imply

$$\|T(t,x) - T(t',x)\| \le \eta$$

b.   For all t and x with $\|x\| \le K$, $T(t,x)$ has a bounded first variation $\delta T(t,x,y)$ such that:

i. $R(t,x,y) = [T(t, x + y) - T(t,x) - \delta T(t,x,y)]/\|y\|$ approaches zero as y approaches zero, uniformly in t and locally uniformly in x; that is, for all $\epsilon > 0$ and x_0 with $\|x_0\| \le K$ there exist positive numbers α and β such that $\|x - x_0\| \le \beta$ and $\|y\| \le \alpha$ imply

$$\|R(t,x,y)\| \le \epsilon \qquad \text{for all } t$$

ii. $\delta T(t,x,y)$ is continuous in x uniformly in t with modulus $\|y\|$; that is, for every $\epsilon > 0$ and all x_0, with $\|x_0\| \le K$, there exists a number $\gamma > 0$ such that $\|x_0\| \le K$, $\|x\| \le K$, and $\|x - x_0\| \le \gamma$ imply that

$$\|\delta T(t,x,y) - \delta T(t,x_0,y)\| \le \epsilon\|y\|$$

c. $\delta T(t,x,y)$ has an inverse $\delta^{-1}T$ for all x, $\|x\| \le K$, and t; that is, $\delta^{-1}T$ is a mapping such that $\delta T(t,x,y) = z$ if and only if $\delta^{-1}T(t,x,y) = y$. In addition, $\delta^{-1}T$ is uniformly bounded in t; that is, for all t, x, and z there exists a function $\mu(x) > 0$ such that

$$\|\delta^{-1}T(t,x,z)\| \le \mu(x)\|z\|$$

d.   For all t, the mapping $x - T(t,x)$ is completely continuous.

e.   If $T(t,x) = \mathbf{0}$ for some t, $0 \le t \le 1$, and some x such that $\|x\| \le K$, then any solution x for $T(t,x) = \mathbf{0}$ lies in the interior of the sphere $S = \{x:\|x\| \le K\}$.

With the above five properties, we are now ready to give the main result for the method of continuity.

Theorem 2-14   Let:

1.   The known transformation $\bar{T}$ be such that $\bar{T}x = \mathbf{0}$ for $\|x\| \le K$ if and only if $x = \mathbf{0}$.

2. Tx be another transformation connected to $\bar{T}x$ by a family of transformations $T(t,x) = 0$ defined for $0 \leq t \leq 1$, where $T(0,x) = \bar{T}x$ and $T(1,x) = Tx$.

Then, if assumptions a to e hold, there exists a function $x(t)$, defined and continuous on the interval $0 \leq t \leq 1$, with $\|x(t)\| \leq K$, and such that $T(t,x) = 0$, with $\|x\| \leq K$, if and only if $x = x(t)$.

The proof of the theorem is very long, and only a brief outline will be given here. The approach is in three steps: (1) to show that $T(t,x) = 0$ has a solution $x \, \epsilon \, S$ for all t, $0 \leq t \leq 1$; (2) to show that $T(t,x) = 0$ has a continuous solution $x(t)$, in $0 \leq t \leq 1$, which is inside S and which is locally unique in the sense that there is a real function $\gamma(t) > 0$ such that $T(t,x) = 0$ and $\|x - x(t)\| \leq \gamma(t)$ only if $x = x(t)$; and (3) to show that for all t, $x(t)$ is the unique solution of $T(t,x) = 0$ with $\|x\| \leq K$.

For applications see the work of Friedrichs.

THE METHOD OF LEAST SQUARES

The least-squares criterion is used in several approximation techniques when applied to operator equations. The text by Linnik considers a variety of problems posed in a least-squares setting, in which one is given a set of numbers

$$x_{ki} \qquad k = 1, 2, \ldots, n; \, i = 1, 2, \ldots, n$$

and $\qquad w_i > 0 \qquad$ and $\qquad \eta_i \qquad i = 1, \ldots, n$

and needs to find numbers $a_1, a_2, \ldots, a_n$ which minimize the expression

$$Q = \sum_{i=1}^{N} w_i \left(\eta_i - \sum_{k=1}^{n} a_k x_{ki} \right)^2 \tag{2-78}$$

and satisfy some supplementary constraining conditions

$$F_j(a_1, a_2, \ldots, a_n) = 0 \qquad j = 1, 2, \ldots, m$$

where the F_j's are known functions of n arguments. Minimization techniques of the calculus are frequently used.

Sometimes η_i and x_{ki} are random variables, in which case the quantities $a_1, a_2, \ldots, a_n$, obtained by the algebraic approach of the method of least squares, are also random variables. Such a probabilistic use of the method gives additional substance and meaning to its usefulness.

Linnik[29] applies the method of least squares to problems of statistical estimation by constructing a likelihood function, using the method of maximum likelihood, and casting this function into a least-squares problem.

If quantities $y_r = \sum_{j=1}^{n} a_j x_{rj}$ $(r = 1, \ldots, N)$ $N \geq n$ are measured with errors Δ_r and $a_1, \ldots, a_n$ are sought, one uses the observed values $L_r = y_r + \Delta_r$, and the least-squares criterion is to find $a_j, j = 1, \ldots, n$, which minimize

$$\sum_{r=1}^{N} p_r \left(L_r - \sum_{j=1}^{n} a_j x_{rj} \right)^2 \tag{2-79}$$

where p_r are weights indicating the relative accuracies of the observations. If the accuracies of observation are the same, $p_r = 1$. In this case suppose that $\tilde{a}_j$ yields the minimum, and write

$$\tilde{L}_r = \sum_{j=1}^{n} \tilde{a}_j x_{rj} \qquad L - \tilde{L} = \epsilon$$

where $\epsilon = (L_1 - \tilde{L}_1, \ldots, L_n - \tilde{L}_n)$ is a vector of differences. The problem then is to minimize

$$\sum_{i=1}^{n} (L_i - \tilde{L}_i)^2 \qquad \text{or} \qquad \sum_{i=1}^{n} \epsilon_i{}^2 = (\epsilon, \epsilon)$$

In chap. 6 of "NM" a useful application of this idea is given. It derives from the concept of an element of best approximation.

If E is a normed linear space and E_1 is a finite-dimensional subspace, then for any $x \,\epsilon\, E$ there exists an element of best approximation $y \,\epsilon\, E_1$ in the sense that $\|x - y\|$ has the smallest value. This theorem is well known. Since the subspace E_1 can be generated by a finite collection of elements, and since any $y \,\epsilon\, E_1$ is a linear combination of these elements, the problem becomes one of choosing the coefficients in the linear combination which minimize $\|x - y\|$.

For an element $x \,\epsilon\, H$, where H is Hilbert space, if L is a finite-dimensional subspace, the coefficients of the linear combination are the Fourier coefficients of x with respect to the basis generating L. This results from minimizing $\|x - y\|^2 = (x - y, \, x - y) = \|y\|^2 - 2(x,y) + \|x\|^2$. Thus if the orthonormal basis elements are $z_1, \ldots, z_n$, and if $y = \sum_{k=1}^{n} \beta_k z_k$, then $\|y\|^2 = \sum_{k=1}^{n} \beta_k{}^2$. If also $(z_k,x) = \alpha_k$ are the Fourier coefficients of x, it is not difficult to substitute and show that $\|x - y\|^2$ is minimum for $\beta_k = \alpha_k$.

SEMIGROUPS (*FOR LINEAR UNBOUNDED OPERATORS*)

We now give an application of ideas covered in depth by Hille and Phillips (Ref. 28 of Chap. 1). A useful treatment of the resolvent is found in the theory of a one-parameter semigroup of linear transformations $T(\xi)$, $0 \leq \xi < \infty$, of a Banach space X into itself. $T(\xi)$ is a semigroup if $T(\xi_1 + \xi_2)x = [T(\xi_1)T(\xi_2)]x$, $0 < \xi_1, \xi_2 < \infty$. To illustrate this treatment let it be required to find all solutions of $t^2(dx/dt) + \lambda x = y$ in the space $C[0,\infty]$ of bounded continuous functions. We shall regard this equation in the form $(\lambda I - A)x = y$. We first examine the homogeneous equation with $y(t) \equiv 0$. The resulting equation has the solution $x = ae^{\lambda/t}$, which is in $C[0,\infty]$ for Re $(\lambda) < 0$ and for $\lambda = 0$. Actually λ is also in the spectrum for Re $(\lambda) = 0$, which follows from the ensuing discussion. We ask whether there is a semigroup $T(\xi)$ of which the unbounded operator $A = -t^2(d/dt)$ is the infinitesimal generator, i.e., satisfies

$$\frac{d}{d\xi} T(\xi)x = A T(\xi)x = T(\xi)Ax$$

for all x such that $t^2x''(t)$ is in $C[0,\infty]$. We find that if

$$T(\xi)x = x[t/(1 + \xi t)]$$

then indeed $(d/d\xi)x[t/(1 + \xi t)] = x'[t/(1 + \xi t)](-t^2)/(1 + \xi t)^2$. This is $T(\xi)$ acting on $Ax = -t^2x'(t)$, and hence $A = -t^2(d/dt)$. To show that $T(\xi)$ defines a semigroup we apply $T(\xi_2)$ to

$$T(\xi_1)x(t) = x\left[\frac{t}{1 + \xi_1 t}\right]$$

We obtain $x\left[\dfrac{t/(1 + \xi_2 t)}{1 + \xi_1(t/1 + \xi_2 t)}\right] = x\left[\dfrac{t}{1 + (\xi_1 + \xi_2)t}\right] = T(\xi_1 + \xi_2)x$

Formally, from $dT/d\xi = AT$, we can expect that $T = \exp(\xi A)$ in some sense. We want the solution of $(\lambda I - A)x = y$, which is

$$x = (\lambda I - A)^{-1}y$$

Since

$$\int_0^\infty e^{-p\xi} d\xi = \frac{1}{p}$$

for Re $(p) > 0$, we have formally,

$\int_0^\infty \exp[-\xi(\lambda I - A)]x \, d\xi$

$$= \int_0^\infty e^{-\xi\lambda}T(\xi)x \, d\xi = (\lambda I - A)^{-1}x \equiv R(\lambda,A)x \quad (2\text{-}80)$$

the resolvent of A. This holds for Re $(\lambda) > \lim_{\xi \to \infty} \log \|T(\xi)\|/\xi \equiv \sigma_0$. For our problem this gives $\int_0^\infty e^{-\xi\lambda}y[t/(1 + \xi t)] \, d\xi$ for Re $(\lambda) > 0$. Thus the resolvent is in the right half-plane, and since the spectrum is closed, it occupies the left half-plane and the rest of the imaginary axis. Thus

Re $(\lambda) \leq 0$ for λ in the spectrum. After simplification of the resolvent [put $s = t/(1 + \xi t)$ and $r = 1/s$], we have the resolvent acting on a function y: $e^{\lambda/t} \int_{1/t}^{\infty} e^{-\lambda r} y(1/r) \, dr$, Re $(\lambda) > 0$, which is the solution of the nonhomogeneous equation.

The same approach can be applied to the hyperbolic partial differential equation

$$x \frac{\partial^2 U}{\partial x^2} + \frac{\partial^2 U}{\partial t^2} = 0$$

with initial values of $U(x,t)$ and $\partial U(xt)/\partial t$ prescribed at $t = 0$.

This equation can be transformed to the parabolic type by writing

$$A = \begin{pmatrix} 0 & I \\ -x \dfrac{\partial^2}{\partial x^2} & 0 \end{pmatrix}$$

which operates on the vectors $F = [f_1(x), f_2(x)]$ with components belonging to $C[-\infty, 0]$. We then solve an initial-value (Cauchy) problem in the equation

$$AV = \frac{\partial}{\partial t} V \qquad V = (U, U_t)$$

A is then regarded as the infinitesimal generator of a semigroup determined by the last equation. Thus we should have our solution if we determined the semigroup. Here one finds expressions of the resolvent for $AF - \lambda F = -G$, where $G = [g_1(x), g_2(x)]$. The spectrum is found to occupy the imaginary axis and to consist entirely of the continuous spectrum. Now determination of the semigroup is obtained by taking the inverse transform of (2-80), which relates the semigroup and the resolvent. We have

$$\lim_{\omega \to \infty} \frac{1}{2\pi i} \int_{\delta - i\omega}^{\delta + i\omega} e^{\lambda t} R(\lambda; A) x \, dA = T(t)x$$

where $\delta > \sigma_0$, $t > 0$, $x \, \epsilon \, D[A]$. See Ref. 59 of Chap. 1 for details.

REFERENCES

1. Altman, M.: Concerning the Method of Tangent Hyperbolas for Operator Equations, *Bull. Acad. Polon. Sci. Sér. Sci. Math. Astronom. Phys.*, vol. 9, pp. 633–637, 1961.

2. ———: Iterative Methods of Higher Order, *Bull. Acad. Polon. Sci. Sér. Sci. Math. Astronom. Phys.*, vol. 9, pp. 63–68, 1961.

3. ———: Concerning Approximate Solutions of Non-linear Functional Equations, *Bull. Acad. Polon. Sci. Sér. Sci. Math. Astronom. Phys.*, vol. 5, pp. 461–465, 1957.

4.   Bartle, R. G.: Newton's Method in Banach Spaces, *Proc. Amer. Math. Soc.*, vol. 6, pp. 827–831, 1955.

5.   Birman, M. S.: Some Estimates for the Method of Steepest Descents, *Uspehi Mat. Nauk.*, vol. 5, pp. 152–155, 1950.

6.   Brauer, F.: A Note on Uniqueness and Convergence of Successive Approximations, *Canad. Math. Bull.*, vol. 2, pp. 5–8, 1959.

7.   ——— and S. Sternberg: Local Uniqueness, Existence in the Large and the Convergence of Successive Approximations, *Amer. J. Math.*, vol. 80, pp. 421–430, 1958.

8.   ———: Some Results on Uniqueness and Successive Approximations, *Canad. J. Math.*, vol. 11, pp. 527–533, 1959.

9.   Dieudonné, J.: Sur la Convergence des approximations successives, *Bull. Sci. Math.*, ser. 2, vol. 69, pp. 62–72, 1945.

10.   Faddeev, D. K., and V. K. Faddeeva: "Computational Methods of Linear Algebra," W. H. Freeman and Company, San Francisco, 1963.

11.   Freudenstein, F., and B. Roth: Numerical Solution of Systems of Nonlinear Equations, *J. Assoc. Comput. Mach.*, vol. 10, no. 4, pp. 550–556, 1963.

12.   Fridman, V. M.: The Method of Minimal Iterations with Minimum Errors for a System of Linear Algebraic Equations with Symmetric Matrices," *Z. Vycisl. Mat. i. Mat. Fiz.*, vol. 2, pp. 341–342, 1962.

13.   ———: An Iteration Process with Minimum Errors for Non-linear Operator Equations," *Dokl. Akad. Nauk SSSR*, vol. 139, pp. 1063–1066, 1961.

14.   Friedrichs, K. O.: "Functional Analysis and Applications," New York University Press, New York, 1953.

15.   Ghinea, M.: Sur la résolution des équations operationnelles dans les espaces de Banach, *Rev. Française Traitement Information*, vol. 8, no. 1, pp. 3–22, 1965.

16.   Hille, E.: "Analysis," vol. I, The Blaisdell Publishing Company, New York, 1964.

17.   Kadec, M. I.: A Method of Metric Projections in the Theory of Non-linear Operators, Funkcionalnyi Analiz i ego Premenenie (Trudy 5 Konf. po Funkcionalnomu Analizu i ego Primenenyu), *Izv. Akad. Nauk Azerbaidzan, SSR*, 1961.

18.   Kantorovich, L. V.: Approximate Solution of Functional Equations, *Uspehi Mat. Nauk*, ser. 11, vol. 6, no. 72, pp. 99–116, 1956.

19.   ———: The Method of Successive Approximations for Functional Equations, *Acta Math.*, vol. 71, pp. 63–97, 1939.

20.   ———: and V. I. Krylov, "Approximate Methods of Higher Analysis," Interscience Publishers, Inc., New York, 1958.

21.   Khamis, S.: A Note on Some Iterative Techniques for the Solution of Numerical Equations, *Univ. Nac. Tucumán Rev. Ser. A*, vol. 13, nos. 1 and 2, pp. 80–84, 1962.

22.   Kivistik, L. A., and A. Ustaal: Some Convergence Theorems for Iteration Processes with Minimum Residuals, *Tartu Riikl, Ül. Toimetised*, no. 129, pp. 382–393, 1962.

23.   ———: A Modification of the Iterative Method with Minimum Residuals for the Solution of Non-linear Operator Equations, *Dokl. Akad. Nauk SSSR*, vol. 136, pp. 22–25, 1961.

24.   Koselev, A. I.: Convergence of the Method of Successive Approximations for Quasi-linear Elliptic Equations, *Dokl. Akad. Nauk SSSR*, vol. 142, pp. 1007–1010, 1962.

25.   Krasnosel'skii, M. A.: Two Remarks on the Methods of Successive Approximations, *Uspehi Mat. Nauk*, n.s., vol. 10, no. 1, pp. 123–127, 1955.

26.   ———: "Topological Methods in the Theory of Non-linear Integral Equations," Pergamon Press and The Macmillan Company, New York, 1964.

27. La Salle, J.: Uniqueness Theorems and Successive Approximation, *Ann. of Math.*, vol. 50, pp. 722–730, 1949.

28. Lindenstrauss, J.: On Nonlinear Projections in Banach Spaces, *Michigan Math. J.*, vol. 11, no. 3, September, 1964.

29. Linnik, Yu. V.: "Method of Least Squares and Principles of the Theory of Observations," Pergamon Press, New York, 1961.

30. Lorch, E. R.: "Spectral Theory," Oxford University Press, New York, 1961.

31. Lotkin, M.: The Solution by Iteration of Nonlinear Integral Equations, *J. Math. and Phys.*, vol. 33, pp. 346–355, 1955.

32. Mann, W. Robert: Mean Value Methods in Iteration, *Proc. Amer. Math. Soc.*, vol. 4, pp. 506–510, 1953.

33. McGill, R., and P. Kenneth: Solution of Variational Problems by Means of a Generalized Newton-Raphson Operator, *AIAA J.*, vol. 2, no. 10, pp. 1761–1766, 1964.

34. Miklin, S. G.: "Direct Methods in Mathematical Physics," Moscow-Leningrad, 1950.

35. Nashed, M. Z.: On General Iterative Methods for the Solutions of a Class of Nonlinear Operator Equations, *Math. Comp.*, vol. 19, no. 89, pp. 14–24, 1965.

36. ———: The Convergence of the Method of Steepest Descents for Nonlinear Equations with Variational or Quasi-variational Operators, *J. Math. Mech.*, vol. 13, no. 5, pp. 765–794, 1964.

37. Petryshyn, W. V.: Direct and Iterative Methods for the Solution of Linear Operator Equations in Hilbert Space, *Trans. Amer. Math. Soc.*, vol. 105, pp. 136–175, 1962.

38. Picard, E.: "Traité d'analyse," Gauthier-Villars, Paris, 1925.

39. Pol'skii, N. I.: Projective Methods in Applied Mathematics, *Dokl. Akad. Nauk SSSR*, vol. 143, pp. 787–790, 1962.

40. Rall, L. B.: An Application of Newton's Method to the Solution of a Nonlinear Integral Equation, *Oregon State College O.R. Project* 452, July, 1955.

41. ———: Error Bounds for Iterative Solutions of Fredholm Integral Equations, *Pacific J. Math.*, vol. 5, pp. 977–986, 1955.

42. Rasulov, M. L.: The Residue Method of Solving Mixed Problems and Some Related Expansion Formulas, *Dokl. Akad. Nauk SSSR*, vol. 120, pp. 33–36, 1958.

43. Rothe, E. H.: Gradient Mappings and Extrema in Banach Spaces, *Duke Math. J.*, vol. 15, pp. 421–431, 1948.

44. Saaty, T. L., and J. Bram: "Nonlinear Mathematics," McGraw-Hill Book Company, New York, 1964.

45. Schechter, S.: Iteration Methods for Nonlinear Problems, *Trans. Amer. Math. Soc.*, vol. 94, no. 1, pp. 179–189, 1962.

46. Vainberg, M. M.: On the Convergence of the Method of Steepest Descents for Nonlinear Equations, *Dokl. Akad. Nauk SSSR*, vol. 130, pp. 9–12, 1960.

47. Viswanatham, B.: The General Uniqueness Theorem and Successive Approximations, *J. Indian Math. Soc.*, vol. 5, pp. 69–73, 1951.

48. Wintner, A.: On the Convergence of Successive Approximations, *Amer. J. Math.*, vol. 68, pp. 13–19, 1940.

THREE

FUNCTIONAL EQUATIONS

3-1 INTRODUCTION

Among the simplest functional equations to which a student becomes exposed early in his study of mathematics is that defining an even function; that is, $f(x) = f(-x)$, which is satisfied, for example, by $f(x) = \cos x$, $f(x) = x^2$, or $f(x) = 1$. Similarly, $f(-x) = -f(x)$ defines an odd function, illustrations of which are $f(x) = x$, $f(x) = \sin x$, and $f(x) = \sinh x$. Another familiar example is the functional relation $f(x + T) = f(x)$, which defines a periodic function with period T; this equation is satisfied for example, by $f(x) = \sin(2\pi x/T)$. Perhaps the earliest work in functional equations was that pursued by Euler in his theorem concerning homogeneous equations: A necessary condition that a function with a continuous derivative be homogeneous of degree k

$$f(\lambda x) = \lambda^k f(x)$$

is

$$x \frac{df(x)}{dx} = kf(x)$$

The student can verify this by differentiating the defining equation with respect to λ and setting $\lambda = 1$.

When a functional relationship is given, the object is, of course, to seek its solutions. In order to obtain a solution, the functions must often be restricted to a specified nature (analytic, bounded, continuous, convex, differentiable, entire, exponential, measurable, meromorphic, monotone, periodic, etc.). The functions themselves may be complex-valued, and the variables may also be complex, in which case one seeks analytic or meromorphic solutions. It often happens that a certain equation has only a trivial solution for a given class of functions. For example, it is known that the only continuous solution of

$$2f(z) = f(x) + f(y)$$

with
$$z = \frac{x - y}{\log(x/y)}$$

where x and y are real, is the identically constant solution $x = y = c$. On the other hand, an equation may have many solutions, and uniqueness is sometimes obtained by imposing side conditions.

The converse problem has also received wide attention, viz., what is the functional relationship which characterizes a given function, e.g., the sine function?[29,37]

Hille[19] points out that the concept of a functional equation is still being debated and that even if one agrees to exclude differential, difference, and integral equations, the problem of classification is difficult, and every functional equation requires its own mode of attack. However, he also states that progress in the field is noticeable.

We shall exemplify some classes of functional equations and inequalities, seek solutions, and demonstrate some of the elementary manipulations used to derive these solutions. The method of attack will involve algebraic operations, calculus methods, complex-variable techniques, series expansions, substitutions, etc. The definitive work on functional equations is that by Aczel.[1]

3-2   EXAMPLES OF FUNCTIONAL EQUATIONS

Some functional equations are generalized statements of known relations among trigonometric or hyperbolic functions. For example, the relation

$$\sin (x + y) = \sin x \cos y + \cos x \sin y$$

can be written in the general form

$$f(x + y) = f(x)g(y) + g(x)f(y) \tag{3-1}$$

Another example is furnished by Truesdell,[36] who attempts to unify by a single theory the special functions of analysis. The unified theory revolves about the functional (differential-difference) equation

$$\frac{\partial F(z,\alpha)}{\partial z} = F(z,\, \alpha + 1) \tag{3-2}$$

An interesting approach to the solution of Eq. (3-2) utilizes the Dirac delta function.† Thus proceeding formally, we assume a solution of the form

$$F(z,\alpha) = \iint K(x,y)e^{zx}e^{y\alpha}\, dy\, dx$$

where K is unknown. Then

$$\frac{\partial F}{\partial z} = \iint K(x,y)xe^{zx}e^{y\alpha}\, dy\, dx \tag{3-3}$$

and using our functional equation we have

$$\iint K(x,y)e^{zx}e^{y\alpha}(x - e^y)\, dy\, dx = 0$$

This expression is satisfied if we let $K(x,y) = a(x,y)\delta(x - e^y)$, where δ is the delta function. If we substitute this in the expression for $F(z,\alpha)$ and put $w = e^y$, we have

$$F(z,\alpha) = \iint a(x, \log w)\delta(x - w)e^{zx}w^{\alpha-1}\, dw\, dx$$
$$= \int b(x)e^{zx}x^{\alpha-1}\, dx \qquad b(x) = a(x, \log x)$$

Alternatively, we can start with a generalized form of this last expression

$$F(z,\alpha) = \int_{x_0}^{x_1} G(z,x)x^{\alpha-1}\, dx$$

Substitution in the functional equation yields

$$\int_{x_0}^{x_1} x^{\alpha-1}\left[\frac{\partial G(z,x)}{\partial z} - xG(z,x)\right] dx = 0$$

and hence the equation can be solved if the equation

$$\frac{\partial G}{\partial z} - xG = 0$$

has a solution.

Our third example is a functional equation which recently has received intensive study. It occurs in the theory of dynamic programming, a special branch of optimization, and has the form

$$f(x) = \max_{0 \le y \le x} \{g(y) + h(x - y) + f[ay + b(x - y)]\}$$

where g and h are known functions. It is obtained as a limiting form for the functional-difference equation

$$f_n(x) = \max_{0 \le y \le x} \{g(y) + h(x - y) + f_{n-1}[ay + b(x - y)]\}$$

---

† The Dirac delta function $\delta(t)$ is defined by $\delta(t) = 0,\ t \ne 0;\ \int_{-\infty}^{\infty} \delta(t)\, dt = 1.$

Another equation in dynamic programming has the integrofunctional form

$$f(s) = \sup_a [g(s,a) + \int f(t) \, dP(t,a,s)]$$

Some of the most important functional equations are known by the names of the men who studied them.

Abel's equation: $\qquad\qquad f[g(x)] = f(x) + \alpha$

where the constant α and $g(x)$ are given.

Cauchy's equation: $\qquad f(x + y) = f(x) + f(y)$
Pythagoras's equation: $\quad |f(x + iy)|^2 = |f(x)|^2 + |f(iy)|^2$

studied by Hille[20] and also called a pythagorean equation.

Poisson's equation: $\; f(x + y) + f(x - y) = 2f(x)f(y)$
Schröder's equation: $\qquad\quad f[g(x)] = \alpha f(x)$

where the constant α and $g(x)$ are given.

Jensen's equation: $\quad f[\frac{1}{2}(x + y)] = \frac{1}{2}[f(x) + f(y)]$
Pexider's equation: $\qquad f[x \oplus y] = g(x) \oplus g(y)$

where $\oplus$ is an associative operation, such as $+$ or $\times$, and f and g are unknown functions.

Examples of functional inequalities are the inequality defining convex functions, $f[(x + y)/2] \leq f(x/2) + f(y/2)$, or more generally $f(\theta x + (1 - \theta)y) \leq \theta f(x) + (1 - \theta)f(y)$ [recall that a necessary and sufficient condition for the convexity of a twice-differentiable function $f(x)$ is that $f''(x) \geq 0$] and the inequality defining subadditive functions

$$f(x + y) \leq f(x) + f(y)$$

Subadditivity can also be defined on a subset S of the reals if the inequality holds for all x and y in S such that $x + y$ is also in S. The function $f(x) = |x|$ is subadditive. An interesting example of subadditivity concerns a conjecture as to the distribution of prime numbers. If $\pi(x)$ denotes the number of primes equal to or less than x, is $\pi(x)$ subadditive? The definition of a monotone function also yields a functional inequality $f(x) \geq f(y)$, $x \geq y$.

A functional equation may have a quite general form; for example,

$$f[G(x,y)] = F[f(x),f(y)] \qquad\qquad (3\text{-}4)$$

Cauchy's equation itself is a special case of the more general equation

$$f(x + y) = F[f(x),f(y)] \qquad\qquad (3\text{-}5)$$

which is also a generalization of $f(x + y) = f(x) + f(y) + f(x)f(y)$. Jensen's equation is a special case of

$$f[\tfrac{1}{2}(x + y)] = F[f(x),f(y)] \tag{3-6}$$

Functional equations occur with difference, differential, and integral forms, as we shall see in later chapters. Let us briefly illustrate these types of equations. The functional equation

$$f(x) = F[x,g(x)]$$

can be reduced to a functional-difference equation. Thus, if we suppose that $g(x)$ is a unit periodic, i.e., it satisfies $g(x + 1) = g(x)$, then

$$f(x + 1) = F[x + 1, g(x)]$$

Hence, $g(x)$ can be eliminated in both expressions, yielding a first-order functional-difference equation

$$f(x + 1) = H[x,f(x)]$$

An example of a functional-differential equation is

$$f(cx) = cf(x)f'(x)$$

whose solutions for $c = 2$ are 0, x, $a^{-1} \sin ax$, and $ae^{x/2a}$. An integrofunctional equation is illustrated by

$$F\{x,\varphi(x),\varphi[f(x)], \int_{x_0}^{x} K_n(x,s,\varphi(s),\varphi[f(s)]) \, ds\} = 0$$

On the other hand,

$$\frac{d^2f}{dx^2} = G\{x,f(g(x)), \int_{a}^{b} F[x,y,f(h(y))] \, dy\}$$

is a integrodifferential functional equation. Of course, it is not difficult to exemplify a functional-difference–differential-integral equation.

As an illustration of manipulations with functional equations,[39] consider the problem of finding every real-valued function satisfying the condition

$$\int_{0}^{x} f(t) \, dt = x \sqrt{f(0)f(x)} \qquad x > 0$$

that is, the average of f over any closed interval $[0,x]$ is the geometric mean of its end-point value. If $f(0) = 0$, then $f(x) \equiv 0$ is a trivial solution. If $f(0) \neq 0$, let $g(x) = \sqrt{f(x)/f(0)}$. Then

$$\int_{0}^{x} g^2(t) \, dt = xg(x)$$

Hence
$$g^2(x) = xg'(x) + g(x)$$

whose solutions, upon separating variables, are found to be

$$g(x) = \begin{cases} \dfrac{1}{1 - cx} & 0 \le x < \dfrac{1}{c}, c > 0 \\[2ex] \dfrac{1}{1 + cx} & 0 \le x \le \infty, c \ge 0 \end{cases}$$

and $g(x) \equiv 1$.   Thus,

$$f(x) = \frac{f(0)}{(1 \pm cx)^2} \qquad \text{and} \qquad f(x) \equiv f(0)$$

are the solutions.

As another example of the methods used in solving a functional equation, consider[12]

$$f(x,n_1)f(x,n_2) = f(x, n_1 + n_2 + c) \tag{3-7}$$

where n_1, n_2, and c are nonnegative integers.   In (3-7), put $n_1 = n - 1$, $n_2 = n + 1$ to obtain

$$f(x, n - 1)f(x, n + 1) = f(x, 2n + c) = [f(x,n)]^2 \tag{3-8}$$

Now if $f(x,0) = 0$, then $f(x,n) = 0$ for all n.   If $f(x,0) \neq 0$, no $f(x,n)$ can vanish, and (3-8) can be written as

$$\frac{f(x, n + 1)}{f(x,n)} = \frac{f(x,n)}{f(x, n - 1)} = \cdots = \frac{f(x,1)}{f(x,0)} \equiv F(x)$$

Thus $f(x,n) = f(x,0)[F(x)]^n$

But, from (3-8), $[f(x,0)]^2 = f(x,c) = f(x,0)[F(x)]^c$

and hence the general solution is given by

$$f(x,n) = [F(x)]^{n+c}$$

Again, suppose we wish to find a solution of

$$\frac{f(x) - f(y)}{x - y} = f'\left(\frac{x + y}{2}\right) \tag{3-9}$$

or $f(x) - f(y) = (x - y)f'\left(\dfrac{x + y}{2}\right)$

Upon differentiating with respect to x and then with respect to y we have

$$0 = \frac{x - y}{4} f'''\left(\frac{x + y}{2}\right)$$

or $f'''(x) = 0$

and therefore $f(x) = ax^2 + bx + c$.

As another example let us find all *analytic* functions f such that

$$f(x + y)f(x - y) = [f(x) + f(y)][f(x) - f(y)]$$

Let $x = y$. Then $f(0) = 0$ or $f(x) \equiv 0$, which we exclude. Operating on the equation with $\partial^2/(\partial x\,\partial y)$, we have

$$f''(x + y)f(x - y) = f(x + y)f''(x - y) \tag{3-10}$$
$$f''(x) = \pm a^2 f(x)$$

and $f(x) = c \sin ax,\ cx,\ c \sinh ax$. The solution cx is a limiting form of $(c/a) \sin ax = cx + o(a^2 x^3)$ as $x \to 0$.

Functional equations and inequalities can arise in practice in several ways. Some occur in direct physical applications; others are from defining properties of functions; and still others are generalizations of known identities.

For example, one may seek all solutions of the simultaneous functional equations

$$f(x + y) = f(x)g(y) + g(x)f(y)$$
$$f^2(x) + g^2(x) = 1 \tag{3-11}$$

which are generalizations of known trigonometric identities. Clearly $f(x) = \sin x,\ g(x) = \cos x$ is one such solution. Are there others? What are other solutions of the first equation alone? One may take other generalizations of trigonometric, hyperbolic, or even elliptic identities. To find periodic solutions of

$$f(x + T) - \lambda f(x) = g(x)$$

where $g(x)$ is periodic of period T, is yet another example.

There are few physical problems whose formulation directly leads to functional equations. Often it is only after a problem has been specifically formulated that one realizes that the functions used are solutions of known functional equations. Then the functional equation is written down, and some justification leading to it is developed. For example, a physical argument is used to derive the law of the parallelogram of forces, which leads to Cauchy's equation. This derivation is due to Darboux and is given in Ref. 27, along with other justifications.

Cauchy's equation is used to establish the fact that if a correspondence between two series of elements has a harmonic ratio, then the correspondence is projective. Other equations associated with Cauchy are

$$f(x + y) = f(x)f(y) \qquad f(xy) = f(x)f(y) \qquad f(xy) = f(x) + f(y)$$

The first equation is used to establish a projective definition of angle, due to Laguerre, and also to define semigroups of transformations.

Another example occurs in information theory. The entropy of n mutually exclusive events is defined by $-\sum_{i=1}^{n} p_i \log p_i$, where p_i is the probability of occurrence of the ith event. Since $f(x) = \log x$ is a solution of $f(xy) = f(x) + f(y)$, we note that this is a function whose value on the product of probabilities of events is equal to the sum of its values on the probabilities of the individual events. Shannon used this function since it agrees with our intuitive notion that the information content of two independent events should be the sum of the information in each.

As a final example of applications, the equation

$$f(x + y) + f(x - y) = 2f(x)f(y)$$

has been used to derive the parallelogram-of-forces law, to define circular functions, and to show their relation to spherical trigonometry, plane euclidean and noneuclidean trigonometries, and statics.†

Exercise 3-1   Determine whether the functional equation

$$f(xy) = f(x)f(y) + 1$$

has a real solution.   *Hint:* Let $x = y = 0$ and solve the resulting quadratic in $f(0)$.

Exercise 3-2   Solve the equation

$$f(x + y) = f(x)f\left(\frac{\pi}{2} - y\right) + f\left(\frac{\pi}{2} - x\right)f(y)$$

with $f(0) = 0$.

Exercise 3-3   Show that

$$f(x) = \frac{1}{2h} \int_{x-h}^{x+h} f(t)\, dt$$

has the solution $f(x) = bx + c$ by first differentiating with respect to x and then twice with respect to h and solving the resulting equation, $f''(x + h) - f''(x - h) = 0$.

Exercise 3-4   Find a solution of $\ln (x + y) = \ln x \ln y$ by putting $y = rx$ and solving the resulting quadratic in $\ln x$.

3-3   CONTINUOUS, DISCONTINUOUS, AND MEASURABLE SOLUTIONS—CAUCHY'S ADDITIVE EQUATION

One of the simplest functional equations, and perhaps the oldest, is Cauchy's equation

$$f(x + y) = f(x) + f(y) \tag{3-12}$$

† See *Amer. Math. Monthly*, vol. 32, p. 428, 1925.

which the reader will recognize as linear, since f distributes with respect to its argument. We shall show, for a variety of assumptions about f, that $f(x) = cx$ is the only continuous solution of (3-12). For example, if we assume that f is *differentiable* with respect to x, we have

$$f'(x + y) = f'(x)$$

Hence f' is periodic with arbitrary period y, and consequently $f'(x) = c$, and $f(x) = cx$. We shall then give a nonmeasureable solution (due to Hamel) which can be constructed if we permit discontinuous solutions. The treatment of (3-12) will be somewhat typical of the methods used for real functional equations. Let $y = 0$ in (3-12). We have

$$f(x) = f(x + 0) = f(x) + f(0)$$

from which we have $f(0) = 0$. Also

$$0 = f(0) = f(x - x) = f(x) + f(-x)$$
and hence
$$f(-x) = -f(x) \tag{3-13}$$

and the solution must be an odd function. Now, if n is a positive integer, then by induction

$$f(nx) = f[x + (n - 1)x] = f(x) + f[(n - 1)x]$$
$$= f(x) + (n - 1)f(x) = nf(x) \tag{3-14}$$

On the other hand, using (3-13) for negative integers, we find that (3-14) is still valid. In fact, from

$$f(x) = f\left(n \frac{x}{n}\right) = nf\left(\frac{x}{n}\right)$$
it follows that
$$f\left(\frac{m}{n} x\right) = mf\left(\frac{x}{n}\right) = \frac{m}{n} f(x) \tag{3-15}$$

and therefore (3-14) is valid for any rational number.

We now determine the form of f, assuming continuity for every x.

Theorem 3-1   If $f(x)$ is continuous and satisfies (3-12), then $f(x) = xf(1)$.

Proof: If $s_k (k = 1, 2, \ldots)$ is a sequence of rationals such that $\lim_{k \to \infty} s_k = x$, then, since $f(x)$ is continuous, $\lim_{k \to \infty} f(s_k) = f(x)$. But from (3-15) $f(s_k) = s_k f(1)$. Hence taking $\lim_{k \to \infty}$ on both sides gives $f(x) = xf(1)$. If we define $f(1) \equiv c$, then $f(x) = cx$. We now prove the stronger version.

Theorem 3-2   If f is continuous at one point x_0 and satisfies (3-12), then $f(x) = xf(1)$ for all x.

Proof: We show that $f(x)$ is continuous everywhere. If x_1 is another point, then

$$
\begin{aligned}
\lim_{x \to x_1} f(x) &= \lim_{x \to x_1} f(x - x_1 + x_1 - x_0 + x_0) \\
&= \lim_{x \to x_1} f(x - x_1 + x_0) + f(x_1 - x_0) \\
&= \lim_{x - x_1 + x_0 \to x_0} f(x - x_1 + x_0) + f(x_1 - x_0) \\
&= f(x_0) + f(x_1 - x_0) = f(x_1)
\end{aligned}
$$

Exercise 3-5 Show that if $f(x)$ satisfies (3-12) and is bounded in the neighborhood of a point x_0, then it is continuous at the origin and hence everywhere, so that $f(x) = xf(1)$. Show that this is also true if $f(x)$ is just bounded from above in a neighborhood of x_0.

Definition: The limit superior and right limit superior of the function $f(x)$ at point y are defined respectively as follows:

$$
\overline{\lim_{x \to y}} f(x) = \inf_{\delta > 0} \sup_{|x - y| < \delta} f(x)
$$

$$
\overline{\lim_{x \to y+}} f(x) = \inf_{\delta > 0} \sup_{(x - y) < \delta} f(x)
$$

Definition: f is upper semicontinuous at x_0 if

$$
f(x_0) \geq \overline{\lim_{x \to x_0}} f(x)
$$

that is, given $\epsilon > 0$, there is a $\delta > 0$ such that $|x - x_0| < \delta$ implies $f(x) < f(x_0) + \epsilon$.

Definition: f is upper semicontinuous on the right at x_0 if

$$
f(x_0) \geq \overline{\lim_{x \to x_0+}} f(x) \tag{3-16}
$$

Exercise 3-6 Prove that $f(x) = xf(1)$ is a solution of (3-12) if f is upper semicontinuous on the right at x_0.

Exercise 3-7 Prove that if $f(x)$ is the pointwise limit of a sequence of continuous functions, then $f(x) = xf(1)$.

Theorem 3-3 If in addition to (3-12) $f(x)$ also satisfies

$$
f(xy) = f(x)f(y)
$$

then $f(x) = xf(1)$.

Proof: For $x > 0$, we can write $f(x) = f(x^{1/2})f(x^{1/2}) = f^2(x^{1/2})$, and hence $f(x) \geq 0$ for $x > 0$. Therefore, from (3-13), for values of x in a neighborhood of $x = -1$ we have $f(x) \leq 0$, and hence $f(x)$ is bounded above, and the proof is complete. Note that since $f(1) = 1$, we have $f(x) = x$ for all x.

Example: Obtain a solution of

$$f(x + y) = f(x) + f(y) + f(x)f(y) \tag{3-17}$$

Let $g(x) = f(x) + 1$; then if $f(x) \not\equiv -1$,

$$g(x + y) = g(x)g(y)$$

Since $g(x) = g^2(x/2)$, $g(x) \geq 0$. If $g(x_0) = 0$, then $g(x) \equiv 0$, $f(x) \equiv -1$. Then if we write $h(x) = \log g(x)$ for $g(x) \neq 0$, we have

$$h(x + y) = h(x) + h(y) \tag{3-18}$$

and hence $f(x) = -1$ and $f(x) = e^{cx} - 1$ are the only continuous solutions. The constant c is given by $\log [f(1) + 1]$.

We have thus successively relaxed the restrictions on the solution of Cauchy's equation and each time arrived at a continuous, differentiable solution. In the complex plane, however, a continuous solution is not necessarily analytic. Thus

$$f(w) = au + bv \qquad \text{for } w = u + iv$$

is a continuous solution, but it is analytic if and only if $b = ia$.

Finally, we show that a discontinuous solution can be obtained if the requirement for boundedness on a set of positive measure is removed.

Theorem 3-4   There exists an $f(x)$ such that $f(x + y) = f(x) + f(y)$ but is not of the form $f(x) = cx$.

Proof: We shall need the idea of a *Hamel basis*. A set of real numbers S forms a Hamel basis for the real numbers if:

1.   The relation $\sum\limits_{i=1}^{n} a_i r_i = 0$ with r_i in S and a_i rational implies $a_1 = \cdots = a_n = 0$; that is, any finite subset of S is linearly independent over the rationals.

2.   For every real number x, there are a finite number of elements $r_1, \ldots, r_k$ in S and rationals $b_1, \ldots, b_k$ such that

$$x = \sum_{j=1}^{k} b_j r_j \tag{3-19}$$

A Hamel basis is nondenumerable. Note that the representation of x in (3-19) is unique as a consequence of condition 1. Such a basis can be shown to exist by making use of Zorn's lemma or, equivalently, by transfinite induction.

Now let S be a Hamel basis, and for each r in S choose $f(r)$ as an arbitrary real number. Then for any real x, the representation (3-19) implies that

$$f(x) = \sum_{j=1}^{k} b_j f(r_j)$$

The question is whether $f(x)$ satisfies (3-12). The answer is in the affirmative, for if

$$x = \sum_{k=1}^{n} a_k r_k \qquad y = \sum_{k=1}^{n} b_k r_k$$

(note that some of the coefficients may be zero, but we use n terms in both cases), then

$$x + y = \sum_{k=1}^{n} (a_k + b_k) r_k$$

and from the definition of f we have

$$f(x + y) = \sum_{k=1}^{n} (a_k + b_k) f(r_k) = f(x) + f(y)$$

To exhibit an f that is *discontinuous*, let $f(r_1) = 1$, for a particular $r_1 \, \epsilon \, S$, $f(r) = 0$ for $r \neq r_1$, $r \, \epsilon \, S$. If f is continuous, then we know from Theorem 3-1 that there is a c such that $f(r) = cr$, and hence $f(r)/f(r_1) = r/r_1$. But the left side is zero for the above assignment of values for f with $r \neq r_1$, while the right side can never be equal to zero (since a basis element must be nonzero). This contradiction shows that f is discontinuous.

Exercise 3-8   Use the above results to show that in the representation of the numbers on any finite interval by means of a Hamel basis every element of the basis is used and that the coefficients used in the representation are unbounded.

Theorem **3-5**   For Cauchy's equation, continuity and measurability are equivalent.

Proof: If f is measurable and satisfies (3-12), then f is bounded on every bounded interval. Let $E_n = \{x : x \, \epsilon \, I, \, |f(x)| \leq n\}$ for integral n. Then $E_1 \subset E_2 \subset \cdots \subset \bigcup_{n=1}^{\infty} E_n = I$. Therefore there is an n for which E_n has positive measure $\mu(E_n)$. Now suppose there is an interval $I = (-A, A)$ on which f is not bounded. Choose a sequence y_k in I so that $f(y_k) > 2n + f(y_{k-1})$, for fixed n, and define

$$F_k = \{z : |z - y_k| < A, \, |f(z - y_k)| \leq n\} \qquad \{z : z = y_k + x, \, x \, \epsilon \, E_n\}$$

If we now let

$$G_k = \{z : |z| < 2A, \, f(y_k) - n \leq f(z) \leq n + f(y_k)\}$$

we see that $F_k \subset G_k$. Now pick an integer $j > k$. It follows, from the iteration procedure by which $f(y_k)$ was obtained, that $-f(y_j) + 2n + f(y_k) < 0$. If $z_j \, \epsilon \, G_j$, then $f(y_j) - n \leq f(z_j)$ by the definition of G_j. Adding the two inequalities and making use of the definition of G_k, we find, for $z_k \, \epsilon \, G_k$, that $f(z_k) \leq n + f(y_k) < f(z_j)$. Therefore $z_j \, \epsilon \, G_k$, $k \neq j$, and $G_k \cap G_j = 0$. It follows that $F_k \cap F_j = 0$ for $k \neq j$, so that $\mu(\overset{\infty}{\underset{k=1}{\cup}} F_k) = \overset{\infty}{\underset{k=1}{\sum}} \mu(F_k) \leq 4A$, because $F_k \subset G_k \subset (-2A, 2A)$. The latter inequality implies that $\mu(F_k) = 0$ for every k. Finally, we have $\mu(F_k) = \mu(E_n)$ for every k (because F_k was constructed by shifting E_n without changing its size), so that $\mu(E_n) = 0$, a contradiction. Therefore, f is bounded on every finite interval and hence is continuous.

Another method of proving the continuity of f is as follows:

$$\int_0^1 f(x + y) \, dy = f(x) + \int_0^1 f(y) \, dy$$

or
$$f(x) = \int_x^{x+1} f(t) \, dt - \int_0^1 f(y) \, dy$$

We conclude that f is continuous since it is the integral of a measurable bounded function. That measurability follows from continuity is a well-known fact of measure theory. Thus, for Cauchy's equation, continuity and measurability are equivalent.

We now use the foregoing ideas to study solutions of the system

$$\begin{aligned} \varphi(x + y) &= \varphi(x)\varphi(y) - \psi(x)\psi(y) \\ \psi(x + y) &= \varphi(x)\psi(y) + \psi(x)\varphi(y) \end{aligned} \qquad (3\text{-}20)$$

where φ and ψ are real-valued and measurable functions. We want to determine all solutions of this pair of equations. Let

$$F(x) = \varphi(x) + i\psi(x) \qquad (3\text{-}21)$$

Then $F(x)$ is complex-valued, and (3-20) reduces to the single equation

$$F(x + y) = F(x)F(y) \qquad (3\text{-}22)$$

Thus
$$\log |F(x + y)| = \log |F(x)| + \log |F(y)| \qquad (3\text{-}23)$$

We see that $f(x) = \log |F(x)|$ is measurable (since all the foregoing transformations preserve measurability) and satisfies Cauchy's equation. Thus, from Theorems 3-1 and 3-5, $\log |F(x)| = cx$ for some real c, and hence

$$|F(x)| = \exp (cx) \qquad (3\text{-}24)$$

Therefore $F(x)$ is bounded and measurable on every bounded interval. We now show that $F(x)$ is continuous and differentiable. Since F is bounded and measurable, it is integrable on every finite interval. There is some $\alpha > 0$ such that $\int_0^\alpha F(y)\,dy \neq 0$, for otherwise $F(y) = 0$ almost everywhere, contradicting (3-24). Now using (3-22), we can write

$$\int_0^\alpha F(x + y)\,dy = F(x) \int_0^\alpha F(y)\,dy$$

or

$$F(x) = \frac{\int_x^{x+\alpha} F(z)\,dz}{\int_0^\alpha (Fy)\,dy}$$

This exhibits $F(x)$ as a continuous function of x and therefore as the integral of the continuous function $F(z)$. It follows that $F(x)$ is differentiable, and we have

$$F'(x) = \frac{F(x + \alpha) - F(x)}{\int_0^\alpha F(y)\,dy}$$

If we let $\int_0^\alpha F(y)\,dy = \beta^{-1}$ and use (3-22), we get

$$F'(x) = \beta F(x)[F(\alpha) - 1]$$
or
$$F'(x) = \gamma F(x)$$

from which $F(x) = \exp(\gamma x)$ (with $\gamma = \beta[F(\alpha) - 1]$). Now, from (3-24), it follows that $\gamma = c + ib$ with b real, and from (3-21) we get

$$\varphi(x) = e^{cx} \cos bx$$
$$\psi(x) = e^{cx} \sin bx$$

Remark: Functional equations also occur in vector and matrix form. Thus Cauchy's equation

$$F(x + y) = F(x) + F(y)$$

where F is an m-dimensional vector function of the n-dimensional vectors x and y, has the solution

$$F(x) = Cx$$

where C is an $(m \times n)$-dimensional constant matrix.

The matrix equation

$$F(X \cdot Y) = F(X)Y$$

where F, X, and Y are square matrices of the same order, is easily solved by setting $X = I$, the identity matrix, giving

$$F(Y) = F(I)Y = CY$$

Exercise 3-9   Using $f(x + y) = f(x)f(y)$, $f(0) \neq 0$, and defining[33]

$$s(x) = \frac{f(x) - f(-x)}{2k} \qquad c(x) = \frac{f(x) + f(-x)}{2} \qquad k \neq 0$$

show that the following hold:

$$c(x) + ks(x) = f(x)$$
$$c(x) - ks(x) = f(-x)$$
$$c^2(x) - [ks(x)]^2 = f(0)$$
$$[c(x) + ks(x)]^n = f^n(x) = f(nx) = c(nx) + ks(nx)$$

The last relation is known as the *generalized De Moivre theorem*.

Let $f(x) = e^{ix}$ and $k = ni$, and show that the third relation becomes

$$X^2 + n^2Y^2 = 1$$

where $X = c(x)$, $Y = s(x)$. If X and Y are interpreted as rectangular coordinates, we have elliptical trigonometry, and if $n = 1$, we have ordinary circular trigonometry. If $f(x) = e^x$ and $k = 1$, we have the theory of hyperbolic functions. Briefly consider the case where $f(x)$ is a matrix $A(x)$. Then $s(x)$ and $c(x)$ will be matrices. Also consider generalizations starting with

$$f(x_1, \ldots, x_n)f(y_1, \ldots, y_n) = f(x_1 + y_1, \ldots, x_n + y_n), f(0, \ldots, 0) \neq 0$$

Define $s(x_1, \ldots, x_n)$, $c(x_1, \ldots, x_n)$ as before. Then let

$$f(x_1, \ldots, x_n) = a_1^{x_1} \cdots a_n^{x_n}, a_i > 0, i = 1, \ldots, n$$

Exercise 3-10 Show that the only continuous solutions of[5]

$$f(x + y) = f(x) + f(y) + a(1 - A^x)(1 - A^y)$$

where a and A are real numbers, $A > 0$, are of the form $f(x) = kx - a(1 - A^x)$. *Hint:* Differentiate with respect to x and then with respect to y. Let $u = x + y$; integrate twice with respect to u to get $f(u) = aA^u - au \ln A - a$. Let f_1 be any other continuous solution; then $g = f - f_1$ is also continuous. Compute $g(x + y)$, and show that g satisfies Cauchy's equation and hence $g(x) = cx$, and finally obtain the desired solution.

Exercise 3-11 Show that the only continuous solutions of

$$f(x + y) = A^x f(y) + A^y f(y)$$

where A is a positive real number, are of the form $f(x) = kxA^x$, where k is a real number. *Hint:* Put $g(x) = A^{-x}f(x)$ and show that g satisfies Cauchy's equation.

3-4 SOME GENERALIZATIONS

If f is continuous, and if there are g and h such that[13]

$$f(x + y) = g(x) + h(y) \qquad (3\text{-}25)$$

then all *continuous* solutions of (3-25) are of the form

$$f(x) = ax + b \qquad (3\text{-}26)$$

***Theorem* 3-6** f has property (3-25) if and only if it has the property

$$f(x + y) = f(x) + f(y) - f(0) \tag{3-27}$$

Proof: If f has property (3-25), let $y = 0$ to obtain

$$f(x) = f(x + 0) = g(x) + h(0)$$

and let $x = 0$ to obtain

$$f(y) = f(0 + y) = g(0) + h(y)$$

Addition yields (3-27).   Conversely, if we let

$$g(x) = f(x) \qquad h(x) = f(x) - f(0)$$

in (3-27), we obtain (3-25).

Exercise 3-12   Show by induction that if f has property (3-25), then for any real x and integer n

$$f(nx) = nf(x) - (n - 1)f(0) \tag{3-28}$$

***Theorem* 3-7**   If f has property (3-25), and if m/n, $n \neq 0$, is a rational number, then $f(m/n) = [f(1) - f(0)](m/n) + f(0)$.

Proof: From (3-28) we have for $x = m/n$

$$f(m) = nf\left(\frac{m}{n}\right) - (n - 1)f(0)$$

For the case $n = m$ we obtain

$$f(m) = mf(1) - (m - 1)f(0)$$

Equating the two expressions for $f(m)$, we obtain the result.

Exercise 3-13   Show that if f with property (3-25) is continuous at a point, then it is continuous everywhere.

Exercise 3-14   Show that if f is continuous, then it has property (3-25) if and only if

$$f(x) = [f(1) - f(0)]x + f(0)$$

for any real x.

Exercise 3-15   Use the Hamel-basis argument to produce a discontinuous solution.   *Hint:* Define f arbitrarily on $S \cup \{0\}$; that is, write

$$f(x) = \sum_{k=1}^{n} a_k f(r_k) - \left(\sum_{k=1}^{n} a_k - 1 \right) f(0)$$

Imitate the case previously given.

3-5   MEASURABLE AND BOUNDED SOLUTIONS—A
GENERALIZATION OF AN EQUATION DUE TO BANACH

The functional equation

$$f(x + y) + f(x - y) = \varphi(2y)f(x) \qquad (3\text{-}29)$$

in which φ is given and f is unknown, has been treated by Kaczmarz.[23] The problem was posed by Banach, who asked for all solutions that are functions of a real variable having the property analogous to that of harmonic functions; e.g., in two dimensions,

$$g(a,b) = \frac{1}{2\pi r} \int_c g(x,y) \, ds$$

with c a circle with center at $[a,b]$ and radius r.   If f is not identically zero and satisfies (3-29), then φ must satisfy

$$\varphi(x + y) + \varphi(x - y) = \varphi(x)\varphi(y) \qquad (3\text{-}30)$$

This is proved by taking a point x_1 for which $f(x_1) \neq 0$; then, with $x = x_1$ in (3-29) and y replaced successively by $(x + y)/2$ and $(x - y)/2$, (3-29) yields

$$f\left(x_1 + \frac{x + y}{2}\right) + f\left(x_1 - \frac{x + y}{2}\right) = \varphi(x + y)f(x_1)$$

$$f\left(x_1 + \frac{x - y}{2}\right) + f\left(x_1 - \frac{x - y}{2}\right) = \varphi(x - y)f(x_1)$$

Also, with y replaced by $y/2$ and x replaced successively by $x_1 + x/2$ and $x_1 - x/2$, (3-29) yields

$$f\left(x_1 + \frac{x}{2} + \frac{y}{2}\right) + f\left(x_1 + \frac{x}{2} - \frac{y}{2}\right) = \varphi(y)f\left(x_1 + \frac{x}{2}\right)$$

$$f\left(x_1 - \frac{x}{2} + \frac{y}{2}\right) + f\left(x_1 - \frac{x}{2} - \frac{y}{2}\right) = \varphi(y)f\left(x_1 - \frac{x}{2}\right)$$

If we sum the first two equations and then the second two, we see that the left members of the two sums are the same, so that

$$f(x_1)[\varphi(x + y) + \varphi(x - y)] = \varphi(y)\left[f\left(x_1 + \frac{x}{2}\right) + f\left(x_1 - \frac{x}{2}\right)\right]$$

Finally, we apply (3-29) to the last bracketed quantity and cancel $f(x_1) \neq 0$ to get (3-30).

If f is measurable, then by taking x fixed with $f(x) \neq 0$, (3-29) shows that φ is also measurable.   Furthermore, we shall show that if φ vanishes almost everywhere, it vanishes everywhere.   For let z be given, let

$x + y = z$ and eliminate y in (3-30). We get

$$\varphi(z) = \varphi(x)\varphi(z - x) - \varphi(2x - z)$$

If $N = \{u:\varphi(u) \neq 0\}$, then, by hypothesis, N is a set of measure zero. Similarly $M = \{2u - z:\varphi(2u - z) \neq 0\}$ is a set of measure zero. Hence we may take x such that $x \notin N$ and $2x - z \notin M$. Then

$$\varphi(x) = \varphi(2x - z) = 0$$

so that $\varphi(z) = 0$.

We now prove that if f is a *measurable* solution of (3-29), then f is *bounded* on every bounded interval. First let us observe that if f is not identically zero, then neither is φ; for otherwise (3-29) gives

$$f(x + y) + f(x - y) = 0$$

for all x, y, which implies that $f(u) + f(v) = 0$ for all u, v. From the previous paragraph, we conclude that there is a positive integer n such that

$$\mu(\{y:|\varphi(2y)| > n^{-1}, |y| \leq n\}) \geq n^{-1} \tag{3-31}$$

in which $\mu(E)$ denotes the Lebesgue measure of the measurable set E; we let

$$E_n = \{y:|\varphi(2y)| > n^{-1}, |y| \leq n\} \tag{3-32}$$

[If (3-31) were false, then for every n, we should have $\mu(E_n) \leq n^{-1}$; but $E_1 \subset E_2 \subset \cdots$, and $\bigcup_{m=1}^{\infty} E_m = \{y:\varphi(2y) \neq 0\}$, a set of positive measure.]

Now suppose that there is a bounded interval of the form $I = (-A,A)$ on which f is not bounded. Then for each $k = 1, 2, \ldots$ there is an x_k in I such that $|f(x_k)| > k$. From (3-29) and (3-32) it follows that if $y \in E_n$, then either $|f(x_k + y)| > \frac{1}{2}kn^{-1}$ or $|f(x_k - y)| > \frac{1}{2}kn^{-1}$. We let

$$A_k = \{y:y \in E_n, |f(x_k + y)| > \frac{1}{2}kn^{-1}\}$$
$$B_k = \{y:y \in E_n, |f(x_k - y)| > \frac{1}{2}kn^{-1}\}$$

We have $E_n = A_k \cup B_k$ for each k, so that

$$n^{-1} < \mu(A_k) + \mu(B_k) \tag{3-33}$$

If F is any set of reals and u is real, then $F + u$ is the set $\{z + u:z \in F\}$ and $u - F$ denotes the set $\{u - z:z \in F\}$. So

$$A_k + x_k = \{y + x_k:y \in A_k\}$$
$$= \{v:|v - x_k| \in E_n, |f(v)| > \frac{1}{2}kn^{-1}\}$$

Since $|v| \leq |v - x_k| + |x_k|$, $v \, \epsilon \, A_k + x_k$ implies that $|v| \leq n + A$, from (3-32) and the fact that $x_k \, \epsilon \, I$. It follows that

$$A_k + x_k \subset \{v : |v| \leq n + A, |f(v)| > \tfrac{1}{2}kn^{-1}\}$$

We denote the last set by D_k. Then $D_1 \supset D_2 \supset \cdots$ and $\overset{\infty}{\underset{1}{\cap}} D_k = \emptyset$, the empty set. The fact that the D_k's are bounded now yields $\mu(D_k) \to 0$ as $k \to \infty$. It follows that $\mu(A_k) = \mu(A_k + x_k) \to 0$ as $k \to \infty$. Similarly, we could start with $x_k - B_k$ and conclude that $\mu(B_k) \to 0$. This contradicts (3-33); therefore f is bounded on every finite interval $I = (-A, A)$.

We can now show that every measurable solution of (3-29) is continuous. First let us observe that from (3-30) it follows that

$$\varphi(x + y) + \varphi(x - y) = \psi(2y)\varphi(x)$$

where $\psi(2y) = \varphi(y)$. Therefore φ itself satisfies (3-29). If f is a measurable solution of (3-29), we have shown that φ is also measurable. Thus φ, being a measurable solution of (3-29), is also bounded on every finite interval. If f is not identically zero, then there is an a such that $\int_0^a \varphi(2y)\, dy \neq 0$, for otherwise $\varphi(2a) = 0$, for almost every a, which we proved was impossible if $f \not\equiv 0$. From (3-29) we obtain

$$\int_0^a f(x + y)\, dy + \int_0^a f(x - y)\, dy = f(x) \int_0^a \varphi(2y)\, dy$$
$$= \int_x^{a+x} f(y)\, dy + \int_{x-a}^x f(y)\, dy \quad (3\text{-}34)$$

and since $\int_0^a \varphi(2y)\, dy \neq 0$, we see that $f(x)$ is continuous. The last equation and the fact that f is continuous show also that

$$f(x + a) - f(x - a) = f'(x) \int_0^a \varphi(2y)\, dy$$

so that f' exists for every x and is a continuous function. We see that all the derivatives of f exist and are continuous.

Now we can consider solutions of (3-29). By subtracting $2f(x)$ from both sides of (3-29) we get, for $y \neq 0$,

$$\frac{f(x + y) - 2f(x) + f(x - y)}{y^2} = \frac{\varphi(2y) - 2}{y^2} f(x) \quad (3\text{-}35)$$

As $y \to 0$, the left member of (3-35) converges to $f''(x)$. Therefore the right member converges, so that we can set

$$\lambda \equiv \lim_{y \to 0} \frac{\varphi(2y) - 2}{y^2}$$

and obtain

$$f''(x) = \lambda f(x), \quad (3\text{-}36)$$

According as $\lambda < 0, \lambda = 0,$ or $\lambda > 0$, we have $f(x) = A \cos kx + B \sin kx$, or $f(x) = Ax + B$, or $f(x) = A \cosh kx + B \sinh kx$, for some A, B, where $k = |\lambda|^{\frac{1}{2}}$. By substituting back in (3-29) we find that $\varphi(2y)$ must be $2 \cos ky$, or 2, or $2 \cosh ky$, respectively; and conversely, if $\varphi(2y) = 2 \cos ky$, for example, then for all $A, B, f(x) = A \cos kx + B \sin kx$ does satisfy (3-29),

Exercise 3-16   Show by replacing x by $x - y$ and y by $2y$ that the equation

$$f(x) + f(x + y) = \varphi(y)f\left(x + \frac{y}{2}\right)$$

is equivalent to (3-29).

3-6   ANALYTIC SOLUTIONS OF A GENERALIZATION OF A TRIGONOMETRIC IDENTITY

Let us seek all *analytic* solutions of the equation

$$f(x + y) = f(x)g(y) + f(y)g(x) \tag{3-37}$$

This equation is a special case of

$$f(x + y) = F_1(x)G_1(y) + F_2(y)G_2(x)$$

If we differentiate (3-37) twice with respect to y and put $y = 0$, we have

$$f(x) = \alpha f(x) + \beta g(x)$$
$$f'(x) = \gamma f(x) + \delta g(x)$$
$$f''(x) = \epsilon f(x) + \zeta g(x)$$

Thus there are constants c_0, c_1, c_2 not all zero such that

$$c_2 f''(x) + c_1 f'(x) + c_0 f(x) = 0$$

from which we have
or

$$f(x) = ae^{\lambda_1 x} + be^{\lambda_2 x}$$
$$f(x) = (cx + d)e^{\lambda x} \tag{3-38}$$

Now if $f(0) \neq 0$, then from (3-37) with $x = 0$ we see that g is a constant multiple of f. Thus we have $f(x + y) = Af(x)f(y)$, which has the solution $f(x) = f(0) \exp (rx)$. Substituting this $f(x)$ and $g(x) = g(0) \exp (rx)$ into (3-37), we find that $g(0) = \frac{1}{2}$, so that $g(x) = \frac{1}{2} \exp (rx)$.

Exercise 3-17   Give the solution if $f(0) = 0$.

Exercise 3-18   It is known[32] that all real continuous solutions, defined for all $x > 0$, of the system of functional equations

$$f(xy) = g(x)^{p(y)}h(y)^{q(x)}$$
$$r(xy) = p(x)q(y)$$

are of the form

$$p(x) = c_1 x^\beta \qquad q(x) = c_2 x^\beta \qquad r(x) = c_1 c_2 x^\beta$$
$$g(x) = (k_1 x^\alpha)^{z^\beta / c_1} \qquad h(x) = (k_2 x^\alpha)^{z^\beta / c_2} \qquad f(x) = (k_1 k_2 x^\alpha)^{z^\beta}$$

where $c_i \neq 0$, $k_i > 0$ $(i = 1, 2)$, and α, β are real constants. Specialize this result to the equation $f(xy) = f(x)f(y)$ and to

$$f(xy) = [f(x)]^{y^\beta}[f(y)]^{x^\beta}$$

Also specialize it to the equation

$$F(xy) = p(y)G(x) + q(x)H(y)$$
$$r(xy) = p(x)q(y)$$

by writing

$$e^{F(xy)} = e^{G(x)p(y)}e^{H(y)q(x)}$$

3-7 A CONTINUOUS STRICTLY INCREASING SOLUTION[35]

We now consider the problem of monotone solutions of the functional equation

$$\varphi[F(x,y)] = \varphi(x) + \varphi(y) \tag{3-39}$$

where F is a polynomial.

Theorem 3-8 If $F(x,y)$ is a polynomial, of degree greater than unity, for which (3-39) has a continuous strictly increasing solution, then $F(x,y)$ must be of the form

$$F(x,y) = \frac{(ax + b)(ay + b) - b}{a} \qquad a \neq 0$$

Proof: Since by hypothesis a continuous strictly increasing solution $\varphi(x)$ of (3-39) exists, $\varphi^{-1}(x)$ must also exist, and hence

$$F(x,y) = \varphi^{-1}[\varphi(x) + \varphi(y)] \tag{3-40}$$

To simplify notation, let us denote $F(x,y)$ by $x \circ y$. Because of (3-40) this operation has the properties:

1. $x \circ y = y \circ x$ (commutativity).
2. $x \circ (y \circ z) = (x \circ y) \circ z$ (associativity).

If $x \circ y$ is a polynomial of degree n in x and of degree m in y, then $n = m$ from 1. The right side of 2 is a polynomial of degree n in z, while the left side is of degree n^2 in z. Thus $n = 1$ and $x \circ y$ is a symmetric polynomial of degree 1 in x, and y and can be written as

$$x \circ y = axy + b(x + y) + c$$

To find conditions on a, b, and c, we substitute in the associative condition and we have:

$$ax(ayz + b(y + z) + c) + b(x + ayz + b(y + z) + c) + c = a(axy$$
$$+ b(x + y) + c)z + b(axy + b(x + y) + c + z) + c \tag{3-41}$$

Equating coefficients of like products of variables, we find everything is an identity except for the coefficients of x and z. In both cases we get $ac + b = b^2$.

If $a = 0$, we find $b = 0$ or 1; if $b = 0$, (3-39) has only the trivial solution $\varphi \equiv 0$. If $b = 1$, we have the slight generalization of Cauchy's equation $\varphi(x) = \alpha(x + c)$.

To get something essentially new, we require $a \neq 0$. Then

$$x \circ y = axy + b(x + y) + c$$

where

$$c = \frac{b^2 - b}{a}$$

and

$$F(x,y) = \frac{a^2xy + ab(x + y) + b^2 - b}{a}$$

3-8　VARIOUS TYPES OF SOLUTION—EVEN, POSITIVE, ENTIRE EXPONENTIAL, BOUNDED, AND PERIODIC

In an interesting paper Green[15] derived various types of solutions of the delay equation

$$f'(x) = f(x + a) \tag{3-42}$$

If f is assumed continuous and $a \neq 0$, then a solution of (3-42) is continuous and has continuous derivatives of all orders; i.e., it is of class C_∞. We now impose restrictions on f in search for specific solutions.

AN EVEN SOLUTION

Let us seek a solution that is an *even* function, i.e., a solution having the property $f(x) = f(-x)$. Differentiating this relation once and using (3-42), we obtain

$$f'(x) = -f'(-x) = -f(-x + a)$$

A second differentiation gives

$$f''(x) = f'(-x + a) = -f'(x - a) = -f(x)$$

The general solution of $f''(x) = -f(x)$ is

$$f(x) = A \sin x + B \cos x$$

for which $a = \pi/2$. Since the requirement is that $f(x)$ be even, we have $f(x) = B \cos x$. An odd-function solution is $f(x) = A \sin x$.

A POSITIVE SOLUTION

The requirement here is that $f(x) \geq 0$ be a solution of (3-42) for all x and for $a > 0$. From this it follows that all derivatives of $f(x)$ are non-negative [since $f^{(n)}(x) = f(x + na) \geq 0$], so that $f(x)$ is an absolutely monotone function.

Now we show that $f(x)$ is an entire function, i.e., that $f(x)$ is analytic for all x. Using Taylor's formula with an integral form of the remainder, and letting $x \geq x_0$ for any fixed x_0 and $h = x - x_0$, we have

$$f(x) = f(x_0) + \sum_{k=1}^{n} \frac{h^k}{k!} f^{(k)}(x_0) + R_n(x)$$

where $\qquad R_n(x) = \int_{x_0}^{x} \frac{(x - y)^n}{n!} f^{(n+1)}(y) \, dy$

Making the substitution $w = (y - x_0)/h$, we obtain the form

$$R_n(h) = \frac{h^{n+1}}{n!} \int_0^1 (1 - w)^n f^{(n+1)}(x_0 + wh) \, dw$$

For fixed w, $f^{(n+1)}$ is monotone in h, since $f^{(n+2)} \geq 0$. Then, putting $g = c - x_0$, $x_0 \leq x \leq c < \infty$, we have

$$0 \leq R_n(h) \leq \frac{h^{n+1}}{n!} \int_0^1 (1 - w)^n f^{(n+1)}(x_0 + wg) \, dw = \left(\frac{h}{g}\right)^{n+1} R_n(g)$$

$$= \left(\frac{h}{g}\right)^{n+1} \left[f(c) - f(x_0) - \sum_{k=1}^{n} \frac{g^k}{k!} f^{(k)}(x_0) \right]$$

$$\leq f(c) \left(\frac{h}{g}\right)^{n+1} \to 0 \qquad \text{as } n \to \infty$$

Because c is arbitrary, we can represent $f(x)$ with Taylor's series for all x. Thus $f(x)$ is an entire function. If we define $f(z) = f(x + iy)$, then we have the representation

$$f(z) = \sum_{k=0}^{\infty} \frac{(z - z_0)^k}{k!} f^{(k)}(z_0)$$

Exercise 3-19 Explain why it is possible to obtain the foregoing expansion for $f(z)$.

Next, we show that $f(x)$ has an exponential lower bound. Using the Maclaurin expansion, we have $f(x) = \Sigma A_n x^n / n!$, where

$$A_n = f^{(n)}(0) = f(na)$$

Since $f(x)$ is a monotone function, $A_{n+1} \geq A_n \geq A_0$. Hence, for $x > 0$, $f(x) \geq \Sigma A_0 x^n / n! = A_0 \exp (x)$, and $A_n = f(na) \geq A_0 \exp (na)$. Now we can write $f(x) \geq \Sigma [A_0 \exp (na)] x^n / n! = A_0 \exp [x \exp (a)]$ and compute a new lower bound for A_n. Repetition gives

$$f(x) \geq A_0 \exp \{x \exp [a \exp (\cdot \cdot \cdot)]\}$$

We let $S_1(a) = \exp (a)$ and $S_n(a) = \exp [a S_{n-1}(a)]$. Since $f(x) \geq A_0$ $\exp [x \exp (S_n)]$, the sequence S_n $(n = 1, 2, \ldots)$ is bounded and non-decreasing; hence it has a limit $S(a)$, which must satisfy

$$S(a) = \exp [a S(a)]$$

Thus, we have $(\log S)/S = a$. The function $(\log S)/S$ attains its maximum at $S = e$, so that $a \leq 1/e$. Thus, in order to obtain a positive solution $f(x)$ of (3-42), we must have $a \leq 1/e$. Then, for $x \geq x_0$, $f(x)$ has the lower bound

$$f(x) \geq f(x_0) \exp [(x - x_0) S(a)]$$

where S is the smaller of the two roots of $(\log S)/S = a$.

The last inequality suggests that one try $f(x) = \exp [x S(a)]$ as a solution, which is considered in the next paragraph.

ENTIRE SOLUTIONS OF EXPONENTIAL TYPE

An exponential-type entire function $f(z)$ is one which satisfies

$$|f(z)| \leq A e^{k|z|}$$

for all values of z and for some constants A and k. If $f(x)$ satisfies (3-42) and also $|f(x)| \leq A e^{k|z|}$ for real x, then clearly $f(z)$ is entire and of exponential type. It is entire since, for any $x > 0$, the remainder in Taylor's formula gives

$$\frac{1}{n!} |x^n f^{(n)}(\theta x)| = \frac{1}{n!} |x^n f(\theta x + na)| \leq A \frac{|x|^n}{n!} e^{k(\theta x + na)} \to 0$$

as $n \to \infty$. It is of exponential type since

$$|f(z)| = \left| \sum f^{(n)}(0) \frac{z^n}{n!} \right| = \left| \sum f(na) \frac{z^n}{n!} \right|$$

$$\leq \sum |f(na)| \frac{|z|^n}{n!} \leq A \sum \frac{[|z| \exp (ka)]^n}{n!}$$

$$= A \exp [|z| \exp (ka)]$$

We now prove that $f(z)$ has the explicit form

$$f(z) = \sum_{k=1}^{n} C_k \exp (s_k z) \tag{3-43}$$

where s_k is a solution of $(\log S)/S = a$. If $a = 1/e$, we have an additional term of the form $Cz \exp(ez)$. The Laplace transform $f^*(s)$, $s = \sigma + i\tau$, of $f(t)$

$$f^*(s) = \mathfrak{L}\{f(t)\} = \int_0^\infty e^{-st}f(t)\, dt$$

converges absolutely for $\sigma > k$. If we split $f^*(s)$ into two integrals, the first on $[0,a]$ and the second on $[a, \infty]$, we obtain

$$f^*(s) = \int_0^a e^{-st}f(t)\, dt + \int_a^\infty e^{-st}f(t)\, dt$$

But,
$$\int_a^\infty e^{-st}f(t)\, dt = \int_a^\infty e^{-st}f'(t-a)\, dt$$
$$= \int_0^\infty e^{-s(u+a)}f'(u)\, du = e^{-as}\mathfrak{L}\{f'(u)\}$$
$$= e^{-as}[sf^*(s) - f(0)]$$

so that
$$f^*(s) = \frac{e^{as}}{s - e^{as}}\left[f(0)e^{-as} - \int_0^a e^{-st}f(t)\, dt\right]$$

This is a meromorphic function (an analytic function whose only singularities except at infinity are poles) whose poles are at the zeros of $s = e^{as}$. Since it is known that the Laplace transform of an exponential-type function is analytic at infinity and vanishes there, there is a circle in the s plane outside which $f^*(s)$ has no poles, and hence the number of its poles is finite. If s_k $(k = 1, \ldots, n)$ are the poles, and if no double root of $s = e^{as}$ is present, then $f^*(s) = \displaystyle\sum_{k=1}^n \frac{C_k}{s - s_k} + h(s)$, where $h(s)$ is an entire function. But $h(s)$ is zero at infinity and is therefore zero everywhere (since a nonzero entire function has an essential singularity at ∞). Inversion, using a circle with center at the origin and radius $> k$, gives

$$f(z) = \sum_{k=1}^n C_k e^{s_k z}$$

If $a = 1/e$, then an extra term of the form $1/(s - e)^2$ is obtained, and its inversion gives rise to a term of the form ze^{ez}.

BOUNDED PERIODIC SOLUTIONS

A bounded solution of (3-42) is of the exponential type and hence has the form (3-43). In order for $f(x)$ to be periodic, we must have $\sigma_k = 0$; that is, we can use only pure imaginary roots $s_k = \sigma_k + i\tau_k = i\tau_k$ of $(\log S)/S = a$. The only pure imaginary root is $S = i$, when $a = \pi/2 \pmod{2\pi}$, so that $f(x)$ has the form $e^{Cx}(A \cos x + B \sin x)$. But $C = 0$, in order for $f(x)$ to be bounded. Consequently, for a bounded periodic solution of (3-42), we have

$$f(x) = A \sin x + B \cos x \qquad a = \frac{\pi}{2} \pmod{2\pi}$$

3-9   CONVEX SOLUTION OF $g(x + 1) - g(x) = \log x$

To refresh the reader's memory we recall that a function is weakly convex if it satisfies

$$f\left(\frac{x + y}{2}\right) \leq \frac{f(x) + f(y)}{2}$$

and strongly convex if it satisfies

$$f[\lambda x + (1 - \lambda)y] \leq \lambda f(x) + (1 - \lambda)f(y) \qquad 0 \leq \lambda \leq 1$$

Note that strong convexity implies weak convexity (let $\lambda = \frac{1}{2}$) but not conversely. Strong convexity also implies continuity. For weak convexity, discontinuity at a point implies discontinuity everywhere and unboundedness at every point; and continuity at a point implies continuity everywhere. Weak convexity with either continuity or boundedness implies strong convexity.

Let $f(z)$ be analytic in the half-plane $R(z) > 0$ and satisfy the functional equation[27]

$$f(z + 1) = zf(z) \tag{3-44}$$

If $g(z)$ is defined in $-1 + \epsilon < R(z) < \epsilon$ by $g(z) = f(z + 1)/z$, then since $f(z + 1)$ is analytic in this band, $g(z)$ has a simple pole at $z = 0$ as the only singularity. Thus $g(z)$ is a meromorphic function which coincides with $f(z)$ in $0 < R(z) < \epsilon$ and, because of (3-44), is an analytic extension of $f(z)$. The process can be repeated for $-2 + 2\epsilon < R(z) < -1 + 2\epsilon$ and so on, reaching an arbitrary point of the plane and introducing simple poles only at $0, -1, -2, \ldots$.

The gamma function, defined by the integral

$$\Gamma(z) = \int_0^\infty e^{-x}x^{z-1}\,dx \qquad R(z) > 0$$

satisfies (3-44). To show this we write

$$\frac{d(e^{-x}x^z)}{dx} = ze^{-x}x^{z-1} - e^{-x}x^z$$

and integrate from 0 to ∞ to obtain

$$0 = z\Gamma(z) - \Gamma(z + 1) \tag{3-45}$$

We can obtain the analytic continuation of $\Gamma(z)$ by writing

$$\Gamma(z) = \int_0^1 e^{-x}x^{z-1}\,dx + \int_1^\infty e^{-x}x^{z-1}\,dx$$

The second expression on the right is everywhere defined and analytic. The exponential factor in the integrand of the first expression can be expanded in series and the result integrated, yielding

$$\frac{1}{z} - \frac{1}{z+1} + \frac{1}{2!}\frac{1}{z+2} - \cdots + \frac{(-1)^n}{n!}\frac{1}{z+n} + \cdots$$

which converges for all values of z that are not zero or negative integers.

The general solution of (3-44) is obtained by writing

$$f(z) = \Gamma(z)G(z) \tag{3-46}$$

Applying (3-44) and (3-45), we see that

$$G(z+1) = G(z)$$

that is, $G(z)$ is a unit periodic function.

Exercise 3-20   Let $f(z)$ satisfy (3-44). By writing

$$f(z) = \exp\left[\int_0^z F(t)\,dt\right]$$

where $F(z)$ is the derivative of $\log f(z)$, show that $F(z)$ satisfies $F(z+1) - F(z) = 1/z$. *Hint:* Substitute in (3-44), rearrange, and differentiate.

We now show that $\log[C\Gamma(x)]$ is the only convex solution of the equation

$$g(x+1) - g(x) = \log x \qquad x > 0 \tag{3-47}$$

To prove this fact (see Courant, "Calculus," vol. 2, p. 324), we note first that if $g(x)$ is strongly convex and $0 \le \delta \le h$, then

$$g(x+\delta) + g(x-\delta) \le g(x+h) + g(x-h)$$

This is obtained by adding

$$g(x-\delta) \le \tfrac{1}{2}\left(1 + \frac{\delta}{h}\right)g(x-h) + \tfrac{1}{2}\left(1 - \frac{\delta}{h}\right)g(x+h)$$

$$g(x+\delta) \le \tfrac{1}{2}\left(1 - \frac{\delta}{h}\right)g(x-h) + \tfrac{1}{2}\left(1 + \frac{\delta}{h}\right)g(x+h)$$

each of which follows from strong convexity.

If $\alpha(x)$ and $\beta(x)$ are two continuous strongly convex solutions, then it is clear that

$$\varphi(x) = \alpha(x) - \beta(x)$$

is a continuous periodic function with unit period.

Now

$$\alpha(x + 1) - \alpha(x) = \log x$$
$$\alpha(x) - \alpha(x - 1) = \log (x - 1)$$

and hence

$$\alpha(x + 1) + \alpha(x - 1) - 2\alpha(x) = \log \frac{x}{x - 1}$$

Thus from the first remark if $\delta \leq 1$, we have

$$\alpha(x + \delta) + \alpha(x - \delta) - 2\alpha(x) \leq \log \frac{x}{x - 1}$$

The function $\beta(x)$ satisfies a similar relation and we have

$$|\varphi(x + h) + \varphi(x - h) - 2\varphi(x)| \leq 2 \log \frac{x}{x - 1}$$

As $x \rightarrow \infty$, $\log [x/(x - 1)] \rightarrow 0$ and hence $\varphi(x + h) + \varphi(x - h) - 2\varphi(x) \rightarrow 0$. Since $\varphi(x)$ is periodic, we have

$$\varphi(x + h) + \varphi(x - h) - 2\varphi(x) = 0$$

The only periodic continuous solution of this equation is constant. To see this, one proceeds along lines similar to those used for Cauchy's equation. Thus if $\varphi(1) = \varphi(2) = C$, then $\varphi(3/2) = 1/2[\varphi(1) + \varphi(2)] = C$ and similarly for every dyadic between 1 and 2. Because the dyadics are dense, we have by continuity that $\varphi(x) \equiv C$.

Now we only need to show that $\log \Gamma(x)$ is weakly convex. We have

$$\Gamma(x) = \int_0^\infty e^{-t/2} t^{(x-1+h)/2} e^{-t/2} t^{(x-1-h)/2} dt$$

and applying the Cauchy-Schwartz inequality, we have

$$[\Gamma(x)]^2 \leq \Gamma(x + h)\Gamma(x - h)$$

and hence

$$\log \Gamma(x) \leq 1/2[\log\Gamma(x + h) + \log\Gamma(x - h)] \tag{3-48}$$

Using the foregoing remarks the reader can show that it is also **strongly** convex. The proof is now complete.

Exercise 3-21 Show that the only periodic function which is logarithmically convex is a constant.

Remark: Thielman[34] has proved that the only function which is convex for all $x > K \geq 0$ and satisfies

$$\frac{1}{f(x + a)} = x^p f(x) \qquad x > 0, a > 0, p > 0$$

is

$$f(x) = \left[\frac{\Gamma(x/2a)}{(2a)^{1/2}\Gamma[(x + a)/2a]} \right]^p$$

3-10 THREE EXAMPLES—SERIES EXPANSION, REDUCTION TO SIMPLER FORMS, AND SUCCESSIVE APPROXIMATIONS

We give in outline form a series approach to the solution of a functional equation.   We consider the equation[8]

$$h(x + y)h(x - y) = h^2(x) + h^2(y) - \alpha^2 \qquad \alpha \neq 0$$

where $h(x)$ is analytic, and put $x = y = 0$.   We obtain $h(0) = \pm \alpha$.   We then let $\pm h(x) = \alpha f(x)$, obtaining

$$f(x + y)f(x - y) = f^2(x) + f^2(y) - 1 \qquad (3\text{-}49)$$

together with $f(0) = 1$.   Now if we put $x = y$ in (3-49), we obtain

$$f(2x) = 2f^2(x) - 1 \qquad (3\text{-}50a)$$

Using (3-50a) and a similar equation for y in (3-49), we have the form

$$f(x + y)f(x - y) = \frac{f(2x) + f(2y)}{2} \qquad (3\text{-}50b)$$

Setting $x + y = u$, $x - y = v$, $2x = u + v$, and $2y = u - v$ in (3-50b), we obtain

$$f(u + v) + f(u - v) = 2f(u)f(v)$$

since the constant of integration is zero because $f(0) = 1$.

We can see that $f(x)$ is even by letting $y = -x$ in (3-50b) and using $f(0) = 1$.   Since $f(x)$ is even and regular in the neighborhood of the origin, we can write

$$f(x) = 1 + \frac{ax^2}{2!} + \frac{bx^4}{4!} + \frac{cx^6}{6!} + \frac{dx^8}{8!} + \cdots$$

and substitute in (3-50a) to determine the coefficients.   We obtain $b = a^2$, $c = a^3$, $d = a^4$, ...   With $k^2 = a$ we have

$$f(x) = \tfrac{1}{2}(e^{kx} + e^{-kx}) = \cosh kx$$

from which we obtain

$$h(x) = \pm \frac{\alpha}{2} (e^{kx} + e^{-kx}) \qquad \alpha \neq 0$$

Exercise 3-22   Obtain the solution if $\alpha = 0$.

Reference 10 gives the most general solution of

$$2^{n/2}f_n(x \sqrt{2}) = \sum_{r=0}^{n} \binom{n}{r} f_r(x) f_{n-r}(x) \qquad (3\text{-}50c)$$

analytic at the origin.   The function

$$g(x,y) = \sum_{n=0}^{\infty} \frac{f_n(x)y^n}{n!} \qquad |y| < 1 \qquad (3\text{-}50d)$$

is continuous in some region about the origin.   Now

$$g(\sqrt{2}\,x,\,\sqrt{2}\,y) = \sum_{n=0}^{\infty} \frac{2^{n/2}f_n(x\,\sqrt{2})y^n}{n!}$$

and substitution in (3-50c) gives

$$g(\sqrt{2}\,x,\,\sqrt{2}\,y) = g^2(x,y)$$

which has the general continuous solution

$$g(x,y) = \exp\,(ax^2 + 2bxy - c^2y^2)$$

with a, b, and c as arbitrary constants.   If substituted in (3-50d), this gives

$$\exp\,(2bxy - c^2y^2) = \sum_{n=0}^{\infty} \frac{e^{-ax^2}f_n(x)y^n}{n!}$$

Expanding the left side and equating the coefficients of y^n yields

$$f_n(x) = c^n e^{ax^2} H_n\left(\frac{bx}{c}\right)$$

which by analytic continuation is the solution for all x.   Here

$$H_n(x) = n! \sum_{i=0}^{[n/2]} \frac{(-1)^i(2x)^{n-2i}}{i!(n-2i)!}$$

are the nth-degree Hermite polynomials defined by

$$\exp\,(2xy - y^2) = \sum_{n=0}^{\infty} \frac{1}{n!}\,y^n H_n(x)$$

THE GENERAL SOLUTION OF $f^{(k)}(x) = f(1/x)$

We define a functional equation to be linear if a linear combination of any two solutions is also a solution.   There is no general theory as to the number of solutions of such an equation.   However, the foregoing equation can be reduced to a linear differential equation of order $2k$ (see Exercise 3-25), with k initial conditions on f and its derivatives up to the $(k-1)$st.   Thus only k arbitrary constants will remain in the general solution.

We begin by assuming a solution of the form[30,31]

$$f(x) = x^m + \lambda x^n$$

where m, n, and λ are arbitrary constants. Substitution yields

$$m(m-1) \cdots (m-k+1)x^{m-k} + \lambda n(n-1) \cdots (n-k+1)x^{n-k}$$
$$- x^{-m} - \lambda x^{-n} = 0$$

Equating powers of x and corresponding coefficients, we see that

$$m + n = k \qquad \lambda = m(m-1) \cdots (m-k+1)$$
$$\lambda n(n-1) \cdots (n-k+1) = 1 \qquad (3\text{-}51)$$

Eliminating λ gives

$$m(m-1)^2 \cdots (m-k+1)^2(m-k) = (-1)^k \qquad (3\text{-}52)$$

Letting $p = m^2 - mk$ in (3-52) and combining m and $(m-k)$, $(m-1)^2$, and $(m-k+1)^2$, etc., we obtain

$$[p + (k-1)]^2[p + 2(k-2)]^2 \cdots [p + \tfrac{1}{2}(k-1)\tfrac{1}{2}(k+1)]^2 p$$
$$= -1 \qquad k \text{ odd}$$
$$[p + (k-1)]^2[p + 2(k-2)]^2 \cdots \left[p + \left(\frac{k}{2}\right)^2\right]^2 p$$
$$= 1 \qquad k \text{ even} \qquad (3\text{-}53)$$

so that m depends on the solutions of a kth-degree equation (the indicial equation) in p. Let $p_1, \ldots , p_k$ be distinct solutions of (3-53) (the case in which some of the roots are the same is left as an exercise). We obtain two solutions of (3-51) for each root p of Eq. (3-53). Associated with each of these two values of m we have a value of λ, and the product of these two values of λ is unity, implying that the two solutions are identical. Thus the general solution is given by

$$f(x) = \sum_{i=1}^{k} a_i(x^{m_i} + \lambda_i x^{k-m_i})$$

where m_i is one of the two roots described above; the corresponding λ_i is obtained from (3-51), and a_i, $i = 1, \ldots , k$, are arbitrary constants.

Exercise 3-23   Let $k = 1$ and obtain the solution. Put $x = e^t$ and obtain

$$f(e^t) = ae^{t/2} \left(\sqrt{3} \cos \frac{\sqrt{3}\,t}{2} + \sin \frac{\sqrt{3}\,t}{2} \right)$$

Replace t by $\log x$ and evaluate a at $x = 1$. Using the identity

$$\sqrt{3} \cos \frac{\sqrt{3}\,y}{2} + \sin \frac{\sqrt{3}\,y}{2} = 2 \cos \left(\frac{\sqrt{3}\,y}{2} - \frac{\pi}{6} \right)$$

obtain

$$f(x) = \frac{2}{\sqrt{3}} f(1) \sqrt{x} \cos \left(\frac{\sqrt{3}}{2} \log x - \frac{\pi}{6} \right)$$

Exercise 3-24   Let $k = 2$, and write down the solution.   Show that if m_{11} and m_{21} correspond to p_1 and m_{12} and m_{22} correspond to p_2, then

$$m_{11} = \tfrac{1}{2}(2 + \sqrt{2 + 2\sqrt{5}}) \qquad m_{12} = \tfrac{1}{2}(2 + \sqrt{2 - 2\sqrt{5}})$$
$$m_{21} = \tfrac{1}{2}(2 - \sqrt{2 + 2\sqrt{5}}) \qquad m_{22} = \tfrac{1}{2}(2 - \sqrt{2 - 2\sqrt{5}})$$

and correspondingly

$$\lambda_{11} = \tfrac{1}{2}(1 + \sqrt{5} + \sqrt{2 + 2\sqrt{5}}) \qquad \lambda_{12} = \tfrac{1}{2}(1 - \sqrt{5} + \sqrt{2 - 2\sqrt{5}})$$

Exercise 3-25   In $f'(x) = f(1/x)$ assume that $f'(x)$ is differentiable.   Obtain the second-order linear differential equation $f''(x) + f(x)/x^2 = 0$.   Solve this equation. What is the relation between its solution and that of the parent functional equation?

Exercise 3-26   Attempt the solution of $f'(x) = af(g(x))$ for some form of the given function $g(x)$ essentially different from x and $1/x$.   Thus, for example, solve $f'(x) = f(1 - x)$ and $f'(x) = f(\omega x)$, where

$$\omega = -\frac{1}{2} + \frac{\sqrt{3}}{2} i \qquad i = \sqrt{-1}$$

Remark: Consider the equation

$$F(\alpha x) - \beta F(x) = G(x)$$

where $\alpha > 0$ and β are arbitrary real constants.   Let

$$x = \alpha^u \qquad F(\alpha^u) = f(u) \qquad G(\alpha^u) = g(u)$$

and our equation becomes

$$f(u + 1) - \beta f(u) = g(u)$$

This is a difference equation which, for $\beta > 0$, yields the general solution for our equation

$$F(x) = \beta^{\alpha \log x} S\left[\frac{G(\alpha^u)}{\beta^{u+1}}\right]$$

for our original equation.   Here $u = \alpha \log x$, and we define

$$S[\varphi(u)] \equiv h(u)$$

where $h(u)$ satisfies the equation

$$h(u + 1) - h(u) = \varphi(u)$$

Remark: Consider the following equation, whose solutions have been studied:

$$f[z, \varphi(g(z))] = \varphi(z)$$

Let ζ be a double point of $g(z)$ [that is, satisfy $g(\zeta) = \zeta$] such that g is regular in a neighborhood of $z = \zeta$ and $|g'(\zeta)| < 1$.   Suppose also that

$f(z,w)$ is regular in a neighborhood of $(z,w) = (\zeta,\beta)$, where β is such that $f(\zeta,\beta) = \beta$ and $|\partial f(\zeta,\beta)/\partial w| < |g'(\zeta)|^{-1}$. Then the above equation has a unique solution $\varphi(z)$ that is regular in a neighborhood of $z = \zeta$ with $\varphi(\zeta) = \beta$. The solution is the limit of a sequence of functions $\varphi_n(z)$ defined by

$$\varphi_{n+1}(z) = f[z,\varphi_n(g(z))]$$

where $\varphi_1(z)$ is an arbitrary function that is regular at $z = \zeta$ with $\varphi_1(\zeta) = \beta$. The proof utilizes the existence of a solution of the simpler equation $f[z,\varphi(z)] = \varphi(sz)$.

3-11 A GENERAL METHOD OF SOLUTION

The foregoing pages have illustrated many of the techniques useful for solving functional equations. By now the reader should be aware that there is no general rule indicating which method should be used in solving a particular equation. Resourcefulness, ingenuity, and occasionally a new technique are required for a successful attack on individual problems. That this situation is changing is illustrated by the following method devised by Vincze.[38] The presentation here essentially follows that reported by Hille.[19]

Let $F_1(z)$, $F_2(z)$, . . . , $F_n(z)$ be functions defined on an arbitrary set S of the complex plane, and let

$$\Delta[F_1(z_1),F_2(z_2), \ldots ,F_n(z_n)] = \Delta[F_1,F_2, \ldots ,F_n] \tag{3-54}$$

be the value of the determinant

$$\begin{vmatrix} F_1(z_1)F_2(z_1) & \cdots & F_n(z_1) \\ F_1(z_2)F_2(z_2) & \cdots & F_n(z_2) \\ \cdot \cdot \cdot \cdot \cdot \cdot \cdot \cdot \cdot \cdot \cdot \cdot \\ F_1(z_n)F_2(z_n) & \cdots & F_n(z_n) \end{vmatrix}$$

Suppose that the following identity holds for the $n(k + 1)$ functions, $F_1, F_2, \ldots F_{kn+n}$, and all points $z_1, z_2, \ldots , z_n$ in S:

$$\sum_{\nu=0}^{k} \Delta[F_{\nu n+1}(z_1),F_{\nu n+2}(z_2), \ldots ,F_{\nu n+n}(z_n)] = 0 \tag{3-55}$$

where $F_0(z)$ is an arbitrary function in S. Then by expanding each determinant in terms of minors of the last column it is easy to show that

$$\sum_{\nu=0}^{k} \Delta[F_{\nu n+1}(z_1),F_{\nu n+2}(z_2), \ldots ,F_{\nu n+n}(z_n),F_0(z_{n+1})] = 0 \tag{3-56}$$

holds for any set of $n + 1$ points $z_1, z_2, \ldots , z_{n+1}$ in S.

The last equation is called by Vincze the *extension by* $F(z)$ of the first.

The use of the method is illustrated by the following example.   Find functions F, G, H, K, and L satisfying

$$F(z_1 \circ z_2) = G(z_1) + H(z_2) + K(z_1)L(z_2) \tag{3-57}$$

where the operation $\circ$ is an associative binary composition on S such that $z_1 \circ z_2 = z_2 \circ z_1$, that is, S is an abelian semigroup with respect to $\circ$.   From the commutativity of the operation $\circ$ we obtain

$$G(z_1) + H(z_2) + K(z_1)L(z_2) = G(z_2) + H(z_1) + K(z_2)L(z_1)$$

and in the notation of Eq. (3-54)

$$\Delta(G,1) + \Delta(1,H) + \Delta(K,L) = 0 \tag{3-58}$$

By choosing as the arbitrary function $F(z) = 1$, (3-58) becomes

$$\Delta(G,1,1) + \Delta(1,H,1) + \Delta(K,L,1) = 0 \tag{3-59}$$

Since the first two determinants have two columns equal to unity, they are each identically zero, giving

$$\Delta(K,L,1) = 0 \tag{3-60}$$

Now, if the points z, z_1, and z_2 are used in the determinant expression and it is then expanded by minors of the first row, it follows that

$$AK(z) + BL(z) + C = 0 \tag{3-61}$$

Thus, the functions K, L, and 1 are linearly dependent.   In fact it is easy to see that, in general, if

$$\Delta[F_1,F_2, \ldots ,F_n] = 0 \tag{3-62}$$

then a relationship similar to (3-61) holds for F_1, F_2, $\ldots$, F_n.   Conversely, if F_1, F_2, $\ldots$, F_n are linearly dependent, then (3-62) follows.

We continue the illustrative problem for the case when the constant A in (3-61) is zero.   For this condition, the function $L(z)$ is a constant. Using this fact in (3-58), we obtain

$$\Delta(G,1) + \Delta(1,H) + \Delta(K,\alpha) = 0 \tag{3-63}$$

which, upon rearranging, becomes

$$\Delta(G,1) - \Delta(H,1) + \Delta(\alpha K,1) = \Delta(G - H + \alpha K, 1) = 0$$
Thus $\qquad G(z) - H(z) + \alpha K(z) = \beta \tag{3-64}$

Upon substituting $H(z)$ from (3-64), (3-57) becomes

$$F(z_1 \circ z_2) = \left[G(z_1) + \alpha K(z_1) - \frac{\beta}{2} \right] + \left[G(z_2) + \alpha K(z_2) - \frac{\beta}{2} \right] \tag{3-65}$$

which is a Pexider equation. An example of this case is the following. If $f(z)$ is a solution of the generalized Cauchy equation

$$f(z_1 \circ z_2) = f(z_1) + f(z_2)$$

then setting

$$F(z) = f(z) + \gamma$$

we obtain

$$G(z) = f(z) - \alpha K(z) + \frac{\beta}{2} + \frac{\gamma}{2}$$

$$H(z) = f(z) - \frac{\beta}{2} + \frac{\gamma}{2}$$

$$K(z) \quad \text{arbitrary}$$

$$L(z) = \alpha$$

The case when L is not a constant is more complicated but involves no new principles.

3-12 FUNCTIONAL INEQUALITIES

The subject of functional inequalities is of great interest in analysis. We illustrate the ideas with a few examples.

Brauer[6] has studied the functional inequality

$$f^n(x) \geq \varphi(x) \qquad -\infty < x < \infty \tag{3-66}$$

where f^n is the nth iterate, or application, of the continuous function f, n is a given positive integer, and $\varphi(x)$ is a continuous strictly increasing function with $\lim_{x \to \infty} \varphi(x) = \infty$ and $\lim_{x \to -\infty} \varphi(x) = -\infty$. We shall illustrate the approach through the following theorems, characterizing solutions of (3-66), proving some and just mentioning others.

Theorem 3-9 If $f(x)$ is a solution of (3-66), then $\lim_{x \to \infty} f(x) = \infty$ or $-\infty$.

Proof: Otherwise there exists a number $a \neq \pm\infty$ and a sequence $\{x_j\}$ such that $\lim_{j \to \infty} x_j = \infty$ and $\lim_{j \to \infty} f(x_j) = a$. For each n,

$$\lim_{j \to \infty} f^n(x_j) = a \lim_{j \to \infty} f^{n-1}(x_j)$$

etc., a contradiction of the fact that f satisfies (3-66).

Theorem 3-10 If $f(x)$ is a solution of (3-66) and $\lim_{x \to \infty} f(x) = -\infty$, then $\lim_{x \to -\infty} f(x) = \infty$.

Theorem 3-11 If $f(x)$ is a solution of (3-66) for an odd value of n, then $\lim_{x \to \infty} f(x) = \infty$.

Proof: If not, then $\lim_{x \to \infty} f(x) = -\infty$ and $\lim_{x \to -\infty} f(x) = \infty$, and there is a fixed point c of f. The interval $[c, \infty)$ is mapped onto an interval containing $(-\infty, c]$, and the interval $(-\infty, c]$ is mapped onto an interval containing $[c, \infty)$ since all odd iterates of f map $[c, \infty)$ onto an interval containing $(-\infty, c]$. But $\varphi(x) \to \infty$, and hence f cannot be a solution for odd n.

Theorem 3-12   If $f(x)$ is a solution of (3-66) such that $\lim_{x \to \infty} f(x) = \infty$, then $f(x) \geq \varphi(x)$ whenever $f(x) < x$. In particular, this is true if f satisfies (3-66) for n odd.

Proof: One can show by the mean-value theorem and by induction that if $\limsup_{x \to \infty} f(x) = \infty$ and x_0 is a point such that $f(x_0) < x_0$, then for each integer k there is a point $x_k \geq x_0$ such that $f^k(x_k) = x_0$. If $f(x_0) < x_0$, there is an x_{n-1} such that $x_{n-1} > x_0$ and $f^{n-1}(x_{n-1}) = x_0$. Thus, if $f(x) < \varphi(x_0)$, $f^n(x_{n-1}) = f(x_0) < \varphi(x_0) < \varphi(x_{n-1})$, and f is not a solution of (3-66).

Corollary: If $f(x)$ is a solution of $f^n(x) \geq x$ for n odd, then $f(x) \geq x$. One can also show that if $f(x)$ is a solution of (3-66) for n even, then $f^2(x) \geq \varphi(x)$ whenever $f^2(x) < x$. In this case $\varphi(x) = x$; then either $f(x) \geq x$ or $f^2(x) = x$.

Hille[19] has classified functional inequalities of the form

$$f(t) \leq T[f(t)] \qquad 0 \leq t \leq a \tag{3-67}$$

where T maps a subset G of the class F of real functions on $(0,a)$ onto itself, as follows: (1) *trivial* if the subset G on which (3-67) holds is equal to F, (2) *restrictive* if G is proper and nonempty, (3) *determinative* if G is a single element, and (4) *absurd* if G is empty. Consider the determinative inequality

$$0 \leq y(t) \leq \alpha \int_0^t y(s)\, ds \qquad 0 \leq t \leq a, \alpha > 0$$

where y is bounded and measurable on $(0,a)$. If we denote the integral by $Y(t)$, then $Y(0) = 0$ and almost everywhere

$$Y'(t) = y(t) \leq \alpha Y(t)$$

so that we can write

$$\frac{d}{dt}[Y(t)e^{-\alpha t}] \leq 0$$

From $Y(0) = 0$ we have

$$Y(t)e^{-\alpha t} \leq 0$$

and hence $Y(t) = 0$ for all t, and $y(t) = 0$.

We now prove and apply an interesting theorem regarding another functional inequality.[17]

Theorem 3-13   If $f(z)$ is an entire function of $z = x + iy$ which satisfies the functional inequality

$$\left| f\left(\frac{x+y}{2}\right) \right| \leq \frac{|f(x)| + |f(y)|}{2} \tag{3-68}$$

then every solution of (3-68) has one of three possible forms: $f(z) =$ constant, or $f(z) = (\alpha z + \beta)^n$, or $f(z) = \exp(\alpha z + \beta)$, where α and β are arbitrary complex numbers and n is an arbitrary natural number.

Proof: Suppose $f(z) \not\equiv 0$, and let $z = s + it$, $\varphi(s,t) = |f(z)|$. Let τ be a real parameter. Let us examine the function

$$F(\tau) = \varphi(a + h\tau, b + k\tau) + \varphi(a - h\tau, b - k\tau)$$

where a, b, h, k are arbitrary real constants and $f(a + ib) \neq 0$. By (3-68) $F(\tau)$ has a minimum $2\varphi(a,b)$ at $\tau = 0$. Thus $F''(0) \geq 0$. But

$$F''(0) = 2[h^2\varphi_{ss}(a,b) + 2hk\varphi_{st}(a,b) + k^2\varphi_{tt}(a,b)] \geq 0$$

and because h and k are arbitrary, we have

$$\varphi_{st}{}^2(a,b) - \varphi_{ss}(a,b)\varphi_{tt}(a,b) \leq 0 \tag{3-69}$$

Now since $f(a + ib) \neq 0$, with a, b arbitrary, there exists a regular branch $g(z)$ of $\sqrt{f(z)}$ in a neighborhood U of $z = \gamma = a + ib$. Thus $\varphi = |g^2|$. By the Cauchy-Riemann equations we have

$$\varphi_{st}{}^2(a,b) - \varphi_{ss}(a,b)\varphi_{tt}(a,b) = 4[|g(\gamma)g''(\gamma)|^2 - |g'(\gamma)|^4]$$

Exercise 3-27 Verify this result.

From (3-69) this becomes

$$|g(\gamma)g''(\gamma)| \leq |g'(\gamma)|^2 \tag{3-70}$$

Because $f(z) = g^2(z)$ in U, we have

$$|f(\gamma)f''(\gamma)| \leq |f'(\gamma)|^2$$

and this holds at every point z; that is,

$$|f(z)f''(z)| \leq |f'(z)|^2$$

Let
$$G(z) = \frac{f(z)f''(z)}{[f'(z)]^2}$$

Now, we show that $G(z)$ is a constant. We must consider two cases. If $f'(z) \neq 0$, then $G(z)$ is analytic (in fact entire). Since $|G(z)| \leq 1$, by Liouville's theorem (a function which is analytic for all finite values of z and is bounded is constant), $G(z)$ is constant. If $f'(z)$ vanishes for some z, then $G(z)$ may be meromorphic, and because $|G(z)| \leq 1$, it maps the complex plane into the interior of the unit disk. This violates Picard's theorem [if $G(z)$ is meromorphic, $G(z) = a$ has solutions for all a with the possible exception of two points; this is true for all nonconstant $G(z)$] unless $G(z)$ is constant. Thus, in either case, $G(z)$ is some constant

A with $|A| \leq 1$. Consequently,

$$f(z)f''(z) = A[f'(z)]^2$$

and solving this differential equation proves the assertion of the theorem.

One can show that if $f(z)$ is meromorphic and satisfies (3-68) for $|z| < \infty$, then it is entire.

Exercise 3-28 Using this theorem, show that all nontrivial solutions of $f[(x + y)/2] = \frac{1}{2}[f(x) + f(y)]$ have the form $f(z) = \alpha z + \beta$ and of $f(x + y) = f(x) + f(y)$ have the form $f(z) = \alpha z$.

Consider the functional equation

$$f(x + y) = f(x)f(y)$$

We have

$$\left| f\left(\frac{x + y}{2}\right) \right| = \left| f\left(\frac{x}{2}\right) \right| \left| f\left(\frac{y}{2}\right) \right| \leq \frac{|f^2(x/2)| + |f^2(y/2)|}{2}$$
$$= \frac{|f(x)| + |f(y)|}{2}$$

Hence, $f(z)$ satisfies the condition of Theorem 3-13, and $f(z) = 0$, or $f(z) = \exp(\alpha z)$, where α is an arbitrary constant.

Exercise 3-29 Consider $f(xy) = f(x)f(y)$. Let $g(z) = f(e^z)$. Using (3-70), we have $g(x + y) = g(x)g(y)$. Show that $f(z) = 0$, 1, or z^n, where n is arbitrary.

Exercise 3-30 Consider

$$|f(x + y)| + |f(x - y)| = 2|f(x)| + 2|g(y)|$$

Using

$$|f(x + y)| + |f(x - y)| \geq 2|f(x)|$$

derive (3-68) and show that

$$f(z) = (\alpha z + \beta)^2 \qquad g(z) = \alpha^2 z^2$$

Then show that for

$$|f(x + y)| + |f(x - y)| = 2|f(x)| + 2|f(y)|$$

we have

$$f(z) = \alpha z^2$$

We now give an example, developing some properties of a function from a functional inequality which it satisfies.[26] Let f be a real-valued measurable function defined for all real x, and let $G(x,y) \geq 0$ be continuous and real. We show that if

$$|f(x + y)| \leq G[f(x),f(y)]$$

for all x and y, then f is bounded on bounded sets. Note that since f is measurable, $h(x) \equiv |f(x)| + |f(-x)|$ is measurable. Thus, there is a set of positive measure on which h is bounded. The hypotheses imply that $f(x - y)$ is bounded for all $(x,y) \epsilon S$. Since S has positive measure, it is known that the set $\{z:z = x - y \text{ and } (x,y) \epsilon S\}$ is a neighborhood of the origin. Thus $f(z)$ is bounded for all z in some neighborhood U of the origin. By induction, for an arbitrary positive integer n, $f(nz)$ is bounded for all $z \epsilon U$. Thus f is bounded on any bounded set.

SUBADDITIVITY AND SUPERADDITIVITY

A real-valued function is said to be superadditive on an interval $I = (0,a)$ if it satisfies the inequality

$$f(x + y) \geq f(x) + f(y)$$

whenever x, y, and $x + y$ are in I. An example is $f(x) = x^2$ on $I(0, \infty)$. It is subadditive if $\leq$ holds. An example is $f(x) = -x^3$ on the same interval.[28]

Theorem 3-14 If $f'(x)$ exists on $I = (a, \infty)$, $a \geq 0$, then $f(x)/x$ is decreasing [and hence $f(x)$ is subadditive] on I if and only if $f'(x) < f(x)/x$ on I.

Proof: We have

$$\frac{xf'(x) - f(x)}{x} < 0$$

and hence

$$\frac{d}{dx}\frac{f(x)}{x} = \frac{xf'(x) - f(x)}{x^2} < 0$$

For the converse, the steps of this proof are reversed

$$f(x + y) = x\frac{f(x + y)}{x + y} + y\frac{f(x + y)}{x + y} \leq x\frac{f(x)}{x} + y\frac{f(y)}{y} = f(x) + f(y)$$

Exercise 3-31 Prove superadditivity by assuming that $f'(x) > f(x)/x$ on $I(a, \infty)$, $a \geq 0$.

Exercise 3-32 Show that if $f(x)$ is subadditive and $f'(x)$ exists on (a, ∞) and $f(x) + f(-x) \leq 0$ for all $x \epsilon I \equiv (a, \infty)$, $a \geq 0$, then $f'(x)$ is nonincreasing on I. *Hint:* Given $x_1, x_2 \epsilon I$, $h > 0$, there exists an $\epsilon > 0$ such that

$$f'(x_1 + x_2) - \epsilon < \frac{f(x_1 + x_2 + h) - f(x_1 + x_2)}{h}$$

with $\lim_{h \to 0} \epsilon = 0$. To complete the argument add x_1 and subtract it from the argument of the first term on the right, apply subadditivity appropriately, and from the fact that $f(-x_1) < -f(x_1)$ obtain on the right

$$\frac{1}{h}[f(x_1 + h) - f(x_1)]$$

and let $h \to 0$.

3-13 OPTIMIZATION AND FUNCTIONAL EQUATIONS[3]

In recent years functional equations have come to play a significant role in the theory of optimization. A field that makes extensive use of functional equations as a part of an optimization scheme subject to constraints is dynamic programming, developed mostly by Bellman. We give an example.

Suppose that it is desired to divide an amount of effort x into two parts, y and $x - y$, with returns from y given by $g(y)$ and from $x - y$ given by $h(x - y)$. After a part of the original amount x has been expended, there remains an amount of effort equal to $ay + b(x - y)$, where $0 < a$ and $b < 1$. The latter amount is again divided into two parts with the same return functions. In this manner the process is continued in several stages. How should this allocation of effort be made at each stage in order for the total return to be maximized for a finite number of stages? Formally the problem is described by taking

$$R_1(x,y) = g(y) + h(x - y)$$

as the total return from the first stage,

$$R_2(x,y_1,y_2) = g(y_1) + h(x_1 - y_1) + g(y_2) + h(x_2 - y_2)$$

where $x_1 = x$, $x_2 = ay_1 + b(x_1 - y_1)$ and $0 \le y_1 \le x_1$, as the total return from two stages, and

$$R_N(x,y_1 \ldots ,y_N) = \sum_{i=1}^{N} [g(y_i) + h(x_i - y_i)]$$

as the total return from successive allocations y_1, y_2, $\ldots$, y_N, where $x_1 = x$, $x_{i+1} = ay_i + b(x_i - y_i)$ $(i = 1, \ldots , N)$,   and $0 \le y_i \le x_i$ $(i = 1, \ldots , N)$. Any choice of y_i $(i = 1, \ldots , N)$ is called a *policy*, and a choice which maximizes R_N is called an *optimum policy*. For an optimum policy, R_N depends only on the original effort x and on the number of stages N.

In order to derive the functional equation of this problem, let $f_N(x)$ be the total return from any N stages, using an optimum policy and starting with an initial amount x. Note that for the first stage, x is divided into y and $x - y$, and then $g(y) + h(x - y)$ is computed. The remaining effort $ay + b(x - y)$ is used optimally for the remaining $N - 1$ stages, no matter what the return $f_{N-1}[ay + b(x - y)]$ of y is. Therefore

$$R_N(x,y) = g(x) + h(x - y) + f_{N-1}[ay + b(x - y)]$$

Thus, by definition, one has

$$f_N(x) = \max_{0 \le y \le x} R_N(x,y) = \max_{0 \le y \le x} \{g(y) + h(x - y) + f_{N-1}[ay + b(x - y)]\}$$

which is the basic functional equation. For N large, f_N and f_{N-1} are approximately the same, and can be denoted by a function f which is independent of N. This gives the simpler equation

$$f(x) = \max_{0 \le y \le x} \{g(y) + h(x - y) + f[ay + b(x - y)]\} \qquad (3\text{-}71)$$

Existence and uniqueness theorems are available for this equation.

With the requirement that $f(0) = 0$, (3-71) can be solved by successive approximations as follows. Let

$$y = \begin{cases} 0 & \text{if } \dfrac{g(x)}{1 - a} \ge \dfrac{h(x)}{(1 - b)x} \\[2ex] x & \text{if } \dfrac{g(x)}{1 - a} < \dfrac{h(x)}{(1 - b)x} \end{cases}$$

Otherwise, choose

$$\frac{g(y)}{(1 - a)y} = \frac{h(x - y)}{(1 - b)(x - y)}$$

Let $f_0(x) = f_g(x)$ be the return calculated by recurrence, using one or the other of these policies. Successive approximations can now be computed using the functional equation in N stages. This process leads to improved approximations at each step, since $f_i \le f_j$ for $i \le j$. By making assumptions on the properties of g and h, it is possible to prescribe methods for computing

$$y_N = y_N(x) \qquad \text{and} \qquad f_N(x)$$

For example, if both g and h are strictly convex in x, an optimum policy requires that $y = 0$ or that $y = x$. Several applications to bottleneck and other types of problems are available.

Exercise 3-33 The equation

$$f(x_3, x_1 + x_2) - f(x_1, x_2 + x_3) + f(x_2, x_1) - f(x_2, x_3) = 0$$

has the general solution

$$f(x,y) = F(x + y) - F(x) - F(y) + 2c(x) + c(y)$$

where F is arbitrary and c satisfies Cauchy's equation $c(x + y) = c(x) + c(y)$. Verify this solution.

Exercise 3-34 The most general complex solution of

$$f(x + y) + f(x - y) = 2f(x)f(y)$$
$$f(x) = \frac{e(x) + e(-x)}{2}$$

is

where $e(x + y) = e(x)e(y)$. Verify this solution and give applications of the result to known identities.

Exercise 3-35 Show that

$$f[xy - (1 - x^2)^{1/2}(1 - y^2)^{1/2}] = f(x) + f(y)$$

is satisfied by $f(x) = c \cos^{-1} x$ and that

$$f[xy - (x^2 - 1)^{1/2}(y^2 - 1)^{1/2}] = f(x) + f(y)$$

is satisfied by $f(x) = c \cosh^{-1} x$.

Exercise 3-36 Show that the only solutions of Poisson's equation

$$f(x + y) + f(x - y) = 2f(x)f(y)$$

other than $f(x) = 0$ satisfy the algebraic-addition theorem

$$[f(x + y)]^2 - 2f(x)f(y)f(x + y) + [f(x)]^2 + [f(y)]^2 - 1 = 0$$

Note that the converse holds; i.e., solutions of the last equation other than $f(x) = -\frac{1}{2}$ satisfy the original equation.

Exercise 3-37 Find solutions of the system of functional equations

$$f(x + y) + f(x - y) = 2f(x)f(y)$$
$$f(x + y) = f(x)f(y) - g(x)g(y)$$

when both f and g are unknown. Note that $f(x) = \cos ax$ is the only nonconstant even solution.

Exercise 3-38 The function

$$f(x) = \sum_{n=0}^{\infty} (-1)^n x^{2n}$$

is the only analytic solution of the functional equation

$$f(x) + f(x^2) = x \qquad 0 \leq x < 1$$

There are infinitely many other solutions (even continuous) in $0 < x < 1$ but none continuous at $x = 0$, and none is known to be continuous at $x = 1$. Verify that the above solution satisfies the equation.

Exercise 3-39 Show that the differentiable real (nonzero) solutions of $f(x + yf(x)) = f(x)f(y)$ are of the form $f(x) = 1 + ax$, where a is an arbitrary real number.

Exercise 3-40 Show that $f(x) = c \log x$ is the only continuous solution which satisfies

$$f(xy) = f(x) + f(y)$$

Hint: Start with $f(mn) = f(m) + f(n)$, where m and n are integers. Use $f(x^n) = nf(x)$.

Exercise 3-41 Verify that $f(x) = u(x) + b$, $\varphi(x,y) = u(xy) - u(x)$ is a solution of

$$f(xy) - f(x) = \varphi(x,y)$$

It is known that this is the general solution, measurable in each variable. Here u is measurable and b is constant.

Exercise 3-42 Show that $ax + by$ is the general form of the solution of

$$f(x_1 + \cdots + x_n, y_1 + \cdots + y_n) = \sum_{k=1}^{n} f(\lambda x_k, \mu y_k)$$

with $\lambda = \mu = 1$. The solution is zero otherwise.

Exercise 3-43 Find a solution of the equation

$$f(x + y)f(x - y) = f^2(x) - f^2(y)$$

Is this the most general solution of its kind?

Exercise 3-44 If $f(x)$ is analytic and satisfies[22]

$$f(x + y) = F[f(x),f(y)] \tag{3-72}$$

then it also satisfies

$$f(2x) = F[f(x),f(x)] \tag{3-73}$$

To show that an analytic solution $\varphi(x)$ of (3-72) satisfies (3-73) consider

$$\varphi(x + y) - F[\varphi(x),\varphi(y)]$$

Put $y = x + h$ and expand in Maclaurin series in powers of h. Note that in order for a solution to (3-72) to exist it is necessary that $F(f(x),f(y)) = F(f(y),f(x))$, and so, for example,

$$\frac{dF}{dh}\bigg|_{h=0} = \frac{1}{2}\frac{d}{dx}F[\varphi(x),\varphi(x)]$$

$$\frac{d^2F}{dh^2}\bigg|_{h=0} = \frac{1}{2^2}\frac{d^2}{dx^2}F[\varphi(x),\varphi(x)]$$

etc., and

$$\frac{d\varphi(2x + h)}{dh}\bigg|_{h=0} = \frac{1}{2}\frac{d\varphi(2x)}{dx}$$

. . . .

Form the appropriate differences of these derivatives and, using the fact that $\varphi(x)$ satisfies (3-73), obtain the result.

REFERENCES

1. Aczel, J.: "Vorlesungen über Funktionalgleichungen und ihre Anwendungen," Birkhäuser Verlag, Basel, 1961.

2. Beckenbach, E., and R. Bellman: "Inequalities," Springer-Verlag OHG, Berlin, 1961.

3. Bellman, R.: "Dynamic Programming," Princeton University Press, Princeton, N.J., 1957.

4. Blackwell, D.: On the Functional Equation of Dynamic Programming, *J. Math. Anal. Appl.*, vol. 2, pp. 273–276, 1961.

5. Boswell, R. D.: On Two Functional Equations, *Amer. Math. Monthly*, vol. 66, p. 716, October, 1959.

6. Brauer, G.: A Functional Inequality, *Amer. Math. Monthly*, vol. 68, pp. 638–642, 1961.

7. Carlitz, L.: Some Functional Equations, *Amer. Math. Monthly*, vol. 68, no. 8, pp. 753–756, October, 1961.

8. Carmichael, R. D.: On Certain Functional Equations, *Amer. Math. Monthly*, vol. 16, pp. 180–183, 1909.

9. Chaundy, T. W., and J. B. McLeod: On a Functional Equation, *Quart. J. Math. Oxford Ser.*, vol. 9, pp. 202–206, 1958.

10. Danese, A. E.: Solution of a Functional Equation Posed by L. Carlitz, *Amer. Math. Monthly*, vol. 61, p. 598, September, 1954.

11. Flett, T. M.: Continuous Solutions of the Functional Equation $f(x + y) + f(x - y) = 2f(x)f(y)$, *Amer. Math. Monthly*, vol. 70, no. 4, pp. 392–397, April, 1963.

12. Franklin, P.: Two Functional Equations with Integral Arguments, *Amer. Math. Monthly*, vol. 38, pp. 154–157, 1931.

13. Friedman, D.: The Functional Equation $f(x + y) = g(x) + h(y)$, *Amer. Math. Monthly*, vol. 69, no. 8, pp. 769–772, October, 1962.

14. Grant, J. D.: Doubly Homogeneous Functional Equations, *Amer. Math. Monthly*, vol. 36, pp. 267–273, 1929.

15. Green, J. W.: On the Solution of the Equation $f'(x) = f(x + a)$, *Math. Mag.*, vol. 26, pp. 117–120, 1952–1953.

16. Hardy, G. H., J. E. Littlewood, and G. Polya: "Inequalities," Cambridge University Press, New York, 1934.

17. Haruki, H.: On the Functional Inequality $\left| f\left(\dfrac{x + y}{2}\right) \right| \le \dfrac{|f(x)| + |f(y)|}{2}$, *J. Math. Soc. Japan*, vol. 16, pp. 39–41, 1964.

18. Herschfeld, A.: On Bell's Functional Equations, *Amer. Math. Monthly*, vol. 38, pp. 395–396, 1931.

19. Hille, E.: Functional Equations, chap. 1 in "Lectures on Modern Mathematics," vol. 3, edited by T. L. Saaty, John Wiley & Sons, Inc., New York, 1965.

20. ———: A Pythagorean Functional Equation, *Ann. of Math.*, ser. 2, vol. 24, pp. 174–180, 1923.

21. Jones, G. S.: Fundamental Inequalities for Discrete and Discontinuous Functional Equations, *RIAS Tech. Rep.* 63-11, April, 1963.

22. Kaba, M.: On the Functions Which Have a Given Algebraical Addition Theorem, *Amer. Math. Monthly*, vol. 13, pp. 181–183, 1906.

23. Kaczmarz, S.: Sur l'équation fonctionnelle $f(x) + f(x + y) = \varphi(y)f(x + y/2)$, *Fund. Math.*, vol. 6, pp. 122–129, 1924.

24. Kamke, E.: "Theory of Sets," Dover Publications, Inc., New York, 1950.

25. Kuczma, M., and K. Szymiczek: On Periodic Solutions of a Functional Equation, *Amer. Math. Monthly*, vol. 70, no. 8, 1963.

26. Newman, D. J.: Solution to "Bounded Functions" Problem 4879 Proposed by L. J. Wallen, *Amer. Math. Monthly*, vol. 68, no. 1, p. 72, 1961.

27. Picard, M. E.: "Leçons sur quelques équations fonctionnelles," Gauthier-Villars, Paris, 1950.

28. Rathore, S. P. S.: On Subadditive and Superadditive Functions, *Amer. Math. Monthly*, vol. 72, pp. 653–654, June–July, 1965.

29. Rosenbaum, R. A., and S. L. Segal: A Functional Equation Characterising the Sine, *Math. Gaz.*, vol. 44, nos. 347–350, pp. 97–105, 1960.

30. Sarma, P. N.: On the Differential Equation $f^{(n)}(x) = f(x^{-1})$, *Math. Student*, vol. 10, pp. 173–174, 1942.

31. Silberstein, L.: Solutions of the Equation $f'(x) = f(x^{-1})$, *Philos. Mag.*, vol. 30, pp. 185–187, 1940.

32. Thielman, H. P.: A Pair of Functional Equations, *Amer. Math. Monthly*, vol. 57, no. 8, October, 1950.

33. ———: A Generalization of Trigonometry, *Nat. Math. Mag.*, vol. 11, no. 8, May, 1937.

34. ———: On the Convex Solution of a Certain Functional Equation, *Bull. Amer. Math. Soc.*, vol. 47, pp. 118–120, 1941.

35. ———: A Note on a Functional Equation, *Amer. J. Math.*, vol. 73, pp. 482–484, 1951.

36. Truesdell, C.: "An Essay toward a Unified Theory of Special Functions," Annals of Mathematics Studies 18, Princeton University Press, Princeton, N.J., 1948.

37. Vaughan, H. E.: Characterization of the Sine and Cosine, *Amer. Math. Monthly*, vol. 62, pp. 707–713, 1955.

38. Vincze, E.: Eine allgemeine Methode in der Theorie der Funktionalgleichungen, *Publ. Math. Debrecen*, vol. 9, pp. 149–163, 314–323, 1962.

39. William Lowell Putnam Competition, *Amer. Math. Monthly*, vol. 70, pp. 712–717, September, 1963.

40. Zarantonello, E. H.: Solving Functional Equations by Contractive Averaging, *Univ. Wisconsin Math. Res. Center Tech. Sum. Rep.* 160, June, 1960.

FOUR

NONLINEAR DIFFERENCE EQUATIONS

4-1 INTRODUCTION

Difference equations and their associated operators not only play a role in their own right as direct mathematical models of physical phenomena but also provide the field of numerical analysis with powerful tools. By approximating differential equations with difference equations, and thereby obtaining numerical solutions, we can study approximate numerical solutions of nonlinear differential and integral equations for which there are no known closed forms.

Difference equations also occur in combined form with differential equations, yielding rich models, particularly in the field of stochastic processes. As an illustration, the solution of a linear differential-difference equation is carried out in detail in this chapter.

We shall restrict the discussion to ordinary difference equations, i.e., equations with a single independent variable. The independent variable may be discrete (a general nonnegative integer-valued variable) or continuous. In the latter case the equation is sometimes called a *functional difference equation*. A chart for difference equations can be developed along lines similar to those given for differential equations on page 177 of "NM."

Since no general methods are available for the solution of nonlinear difference equations, we follow the method adopted in other chapters and often elucidate techniques by specific examples which are applicable to a general class of problems.

First we introduce the operator notation peculiar to difference equations, some of which is to be used in the remainder of the chapter.

NOTATION AND OPERATORS FOR DIFFERENCE EQUATIONS

Let $f(x_0)$, $f(x_1)$, $f(x_2)$, . . . be given values of $f(x)$ and $x_i - x_{i-1} = h$, $i = 1, 2, \ldots$. Then the definitions of various operators with respect to the interval h together with its discrete analogue are

The shifting operator:

$$Ef(x) = f(x + h) \qquad Ey_n = y_{n+1} \tag{4-1}$$

The forward-difference operator:

$$\Delta f(x) = f(x + h) - f(x) \qquad \Delta y_n = y_{n+1} - y_n \tag{4-2}$$

Thus
$$\Delta \equiv E - 1 \tag{4-3}$$

which is to be interpreted as $\Delta y_n = (E - 1)y_n$.

The backward-difference operator:

$$\nabla f(x) = f(x) - f(x - h) \qquad \nabla y_n = y_n - y_{n-1} \tag{4-4}$$

Thus†
$$\nabla = 1 - \frac{1}{E} = \frac{E - 1}{E} \tag{4-5}$$

Example:

$$\left[\frac{E - 1}{E}\right] f(x) = \frac{1}{E}\left[f(x + h) - f(x)\right] = f(x) - f(x - h) = \nabla f(x)$$

The central-difference operator:

$$\delta f(x) = f\left(x + \frac{h}{2}\right) - f\left(x - \frac{h}{2}\right) \qquad \delta y_n = y_{n+\frac{1}{2}} - y_{n-\frac{1}{2}} \tag{4-6}$$

Thus
$$\delta = E^{\frac{1}{2}} - E^{-\frac{1}{2}} \tag{4-7}$$

† The powers of the various operators are defined in a natural manner, for example, $E^2 f = E(Ef) = E(f(x + h)) = f(x + 2h)$, and the definition is extended to arbitrary real-number exponents, where applicable, in exactly the same manner as for powers of real numbers.

The mean-difference operator:

$$\mu f(x) = \tfrac{1}{2}\left[f\left(x + \frac{h}{2}\right) + f\left(x - \frac{h}{2}\right) \right] \qquad \mu y_n = \tfrac{1}{2}[y_{n+\frac{1}{2}} + y_{n-\frac{1}{2}}] \quad (4\text{-}8)$$

Thus

$$\mu = \tfrac{1}{2}[E^{\frac{1}{2}} + E^{-\frac{1}{2}}]$$

The differential operator:

$$Df(x) = \lim_{\alpha \to 0} \frac{(E^\alpha - 1)f(x)}{\alpha} = \lim_{\alpha \to 0} \frac{f(x + \alpha) - f(x)}{\alpha} = f'(x) \quad (4\text{-}9)$$

Thus from the series expansion

$$Ef(x_k) = f(x_k + h) = f(x_k) + \frac{h}{1!} f'(x_k) + \cdots + \frac{h^n}{n!} f^{(n)}(x_k) + \cdots$$

$$= \left[1 + \frac{hD}{1!} + \cdots + \frac{(hD)^n}{n!} + \cdots \right] f(x_k) = e^{hD}f(x_k)$$

we have $E = e^{hD}$, or $E = e^D$ in the discrete case, and

$$hD = \log E = \log (1 + \Delta) = \Delta - \tfrac{1}{2}\Delta^2 + \tfrac{1}{3}\Delta^3 - \cdots$$

For example,

$$Df(x_k) = \frac{df}{dx}\Big|_{x = x_k} = \frac{1}{h} \left(\Delta f(x_k) - \tfrac{1}{2}\Delta^2 f(x_k) + \tfrac{1}{3}\Delta^3 f(x_k) - \cdots \right) \quad (4\text{-}10)$$

There are several formulas for numerical differentiation. Well-known expressions are

$$D^{(n)} = \left[\sum_{k=1}^{\infty} (-1)^{k-1} \frac{\Delta^k}{k} \right]^n \quad (4\text{-}11)$$

and

$$D^{(2m)} = \left(2 \sinh^{-1} \frac{\delta}{2} \right)^{2m} \quad (4\text{-}12)$$

For interpolation we have Newton's forward formula

$$E^n = (1 + \Delta)^n = 1 + \sum_{k=1}^{n} \binom{n}{k} \Delta^k \quad (4\text{-}13)$$

and backward formula

$$E^{-n} = (1 + \nabla)^n = 1 + \sum_{k=1}^{n} \binom{n}{k} \nabla^k \quad (4\text{-}14)$$

where

$$\binom{n}{k} = \frac{n(n - 1) \cdots (n - k + 1)}{k!}$$

Now the inverse of D, that is, D^{-1}, is the integral operator. The inverse of Δ is defined as follows:

$$S \equiv \Delta^{-1} = (e^D - 1)^{-1} = \sum_{n=0}^{\infty} \frac{B_n}{n!} D^{n-1} \qquad (4\text{-}15)$$

where

$B_{2n+1} = 0,\ n > 0,\ B_0 = 1,\ B_1 = -\tfrac{1}{2},\ B_2 = \tfrac{1}{6},\ B_4 = -\tfrac{1}{30},\ B_6 = \tfrac{1}{42},$

etc. The coefficients B_n are known as the *Bernoulli numbers*.[24] From this we easily obtain the Euler-Maclaurin formula for numerical integration

$$D^{-1}f(x) = \int f(x)\,dx = \Delta^{-1}f(x) - B_1 f(x) - \frac{B_2}{2!} Df(x) - \frac{B_4}{4!} D^3 f(x)$$
$$- \frac{B_6}{6!} D^5 f(x) - \cdots \qquad (4\text{-}16)$$

From the definition of D in terms of Δ in (4-10) we obtain Newton's integration formula

$$\int f(x)\,dx = \Delta^{-1}f(x) + \tfrac{1}{2}f(x) - \tfrac{1}{12}\Delta f(x) + \tfrac{1}{24}\Delta^2 f(x) - \tfrac{19}{720}\Delta^3 f(x) + \cdots$$
$$(4\text{-}17)$$

We also have

$$\mu\delta^{-1}f(x) \Big]_a^b = \tfrac{1}{2}f(b) + f(b-1) + \cdots + f(a+1) + \tfrac{1}{2}f(a) \qquad (4\text{-}18)$$

SYSTEMS OF DIFFERENCE EQUATIONS

The system[44]

$$Ex_m(t) = f_m[t, x_1(t), \ldots, x_n(t)] \qquad m = 1, \ldots, n \qquad (4\text{-}19)$$

defined on $t_0 \leq t \leq t_1$, can be written as a single vector difference equation

$$EX = F(X, t) \qquad (4\text{-}20)$$

where the vectors X and F are given by $X = [x_1, \ldots, x_n]$ and where $F = (f_1, \ldots, f_n)$. If an arbitrary nth-order difference equation

$$\varphi(t, x, Ex, E^2x, \ldots, E^n x) = 0 \qquad (4\text{-}21)$$

can be solved to yield

$$E^n x = f(t, x, Ex, E^2x, \ldots, E^{n-1}x) \qquad (4\text{-}22)$$

then by putting

$$x = x_1 \quad Ex = Ex_1 = x_2 \quad E^2x = Ex_2 = x_3, \ldots, E^{n-1}x = Ex_{n-1} = x_n$$

F becomes $F = [x_2, \ldots, x_n, f]$ and again,

$$EX = F(X, t) \tag{4-23}$$

This reduction can be effected under appropriate assumptions for an arbitrary system of simultaneous difference equations.

If we assume that at $t = t_0$, $X = X_0 = [x_1(t_0), \ldots, x_n(t_0)]$, then since $EX_0 = F(X_0, t_0)$, we have

$$EX_0 = [x_1(t_0 + 1), \ldots, x_n(t_0 + 1)] \equiv X_1 = F(X_0, t_0) \tag{4-24}$$

Also

$$EX_1 = [x_1(t_0 + 2), \ldots, x_n(t_0 + 2)] \equiv X_2 = F(X_1, t_0 + 1) \tag{4-25}$$

giving X_2. Similarly we have

$$EX_k \equiv X_{k+1} = F(X_k, t_0 + k) \qquad \text{for } t_0 + k \leq t \leq t_0 + k + 1 \tag{4-26}$$

where k is an arbitrary positive integer. In order to extend this solution to arbitrary values of t and not just to $t_0 + k$, $k \geq 0$ an integer, it is necessary to know the value of X over an interval of unit length. Thus if we are given

$$X = X_0 \equiv [f_1(t), \ldots, f_n(t)] \qquad t_0 \leq t \leq t_0 + 1 \tag{4-27}$$

then $$EX_0 = F(X_0, t) = X_1 \tag{4-28}$$

defines X for $t_0 + 1 \leq t \leq t_0 + 2$, and more generally

$$EX_k = F(X_k, t + k) = X_{k+1} \tag{4-29}$$

defines X for $t_0 + k + 1 \leq t \leq t_0 + k + 2$ for arbitrary positive integer k. Then X is determined on $t_0 \leq t \leq t_0 + k$.

There is only one solution of the system (4-23), since if we have two solutions of the equation X and Y, whose initial prescriptions on $t_0 \leq t \leq t_0 + 1$ are the same, then on this interval we have

$$EX = F(X, t) = F(Y, t) = EY \qquad t_0 \leq t \leq t_0 + 1 \tag{4-30}$$

and $$E^2 X = F(EX, t + 1) = F(EY, t + 1) = E^2 Y$$
$$t_0 + 1 \leq t \leq t_0 + 2 \tag{4-31}$$

and if $$E^k X = E^k Y \tag{4-32}$$

for $t_0 + k - 1 \leq t \leq t_0 + k$, then

$$E^{k+1} X = F(E^k X, t + k) = F(E^k Y, t + k) = E^{k+1} Y \tag{4-33}$$

for $t_0 + k \leq t \leq t_0 + k + 1$. We have:

Theorem 4-1 The equation $EX = F(X,t)$, where F exists for all X and t and where $X(t) = X_0 = [f_1(t), \ldots, f_n(t)]$ is defined for $t_0 \leq t \leq t_0 + 1$, has a unique solution defined for $t_0 \leq t \leq t_0 + k$ with positive integer k.

Exercise 4-1 Show that an explicit expression for the nth term of a sequence $\{y_n\}$ defined by

$$y_n = y_{n-1} + y_{n-2} + a_n \qquad n = 1, 2, \ldots$$

with y_0 and y_1 given and the explicit expression for the nth term of the sequence $\{a_n\}$ known, is

$$y_n = f_n y_1 + f_{n-1} y_0 + f_{n-1} a_2 + f_{n-2} a_3 + \cdots + f_1 a_n$$

where f_j denotes the jth term of the Fibonacci sequence $1, 1, 2, 3, 5, 8, \ldots$, or explicitly

$$f_j = \tfrac{1}{2}[(1 + \sqrt{5})^{j-1} + (1 - \sqrt{5})^{j-1}]$$

4-2 LINEAR DIFFERENCE EQUATIONS

CONSTANT COEFFICIENTS

This section contains a brief review of the methods used for solving the general linear Nth-order difference equation with constant coefficients

$$\sum_{n=0}^{N} a_n y_{n+k} = b_k \qquad a_N \neq 0 \tag{4-34}$$

in which k may assume any nonnegative integral value. Several methods for the solution of this equation are available.

1. A solution to the homogeneous equation $(b_k = 0, k = 0, 1, \ldots)$ can be obtained by assuming that it has the form

$$y_m = A\lambda^m \tag{4-35}$$

where A and λ are constants, which, when substituted in (4-34), yields

$$A\lambda^k \left(\sum_{n=0}^{N} a_n \lambda^n \right) = 0 \qquad \text{for } b_k = 0 \tag{4-36}$$

If $\lambda_1, \lambda_2, \ldots, \lambda_N$ are the N roots (assumed distinct, see later for the case of multiple roots) of the Nth-degree polynomial in parentheses, then

$$y_m = \sum_{i=1}^{N} A_i \lambda_i^m \qquad m = 0, 1, \ldots \tag{4-37}$$

is the general solution of the homogeneous equation, where A_i must be determined from the N initial conditions $y_0, y_1, \ldots, y_{N-1}$.

In order for the solution to be stable it is necessary that y_m remain bounded as $m \to \infty$. For the present, this condition may be taken as a definition of stability. It is apparent from (4-37) that this will be true if

$$|\lambda_i| \leq 1 \qquad i = 1, 2, \ldots, N$$

2. The nonhomogeneous equation can be solved by iteration in the following manner. Consider the set of equations

$$
\begin{aligned}
y_0 &= y_0 \\
y_1 &= y_1 \\
&\cdot\ \cdot\ \cdot\ \cdot\ \cdot \\
y_{N-1} &= y_{N-1} \\
y_N &= \sum_{n=0}^{N-1} a_n y_n + \beta_0
\end{aligned}
\qquad (4\text{-}38)
$$

where the last equation involves a redefinition of the coefficients a_n and b_k for the case $k = 0$ (or simply assume that $a_N = -1$; $\beta_0 = -b_0$).

The system (4-38) can be written in vector form

$$EY_0 = Y_1 = AY_0 + B_0 \qquad (4\text{-}39)$$

where Y_1, Y_0, and B_0 are the transposes of the vectors $(y_N, y_{N-1}, \ldots, y_1)$, $(y_{N-1}, y_{N-2}, \ldots, y_0)$, and $(\beta_0, 0, \ldots, 0)$, respectively, and A is the matrix

$$
\begin{pmatrix}
a_{N-1} & a_{N-2} & \cdots & a_1 & a_0 \\
1 & 0 & \cdots & 0 & 0 \\
0 & 1 & \cdots & 0 & 0 \\
\cdot & \cdot & \cdots & \cdot & \cdot \\
0 & 0 & & 1 & 0
\end{pmatrix}
$$

Now using the shifting operator E, we obtain

$$
\begin{aligned}
Y_2 &= EY_1 = E(AY_0 + B_0) = AEY_0 + EB_0 \\
&= A(AY_0 + B_0) + EB_0 = A^2 Y_0 + AB_0 + B_1
\end{aligned}
$$

and in general

$$Y_k = A^k Y_0 + (A^{k-1}B_0 + A^{k-2}B_1 + \cdots + AB_{k-2} + B_{k-1}) \qquad (4\text{-}40)$$

where $B_s = EB_{s-1} = (E\beta_{s-1}, 0, \ldots, 0) = (\beta_s, 0, \ldots, 0)$, $\beta_{s-1} = -b_{s-1}$, $\beta_s = -b_s$.

We can compute any power of A using the following formula, which holds for entire functions of a matrix,

$$f(A) = \sum_{i=1}^{d} \sum_{p=0}^{p_i-1} \frac{(A - \lambda_i I)^p}{p!} f^{(p)}(\lambda_i) Z(\lambda_i) \qquad (4\text{-}41)$$

where d is the number of distinct characteristic roots λ_i of A, p_i is the multiplicity of the ith root λ_i (of the characteristic equation $|\lambda I - A| = 0$), $f^{(p)}(\lambda_i)$ is the pth-order formal derivative of f evaluated at λ_i, and $Z(\lambda_i)$ are complete orthogonal idempotent matrices of the transpose matrix A^T; that is, they have the properties

$$\sum_{i=1}^{d} Z(\lambda_i) = I \qquad Z(\lambda_i)Z(\lambda_j) = 0 \qquad i \neq j$$

$$Z^2(\lambda_i) = Z(\lambda_i)$$

(4-42)

where I and $\mathbf{0}$ are the identity and null matrices, respectively. If λ_i, $i = 1, \ldots, N$, are all distinct, then (see "NM," chap. 4)

$$A^m = \sum_{i=1}^{N} \lambda_i{}^m Z(\lambda_i)$$

where

$$Z(\lambda_i) = \frac{\displaystyle\prod_{j \neq i} (\lambda_j I - A)}{\displaystyle\prod_{j \neq i} (\lambda_j - \lambda_i)}$$

For asymptotic stability we are concerned with the solution Y_k as $k \to \infty$, and this we examine later in the section.

Exercise 4-2[1] Solve explicitly for x_i given $x_1 = 2$, $x_2 = 3$, $x_{2n} = x_{2n-1} + 2x_{2n-2}$, $x_{2n+1} = x_{2n} + x_{2n-1}$.

1. By writing $x_{2n+1} = 2x_{2n-2} + 2x_{2n-1}$ and assuming that $x - 2 = -\frac{1}{2}$, $x_{-1} = \frac{3}{2}$, $x_0 = \frac{1}{2}$, form the matrix equation $X_n = AX_{n-1}$, $n = 0, 1, \ldots$, where $X = \begin{pmatrix} x_{2n} \\ x_{2n+1} \end{pmatrix}$, $A = \begin{pmatrix} 2 & 1 \\ 2 & 2 \end{pmatrix}$. Thus $X_n = A^{n+1}X_{-1}$. Since A satisfies its characteristic equation, we have $A^{n+1} = \alpha_n I + \beta_n A$ for some scalars α_n and β_n. Compute the characteristic values of A and find $\alpha_n + \beta_n$ by substituting in the last equation. Thus show that

$$X_n = \begin{pmatrix} \alpha_n + 2\beta_n & \beta_n \\ 2\beta_n & \alpha_n + 2\beta_n \end{pmatrix} X_{-1}$$

and hence that $T_{2n} = (\beta_n - \alpha_n)/2$, $T_{2n+1} = 2\beta_n + 3\alpha_n/2$.

2. By an alternative method.

3. The third method of solution is the most widely used, especially in the design of sampled-data systems, where a modification known as the *z transformation*[38] yields an almost complete analogy between the solution of linear differential equations by the Laplace transform and linear difference equations by the *z* transform.

We shall discuss the use of the generating function first, since a somewhat briefer explanation can be given for it. (The *z* transform can be formed from the generating function by making the substitution $z = 1/t$.)

Let t be a complex variable, and let

$$F(t) = \sum_{m=0}^{\infty} t^m y_m \tag{4-43}$$

$F(t)$ is called the *generating function* of the sequence $\{y_m\}$. Through its use Eq. (4-34) can be transformed into a rational algebraic equation. This equation in $F(t)$ can then be inverted, and the solution for y_m obtained.

Thus if (4-43) is differentiated j times with respect to t and the derivative evaluated at $t = 0$, one obtains

$$y_j = \frac{1}{j!} \frac{d^j F}{dt^j} \Big|_{t=0} \tag{4-44}$$

as an inversion formula for y_j in terms of $F(t)$.

Exercise 4-3   Apply the generating-function technique to the system $\rho p_0 = p_1$, $(1 + \rho)p_n = p_{n+1} + \rho p_{n-1}; n \geq 1 \, \rho < 1$.

In order to solve (4-34) multiply by t^k and sum.   This gives

$$\sum_{k=0}^{\infty} t^k \sum_{n=0}^{N} y_{n+k} a_n = \sum_{k=0}^{\infty} \sum_{n=0}^{N} t^{-n} a_n t^{n+k} y_{n+k}$$

$$= \sum_{n=0}^{N} t^{-n} a_n \left(\sum_{r=0}^{\infty} t^r y_r - \sum_{s=0}^{n-1} t^s y_s \right) \equiv B(t) \tag{4-45}$$

where the last sum is taken as zero when $n = 0$.

Upon rearranging and using (4-43), we obtain

$$F(t) \sum_{n=0}^{N} t^{-n} a_n = \sum_{n=0}^{N} t^{-n} a_n \sum_{s=0}^{n-1} t^s y_s + B(t) \tag{4-46}$$

and therefore

$$F(t) = \frac{\displaystyle\sum_{n=0}^{N} \sum_{s=0}^{n-1} a_n t^{s-n} y_s + B(t)}{\displaystyle\sum_{n=0}^{N} t^{-n} a_n} = \frac{\displaystyle\sum_{n=0}^{N} t^{N-n} a_n \sum_{s=0}^{n-1} t^s y_s + t^N B(t)}{\displaystyle\sum_{n=0}^{N} t^{N-n} a_n} \tag{4-47}$$

If the initial values $y_0, y_1, \ldots, y_{N-1}$ are given, then that part of (4-47) involving these known y's is the ratio of an $(N-1)$st-degree polynomial and an Nth-degree polynomial, which can be expressed in partial fractions as a sum of terms of the form

$$\frac{C}{(t-r)^k}$$

where r is a root of

$$\sum_{n=0}^{N} a_n t^{N-n} = 0 \qquad (4\text{-}48)$$

with a multiplicity at least equal to k. Here r is the reciprocal of λ.

Using (4-44), we obtain

$$y_j = (-1)^k \frac{(k+j-1)!}{(k-1)!j!} \frac{1}{r^{k+j}} = (-1)^k r^{-(k+j)} \binom{k+j-1}{j} \qquad (4\text{-}49)$$

Therefore the general solution consists of a sum of terms like (4-49).

From Eq. (4-49) it is seen that the solution is asymptotically stable ($\lim_{j \to \infty} y_j = 0$) if and only if $|r| > 1$ for all roots.

If we substitute $z = 1/t$ in (4-48), then it becomes

$$\sum_{k=0}^{N} z^{-k} a_{N-k} = 0 \qquad (4\text{-}50)$$

and the condition on the roots s_i, $i = 1, \ldots, N$, of (4-50) for asymptotic stability becomes

$$|s_i| < 1 \qquad (4\text{-}51)$$

since if s_i is a root of (4-50), then $1/s_i$ is a root of (4-48). Here s_i corresponds to λ_i, previously discussed.

Three useful sufficiency tests for asymptotic stability, often loosely called *stability*, follow.[36]

Gauss[13] has shown that there are no roots of (4-50) outside the circle $|\lambda| = \max(1, \alpha\sqrt{2})$, where α is the sum of the positive a_i. Therefore the system is stable if $\alpha \leq 1/\sqrt{2}$.

Walsh[49] has shown that the roots lie within the circle

$$|\lambda| = \sum_{k=1}^{N} |a_{N-k}|^{1/k} \qquad (4\text{-}52)$$

and hence we have stability if this sum is less than or equal to 1.

Finally, Kojima[27] has shown that the roots are inside the circle defined by

$$|\lambda| \leq \max\left[\left|\frac{a_0}{a_1}\right| \quad 2\left|\frac{a_j}{a_{j+1}}\right| \right] \qquad j = 1, 2, \ldots, N-1 \qquad (4\text{-}53)$$

Exercise 4-4 Show that if $\sigma_i = \Sigma \lambda_j \lambda_k \cdots \lambda_r$, where the i of the characteristic roots of (4-36) appear in each product and there are $\binom{N}{i}$ terms in the sum, then a necessary condition for stability is

$$|\sigma_i| \leq \binom{N}{i}$$

Remark: In Ref. 33 Schwartz's lemma [if $f(z)$ is an analytic function, regular for $|z| \leq R$, and $|f(z)| \leq M$ for $|z| = R$, and $f(0) = 0$, then $|f(re^{i\theta})| \leq Mr/R$, $0 \leq r \leq R$] is applied to the study of the location of zeros of the polynomial $f(z) = \sum_{n=0}^{N} a_n z^{N-n}$.   We develop some of the ideas below.

Exercise 4-5   Let $M = \max_{|z|=1} |a_1 z^{N-1} + a_2 z^{N-2} + \cdots + a_n|$.   Using Schwartz's lemma show that $|a_1 z + \cdots + a_n z^N| \leq M|z|$ for $|z| \leq 1$.

Exercise 4-6   Show that all the zeros of $f(z)$ lie in $|z| \leq M/|a_0|$ if $|a_0| \leq M$. Consider $g(z) = a_0 + a_1 z + \cdots + a_n z^N$.   Then for $|z| \leq 1$,

$$|g(z)| \geq |a_0| - |a_1 z + \cdots + a_n z^N| \geq |a_0| - M|z|$$

and hence $|g(z)| > 0$, that is, $g(z)$ does not vanish, if $|z| < |a_0|/M$.   Thus the zeros of $g(z)$ lie in $|z| \geq |a_0|/M$.   But $f(z) = z^N g(1/z)$, and hence the result follows.   This bound is better than the traditional bound $\sum_{n=1}^{N} \frac{|a_n|}{|a_0|}$.   Note that the polynomial $f(z)$ $= -nz^N + z^{N-1} + \cdots + z + 1$ yields the limit in the above theorem.
Also show that if $|a_0| \geq M$, then the zeros of $f(z)$ lie in $|z| \leq 1$.

Exercise 4-7   As a corollary to the theorem of the last exercise show that if $a_N \geq 0$, $n = 1, \ldots, N$, and $|a_0| \leq a_1 + \cdots + a_N$, then all the zeros of $f(z)$ lie in $|z| \leq (a_1 + \cdots + a_N)/|a_0|$.

In the same reference the following results are also found.
1.   If r is the modulus of the zeros of largest modulus of $f(z)$, and if $M' = \max_{|z|=r} |a_0 z^{N-1} + \cdots + a_{N-1}|$, then all the zeros of $f(z)$ lie in $r|a_n|/M' \leq |z| \leq r$ if $|a_n| \leq M'$.   A corollary to this is that if $a_n \geq 0$, $n = 1, \ldots, N-1$, $a_0 > 0$, and $|a_n| \leq a_0 r^N + a_1 r^{N-1} + \cdots + a_{N-1} r$, then all zeros of $f(z)$ lie in

$$\frac{r|a_N|}{a_0 r^N + \cdots + a_{N-1} r} \leq |z| \leq r$$

2.   If $a_0 \geq a_1 \geq \cdots \geq a_{N-1} \geq a_N \geq 0$, then all zeros of $f(z)$ lie in

$$\frac{a_N}{a_0 + \cdots + a_{N-1}} \leq |z| \leq 1$$

3.   All zeros of $f(z)$ whose real parts are positive lie in

$$|z| \leq \max \left(\frac{a_1}{a_0}, \frac{a_2}{a_1}, \ldots, \frac{a_{N-1}}{a_{N-2}}, \frac{a_N}{a_{N-1}} \right)$$

4.   If $a_0 \geq a_1 \geq \cdots \geq a_N > 0$, then the number of zeros of $f(z)$ in $|z| \leq \frac{1}{2}$ does not exceed

$$1 + \frac{1}{\ln 2} \ln \frac{a_0}{a_r}$$

ARBITRARY COEFFICIENTS
First-order Equations

The linear first-order difference equation with variable coefficients is given by[8,32]

$$a(x)y(x + 1) + b(x)y(x) = c(x) \qquad (4\text{-}54)$$

We solve the homogeneous equation obtained by putting $c(x) = 0$ and divide by $a(x)$, assuming $a(x)$ is nowhere zero, to obtain

$$y(x + 1) = p(x)y(x)$$

Taking logarithms and summing the function on the right, we have

$$\ln y(x) = \sum_{k=0}^{N-1} \ln p(k + \theta) + \ln y(\theta) \qquad (4\text{-}55)$$

where $x = N + \theta$, $0 \leq \theta \leq 1$, and N is any positive integer.   Thus

$$y(x) = y(\theta) \exp \sum_{k=0}^{N-1} \ln p(k + \theta) \qquad (4\text{-}56)$$

This solution is completely determined only if $y(x)$ is given in the strip $0 \leq x \leq 1$.

Exercise 4-8   Show that $y(x + 1) = e^{2x}y(x)$ has the solution

$$y(x) = B(x) \exp (x^2 - x)$$

where $B(x)$ is a unit periodic; that is, $B(x) = B(x + 1)$.

The nonhomogeneous equation, on division by $a(x) \neq 0$, takes the form

$$y(x + 1) - p(x)y(x) = q(x) \qquad (4\text{-}57)$$

Let $\bar{y}(x)$ be the solution to the homogeneous case; that is,

$$\bar{y}(x) = y(\theta) \exp \sum_{k=0}^{N-1} \ln p(k + \theta) \qquad x = N + \theta$$

$$0 \leq \theta \leq 1, N \geq 1 \quad (4\text{-}58)$$

We write

$$y(x) = \bar{y}(x)z(x) \tag{4-59}$$

and substitute (4-59) into (4-57) to obtain

$$\bar{y}(x + 1)z(x + 1) - p(x)\bar{y}(x)z(x) = q(x) \tag{4-60}$$

Since $\bar{y}(x + 1) = p(x)\bar{y}(x)$, our equation becomes

$$\bar{y}(x + 1)\, \Delta z(x) = q(x)$$

from which, by dividing by $\bar{y}(x + 1)$ and summing, we have

$$z(x) = \alpha(x) + \sum_{k=0}^{N-1} \frac{q(k + \theta)}{\bar{y}(k + 1 + \theta)} \qquad x = \theta + N, 0 \le \theta \le 1, N \ge 1 \tag{4-61}$$

where $\alpha(x)$ is an arbitrary unit periodic. From $\bar{y}(x)$ and $z(x)$ we obtain the complete solution.

The solution can also be obtained by using the method of continued fractions. See some of the standard references listed at the end of the chapter.

THE GENERAL EQUATION

The general nth-order linear nonhomogeneous difference equation in a single unknown function $y(x)$ is given by

$$a_n(x)y(x + n) + a_{n-1}(x)y(x + n - 1)$$
$$+ \cdots + a_0(x)y(x) = b(x) \tag{4-62}$$

where the coefficients are given. The equation is called homogeneous if $b(x) \equiv 0$.

Let $a_i(x)$, $i = 0, \ldots , n$, be analytic in the finite plane except for poles, and let $y_i(x)$, $i = 1, \ldots , n$, be solutions to the homogeneous equation. Assume for simplicity that $a_n(x) \equiv 1$. If $p_i(x), i = 1, \ldots ,$ n, are arbitrary unit periodics, then

$$\sum_{i=1}^{n} p_i(x)y_i(x) \tag{4-63}$$

is also a solution.

If there are $p_i(x)$ not all identically zero such that (4-63) vanishes identically, then $y_i(x)$, $i = 1, \ldots , n$, are said to be linearly dependent; otherwise they are linearly independent, and are said to form a fundamental system of solutions. The Casorati theorem says that (see exercises below) $y_i(x), i = 1, \ldots , n$, are linearly dependent if and only if the

Casorati determinant (the analogue of the wronskian in differential equations) vanishes.   This determinant is

$$\begin{vmatrix} y_1(x) & y_2(x) & \cdots & y_n(x) \\ y_1(x+1) & y_2(x+1) & \cdots & y_n(x+1) \\ \cdots\cdots\cdots\cdots\cdots\cdots\cdots\cdots\cdots\cdots\cdots\cdots \\ y_1(x+n-1) & y_2(x+n-1) & \cdots & y_n(x+n-1) \end{vmatrix}$$

Exercise 4-9   Prove necessity by equating (4-63) to zero and generating a system of equations by replacing the argument by $x + 1$, $x + 2$, etc., successively.   Thus if the $y_i(x)$, $i = 1, \ldots, n$, form a fundamental system of solutions, every solution $y(x)$ can be expressed by (4-63) with an appropriate choice of $p_i(x)$.

Exercise 4-10   Prove the last statement using each of $y_i(x)$ in (4-62), with $b(x) \equiv 0$, and note that their determinant must vanish, as we assume that $a_i(x)$ do not all vanish at the same point.

Exercise 4-11   Show how the $p_i(x)$ can be determined by equating (4-63) to $y(x)$ and generating a system of equations by replacing the argument by $x + 1$, $x + 2$, $\ldots$, $x + n - 1$.   To solve the nonhomogeneous equation (4-62), write in terms of a fundamental system of solutions $y_i(x)$, $i = 1, \ldots, n$,

$$y(x) = \sum_{i=1}^{n} \alpha_i(x)y_i(x) \tag{4-64}$$

replace x by $x + 1$, write

$$\alpha_i(x+1) = \alpha_i(x) + \Delta\alpha_i(x)$$

substitute in (4-64), and equate to zero the sum of all the terms in $\Delta\alpha_i(x)$.   With this accomplished, replace x in the resulting equation by $x + 1$ (making the argument $x + 2$) and again equate to zero the sum of the terms involving $\Delta\alpha_i(x)$.   If this process is repeated n times, until the argument becomes $x + n$, and in the last equation the sum of the terms in $\Delta\alpha_i(x)$ is equated to $b(x)$, we have n equations from which $\Delta\alpha_i(x)$ are determined.   Thus by taking the inverse of the difference operator Δ we obtain $\alpha_i(x)$ and from (4-64) a solution of (4-62).   Apply this procedure to $y(x + 1) + ay(x) = b$.

4-3   A GENERAL DIFFERENCE EQUATION OF THE FIRST ORDER

Here we shall be concerned with the equation[30]

$$y(x + 1) = f[y(x)]$$

where $f(y(x))$ does not depend on x, except through y.

We shall study the behavior of solutions of this equation for all possible initial values as $x \to \infty$.

Graphically we plot $y(x + 1)$ as a function of $y(x)$, that is,

$$y(x + 1) = f[y(x)]$$

Then for any initial $y(x)$ we obtain $y(x + 1)$, which now must be used again to obtain a new value of $y(x + 2)$. This is accomplished by drawing the straight line $y(x + 1) = y(x)$, so that each $y(x + 1)$ can be used as a new $y(x)$, and hence the horizontal line corresponding to a value of $y(x + 1)$ from the curve intersects the line $y(x + 1) = y(x)$ at a point which, when projected on the $y(x)$ axis, gives the new value of $y(x)$. This is then used to determine a new $y(x + 1)$, etc. Since successive values of $y(x + 1)$ correspond to unit increases in x at each step, we obtain the behavior of $y(x)$ as $x \to \infty$ for any initial value of $y(x)$.

The question is whether these values of $y(x + 1)$ will converge to a point (a,a), which is an intersection point of the straight line with the curve, as the process is repeated, i.e., as $x \to \infty$, or whether it will tend to ∞ or to $-\infty$. Analysis of this behavior leads to useful asymptotic expressions for the unknown function $y(x)$ [easily obtainable from that of $y(x + 1)$], which often is all that one can hope for in a nonlinear problem.

Note that if the curve lies completely above or below the line

$$y(x + 1) = y(x)$$

then the iterations tend to $+ \infty$ or to $-\infty$ as $x \to + \infty$. If the curve and the line intersect, then, depending on whether the initial value is chosen to the left or to the right of the abscissa of the point of intersection, the iterations may tend to an intersection point or to ∞ or $-\infty$ as $x \to \infty$, depending on whether, as $y(x)$ increases, the curve goes from above to below the line at the point of intersection or from below to above the line. See Figs. 4-1 and 4-2.

Let (a,a) be any point such that $f(a) = a$, and let $y(x) = a + u(x)$. Using a series expansion we can then write

$$y(x + 1) = f[y(x)] = f[a + u(x)]$$
$$= f(a) + f'(a)u(x) + \tfrac{1}{2}f''(a)[u(x)]^2 + \cdots$$

Using the first-order approximation, we have

$$y(x + 1) = a + f'(a)u(x)$$

For these successive iterations to converge to a finite limit as $x \to \infty$ we must have

$$|y(x + 1) - a| < |y(x) - a|$$

which implies that $\qquad |f'(a)| < 1$

Thus if

$|f'(a)| < 1$	we have convergence
$|f'(a)| = 1$	convergence is conditional
$|f'(a)| > 1$	we have divergence

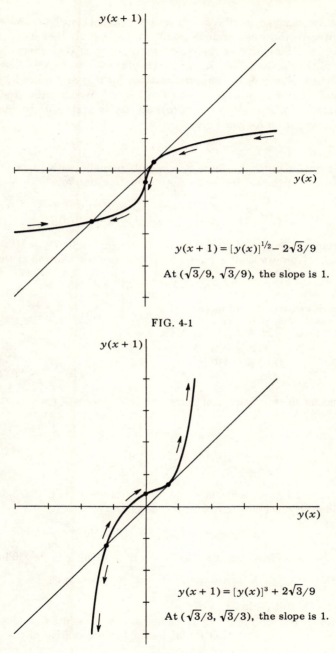

$$y(x + 1) = [y(x)]^{1/2} - 2\sqrt{3}/9$$

At $(\sqrt{3}/9, \sqrt{3}/9)$, the slope is 1.

FIG. 4-1

$$y(x + 1) = [y(x)]^3 + 2\sqrt{3}/9$$

At $(\sqrt{3}/3, \sqrt{3}/3)$, the slope is 1.

FIG. 4-2

If there exists a point on the graph $y(x + 1) = f[y(x)]$ such that everywhere to the right of that point $f[y(x)] > y(x)$, then $y(x) \to +\infty$ as $x \to +\infty$. Similarly, if there exists a point such that everywhere to the left of that point $f[y(x)] < y(x)$, then $y(x) \to -\infty$ as $x \to +\infty$.

We first seek asymptotic approximations for $y(x)$ around finite points (a,a) such that $f(a) = a$ and $0 < |f'(a)| < 1$. We use the first-order approximation $y(x + 1) = a + [f'(a)] \Delta y$, $\Delta y = y(x) - a$, to obtain a first approximation for $y(x)$, given by

$$y(x) = a + A[f'(a)]^x$$

(verify by substitution), where A is a unit periodic with $|A| < 1$. Let $t = A[f'(a)]^x$; then the first-order approximation becomes

$$z(t) = y(x) = a + A[f'(a)]^x = a + t$$

and
$$y(x + 1) = a + A[f'(a)]^{x+1} = a + [f'(a)]t$$

Thus from $y(x + 1) = f[y(x)]$ we have $a + f'(a)t = f[z(t)] = z(kt)$. Let $k = f'(a)$. Now we shall attempt to derive a series solution of the form

$$z(t) = a + t + A_2t^2 + A_3t^3 + \cdots$$

Using series expansion we have

$$
\begin{aligned}
z(a + kt) &= [a + kt + A_2k^2t^2 + A_3k^3t^3 + \cdots] = f[z(t)] \\
&= f[a + t + A_2t^2 + A_3t^3 + \cdots] \\
&= f(a) + f'(a)[t + A_2t^2 + A_3t^3 + \cdots] \\
&\quad + \tfrac{1}{2}[f''(a)][t + A_2t^2 + A_3t^3 + \cdots]^2 + \cdots
\end{aligned}
$$

By equating the coefficients of corresponding powers of t we obtain

$$
\begin{aligned}
a &= f(a) \\
k &= f'(a) \\
A_2k^2 &= \tfrac{1}{2}[f''(a) + 2A_2f'(a)] \\
A_3k^3 &= \tfrac{1}{6}[f'''(a) + 6A_2f''(a) + 6A_3f'(a)]
\end{aligned}
$$

from which all the coefficients can be determined. Furthermore, the ratio between successive terms in the series is

$$(A_{n+1}/A_n)t = (A_{n+1}/A_n)(A)[f'(a)]^x$$

which can be made as small as necessary by making x large. Thus

$$y(x) = a + Ak^x + A_2A^2k^{2x} + A_3A^3k^{3x} + \cdots$$

where A is a unit periodic.

We next develop approximations around finite points (a,a) such that $f(a) = a$ and $f'(a) = 0$. Let $f^{(n)}(a)$ be the first derivative of f at a such that $f^{(n)}(a) \neq 0$. This leads to an approximation of the form

$$u(x + 1) = \frac{1}{n!} [f^{(n)}(a)][u(x)]^n$$

from which we obtain the first approximation for $y(x)$, given by

$$y(x) = a + \frac{A^{n^x}}{[f^n(a)/n!]^{1/(n-1)}} \qquad |A| < 1$$

Let $t = A^{n^x}$ and $B_1 = 1/[f^n(a)/n!]^{1/(n-1)}$. By similar methods we obtain

$$z(t^n) = f[z(t)] \qquad \text{and} \qquad z(t) = a + B_1 t + B_2 t^2 + \cdots$$

and

$$z(t^n) = [a + B_1 t^n + B_2 t^{2n} + \cdots] = f[z(t)]$$
$$= f(a) + \frac{f^n(a)}{n!} [B_1 t + B_2 t^2 + B_3 t^3 + \cdots]^n$$
$$+ \frac{f^{n+1}(a)}{(n+1)!} [B_1 t + B_2 t^2 + B_3 t^3 + \cdots]^{n+1} + \cdots$$

Here B_1 B_2, . . . are coefficients of the expansion and must be determined. We have

$$a = f(a)$$
$$B_1 = \frac{f^n(a)}{n!} B_1{}^n$$

from which

$$B_1 = \frac{1}{[f^n(a)/n!]^{1/(n-1)}}$$

and all other coefficients can be determined. Our expansion becomes

$$z(t) = a + B_1 t + B_2 t^2 + B_3 t^3 + \cdots$$
$$y(x) = a + B_1 A^{n^x} + B_2 A^{2-n^x} + B_3 A^{3-n^x} + \cdots$$

For approximations at (∞, ∞) and $(-\infty, -\infty)$ let $y(x + 1) = p[y(x)]^r$ be the asymptote to the curve $y(x + 1) = f[y(x)]$ at the desired branch. We have two cases to consider.

Case 1: $r = 1$, $p > 1$. Here the first approximation is given by $y(x) = Ap^x$. If we let $t = Ap^x$ and $z(t) = y(x)$, our equation

$$y(x + 1) = f[y(x)]$$

becomes $z(pt) = f[z(t)]$, and the expansion assumes the form

$$z(t) = t + A_0 + \frac{A_1}{t} + \frac{A_2}{t^2} + \cdots$$

which cannot be handled by the general method previously used.

Case 2: $r > 1$. The first approximation is $y(x) = Bc^{r^x}$ where c is a unit periodic and $|c| > |$ B is known. Let $t = c^{r^x}$ and $z(t) = y(x)$.

Then $y(x + 1) = f[y(x)]$ becomes $z(t^r) = f[z(t)]$. The expansion is the same as in case 1 except for the differing value for t. For the approximations at finite points (a,a) such that $f(a) = a$ and $|f'(a)| = 1$ there is no standard pattern of convergence, and hence there is no procedure for approximating $y(x)$ under these conditions.

4-4 SOLUTIONS OF SOME NONLINEAR EQUATIONS

ASYMPTOTIC SOLUTIONS; NONLINEAR EQUATIONS

If $y_1, \ldots, y_{n-1}$ are prime numbers, then $y_n = y_1 y_2 \cdots y_{n-1} + 1$ and $y_{n+1} = y_1 y_2 \cdots y_n + 1$ are relatively prime. Euclid used this type of relation to show that the number of primes is infinite. These two relations yield

$$y_n = \frac{y_{n+1} - 1}{y_n - 1} \tag{4-65}$$

Let $y_n = u_n + \frac{1}{2}$; then we have the nonlinear difference equation

$$u_{n+1} = u_n^2 + \frac{1}{4} \tag{4-66}$$

which is a special case of

$$u_{n+1} = a u_n^2 + b \tag{4-67}$$

and this, on substituting $v_n = a u_n$, becomes

$$v_{n+1} = v_n^2 + ab \tag{4-68}$$

Suppose that $a \neq 0$ (otherwise the problem is trivial); consider the equation $x^2 - x + ab = 0$, and let $r \geq \frac{1}{2}$ and $1 - r$ be its real roots. As $n \to \infty$, if v_n has a limit, it is either r or $1 - r$. Note that if any $v_n = \pm r$ or $\pm(1 - r)$, then $v_{n+i}, i = 1, 2, \ldots$, are all equal to r or $1 - r$, respectively.[14]

It may be that $v_n = r$ or $1 - r$ for all $n = 0, 1, 2, \ldots$. If this is not the case, then let v_n be the first value which is not a root. Then, because of (4-68), $v_n = -r$ if $v_{n+1} = r$, or $v_n = r - 1$ if $v_{n+1} = 1 - r$.

In the first case we obtain

$$v_{n-1}^2 = v_n - ab = -r - ab \geq 0 \qquad r + ab \leq 0 \tag{4-69}$$

if v_{n-1} is to be real.

Substitution of the quadratic $ab = -r^2 + r$ into (4-69) yields

$$r(2 - r) \leq 0 \tag{4-70}$$

Since we have assumed $r \geq \frac{1}{2}$, we have $r \geq 2$ to satisfy (4-70). Similarly, if $v_n = r - 1$, we obtain $r \geq 1$. If these conditions are satisfied, v_0 can be found by successively solving (4-69) for $v_{n-1}, v_{n-2}, \ldots, v_0$. In particular, if $r = 2$, $v_0 = 2 \cos (k\pi/2^m)$ for positive integers k and m.

Finally, suppose the sequence does not become stationary. Then if $\lim\limits_{n \to \infty} y_n$ exists, it equals r or $1 - r$. The limit will exist if and only if $|v_0| < r$ and $\frac{1}{2} \leq r \leq \frac{3}{2}$, and the limit will be $1 - r$. We show that it is not equal to r. The rest of the conditions follow in a similar manner.

Suppose $v_n = r + a$. Then

$$\Delta v_n = v_{n+1} - v_n = a(a + (2r - 1)) \qquad 2r - 1 > 0$$

Then if $a > 0$, $\Delta v_n > 0$, and v_{n+1} is further removed from r. If $1 - r < v_n < r$, so that $1 - 2r < a < 0$, then $\Delta v_n < 0$, and again v_{n+1} is further removed from r. Therefore r cannot be the limit point.

Another asymptotic solution is illustrated by the nonlinear equation[40]

$$u_{n+1}^2 = u_n + a \qquad n = 1, 2, \ldots$$
$$u_1 = \sqrt{a} \qquad (4\text{-}71)$$

where a is complex. It has the asymptotic solution

$$\lim_{n \to \infty} u_n = \frac{1 + \sqrt{1 + 4a}}{2} \qquad (4\text{-}72)$$

Note that $u_{n+1} = \sqrt{a + u_n}$ is an infinite radical with the same period a.

Exercise 4-12 Obtain the foregoing solution. *Hint:* the proof is roughly as follows. If the limit exists, it must be one of the two roots r_1 and r_2 of the quadratic $u^2 - u - a = 0$. The problem is then to show that it is r_1, where r_1 is the root with positive sign before the radical. The argument can be given in several parts: (1) $a \neq -\frac{1}{4}$ is nonreal, and Im $(a) > 0$ or < 0; (2) a is real; (3) $a = -\frac{1}{4}$. To prove (1) define

$$R_n = \frac{u_n - r_1}{u_n - r_2}$$

and hence if $\lim\limits_{n \to \infty} R_n = 0$, then $\lim\limits_{n \to \infty} u_n = r_1$. Since $r_1^2 = r_1 + a$, $r_2^2 = r_2 + a$, we have

$$u_{n+1}^2 - r_1^2 = u_n - r_1 \qquad u_{n+1}^2 - r_2^2 = u_n - r_2$$

Hence one can show by writing $u_n = \alpha_n + i\beta_n$ that

$$\left| \frac{R_{n+1}}{R_n} \right| = \left| \frac{u_{n+1} + r_2}{u_{n+1} + r_1} \right| < 1 - \epsilon$$

for $\epsilon > 0$ and hence $\lim\limits_{n \to \infty} R_n = 0$.

Exercise 4-13 Supply the proof for (2) and (3).

Exercise 4-14 Define $y_0 = 1$, $y_1 = 3$, $y_{n+1} = 2y_n^2 - 1$ for $n \geq 1$ and find†

$$\lim_{n \to \infty} \frac{y_n}{2^n y_0 y_1 \cdots y_{n-1}}$$

† See *Amer. Math. Monthly*, vol. 60, p. 711, 1953.

REDUCTION TO LINEAR EQUATIONS

Some nonlinear difference equations can be reduced to linear equations by a change of variables. Examples of these are

$$y_{n+1} = \frac{y_n}{1 + y_n} \qquad n = 0, 1, \ldots \qquad (4\text{-}73)$$

Thus we obtain a linear equation if we put $u_n = 1/y_n$ provided that $y_0 \neq 0$, $y_n \neq 0$ for every n. The equation

$$y_{n+1} = a - \frac{b}{y_n} \qquad (4\text{-}74)$$

is transformed to a linear equation $u_{n+2} - a u_{n+1} + b u_n = 0$ by setting $y_n = u_{n+1}/u_n$. This substitution can also be used to transform

$$y_{n+1}^2 - 3y_{n+1}y_n + 2y_n^2 = 0 \qquad (4\text{-}75)$$

to the form
$$\left(\frac{u_{n+2}u_n}{u_{n+1}} - 1 \right) \left(\frac{u_{n+2}u_n}{u_{n+1}} - 2 \right) = 0$$

from which $u_n = 1$ or 2. Each of the two values gives rise to a solution of the original equation. The two solutions are y_0 and $y_0 2^n$.

Next consider two concave mirrors[8] A_1 and A_2 facing each other (as these parentheses do) at a distance of α units. Their respective focal lengths are k_1 and k_2. An object is placed on the line segment passing through their respective foci at a distance y_1 from A_1. It forms an image in A_1 at a distance x_1 in front of A_1. This image is now the object for A_2 and forms an image in front of A_2 at a distance y_2 from A_1, etc., for x_n and y_n. We have

$$\frac{1}{y_1} + \frac{1}{x_1} = \frac{1}{k_1} \qquad \frac{1}{\alpha - y_2} + \frac{1}{\alpha - x_1} = \frac{1}{k_2} \qquad (4\text{-}76)$$

and in general

$$\frac{1}{y_n} + \frac{1}{x_n} = \frac{1}{k_1} \qquad \frac{1}{\alpha - y_{n+1}} + \frac{1}{\alpha - x_n} = \frac{1}{k_2} \qquad (4\text{-}77)$$

If we eliminate x_n between these two equations, we obtain a nonlinear difference equation of the general form

$$y_{n+1} = \frac{a y_n + b}{c y_n + d} \qquad (4\text{-}78)$$

which can be reduced to a pair of linear difference equations by writing $y_n = v_n/u_n$ and setting the numerators and denominators equal.

Exercise 4-15 Should one be concerned about common factors in (4-78)? Explain.

Exercise 4-16† Derive the solution of the cyclic set of simultaneous equations

$$x_1 - 2x_n = x_n^2 x_1$$
$$x_{j+1} - 2x_j = x_j^2 x_{j+1}$$

where n is a positive integer $n > 1$, $j = 1, \ldots, n - 1$. *Hint:* Successively substitute $2x_j/(1 - x_j^2)$ for x_{j+1} until only $x_{j+n} = x_j$ appears. Obtain an equation of degree $2^n + 1$ which has $x_j(x_j^2 + 1)$ as a factor, that is, $x_j = 0$, $x_j = \pm i$. Also let $x_j = \tan \theta$; then $x_{j+1} = \tan 2\theta$, $x_{j+n} = x_j = \tan 2^n\theta$, and hence $\theta = k\pi/(2^n - 1)$. The $2^n - 1$ real values of $\tan \theta$ for $k = 0, \ldots, 2^n - 2$, repeating $x_j = 0$ together with $\pm i$, give the complete solution of our equation of degree $2^n + 1$.

DIFFERENCE ANALOGUES TO CLAIRAUT'S AND RICCATI'S DIFFERENTIAL EQUATIONS

In the first-order nonlinear equation

$$F(x, y(x), \Delta y(x)) = 0 \tag{4-79}$$

let $\Delta y(x) = z(x)$. Applying the operator Δ to the resulting equation yields

$$\varphi(x, y(x), z(x), \Delta z(x)) = 0 \tag{4-80}$$

where $\varphi \equiv \Delta F$.

If this equation is independent of $y(x)$, and if we can solve for $z(x)$, we would obtain

$$g(x, z(x), A(x)) = 0 \tag{4-81}$$

where $A(x)$ is an arbitrary unit periodic. As an illustration, consider

$$y(x) = x \, \Delta y(x) + f[\Delta y(x)] \tag{4-82}$$

which is analogous to Clairaut's differential equation. If we put

$$\Delta y(x) = z(x)$$

we have

$$y(x) = xz(x) + f[z(x)] \tag{4-83}$$

which, on applying Δ, becomes

$$0 = (x + 1) \, \Delta z(x) + f(z + \Delta z) - f(z) \tag{4-84}$$

Thus either $\Delta z = 0$, and hence $z = A(x)$, from which we have

$$y(x) = xA(x) + f[A(x)] \tag{4-85}$$

or $\qquad \dfrac{f(z + \Delta z) - f(z)}{\Delta z} = -(x + 1) \tag{4-86}$

from which one can obtain another solution.

† *Amer. Math. Monthly*, vol. 35, p. 211, April, 1928.

Exercise 4-17 Apply the foregoing to

$$y(x) = x \, \Delta y(x) + [\Delta y(x)]^2$$

The difference equation which corresponds to Riccati's equation (see "NM," chap. 4) can be written as

$$y(x)y(x + 1) + a(x)y(x + 1) + b(x)y(x) + c(x) = 0 \qquad (4\text{-}87)$$

On putting

$$y(x) = \frac{z(x + 1)}{z(x)} - a(x) \qquad (4\text{-}88)$$

we have the second-order linear equation

$$z(x + 2) + [b(x) - a(x + 1)]z(x + 1) + [c(x) - a(x)b(x)]z(x) = 0 \qquad (4\text{-}89)$$

As in the case of differential equations, if three particular solutions are known, the solution can be written down.

Exercise 4-18 By putting $y(x) = \tan z(x)$ show that a solution of $y(x + 1)y(x) - a(x)[y(x + 1) - y(x)] + 1 = 0$ is given by

$$y(x) = \tan \left[A(x) + \Delta^{-1} \tan^{-1} \frac{1}{a(x)} \Delta x \right] \qquad (4\text{-}90)$$

Remark: Again, as in the case of ordinary differential equations, one can establish conditions for the equation

$$Q(x,y(x),\Delta y(x)) \, \Delta y(x) + P(x,y(x)) = 0 \qquad (4\text{-}91)$$

to be exact.

TWO EQUATIONS WITH INTEGER SOLUTIONS

In the case of difference equations one may be concerned with obtaining a solution in integers. For example, the following nonlinear difference equation[37]

$$y_{n+1} = \frac{1 + y_n y_{n-1}}{y_{n-2}} \qquad y_1 = y_2 = y_3 = 1$$

has a solution y_n that is an integer. To see this, define the sequence $\{x_n\}$ of integers $x_1 = x_2 = x_3 = 1$, $x_4 = 2$, $x_n = 4x_{n-2} - x_{n-4}$, $n > 4$. Then

$$x_{n+1}x_{n-2} - x_n x_{n-1} = (4x_{n-1} - x_{n-3})x_{n-2} - (4x_{n-2} - x_{n-4})x_{n-1}$$
$$= x_{n-1}x_{n-4} - x_{n-2}x_{n-3}$$

and inductively

$$x_{n+1}x_{n-2} - x_n x_{n-1} = 1$$

or

$$x_{n+1} = \frac{1 + x_n x_{n-1}}{x_{n-2}}$$

from which

$$\{y_n\} = \{x_n\}$$

The classical *regula falsi* method for solving $f(x) = 0$ is given by[45]

$$x_{n+1} = \frac{x_{n-1}f(x_n) - x_n f(x_{n-1})}{f(x_n) - f(x_{n-1})}$$

For $f(x) = x^2 = 0$

$$x_{n+1} = \frac{x_{n-1}x_n}{x_{n-1} + x_n}$$

By definition, the Fibonacci numbers satisfy

$$F_1 = F_2 = 1 \qquad F_{n+2} = F_{n+1} + F_n$$

Thus if $x_0 = 1$, $x_1 = \frac{1}{2}$. Application of the *regula falsi* method using $f(x) = x^2$ gives the reciprocal of the Fibonacci numbers

$$\frac{\dfrac{1}{F_{i+1}} + \dfrac{1}{F_{i+2}}}{\dfrac{1}{F_{i+1}} + \dfrac{1}{F_{i+2}}} = \frac{1}{F_{i+1} + F_{i+2}} = \frac{1}{F_{i+3}}$$

SOLUTION BY SERIES

A first-order difference equation has the general implicit form[43]

$$F(x, y(x), \Delta y(x)) = 0$$

where $\Delta y(x) = y(x + 1) - y(x)$. This form can also be represented by $G(x, y(x), y(x + 1)) = 0$. A canonical form is obtained when one solves for $\Delta y(x)$, obtaining $\Delta y(x) = f(x, y(x))$. Suppose $f(x, y)$ is analytic in the region $D_0: |x| \geq R_0$, $|\arg x| \leq \alpha$, $|y| \leq r_0$, and can be expanded asymptotically as

$$f(x, y) \approx \sum_{j+k=1}^{\infty} a_{jk} x^{-j} y^k$$

Let $a_{01} \neq 0$, put $y = \sum_{j=1}^{\infty} g_j x^{-j}$, and substitute in

$$\sum_{j+k=1}^{\infty} a_{jk} x^{-j} y^k - \Delta y = 0$$

Then equate coefficients of successive powers of x^{-1} and obtain

$$a_{01}g_1 + a_{10} = 0 \qquad a_{01}g_2 + a_{20} + a_{11}g_1 + a_{02}g_1{}^2 = 0, \ldots$$

and generally

$$a_{01}g_n + p_n(g_1, \ldots, g_{n-1}, a_{jk}) = 0 \qquad n = 1, 2, \ldots$$

where p_n is a polynomial in $g_1, \ldots, g_{n-1}$ and depends only on a_{jk}, for which $j + k \leq n$. In this manner the g's are successively determined, giving the expansion for $y(x)$. It can be shown that if $|1 + a_{01}| > 1$, then under specified conditions such a solution exists and is unique. If $|1 + a_{01}| = 1$, we must have $\alpha < \pi/2$. If a_{0m} is such that $a_{0k} = 0$ ($k = 1, \ldots, m - 1$), $a_{0m} \neq 0$, then if $a_{10} \neq 0$ and $m \geq 2$, or if $a_{10} = 0$, $m = 2$, a_{11} arbitrary (or $m \geq 3$, a_{11} is not a negative integer) or, third, if $a_{10} = 0$, $m \geq 3$, a_{11} is arbitrary, $a_{11} \neq (n - m)/(m - 1)^2$, we have m distinct analytic solutions of the form

$$y(x) = g_1{}^p x^{-1/m} + \cdots + g_n{}^p x^{-n/m} \qquad p = 0, \ldots, m - 1$$

The second-order linear difference equation

$$A_1(x)y(x + 2) + A_2(x)y(x + 1) + A_3(x)y(x) = 0$$

where $A_i(x) = \sum_{k=0}^{n} a_{ik}x^{n-k}$, $i = 1, 2, 3$, for the special case in which the characteristic equation $a_{10}\rho^2 + a_{20}\rho + a_{30} = 0$ has distinct roots r_1, r_2 with $|r_1|$, $|r_2| < 1$ and r_1 satisfies $a_{11}\rho^2 + a_{21}\rho + a_{31} = 0$ has the solution $y(x) = \alpha_0 r_1{}^x \sum_{n=1}^{\infty} \alpha_n(x)$, where α_0 is an arbitrary constant and $\alpha_n(x)$ are solutions of nonhomogeneous difference equations with constant coefficients.

Remark: The system

$$y_i(x + 1) = x^{-1} \sum_{j=1}^{n} b_{ij}(x)y_j(x) + f_i(y_1(x), \ldots, y_n(x); p, x)$$

with the conditions

$$f_i(0, \ldots, 0; x) \equiv 0 \qquad i = 1, \ldots n$$

where the f_i are analytic in y_i and p and are continuous in the complex variable x and bounded for all values of $y_i(x)$, p, and x in the domain $G: |y_i(x)| \leq r_i$, $|p| \leq \rho$, $|x| \geq K$ and where the $b_{ij}(x)$ are any continuous and bounded functions of x defined in G, has a solution analytic in p and continuous in x.[43]

4-5 STABILITY OF SOME DIFFERENCE APPROXIMATIONS

We shall examine the behavior of difference approximations to differential equations through some simple examples. We start with the partial differential equation[16]

$$\frac{\partial y(t,x)}{\partial t} = c \frac{\partial^2 y(t,x)}{\partial x^2} \qquad a \leq x \leq b, t \geq 0 \qquad (4\text{-}92)$$

and seek a difference equation (with stable solution) approximating it. First we divide $[a,b]$ into n equal intervals with $\Delta x \equiv (b - a)/n$, and we let $x_j = a + (j + \frac{1}{2}) \Delta x$. Next we define y_j as the average of y over $x_j - \Delta x/2 \leq x \leq x_j + \Delta x/2$, that is,

$$y_j = \frac{c}{\Delta x} \int_{x_j - \Delta x/2}^{x_j + \Delta x/2} y \, dx \qquad (4\text{-}93)$$

We then have

$$\frac{dy_j}{dt} = \frac{c}{\Delta x} \left[\frac{\partial y}{\partial x} \right]_{x_j - \Delta x/2}^{x_j + \Delta x/2} \qquad (4\text{-}94)$$

We approximate $\partial y/\partial x$ by

$$\left[\frac{\partial y}{\partial x} \right]_{x = x_j \pm \Delta x/2} = \pm \frac{1}{\Delta x} [y_{j\pm 1} - y_j] \qquad (4\text{-}95)$$

and hence

$$\frac{dy_j}{dt} = \frac{c}{\Delta x^2} [y_{j+1} + y_{j-1} - 2y_j] \qquad (4\text{-}96)$$

We also divide the time t into finite intervals Δt and let $y_j{}^k \equiv y_j$ at $t = k \Delta t$. Then we introduce an approximation for $dy_j{}^k/dt$, given by

$$\frac{dy_j{}^k}{dt} = \frac{y_j{}^{k+1} - y_j{}^k}{\Delta t} \quad \text{or} \quad \frac{dy_j{}^k}{dt} = \frac{1}{\Delta t} [y_{j-1}^{k+1} - \frac{1}{2}(y_{j-1}^k + y_{j+1}^k)] \quad (4\text{-}97)$$

Alternative approximations are also possible.

We have the resulting difference equations:

$$y_j{}^{k+1} = y_j{}^k + \frac{c \, \Delta t}{\Delta x^2} [y_{j+1}^k + y_{j-1}^k - 2y_j{}^k]$$

$$y_j{}^{k+1} = \frac{1}{2}(y_{j-1}^k + y_{j+1}^k) + \frac{c \, \Delta t}{\Delta x^2} [y_{j+1}^k + y_{j-1}^k - 2y_j{}^k]$$

$$y_j{}^{k+1} = y_j{}^k + \frac{c \, \Delta t}{\Delta x^2} [y_{j+1}^{k+1} + y_{j-1}^{k+1} - 2y_j{}^{k+1}]$$

. . . .

The only a priori requirement on the approximations is that they formally reduce to the original differential equation as Δx and $\Delta t \to 0$.

If we assume that the difference between two solutions can be decomposed into Fourier components of the form

$$A e^{imj\Delta x} e^{ln\Delta t} \qquad i = \sqrt{-1} \tag{4-98}$$

and if this is substituted into the first approximation, we find that it will be a solution if

$$e^{l\Delta t} = 1 - \frac{2c\,\Delta t}{\Delta x^2}\,(1 - \cos(m\,\Delta x)) \tag{4-99}$$

For the Fourier component to remain bounded it is necessary that $|e^{l\Delta t}| \le 1$ for all m. Since $e^{l\Delta t} \le 1$ already, we must have $e^{l\Delta t} > -1$. The worst case is when $\cos m\,\Delta x = -1$, which leads to the requirement

$$\frac{c\,\Delta t}{\Delta x^2} < \frac{1}{2} \tag{4-100}$$

The same argument applied to the second approximation leads to

$$e^{l\Delta t} = \cos m\,\Delta x - \frac{2c\,\Delta t}{\Delta x^2}\,(1 - \cos m\,\Delta x) \tag{4-101}$$

In this case, for $\cos m\,\Delta x = -1$, $e^{l\Delta t} < -1$; so that the second approximation is unconditionally unstable. For further discussion of stability, see below.

Exercise 4-19 Show that the third approximation is unconditionally stable.

We shall illustrate with a second example outlining the steps to be followed

$$\frac{\partial y}{\partial t} = y - \frac{\partial y}{\partial x} + \frac{\partial^2 y}{\partial x^2} \tag{4-102}$$

Solution: Let

$$\Delta x = \frac{b - a}{n} \qquad \text{and} \qquad x_j = a + (j + \tfrac{1}{2})\,\Delta x \tag{4-103}$$

and

$$y_j = \frac{1}{\Delta x} \int_{x_j - \Delta x/2}^{x_j + \Delta x/2} y\,dx \tag{4-104}$$

Then

$$\frac{dy_j}{dt} = \frac{1}{\Delta x} \int_{x_j - \Delta x/2}^{x_j + \Delta x/2} \left[y - \frac{\partial y}{\partial x} + \frac{\partial^2 y}{\partial x^2} \right] dx$$

$$= y_j - \frac{1}{\Delta x}\,[y]_{x_j - \Delta x/2}^{x_j + \Delta x/2} + \frac{1}{\Delta x} \left[\frac{\partial y}{\partial x} \right]_{x_j - \Delta x/2}^{x_j + \Delta x/2} \tag{4-105}$$

If we use the approximations

$$[y]_{x=x_j\pm\Delta x/2} = \tfrac{1}{2}[y_{j\pm1} + y_j] \tag{4-106}$$

$$\left[\frac{\partial y}{\partial x}\right]_{x=x_j\pm\Delta x/2} = \pm\frac{1}{\Delta x}[y_{j\pm1} - y_j] \tag{4-107}$$

we have $\quad \dfrac{dy_j}{dt} = y_j - \dfrac{1}{2\Delta x}[y_{j+1} - y_{j-1}] + \dfrac{1}{\overline{\Delta x}^2}[y_{j+1} - 2y_j + y_{j-1}] \tag{4-108}$

Dividing the time interval yields $dy_j{}^k/dt = 1/\Delta t[y_j{}^{k+1} - y_j{}^k]$.

$$y_j{}^{k+1} = y_j{}^k + \Delta t\, y_j{}^k - \frac{1}{2}\frac{\Delta t}{\Delta x}[y_{j+1}^k - y_{j-1}^k]$$
$$+ \frac{\Delta t}{\overline{\Delta x}^2}[y_{j+1}^k - 2y_j{}^k + y_{j-1}^k] \tag{4-109}$$

$$y_j{}^{n+1} = y_j{}^n + \Delta t\frac{\partial y_j{}^n}{\partial t} + \overline{\Delta t}^2\frac{\partial^2 y_j{}^n}{\partial t^2} + \cdots \tag{4-110}$$

$$y_{j\pm1}^n = y_j{}^n \pm \Delta x\frac{\partial y_j{}^n}{\partial x} + \overline{\Delta x}^2\frac{\partial^2 y_j{}^n}{\partial x^2} \pm \cdots \tag{4-111}$$

Then, in general,

$$\frac{\partial y}{\partial t} - y + \frac{\partial y}{\partial x} - \frac{\partial^2 y}{\partial x^2} = \left[\frac{\overline{\Delta x}^2}{12}\frac{\partial^4 y}{\partial x^4} + \frac{\overline{\Delta x}^4}{360}\frac{\partial^6 y}{\partial x^6} + \cdots\right]$$
$$- \left[\frac{\overline{\Delta x}^2}{6}\frac{\partial^3 y}{\partial x^3} + \frac{\overline{\Delta x}^4}{120}\frac{\partial^5 y}{\partial x^5} + \cdots\right] - \left[\frac{\Delta t}{2}\frac{\partial^2 y}{\partial t^2} + \frac{\overline{\Delta t}^2}{6}\frac{\partial^3 y}{\partial t^3} + \cdots\right]$$
$$\tag{4-112}$$

and the error can be made as small as desired.

Turning now to a nonlinear system, suppose we are given

$$\frac{dy}{dt} = 2z - \frac{\alpha}{2}y|y| \tag{4-113}$$

$$\frac{dz}{dt} = -y \tag{4-114}$$

Let $\qquad\qquad E = \tfrac{1}{2}y^2 + z^2 \tag{4-115}$

Then $\qquad\qquad \dfrac{dE}{dt} = -\dfrac{\alpha}{2}y^2|y| \tag{4-116}$

Centering our attention on $t = k\,\Delta t$, with

$$\frac{dy}{dt} = \frac{y^{k+1} - y^k}{\Delta t}\qquad \frac{dz^k}{dt} = \frac{z^{k+1} - z^k}{\Delta t}\qquad \frac{dE^k}{dt} = \frac{E^{k+1} - E^k}{\Delta t}$$

we have $\qquad\qquad y^{k+1} = y^k + \Delta t\left[2z^k - \dfrac{\alpha}{2}y^k|y^k|\right] \tag{4-117}$

$$z^{k+1} = z^k - y^k\,\Delta t \tag{4-118}$$

From (4-115), (4-117), and (4-118) we have

$$\frac{E^{k+1} - E^k}{\Delta t} = -\frac{\alpha}{2} (y^k)^2 |y^k| + \Delta t \left[\frac{1}{2} \left(2z^k - \frac{\alpha}{2} y^k |y^k| \right)^2 + (y^k)^2 \right] \quad (4\text{-}119)$$

and the first and second terms on the right oppose each other.   Actually $E \to 0$ as $t \to \infty$, and $dE/dt \leq 0$.   Depending on initial values, however, Eq. (4-119) can have either sign or change sign in the problem.   In fact, $E^{k+1} = E^k$ if and only if

$$z^k = \frac{y^k}{2} \left[\frac{\alpha |y^k|}{2} \pm \sqrt{\frac{\alpha}{\Delta t} |y^k| - 2} \right] \quad (4\text{-}120)$$

If $|y^k| < 2\Delta t/\alpha$, there is no real z^k for which $E^{k+1} = E^k$.   In fact, for all z^k, $E^{k+1} > E^k$ if $|y^k| < 2\Delta t/\alpha$, and E^k can grow without bound.   However, if y is not identically zero, then Δt can be chosen small enough so that $|y^k| \geq 2\Delta t/\alpha$ somewhere.   Note that $z = A \sin (t \sqrt{2})$,

$$y = -A \sqrt{2} \cos (t \sqrt{2})$$

is a solution for the system

$$\frac{dy}{dt} = 2z \qquad \frac{dz}{dt} = -y \quad (4\text{-}121)$$

obtained from the original system by setting $\alpha = 0$.   Letting

$$y^{k+1} = y^k + \Delta t \frac{dy^k}{dt} + \frac{\overline{\Delta t}^2}{2} \frac{d^2 y^k}{dt^2} + \cdots$$

$$z^{k+1} = z^k + \Delta t \frac{dz^k}{dt} + \frac{\overline{\Delta t}^2}{2} \frac{d^2 z^k}{dt^2} + \cdots \quad (4\text{-}122)$$

$$E^{k+1} = E^k + \Delta t \frac{dE^k}{dt} + \frac{\overline{\Delta t}^2}{2} \frac{d^2 E^k}{dt^2} + \cdots$$

and substituting these values in (4-117) to (4-119), we obtain

$$\frac{dy}{dt} + \frac{\Delta t}{2} \frac{d^2 y}{dt^2} = 2z - \frac{\alpha}{2} y|y|$$

$$\frac{dz}{dt} + \frac{\Delta t}{2} \frac{d^2 z}{dt^2} = -y \quad (4\text{-}123)$$

$$\frac{dE}{dt} = -\frac{\alpha}{2} y^2 |y| - \frac{\Delta t}{2} \left(y \frac{d^2 y}{dt^2} + 2z \frac{d^2 z}{dt^2} \right)$$

If we seek an approximation to the solution of our system near $\alpha = 0$, we obtain

$$\frac{dE}{dt} = -\alpha A^3 \sqrt{2} \left[\cos^2 (t \sqrt{2}) \right] |\cos (t \sqrt{2})| + 2A^2 \Delta t$$

When the amplitude reaches equilibrium, the average of dE/dt over a cycle will vanish; that is,

$$-\alpha A_\infty^3 \sqrt{2} \left(\frac{4}{3\pi}\right) + 2A_\infty^2 \Delta t = 0$$

Thus
$$A_\infty = \frac{3\pi \Delta t}{2\alpha \sqrt{2}} \qquad (4\text{-}124)$$

Moreover $E = \frac{1}{2}y^2 + z^2 = A^2 \cos^2 (t \sqrt{2}) + A^2 \sin^2 (t \sqrt{2}) = A^2$

from which $\dfrac{dE}{dt} = 2A \dfrac{dA}{dt}$

and, averaging over a quarter cycle, we have

$$\frac{\overline{dA}}{dt} = \frac{1}{2A} \frac{\overline{dE}}{dt} = -\frac{2\alpha \sqrt{2}}{3\pi} A(A - A_\infty)$$

whose solution is

$$A = \frac{A_0 A_\infty}{A_0 + (A_\infty - A_0)e^{-\bar{t}\Delta t}} \qquad (4\text{-}125)$$

Substituting into (4-117), (4-118), and (4-115) yields

$$E = A^2 - \frac{1}{2}A^2 \Delta t \sqrt{2} \sin (2t \sqrt{2}) \qquad (4\text{-}126)$$

When we use (4-124) to (4-126), we see that the stability for any given case depends on A_0.

4-6 STABILITY

THE LYAPUNOV CRITERION FOR DIFFERENCE EQUATIONS

By analogy with the analysis of stability of a system of first-order differential equations one can study the stability of the dynamical system represented by the first-order system of difference equations[25]

$$y_k(t + 1) = Y_k(y_1, \ldots, y_n; t) \qquad k = 1, \ldots, n \qquad (4\text{-}127)$$

where t can take on any integer value. In vector notation we have

$$y(t + 1) = Y(y,t) \qquad (4\text{-}128)$$

By analogy with the existence proof for differential equations this too can be shown to have a unique solution $y(t)$, with $y(t_0) = y^0$, which exists in $t_0 \le t < \infty$. We now transform this system into the disturbed system. Thus if position disturbances $x_k(t_0)$ are introduced at time t_0, and if $y_k(t) = \varphi_k(t)$ is the particular solution at t_0, the position coordinates of the system at t_0 become

$$\bar{y}_k(t_0) = \varphi_k(t_0) + x_k(t_0) \qquad k = 1, \ldots, n \qquad (4\text{-}129)$$

Let $\bar{y}_k(t)$ be the solution of the new problem. We study the equations of disturbed motion

$$x_k(t + 1) = y_k(t + 1) - \varphi_k(t + 1) = Y_k(x_1 + \varphi_1, \ldots, x_n + \varphi_n; t)$$
$$- Y_k(\varphi_1, \ldots, \varphi_n; t) \equiv X_k(x_1, \ldots, x_n; t) \equiv X_k(x, t) \quad (4\text{-}130)$$

If we assume that $\partial Y_k / \partial x_j, j = 1, \ldots, n$, exist and are continuous, we can write

$$X_k(x, t) = \sum_{j=1}^{n} a_{kj}(t) x_j + f_k(x, t) \qquad (4\text{-}131)$$

The vector $f(x, t)$ is continuous in x for $t > t_0$. Thus (4-130) can be written in vector notation as

$$x(t + 1) = A(t)x + f(x, t) \qquad (4\text{-}132)$$

where $A(t)$ is the matrix of the $a_{kj}(t)$. The system $x(t + 1) = A(t)x$ is known as the linear variational system of (4-127) relative to $\varphi_k(t)$. If, as in the autonomous case, we assume that t does not appear explicitly on the right, we have

$$x(t + 1) = Ax + f(x) \qquad (4\text{-}133)$$

where A is a constant matrix.

Definitions of stability and asymptotic stability are identical with those for differential equations (See "NM"). The definition of stability for a system with the above arguments is a special case of the system

$$X(t_{k+1}) = Ax(t_k) + f(x(t_k))$$

A solution $X = 0$ of this system is stable if, given $\epsilon > 0$ and t_0, there is a $\delta(\epsilon, t_0) > 0$ such that $\|X(t_k)\| < \epsilon$, $t_k > t_0$, whenever $\|X(t_0)\| < \delta$. It is asymptotically stable if, in addition, $\lim_{k \to \infty} \|X(t_k)\| \to 0$.

A Lyapunov function $V(x)$ for the difference system (4-133) can also be defined by analogy with differential equations. We define

$$\Delta V(x(t_k)) = V(x(t_{k+1})) - V(x(t_k))$$

We shall illustrate the application of Lyapunov's theory to difference equations by the following theorem.

Theorem 4-2   If $V(x)$ is positive-definite in the whole space, that is, $V(x) > 0$, $x \neq 0$, $V(0) = 0$, and if ΔV is negative-definite, that is, $\Delta V(x) < 0$, $x \neq 0$, $\Delta V(0) = 0$ along a solution, then the solution $X = 0$ of (4-132) is asymptotically stable in the large.

Proof: The proof proceeds exactly as for the continuous case except that here we write

$$V[x_1(t_k), \ldots, x_n(t_k)] = V[x_1(t_0), \ldots, x_n(t_0)]$$
$$+ \sum_{i=0}^{k-1} \Delta V[x(t_i)] \leq V[x(t_0)] - \beta(t_k - t_0)$$

Remark: Consider the nonlinear difference equation

$$x_{n+m} + y_1 x_{n+m-1} + \cdots + y_m x_n = 0 \tag{4-134}$$

in which y_i $(i = 1, \ldots, m)$ depends on x_{n+k} $(k = 0, \ldots, m - 1)$ and on n. If $x_{n+k} = 0$ $(k = 0, \ldots, m - 1)$ is an equilibrium point, then if there is an $\epsilon > 0$ such that $\sum_{i=1}^{m} |y_i| < 1 - \epsilon$, the following can serve as a Lyapunov function:

$$V = \sum_{k=p}^{n+m-1} (1 - \epsilon)|x_k| - \sum_{k=p}^{n+m-2} |y_1| \, |x_k| - \cdots - \sum_{k=p}^{n-1} |y_m| \, |x_k| \qquad n > p$$

Along a solution, $V < 0$, and we have asymptotic stability.

The following example for the analysis of a discrete system is due to O'Shea.[35] Suppose our system of equations is given by

$$X_{k+1} \equiv X(t_{k+1}) = F(X(t_k))$$

of which

$$x_1(t_{k+1}) = x_1{}^2(t_k) \qquad x_2(t_{k+1}) = x_2{}^2(t_k)$$

is a special case. Let us study the stability of this special system. Suppose that $V = x_1{}^2 + x_2{}^2$. Let $X_k = (x_1(t_k), x_2(t_k))$. Then

$$\Delta V(X_k) = V(X_{k+1}) - V(X_k) = V(F(X_{k+1})) - V(F(X_k))$$

and for our example

$$\Delta V = -x_1{}^2(1 - x_1{}^2) - x_2{}^2(1 - x_2{}^2)$$

Exercise 4-20 Derive the last expression for ΔV.

Note that ΔV is negative-definite for $\|X\| < 1$, $X = (x_1, x_2)$, and asymptotic stability is in

$$V = x_1{}^2(t_k) + x_2{}^2(t_{k+1}) < 1$$

Its boundary is $V(X_k) = 1$.

Now form a sequence of Lyapunov functions according to the relation

$$V_n(X_k) = V_{n-1}(F(X_k))$$

with $V_0 \equiv V$. Thus

$$V_n(X_k) = x_1{}^{2n+1} + x_2{}^{2n+1} = 1$$

As $n \to \infty$, the region defined by $V_n(X_k) = 1$ approaches a square circumscribing the unit circle, and asymptotic stability is obtained for X such that $\|X\| = \max |X_i| = 1$. An excellent account of these ideas is given in Ref. 29.

The autonomous system[26]

$$x(t_{k+1}) = h(x(t_k)) \qquad h(0) = 0 \qquad (4\text{-}135)$$

where $x(t)$ is an n-dimensional vector, is asymptotically stable in the large if h can be shown to be a contraction ($\|h(x)\| < \|x\|$) for all $x \neq 0$.

This follows immediately from the observation that

$$V(x) = \|x\| > 0 \qquad x \neq 0 \qquad (4\text{-}136)$$

is a Lyapunov function for x, with

$$\Delta V(x_k) = \|x(t_{k+1})\| - \|x(t_k)\| < 0 \qquad (4\text{-}137)$$

The stability of (4-135) can therefore be determined if a norm for which h is a contraction can be found.

Suppose

$$x_i(t_{k+1}) = \sum_j h_{ij}(x(t_k))x_j(t_k) \qquad (4\text{-}138)$$

and that there are positive constants $c_1, c_2, \ldots, c_n$ such that one of the following holds:

1. $\displaystyle \max_i \left\{ \sum_{j=1}^{n} \frac{c_i}{c_j} |h_{ij}(x)| \right\} < 1.$

2. $\displaystyle \max_j \left\{ \sum_{i=1}^{n} \frac{c_i}{c_j} |h_{ij}(x)| \right\} < 1$ for all x.

Then an appropriate norm in case 1 is

$$\|x\| = \max_i c_i|x_i| \qquad (4\text{-}139)$$

Now

$$\|x(t_{k+1})\| = \max_i \left\{ c_i \left| \sum_{j=1}^{n} h_{ij}(x(t_k))x_j(t_k) \right| \right\}$$

$$\leq \max_i \left\{ \sum_{j=1}^{n} \frac{c_i}{c_j} |h_{ij}(x)||c_j||x_j| \right\}$$

$$\leq \max_i \left\{ \sum_{j=1}^{n} \frac{c_i}{c_j} |h_{ij}(x)| \right\} \max_j \{c_j|x_j|\}$$

Therefore, from condition 1 and (4-139), $\|x(t_{k+1})\| < \|x(t_k)\|$, and h is a contraction.

In case 2 one can show that a similar result is obtained using

$$\|x\| = \sum_{i=1}^{n} c_i|x_i| \qquad (4\text{-}140)$$

Szegö and Kalman[42] have given a sufficient condition for the stability of the system

$$x_{t+1} = A x_t - b\varphi(\sigma_t) \qquad (4\text{-}141)$$

where A and b are given by

$$A = \begin{pmatrix} 0 & 1 & \cdots & 0 \\ \cdots & \cdots & \cdots & \cdots \\ 0 & 0 & \cdots & 1 \\ -\alpha_n & -\alpha_{n-1} & \cdots & -\alpha_1 \end{pmatrix} \qquad b = \begin{pmatrix} 0 \\ \cdots \\ 0 \\ 1 \end{pmatrix} \qquad (4\text{-}142)$$

and φ is a real continuous (nonlinear) function of the real variable σ such that

$$\varphi(0) = 0 \qquad 0 \le \sigma\varphi(\sigma) < \sigma^2 k,\ k < \infty \qquad (4\text{-}143)$$

and $\sigma_t = 2c'x_t,\ t = 0, 1, \ldots$. Here c is a vector, and the vectors b, Ab, A^2b, $\ldots$, $A^{n-1}b$ and c, $A'c$, $\ldots$, $(A')^{n-1}c$ are assumed to be linearly independent. The characteristic values λ_i of A satisfy

$$|\lambda_i| < 1 \qquad i = 1, \ldots, n \qquad (4\text{-}144)$$

Then from the inequality

$$k^{-1} + \mathrm{Re}\,\{c'(e^{j\omega}I - A)^{-1}b\} \ge 0 \qquad \text{for all real } \omega \qquad (4\text{-}145)$$

follows the existence of a Lyapunov function $V(x) = xHx$, where H is a symmetric positive-definite matrix, such that

$$\Delta V(x) = -[\gamma\varphi(\sigma) + q'x]^2 - \varphi(\sigma)[\sigma - k^{-1}\varphi(\sigma)] \qquad (4\text{-}146)$$
$$\Delta V(x_t) = 0 \qquad \text{only if } x_t = 0$$

$\Delta V(x)$ can be rewritten as

$$\Delta V(x) = -[\gamma\varphi(\sigma) + q'x]^2 - \sigma\varphi(0)\left[1 - \frac{\sigma\varphi(\sigma)}{\sigma^2 k}\right]$$

and because of (4-143) this ensures that $\Delta V(x) \le 0$, and therefore the absolute stability of (4-141) follows.

Reference 42 establishes the fact that the inequality (4-145) is a necessary and sufficient condition for the existence of a real number $\gamma \ne 0$, a real vector q, and a real symmetric positive-definite matrix H satisfying:

1. $H - A'HA = qq'$.
2. $A'Hb - c = \gamma q$.
3. $k^{-1} - b'Hb = \gamma^2$.

From (4-141), $V(x) = x'Hx$, and when conditions 1 to 3 are used, Eq. (4-146) is easily obtained.

4-7   DIFFERENTIAL - DIFFERENCE EQUATIONS—AN EXAMPLE[39]

The purpose of this section is to accustom the reader to techniques and concepts which occur in the analysis of a linear system of differential-difference equations.   Later, in Chap. 5, when we treat some nonlinear examples (we have learned from our study of nonlinear differential equations how little we can expect so far as solutions are concerned), we briefly examine questions of stability and asymptotic behavior characteristic of solutions of such systems.

The model developed here will be concerned with the distribution of arrivals, e.g., telephone calls, under appropriate assumptions.   The development examines the properties of the Poisson process.   By making explicit assumptions in keeping with these Poisson properties, the Poisson process is then derived as an arrival distribution.

From $(\lambda t)^n e^{-\lambda t}/n!$ which represents the Poisson process, we note that during a time interval t, the probability of no arrivals is $e^{-\lambda t}$ and that of a single arrival is $\lambda t e^{-\lambda t}$; hence the probability of more than one arrival is

$$
\begin{aligned}
1 - (e^{-\lambda t} + \lambda t e^{-\lambda t}) = 1 - &\left\{ \left[1 - \lambda t + \frac{(\lambda t)^2}{2!} - \cdots \right] \right. \\
&\left. + \lambda t \left[1 - \lambda t + \frac{(\lambda t)^2}{2!} - \cdots \right] \right\} \\
= \frac{(\lambda t)^2}{2} + \cdots &= O(t^2) \qquad (4\text{-}147)
\end{aligned}
$$

a function which behaves like t^2.

Thus if t is small, terms with t^2 are negligible compared with terms without t or with the first power of t.   Hence for small t the probability of more than one arrival is negligible.   As a very desirable property for many practical applications, it makes the Poisson process useful.   The probability of at least one arrival during t is given by

$$
1 - e^{-\lambda t} = \lambda t + O(t^2) \qquad (4\text{-}148)
$$

The probability of no arrivals is $e^{-\lambda t} = 1 - \lambda t + O(t^2)$.   In both these expressions we can ignore the last quantity on the right if we consider only small values of t.

At the risk of some repetition but in conformity with standard methods of procedure, let us assume these properties, i.e., that the probability of a single arrival during a small time interval Δt is $\lambda \, \Delta t$ and that of more than a single arrival during Δt is negligible; then we can derive the Poisson distribution.

We let $P_n(t)$ be the probability that exactly n items have arrived by time t. We first observe that $0 \leq P_n(t) \leq 1$, since the $P_n(t)$ are probabilities. Besides, $\sum_{n=0}^{\infty} P_n(t) = 1$, because a definite number must have arrived by time t, and these alternatives are mutually exclusive. To see what happens during the subsequent small time interval Δt we write

$$P_0(t + \Delta t) = P_0(t)(1 - \lambda \, \Delta t)$$
$$P_n(t + \Delta t) = P_n(t)(1 - \lambda \, \Delta t) + P_{n-1}(t)\lambda \, \Delta t \qquad n \geq 1 \tag{4-149}$$

The first equation gives the probability that there have been no arrivals by time $t + \Delta t$. This probability can be related to the state of the system at time t. Thus, by the law of compound probabilities of two events which occur independently, i.e., one takes the product of the probabilities of these events, it is equal to the probability that nothing had arrived by time t multiplied by the probability that nothing has arrived during Δt. For the case $n \geq 1$ this property (i.e., having the same number in the system as at time t and nothing arriving during Δt) also holds, but in addition there might have been $n - 1$ arrivals during time t, followed by an additional arrival during Δt. The product of these quantities yields the second term on the right. We have not mentioned the possibility of more than one arrival during the small interval Δt since it is negligible and can be shown to vanish in what follows.

On simplifying, transposing $P_n(t)$ to the left, and dividing by Δt, the system of equations for $n \geq 1$ becomes

$$\frac{P_n(t + \Delta t) - P_n(t)}{\Delta t} = -\lambda P_n(t) + \lambda P_{n-1}(t)$$

If we take the limit as $\Delta t \to 0$, then, by definition, the left side is the derivative $P_n'(t) = dP_n(t)/dt$, and the equations become

$$P_0'(t) = -\lambda P_0(t)$$
$$P_n'(t) = -\lambda P_n(t) + \lambda P_{n-1}(t) \qquad n \geq 1 \tag{4-150}$$

These are linear differential equations with respect to t and linear first-order (one shift in subscript) difference equations with respect to n.

We can solve these equations conveniently by the use of a generating function. We define such a function by

$$P(z,t) \equiv \sum_{n=0}^{\infty} P_n(t)z^n = P_0(t) + P_1(t)z + P_2(t)z^2 + \cdots \tag{4-151}$$

It can be seen that $P_n(t)$ is obtained by differentiating $P(z,t)$ n times with respect to z, then dividing by $n!$, and putting $z = 0$. Thus if $P(z,t)$ is known, $P_n(t)$ is easily determined in this manner.

The time origin for a specific study can be chosen anywhere, even after arrivals have actually occurred. Thus it may be that, by $t = 0$, i units have arrived. In that case, $P_n(0)$ is zero if $n \neq i$ and unity if $n = i$. Thus

$$P(z,0) = \sum_{n=0}^{\infty} P_n(0)z^n = P_i(0)z^i = z^i \tag{4-152}$$

Note also that $P(1,t) = 1$ and

$$\frac{\partial P(z,t)}{\partial t} = \frac{\partial}{\partial t} \sum_{n=0}^{\infty} P_n(t)z^n = \sum_{n=0}^{\infty} P_n'(t)z^n \tag{4-153}$$

We use partial differentiation because two variables are involved, z and t.

Now if we multiply the system of differential-difference equations for $n \geq 1$ by z^n and the first equation by z^0 and sum over n, we have for the sum on the left $\partial P(z,t)/dt$ and for the sum of the first terms on the right $-\lambda P(z,t)$.

The second terms on the right, when summed over n, give

$$\sum_{n=1}^{\infty} \lambda P_{n-1}(t)z^n = \lambda P_0(t)z + \lambda P_1(t)z^2 + \cdots$$

If we factor λz out of the terms on the right, they can be written as $\lambda z P(z,t)$. Thus the system has been reduced to the linear differential equation in the generating function

$$\frac{\partial P(z,t)}{\partial t} - \lambda(z - 1)P(z,t) = 0 \tag{4-154}$$

This equation is very similar to the simplest types of ordinary differential equations encountered in elementary calculus courses. Its solution, with z treated as a constant since it is independent of t, is given by

$$P(z,t) = Ce^{\lambda(z-1)t} \tag{4-155}$$

This can be verified by substitution. Note that C may depend on z.

Suppose that at $t = 0$ nothing had arrived; then $P(z,0) = 1$, since $i = 0$. Thus $C = 1$, and we have

$$P(z,t) = e^{\lambda(z-1)t}$$

But as we pointed out, $P_n(t)$ can be obtained by differentiation; hence

$$P_n(t) = \frac{1}{n!} \frac{\partial^n P(z,t)}{\partial z^n} \Big|_{z=0}$$

Thus

$$P_0(t) = e^{-\lambda t}$$

$$P_1(t) = \lambda t e^{-\lambda t}$$

and

$$P_n(t) = \frac{(\lambda t)^n e^{-\lambda t}}{n!} \tag{4-156}$$

which gives the desired Poisson process. It is clear that t is the length of the time interval in which the events occur rather than absolute time.

The foregoing method of derivation is standard for many queueing problems. The equations may be more complicated, but the basic procedure often is the same. The reader should have no difficulty in writing, for $i \neq 0$ and $n \geq i$,

$$P_n(t) = \frac{(\lambda t)^{n-i} e^{-\lambda t}}{(n-i)!}$$

since for this case $P(z,0) = z^i$, C has the value z^i, and

$$P(z,t) = z^i \sum_{n=i}^{\infty} P_n(t) z^{n-i}$$

We now consider a more elaborate model, due to Erlang, using the previous ideas. We suppose that we have a waiting line before one service counter, that the arrivals occur by a Poisson process, and that individuals are taken into service according to a Poisson process with a different parameter. If the server is not free, they must wait for their turn.

If we assume that an operation starts with i items waiting in line, the following equations provide a representation for a Poisson input with parameter μ, exponential holding time with parameter μ, for a first-come first-served single-channel queue:

$$P_n(t + \Delta t) = P_n(t)[1 - (\lambda + \mu) \Delta t] + P_{n-1}(t)\lambda \Delta t \\ + P_{n+1}(t)\mu \Delta t \qquad n \geq 1 \quad (4\text{-}157)$$

$$P_0(t + \Delta t) = P_0(t)(1 - \lambda \Delta t) + P_1(t)\mu \Delta t \qquad n = 0$$

These equations state that the probability of n items in the system at time $t + \Delta t$ equals the probability of n items in the system at time t multiplied by the probability of no arrivals and no departures plus the probability of $n - 1$ items in the system at time t multiplied by the probability of one arrival and no departures plus the probability of $n + 1$ items in the system at time t multiplied by the probability of a single departure and no arrivals.

Note that the probability of no arrivals and no departures is given by $(1 - \lambda \, \Delta t) (1 - \mu \, \Delta t)$. The term involving $(\Delta t)^2$ drops out in forming the differential equation. Thus one can use $1 - (\lambda + \mu) \, \Delta t$. For the remaining two terms above, note also that $\lambda \, \Delta t(1 - \mu \, \Delta t) \cong \lambda \, \Delta t$ and $\mu \, \Delta t(1 - \lambda \, \Delta t) \cong \mu \, \Delta t$.

After transposing and passing to the limit with respect to Δt, we have

$$\frac{dP_n(t)}{dt} = -(\lambda + \mu)P_n(t) + \lambda P_{n-1}(t) + \mu P_{n+1}(t) \qquad n \geq 1$$

$$\frac{dP_0(t)}{dt} = -\lambda P_0(t) + \mu P_1(t) \qquad\qquad\qquad\quad n = 0 \tag{4-158}$$

For many problems time dependence is not the typical situation to be studied. After a long period of operation the system acquires a pattern of behavior which, although describable in terms of probabilities, does not depend on time. Thus as $t \to \infty$ this situation can be expected to occur for many systems. The question of whether such $p_n = \lim_{t \to \infty} P_n(t)$ exist is a main concern of ergodic analysis, where one actually looks for the probabilities p_n which describe the steady or equilibrium state. But one also notes that the requirement that the probabilities $P_n(t)$ do not change with time is the definition of the steady state.

The changing of $P_n(t)$ with respect to t is described by the derivative $P'_n(t)$, and the steady-state condition requires that $P'_n(t) = 0$. Thus essentially we have two ways of obtaining the steady-state probabilities:
1. From $P'_n(t) = 0$, which gives probabilities p_n independent of t.
2. From $\lim_{t \to \infty} P_n(t)$, which also gives the p_n independent of t.

Transient or time-dependent solutions are particularly useful when the traffic-intensity or utilization factor $\rho = \lambda/\mu \geq 1$, since in this case no steady state occurs. Here an expression for the expected number waiting and the expected waiting time will be derived in the steady state with $\lambda/\mu < 1$. If $\lambda/\mu \geq 1$, the number waiting will be infinite.

By setting the time derivatives equal to zero and eliminating time from the above equations, we get, after transposing,

$$(\lambda + \mu)p_n = \lambda p_{n-1} + \mu p_{n+1} \qquad n \geq 1$$
$$\lambda p_0 = \mu p_1 \qquad\qquad\qquad\quad n = 0$$

Let $\rho = \lambda/\mu$; then these equations become

$$(1 + \rho)p_n = p_{n+1} + \rho p_{n-1} \qquad n \geq 1$$
$$p_1 = \rho p_0 \qquad\qquad\qquad\quad n = 0$$

Let $n = 1$ in the first equation. Then $(1 + \rho)p_1 = p_2 + \rho p_0$. Substituting for p_1 from the second equation, this becomes $p_2 = \rho^2 p_0$. Repetition of this process yields $p_n = \rho^n p_0$. Now $\sum_{n=0}^{\infty} p_n = 1$, since the sum gives the total probability that there are $0, 1, 2, \ldots$ items in the system. This total probability must yield certainty, since it accounts for all the possible states of the system. Thus

$$\sum_{n=0}^{\infty} \rho^n p_0 = 1$$

or

$$\sum_{n=0}^{\infty} p_0 \rho^n = p_0 \sum_{n=0}^{\infty} \rho^n = \frac{p_0}{1-\rho} = 1$$

We now return to the problem of deriving $P_n(t)$. We need the following theorem, due to Rouché.

Theorem 4-3   If $f(z)$ and $g(z)$ are analytic inside and on a closed contour C, and if $|g(z)| < |f(z)|$ on C, then $f(z)$ and $f(z) + g(z)$ have the same number of zeros inside C.

A proof of this theorem can be found in any standard work on complex variables.

Again we define the generating function of $P_n(t)$ by

$$P(z,t) = \sum_{n=0}^{\infty} P_n(t) z^n \tag{4-159}$$

which must converge within the unit circle $|z| = 1$, as it is dominated by the geometric series.

On multiplying each equation in $P_n(t)$ by z^{n+1} and summing over n, one has

$$z \frac{\partial P(z,t)}{\partial t} = (1-z)[(\mu - \lambda z)P(z,t) - \mu P_0(t)] \tag{4-160}$$

Note that the initial condition that there are i waiting items at $t = 0$ becomes $P(z,0) = z^i$, since $P_n(0) = 0$, except when $n = i$, in which case $P_i(0) = 1$.

Applying the Laplace transform to this linear first-order partial differential equation and solving for the transform of $P(z,t)$, one has

$$P^*(z,s) = \frac{z^{i+1} - \mu(1-z)P_0^*(s)}{sz - (1-z)(\mu - \lambda z)} \tag{4-161}$$

where typically

$$f^*(s) \equiv \int_0^{\infty} e^{-st} f(t)\, dt \tag{4-162}$$

Note that

$$\int_0^{\infty} e^{-st} \frac{\partial P}{\partial t}\, dt = e^{-st} P \Big]_0^{\infty} + s \int_0^{\infty} e^{-st} P\, dt = -z^i + sP^*$$

In the next few pages a more convenient expression for $P^*(z,s)$ will be found, and the coefficient of z^n in the power-series expansion of $P^*(z,s)$, that is, $P_n^*(s)$, is recognized as the transform of $P_n(t)$. Taking the inverse Laplace transform gives the transient solution $P_n(t)$ sought.

Since $P^*(z,t)$ converges inside and on the unit circle for Re $(s) > 0$, the zeros of the denominator of (4-161) inside and on $|z| = 1$ must coincide with corresponding zeros of the numerator. But the zeros $\alpha_k(s)$ of the denominator are obtained by setting it equal to zero and solving a quadratic in z:

$$\alpha_k(s) = \frac{\lambda + \mu + s \pm [(\lambda + \mu + s)^2 - 4\lambda\mu]^{\frac{1}{2}}}{2\lambda} \qquad k = 1, 2 \quad (4\text{-}163)$$

Here the value of the square root with positive real part is taken; that is, $\alpha_k(s)$ has Re $(s) > 0$. $\alpha_1(s)$ has the positive sign before the radical.

Also by Rouché's theorem, the denominator of P^* has a single zero within the disk $|z| \leq 1$. Here

$$f(z) = (\lambda + \mu + s)z \qquad\qquad g(z) = -(\lambda z^2 + \mu)$$
$$|f(z)| = |\lambda + \mu + s| > |\lambda + \mu| \qquad |g(z)| = |\lambda + \mu|$$

and therefore $|g(z)| < |f(z)|$ on $|z| = 1$.

Since both $f(z)$ and $g(z)$ are analytic inside and on $|z| = 1$, $f(z)$ and $(fz) + g(z)$ have the same number of zeros. The zero must be $\alpha_2(s)$, since $\alpha|_2(s)| < |\alpha_1(s)|$ and there is no zero on $|z| = 1$. Note that

$$\alpha_1 + \alpha_2 = \frac{\lambda + \mu + s}{\lambda} \qquad \alpha_1\alpha_2 = \frac{\mu}{\lambda} \qquad s = -\lambda(1 - \alpha_1)(1 - \alpha_2)$$

The numerator of (4-161) must vanish for $z = \alpha_2(s)$; otherwise, for this value, $P^*(z,s)$ would not exist, as it must. This leads to

$$P_0^*(s) = \frac{\alpha_2^{\,i+1}}{\mu(1 - \alpha_2)}$$

which, when substituted above, gives

$$P^*(z,s) = \frac{z^{i+1} - [(1 - z)\alpha_2^{\,i+1}/(1 - \alpha_2)]}{-\lambda(z - \alpha_1)(z - \alpha_2)} \qquad (4\text{-}164)$$

When the numerator is multiplied through by $1 - \alpha_2$, it simplifies and factors. We have

$$P^*(z,s) = \frac{\begin{aligned}(z - \alpha_2)(z^i + \alpha_2 z^{i-1} + \cdots + \alpha_2^{\,i}) \\ - z\alpha_2(z - \alpha_2)(z^{i-1} + \alpha_2 z^{i-2} + \cdots + \alpha_2^{\,i-1})\end{aligned}}{\lambda\alpha_1(z - \alpha_2)(1 - z/\alpha_1)(1 - \alpha_2)} \qquad (4\text{-}165)$$

Canceling $z - \alpha_2$ in numerator and denominator, subtracting and adding α_2^{i+1}, in order to factor $1 - \alpha_2$ out of the numerator, and then writing

$$(1 - z/\alpha_1)^{-1} = \sum_{k=0}^{\infty} \left(\frac{z}{\alpha_1}\right)^k \text{ yield}$$

$$P^*(z,s) = \frac{1}{\lambda \alpha_1} (z^i + \alpha_2 z^{i-1} + \cdots + \alpha_2{}^i) \sum_{k=0}^{\infty} \left(\frac{z}{\alpha_1}\right)^k$$

$$+ \frac{\alpha_2{}^{i+1}}{\lambda \alpha_1 (1 - \alpha_2)} \sum_{k=0}^{\infty} \left(\frac{z}{\alpha_1}\right)^k \quad (4\text{-}166)$$

Note that $|z/\alpha_1| < 1$.

Now $P_n^*(s)$ is the coefficient of z^n in the Laplace transform of (4-159). It is clear that the contribution to this coefficient of the second term in (4-166), i.e., the coefficient of z^n in that term, is

$$\frac{\alpha_2{}^{i+1}}{\lambda \alpha_1{}^{n+1}(1 - \alpha_2)} = \frac{\alpha_2{}^{i+1}}{\lambda \alpha_1{}^{n+1}} (1 + \alpha_2 + \alpha_2{}^2 + \cdots)$$

$$= \frac{1}{\lambda} \left(\frac{\lambda}{\mu}\right)^{n+1} \sum_{k=n+i+2}^{\infty} \left(\frac{\mu}{\lambda}\right)^k \frac{1}{\alpha_1{}^k}$$

from the fact that $|\alpha_2| < 1$ and that $\alpha_1 \alpha_2 = \mu/\lambda$.

The latter fact gives rise to powers of μ/λ by multiplying each α_2 raised to a power in the numerator by α_1 raised to the same power and of course also multiplying the denominator by this power of α_1. The first term on the right of $P^*(z,s)$ yields the remaining coefficients of z^n on multiplying the left factor by the appropriate power of z in the series and adding the coefficients of z^n.

For example, it is clear that if $n \geq i$, all the terms of $(1/\lambda \alpha_1)(z^i + \alpha_2 z^{i-1} + \cdots + \alpha_2{}^i)$ will contribute to this coefficient. Thus z^i is multiplied by z^{n-i} in the series, and the coefficient in this case is $1/\alpha_1{}^{n-i}$ to be multiplied by the outside factor $1/\lambda \alpha_1$. To this is added the coefficient resulting from multiplying z^{i-1} by z^{n-i+1}, which is $(1/\lambda \alpha_1)(\alpha_2/\alpha_1{}^{n-i+1})$, and so on.

In general, the contribution to this coefficient from multiplying z^{i-m} by z^{n-i+m} is

$$\frac{1}{\lambda \alpha_1} \frac{\alpha_2{}^m}{\alpha_1{}^{n-i+m}} = \frac{1}{\lambda \alpha_1} \frac{\alpha_2{}^m \alpha_1{}^m}{\alpha_1{}^{n-i+2m}} = \frac{(\mu/\lambda)^m}{\lambda \alpha_1{}^{n-i+2m+1}}$$

Hence, for $n \geq i$, we have

$$P_n^*(s) = \frac{1}{\lambda} \left[\frac{1}{\alpha_1{}^{n-i+1}} + \frac{\mu/\lambda}{\alpha_1{}^{n-i+3}} + \frac{(\mu/\lambda)^2}{\alpha_1{}^{n-i+5}} + \cdots + \frac{(\mu/\lambda)^i}{\alpha_1{}^{n+i+1}} \right.$$

$$\left. + \left(\frac{\lambda}{\mu}\right)^{n+1} \sum_{k=n+i+2}^{\infty} \frac{(\mu/\lambda)^k}{\alpha_1{}^k} \right] \quad (4\text{-}167)$$

Now $P_n(t)$, the inverse transform of $P_n^*(s)$, is given by

$$P_n(t) = \frac{1}{2\pi i} \int_{c-i\infty}^{c+i\infty} e^{st} P^*(s)\, ds$$

To obtain the inverse transform, we first use the theorem that if $f^*(s)$ is the Laplace transform of $f(t)$, then $f^*(s + a)$ is the Laplace transform of $e^{-at}f(t)$, a fact which easily follows from the definitions. Also, the inverse transform can be applied to each term on the right side of (4-167).

Exercise 4-21   Prove the last two statements above.

Note from (4-163) that, in α_1, s has the constant $\lambda + \mu$ added to it; consequently in the case of the inverse transform of every term in (4-167) we apply the above theorem to functions whose Laplace transform is of the form $(\alpha_1)^{-\nu}$. Now $[(s + \sqrt{s^2 - 4\lambda\mu})/2\lambda]^{-\nu}$ is the Laplace transform of

$$(2\lambda)^\nu \nu (2\sqrt{\lambda\mu})^{-\nu} t^{-1} I_\nu(2\sqrt{\lambda\mu}\, t) = \nu \left(\sqrt{\frac{\lambda}{\mu}}\right)^\nu t^{-1} I_\nu(2\sqrt{\lambda\mu}\, t)$$

Here $I_\nu(z)$ is the modified Bessel function of the first kind given by

$$I_\nu(z) = \sum_{k=0}^{\infty} \frac{(z/2)^{\nu+2k}}{k!\,\Gamma(\nu + k + 1)} \qquad I_\nu(z) = i^{-\nu} J_\nu(iz)$$

$$I_\nu(z) = \begin{cases} \dfrac{z^\nu}{2^\nu \nu!} & \text{as } z \to 0 \\[2ex] \dfrac{e^z}{\sqrt{2\pi z}} & \text{as } z \to \infty \end{cases}$$

Finally

$$P_n(t) = \frac{e^{-(\lambda+\mu)t}}{\lambda} \left[\left(\sqrt{\frac{\lambda}{\mu}}\right)^{n-i+1} (n - i + 1) t^{-1} I_{n-i+1}(2\sqrt{\lambda\mu}\, t) \right.$$
$$+ \frac{\mu}{\lambda} \left(\sqrt{\frac{\lambda}{\mu}}\right)^{n-i+3} (n - i + 3) t^{-1} I_{n-i+3}(2\sqrt{\lambda\mu}\, t) + \cdots$$
$$+ \left(\frac{\mu}{\lambda}\right)^i \left(\sqrt{\frac{\lambda}{\mu}}\right)^{n+i+1} (n + i + 1) t^{-1} I_{n+i+1}(2\sqrt{\lambda\mu}\, t)$$
$$\left. + \left(\frac{\lambda}{\mu}\right)^{n+1} \sum_{k=n+i+2}^{\infty} \left(\sqrt{\frac{\mu}{\lambda}}\right)^k k t^{-1} I_k(2\sqrt{\lambda\mu}\, t) \right]$$

Substituting the well-known relation

$$\frac{2\nu}{z} I_\nu(z) = I_{\nu-1}(z) - I_{\nu+1}(z)$$

one has

$$
P_n(t) = \frac{e^{-(\lambda+\mu)t}}{\lambda} \left\{ \left(\sqrt{\frac{\lambda}{\mu}}\right)^{n-i+1} \sqrt{\lambda\mu}\, [I_{n-i}(2\sqrt{\lambda\mu}\,t) - I_{n-i+2}(2\sqrt{\lambda\mu}\,t)] \right.
$$

$$
+ \left(\sqrt{\frac{\lambda}{\mu}}\right)^{n-i+1} \sqrt{\lambda\mu}\, [I_{n-i+2}(2\sqrt{\lambda\mu}\,t) - I_{n-i+4}(2\sqrt{\lambda\mu}\,t)]
$$

$$
+ \cdots + \left(\sqrt{\frac{\lambda}{\mu}}\right)^{n-i+1} \sqrt{\lambda\mu}\, [I_{n+i}(2\sqrt{\lambda\mu}\,t) - I_{n+i+2}(2\sqrt{\lambda\mu}\,t)]
$$

$$
+ \left(\frac{\lambda}{\mu}\right)^{n+1} \sum_{k=n+i+2}^{\infty} \left(\sqrt{\frac{\mu}{\lambda}}\right)^{k} \sqrt{\lambda\mu}\, [I_{k-1}(2\sqrt{\lambda\mu}\,t)
$$

$$
\left. - I_{k+1}(2\sqrt{\lambda\mu}\,t)] \right\} \quad (4\text{-}168)
$$

But

$$
\left(\frac{\lambda}{\mu}\right)^{n+1} \sum_{k=n+i+2}^{\infty} \left(\sqrt{\frac{\mu}{\lambda}}\right)^{k} [I_{k-1}(2\sqrt{\lambda\mu}\,t) - I_{k+1}(2\sqrt{\lambda\mu}\,t)]
$$

$$
= \left(\frac{\lambda}{\mu}\right)^{n+1} \left[\left(\sqrt{\frac{\mu}{\lambda}}\right)^{n+i+2} I_{n+i+1}(2\sqrt{\lambda\mu}\,t) + \sqrt{\frac{\mu}{\lambda}} \sum_{k=n+i+2}^{\infty} \left(\sqrt{\frac{\mu}{\lambda}}\right)^{k} I_k(2\sqrt{\lambda\mu}\,t) \right.
$$

$$
\left. + \left(\sqrt{\frac{\mu}{\lambda}}\right)^{n+i+1} I_{n+i+2}(2\sqrt{\lambda\mu}\,t) - \sqrt{\frac{\lambda}{\mu}} \sum_{k=n+i+2}^{\infty} \left(\sqrt{\frac{\mu}{\lambda}}\right)^{k} I_k(2\sqrt{\lambda\mu}\,t) \right]
$$

$$
= \left(\sqrt{\frac{\lambda}{\mu}}\right)^{n-i} I_{n+i+1}(2\sqrt{\lambda\mu}\,t) + \left(\sqrt{\frac{\lambda}{\mu}}\right)^{n-i+1} I_{n+i+2}(2\sqrt{\lambda\mu}\,t)
$$

$$
+ \left(1 - \frac{\lambda}{\mu}\right)\left(\frac{\lambda}{\mu}\right)^{n} \sqrt{\frac{\lambda}{\mu}} \sum_{k=n+i+2}^{\infty} \left(\sqrt{\frac{\mu}{\lambda}}\right)^{k} I_k(2\sqrt{\lambda\mu}\,t)
$$

therefore the second expression of the first term of (4-168) cancels with the first expression of the second term, and so on, and one has for $n \geq i$

$$
P_n(t) = e^{-(\lambda+\mu)t} \left[\left(\sqrt{\frac{\mu}{\lambda}}\right)^{i-n} I_{n-i}(2\sqrt{\lambda\mu}\,t) + \left(\sqrt{\frac{\mu}{\lambda}}\right)^{i-n+1} I_{n+i+1}(2\sqrt{\lambda\mu}\,t) \right.
$$

$$
\left. + \left(1 - \frac{\lambda}{\mu}\right)\left(\frac{\lambda}{\mu}\right)^{n} \sum_{k=n+i+2}^{\infty} \left(\sqrt{\frac{\mu}{\lambda}}\right)^{k} I_k(2\sqrt{\lambda\mu}\,t) \right] \quad (4\text{-}169)
$$

It will now be shown that this is also the solution for $n < i$.

Note from the first expression on the right in Eq. (4-166) that to obtain the coefficient of z^n one commences the multiplication of the series $\sum_{k=0}^{\infty} (z/\alpha)^k$ by z^n in the outside factor $(1/\lambda\alpha_1)(z_i + \alpha_2 z^{i-1} + \cdots + \alpha_2{}^i)$, using all the terms in z to the right of it. Everything from then on remains the same as above. Hence cancellation takes place in the same manner. However, we must show that the first term for the case $n < i$ coincides with the first term of (4-169).

Now the first contribution to the coefficient of z^n in this case is obtained by multiplying the coefficient of z^n in the outside factor by the first term of the series, which is unity. This gives

$$\frac{\alpha_2^{i-n}}{\lambda \alpha_1} = \frac{(\mu/\lambda)^{i-n}}{\lambda \alpha_1^{i-n+1}}$$

which has the inverse transform

$$\frac{e^{-(\lambda+\mu)t}}{\lambda} \left[\left(\frac{\mu}{\lambda}\right)^{i-n} \left(\sqrt{\frac{\lambda}{\mu}}\right)^{i-n+1} (i - n + 1)t^{-1} I_{i-n+1}(2\sqrt{\lambda\mu}\,t) \right]$$

$$= \frac{e^{-(\lambda+\mu)t}}{\lambda} \left\{ \left(\sqrt{\frac{\lambda}{\mu}}\right)^{n-i+1} \sqrt{\lambda\mu}\,[I_{i-n}(2\sqrt{\lambda\mu}\,t) - I_{i-n+2}(2\sqrt{\lambda\mu}\,t)] \right\}$$

It is also known that if n is an integer, $I_{-n}(z) = I_n(z)$, and since the second expression is canceled, as outlined above, the first terms coincide, and the solution is complete in detail.

We again derive an expression for the equilibrium probabilities $p_n = \lim\limits_{t \to \infty} P_n(t)$. We use the relation

$$e^{(x/2)(y+1/y)} = \sum_{n=-\infty}^{\infty} y^n I_n(x) = \sum_{n=0}^{\infty} y^n I_n(x) + \sum_{n=1}^{\infty} y^{-n} I_n(x) \quad (4\text{-}170)$$

and show that the above solution for $\lambda < \mu$ tends to a steady-state limit as $t \to \infty$. We shall simplify the computation by assuming that $i = 0$. For fixed λ and μ as $t \to \infty$ one has

$$I_n(2\sqrt{\lambda\mu}\,t) \sim \frac{\exp(2\sqrt{\lambda\mu}\,t)}{\sqrt{2\pi(2\sqrt{\lambda\mu})\,t}}$$

which is independent of n.

When this expression is substituted for I_n in the first two terms in brackets in (4-169) and the latter is multiplied by the outside factor, each of the resulting two expressions is readily seen to approach 0 as $t \to \infty$, using the fact that $\lambda + \mu > 2\sqrt{\lambda\mu}$ (since $\lambda/\mu < 1$ was assumed). As for the limiting behavior of the last factor of the solution, note that

$$\sum_{k=n+2}^{\infty} \left(\sqrt{\frac{\mu}{\lambda}}\right)^k I_k(2\sqrt{\lambda\mu}\,t) = \sum_{k=0}^{\infty} \left(\sqrt{\frac{\mu}{\lambda}}\right)^k I_k(2\sqrt{\lambda\mu}\,t)$$

$$- \sum_{k=0}^{n+1} \left(\sqrt{\frac{\mu}{\lambda}}\right)^k I_k(2\sqrt{\lambda\mu}\,t) \quad (4\text{-}171)$$

When multiplied by $e^{-(\lambda+\mu)t}$, the finite sum on the right tends to zero as $t \to \infty$, since each term tends to zero separately. The series will now be examined.

When multiplied by $e^{-(\lambda+\mu)t}$, the last sum on the right of (4-170), with $y = \sqrt{\mu/\lambda}$ and $x = 2\sqrt{\lambda\mu}\, t$, tends to zero as $t \to \infty$ when the asymptotic expression of $I_n(x)$ is used, together with the fact that $\lambda/\mu < 1$. On the other hand, the first sum on the right of (4-170) is the desired series of (4-171) under study. Hence, finally,

$$e^{-(\lambda+\mu)t} \sum_{n=0}^{\infty} \left(\sqrt{\frac{\mu}{\lambda}}\right)^k I_k(2\sqrt{\lambda\mu}\, t)$$

$$= e^{-(\lambda+\mu)t} \exp\left[\tfrac{1}{2} t(2\sqrt{\lambda\mu}) \left(\sqrt{\frac{\mu}{\lambda}} + \sqrt{\frac{\lambda}{\mu}}\right)\right] = 1$$

Therefore
$$p_n = \lim_{t \to \infty} P_n(t) = \left(\frac{\lambda}{\mu}\right)^n \left(1 - \frac{\lambda}{\mu}\right)$$

at long last.

4-8 OPTIMIZATION AND DIFFERENCE EQUATIONS

We shall illustrate a problem in which a function is to be minimized subject to constraints given as difference equations.

Problem: A vehicle is to be advanced 1,000 miles from an original position with a minimum expenditure of fuel. The vehicle has a maximum fuel capacity of 500 units, which it uses at the uniform rate of one unit per mile. The vehicle must stock its own storage points from its one tank by making round trips between storage points. Determine (1) the storage points along the route which minimize the fuel consumption for the entire trip, (2) the number of round trips required between each pair of storage points, and (3) the minimum amount of fuel required at the start.

Formulation and solution: If s_i is the amount of fuel stored at the ith storage point $(i = 0, 1, \ldots, n)$, d_i is the distance in miles between the $(i-1)$st and the ith storage points, and k_i is the number of round trips the vehicle makes between these two points, then

$$s_{i-1} = s_i + 2k_i d_i + d_i \qquad i = 1, \ldots, n \qquad (4\text{-}172)$$

Thus the amount of fuel stored at the $(i-1)$st point is equal to the amount stored at the ith point plus the amount consumed along the route in making k_i round trips and a single forward trip.

It is not difficult to see that a minimum use of fuel is made if the vehicle starts each trip loaded to full capacity. For in this case, fewer trips are made than in any other, and hence less fuel is consumed in travel. Also, in order to have no fuel left at the end, it is necessary that the vehicle be loaded with 500 units of fuel at the 500-mile point.   From these two facts it follows that the ith storage point should be located so that the vehicle makes k_{i+1} round trips between this point and the $(i + 1)$st storage point and one last forward trip, always fully loaded and ultimately leaving no fuel behind at the ith storage point.   Working backward, we see that the last statement is valid back to the first storage point but is not possible for the starting point, since the position of that point is predetermined.   Hence, the vehicle will make its last forward trip between the starting point and the first storage point with a load $c \le 500$.   Thus we have

$$s_i = k_i(500 - 2d_i) + 500 - d_i \qquad i = 2, \ldots, n$$
$$s_1 = k_1(500 - 2d_1) + c - d_1 \qquad\qquad\qquad (4\text{-}173)$$

We wish to minimize s_0, given by

$$s_0 = s_1 + 2k_1d_1 + d_1$$

Now, using for s_1 the value given in (4-173), we have

$$s_0 = 500k_1 + c \qquad\qquad (4\text{-}174)$$

Since the vehicle can travel the last 500 miles without stored fuel, in order to minimize s_0 it suffices to put $s_n = 500$ and to place the storage points along the first 500 miles of the route.   Thus,

$$\sum_{i=1}^{n} d_i = 500 \qquad\qquad (4\text{-}175)$$

Now from (4-173) we have

$$d_i = \frac{500k_i + 500 - s_i}{2k_i + 1} \qquad i = 2, \ldots, n \qquad (4\text{-}176)$$

Replacing i by $i + 1$ in (4-172) and substituting for s_i in the expression for d_i, we have

$$d_i = \frac{500k_i + 500 - s_{i+1} - 2k_{i+1}d_{i+1} - d_{i+1}}{2k_i + 1}$$
$$i = 2, \ldots, n; \; k_{n+1} \equiv 0 \quad (4\text{-}177)$$

Finally, replacing s_{i+1} by its equal from (4-173) and simplifying yields

$$d_i = \frac{500(k_i - k_i + 1)}{2k_i + 1} \qquad i = 2, \ldots, n; \; k_{n+1} \equiv 0 \quad (4\text{-}178)$$

Similarly, we have

$$d_1 = \frac{500(k_1 - k_2) + c - 500}{2k_1 + 1}$$

Since $d_i > 0$, it follows that $k_i > k_{i+1}$. Thus we wish to minimize

$$s_0 = 500k_1 + c \qquad (4\text{-}179)$$

subject to the constraint

$$\sum_{i=1}^{n} d_i = \left(500 \sum_{i=1}^{n} \frac{k_i - k_{i+1}}{2k_i + 1}\right) + \frac{c - 500}{2k_1 + 1} = 500$$

This can be written as

$$\sum_{i=1}^{n} \frac{k_i - k_{i+1}}{2k_i + 1} - \frac{1 - c/500}{2k_1 + 1} = 1 \qquad k_{n+1} \equiv 0 \qquad (4\text{-}180)$$

Since $k_1 > k_2 > \cdots > k_i > \cdots > k_n$, the second term on the left is less than the least value which any term in the sum can assume, that is,

$$\frac{k_i - k_{i+1}}{2k_i + 1} > \frac{1}{2k_1 + 1} \qquad i = 2, \ldots, n$$

Simply stated, this relation requires a choice of the k_i for which

$$\sum_{i=1}^{n} \frac{k_i - k_{i+1}}{2k_i + 1}$$

exceeds unity with a minimum k_1.

It will now be shown that the minimum value of k_1 is obtained by taking $k_i - k_{i+1} = 1$ $(i = 1, \ldots, n)$ and $k_n = 1$. Suppose that for $i = i_0$, $k_{i_0} - k_{i_0+1} = m$, an integer > 1. Then for the i_0th term, one has

$$\frac{k_{i_0} - k_{i_0+1}}{2k_{i_0} + 1} = \frac{m}{2(k_{i_0+1} + m) + 1}$$

where the denominator jumps from $2k_{i_0+1} + 1$ to $2(k_{i_0+1} + m) + 1$ when i decreases from $i_0 + 1$ to i_0.

Now, by taking $k_{i_0} - k_{i_0+1} = 1$, this jump can be replaced by a gradually increasing sum of terms

$$\frac{1}{2(k_{i_0+1} + 1) + 1} + \frac{1}{2(k_{i_0+1} + 2) + 1} + \cdots + \frac{1}{2(k_{i_0+1} + m) + 1}$$

This sum is greater than the previous expression, which is m times the smallest term in the sum. It follows that a minimum k_1 is attained by taking $k_i - k_{i+1} = 1$, that is, by using unit differences. These expressions with unit differences will determine n, the number of storage points. When the number of storage points is decreased, then the number of round trips between two adjacent points must be increased, and the above argument shows that this is inefficient. Again, if the number of storage points is increased, one automatically increases k_1, which is also inefficient.

Furthermore, by taking $k_n = 1$, one obtains a more rapidly increasing sum from the monotonically increasing denominators. Because of the monotone property of the k_i, one chooses n such that $k_n = 1$, $k_{n-1} = 2$, ..., in such a manner as to produce k_2 with

$$\sum_{i=2}^{n} \frac{k_i - k_{i+1}}{2k_i + 1} < 1 < \sum_{i=1}^{n} \frac{k_i - k_{i+1}}{2k_i + 1}$$

In this case, $k_2 = 6$. The corresponding d_i are then calculated, working backward from the 500-mile point. The distance which remains is taken as d_1, with a corresponding $k_1 = 7$ round trips. This choice of k_1 minimizes s_0, and the corresponding choice of d_i satisfies the distance constraint.

Note that if the original total difference is chosen appropriately, the d_{n-i} go as the harmonic series with odd denominators. A consideration of the slowness of divergence of this series casts some light on the economic difficulty of importing into and improving backward countries.

REFERENCES

1. A Difference Equation: *Amer. Math. Monthly*, vol. 72, no. 9, pp. 1024–1025, November, 1965.

2. Batchelder, P. M.: "An Introduction to Linear Difference Equations," Harvard University Press, Cambridge, Mass., 1927.

3. Baumol, W. J.: Topology of Second Order Linear Difference Equations with Constant Coefficients, *Econometrica*, vol. 26, April, 1958.

4. Birkhoff, G. D.: General Theory of Linear Difference Equations, *Trans. Amer. Math. Soc.*, vol. 12, pp. 243–284, April, 1911.

5. ———: Formal Theory of Irregular Linear Difference Equations, *Acta Math.*, vol. 54, pp. 205–246, 1930.

6. ——— and W. J. Trjitzinsky: Analytic Theory of Singular Difference Equations, *Acta Math.*, vol. 60, pp. 1–89, 1932.

7. Bleich, F. R., and E. Melan: "Die gewöhnlichen und partiellen Differenzen-Gleichungen der Baustatik," Springer-Verlag OHG, Berlin, 1927.

8. Boole, G.: "Finite Differences," Macmillan and Company, London, 1880.

9. Brand, L.: A Sequence Defined by a Difference Equation, *Amer. Math. Monthly*, vol. 62, p. 489, 1955.

10. Buchanan, M. L.: A Necessary and Sufficient Condition for Stability of Difference Schemes for Second-order Initial Value Problems, *Adelphi College Dept. Math. AGM Rep.* 104, February, 1962.

11. Fort, T.: "Finite Differences," Oxford University Press, Fair Lawn, N.J., 1948.

12. Forsythe, G. E., and W. R. Wasow: "Finite-difference Methods for Partial Differential Equations," John Wiley & Sons, Inc., New York, 1960.

13. Gauss, C. F.: "Werke," vol. 3, Göttingen, 1866.

14. Goheen, H.: A Quadratic Recurrence Problem 4601, *Amer. Math. Monthly*, vol. 62, p. 736, December, 1955.

15. Goldberg, S.: "Introduction to Difference Equations," John Wiley & Sons, Inc., New York, 1958.

16. Harlow, F. H.: Stability of Difference Equations: Selected Topics, pp. 1–39, Los Alamos Scientific Laboratory, 1960.

17. Harris, W. A., Jr., and Y. Sibuya: Asymptotic Solutions of Systems of Nonlinear Difference Equations, *Arch. Rational Mech. Anal.*, vol. 15, pp. 377–395, 1964; *Math. Reviews*, September, 1964.

18. Henrici, P.: "Discrete Variable Methods in Ordinary Differential Equations," John Wiley & Sons, Inc., New York, 1962.

19. Hildebrand, F. B.: "Introduction to Numerical Analysis," McGraw-Hill Book Company, New York, 1956.

20. Horn, J.: Über nichtlineare Differenzgleichungen, *Arch. Math. u. Phy.*, vol. 25, pp. 137–148, 1916.

21. ———: Über eine nichtlineare Differenzgleichung, *Jber. Deutsch. Math.-Verein.*, vol. 26, pp. 230–251, 1917.

22. ———: Zur Theorie der nichtlinearen Differenzengleichungen, *Math. Z.*, vol. 1, pp. 80–114, 1918.

23. Householder, A. S.: "Principles of Numerical Analysis," McGraw-Hill Book Company, New York, 1953.

24. Jordan, C.: "Calculus of Finite Differences," Chelsea Publishing Company, New York, 1950.

25. Kadota, T. T.: Asymptotic Stability of Some Nonlinear Feedback Systems, *Electronic Research Laboratory*, ser. 60, no. 264, pp. 1–84, Jan. 4, 1960.

26. Kalman, R. E., and J. E. Bertram: Control System Analysis and Design via the Second Method of Lyapunov, II: Discrete Time Systems, *Trans. ASME Ser. D J. Basic Engrg.*, June, 1960.

27. Kojima, T.: On a Theorem of Hadamard's and Its Applications, *Tohoku Math. J.*, vol. 5, p. 58, 1914.

28. Lancaster, O. E.: Nonlinear Algebraic Difference Equations, *Amer. J. Math.*, pp. 187–209, 1939.

29. Lefferts, E. J.: "A Guide of the Application of the Liapunov Direct Method to Flight Control Systems," Martin Marietta Corporation, Contract NAS 2-1777, Baltimore, Md., for NASA, Washington, D.C., April 1965.

30. Levy, H., and F. Lessman: "Finite Difference Equations," The Macmillan Company, New York, 1961.

31. Markov, A. A.: "Differenzenrechnung," Teubner, Leipzig, 1896.

32. Milne-Thompson, L. M.: "The Calculus of Finite Differences," Macmillan & Co., Ltd., London, 1951.

33. Mohammad, Q. G.: On the Zeros of Polynomials, *Amer. Math. Monthly*, vol. 72, pp. 631–633, June–July, 1965.

34. Norlund, N. E.: "Leçons sur les équations linéaires aux différences finies," Gauthier-Villar, Paris, 1929.

35. O'Shea, R. P.: Approximation of the Asymptotic Stability Boundary of Discrete Time Control Systems, Using an Inverse Transformation Approach, presented at 1964 JACC, June 25–29, Stanford University.

36. Pfouts, R. W., and C. E. Ferguson: A Matric General Solution of Linear Difference Equations with Constant Coefficients, *Math. Mag.*, vol. 33, pp. 119–127, January–February, 1960.

37. Pietenpol, J. L.: Solution to a Nonlinear Difference Equation Proposed by M. S. Klamkin and D. J. Newman, *Amer. Math. Monthly*, vol. 68, no. 4, p. 379, 1961.

38. Ragazinni, J. R., and L. A. Zadeh: The Analysis of Sampled-data Systems, *Trans. AIEE*, vol. 71, pt. 2, 1952.

39. Saaty, T. L.: "Elements of Queueing Theory," McGraw-Hill Book Company, New York, 1961.

40. Schuske, G., and W. J. Thron: On Periodic Infinite Radicals, University of Colorado, AFOSR-546, pp. 1–8, n.d.

41. Stiefel, E. L.: "An Introduction to Numerical Mathematics," Academic Press Inc., New York, 1963.

42. Szegö, G., and Rudolph Kalman: Sur la stabilité absolue d'un Système d'équation aux différences finies, *C. R. Acad. Sci. Paris*, vol. 257, pp. 388–390, 1963.

43. Tanaka, S.: On Asymptotic Solutions of Nonlinear Difference Equations of the First Order, *Mem. Fac. Sci. Kyushu Univ. Ser. A*, vol. 6–7, pp. 107–127, 1951–1953, vol. 10–11, pp. 45–83 and 167–184, 1956–1957.

44. Tauber, S.: Existence and Uniqueness Theorems for Solutions of Difference Equations, *Amer. Math. Monthly*, vol. 71, no. 8, pp. 859–862, 1964.

45. Thoro, D.: Regula Falsi and the Fibonacci Numbers, *Amer. Math. Monthly*, vol. 70, no. 10, p. 869, 1963.

46. Trjitzinsky, W. T.: Nonlinear Difference Equations, *Compositio Math.*, vol. 5, pp. 1–66, 1937–1938.

47. Vidal, P., and S. Wengrzyn: Sur la stabilité asymptotique d'une équation non linéaire aux différences finies d'ordre *m*, *C. R. Acad. Sci. Paris*, vol. 256, pp. 2781–2783, 1963; *Math. Reviews*, September, 1964.

48. ———— and ————: Sur la stabilité asymptotique d'une équation non linéaire aux differences finies du deuxième ordre, *C. R. Acad. Sci. Paris*, vol. 256, pp. 1672–1674, 1963; *Math. Reviews*, September, 1964.

49. Walsh, J. L.: An Inequality for the Roots of an Algebraic Equation, *Ann. of Math.*, vol. 25, pp. 285–286, 1924.

50. Zeitlin, D.: On Solutions of Homogeneous Linear Difference Equations with Constant Coefficients, *Amer. Math. Monthly*, vol. 68, no. 2, February, 1961.

FIVE

DELAY–DIFFERENTIAL EQUATIONS

5-1 INTRODUCTION

In this chapter we discuss delay-differential equations, a topic which is in a rapid state of development. It was the Russian mathematician Krasovskii who found an accommodation for delay-differential equations as operators in function spaces. It is worth noting that the theory of delay-differential equations is not just a simple extension of the theory of ordinary differential equations; in some respects it has required new ideas and a novel approach. Some topics have been mentioned in the literature without significant results. In other cases research has unfolded some interesting methods.

A relatively simple example of a delay-differential equation is the linear nonhomogeneous equation

$$\ddot{x}(t) + a(t)\dot{x}(t - \tau) + b(t)x(t - \tau) = f(t) \qquad (5\text{-}1)$$

where τ is constant and positive. Here the second derivative of the unknown function $x(t)$ is seen to depend upon the given function $f(t)$ and upon the "past history" of $x(t)$. The variable t denotes time.

A more complex and general form permits delays which are functions of t. Thus a generalization of the above equation is

$$\ddot{x}(t) + a(t)x(t - \tau_1(t)) + b(t)x(t - \tau_2(t))$$
$$= f(t) \qquad \tau_i(t) \geq 0, \, i = 1, 2 \quad (5\text{-}2)$$

An even more general expression, as a first-order differential equation, is

$$\dot{x}(t) = f(t, x(t), x(g_1(t), \ldots, g_m(t)))$$

where f and $g_i(t)$, $i = 1, \ldots, m$, are given for $t \geq t_0$, with $\alpha \leq g_j(t) \leq t$, α a constant. Driver[14] has studied a more general equation given by

$$\dot{x}(t) = f(t, x(t), x(g_2(t, x(t))), \ldots, x(g_m(t, x(t))))$$

which arises in a two-body problem in classical electrodynamics.

Generally we can expect both advances and delays to occur in the arguments, and we refer to *equations with deviating or perturbed arguments* or *functional difference equations*. Sometimes these equations can also be classed as integrodifferential equations (a topic discussed later), an example of which is

$$\dot{x}(t) = \int_\alpha^t x(s) \, dF(s, t) \qquad t \geq t_0$$

which involves a Stieltjes integral, with $F(s, t)$ being a given function of bounded variation in s for $t \geq t_0 \geq \alpha$.

Differential equations with deviating arguments in which the arguments of the highest-order derivative of the unknown function are identical and assume values that are never less (more) than those of the unknown function and its remaining derivatives are called *equations of delay, retarded* or *lagging* (*advanced* or *leading*) *type*. An example of a delay equation is

$$\dot{x}(t + \tau) = a(t)x(t) + f(t) \qquad \tau > 0, \, t \geq t_0$$

An illustration of the idea of lag is given by a physical system whose acceleration depends upon its velocity and on its position at earlier instants.

Exercise 5-1 Show that

$$\dot{x}(t) = x(t) - x(te^{-t}) \qquad t > 0, \, t_0 = 0$$

is a delay equation.

Exercise 5-2   Illustrate an equation of advanced type.

Example: A system of equations

$$\dot{x}(t) = -ax(t) + b[x_0 - x(t - \tau)] \qquad 0 \le t \le T$$
$$\dot{x}(t) = b[x_0 - x(t - \tau)] \qquad\qquad t > T$$

where a, b, x_0, and τ are positive constants, $0 \le T \le \infty$, and $x(t) = x_0$, $t < 0$, arises in the formulation of the biological reaction phenomena to x-rays.

A general (nonlinear) form of this system, in which both the coefficients and the retardation parameters are functions of time, is

$$\dot{x}_i(t) = X_i(t, x_j(t), x_j(t - \tau_{ij}(t))) + f_i(t) \qquad i,j = 1, \ldots, n$$

where $X_i(t,0,0) = 0$ for $t \ge 0$.

Example: A quasi-linear differential-difference equation [or system if $x(t)$ is a vector] containing a small parameter ϵ is given by

$$\dot{x}(t + 1) = ax(t + 1) + bx(t) + f(x(t + 1), x(t), u(t), t, \epsilon)$$

where $u(t)$ is a given function. This form is suggestive of the use of perturbation methods to obtain solutions near the solutions of the linear equation, in which one puts $\epsilon = 0$. In this manner one can also determine the boundedness, uniqueness, and existence of periodic solutions.

All other equations are called *neutral*. We have the following examples:

$$\dot{x}(t) = f(t, x(t - \tau), x(t + \tau)) \qquad \tau > 0 \qquad (5\text{-}3)$$
$$\dot{x}(t) = ax(t) + bx(t - \tau) + c\dot{x}(t - \tau) \qquad \tau > 0 \qquad (5\text{-}4)$$

This last equation suggests the richness of equations of the neutral type, for here $\dot{x}(t)$ depends on $\dot{x}(t - \tau)$. We shall devote more time to equations of the delay type since considerably more information is available for them than for the other two types.

A fundamental problem in the study of delay-differential equations is the initial-value problem. For example, if we wish to solve Eq. (5-2) for all $t \ge t_0$ with t_0 given, then it may be that $t - \tau_2(t) < t_0$ for some values of $t > t_0$. In that case $x(t - \tau_2(t))$ would have to be defined for all $t > t_0$ for which $t - \tau_2(t) < t_0$. If we thus specify $x(t - \tau_2(t))$ by a function $\varphi(t - \tau_2(t))$ for all those $t > t_0$ for which $t - \tau_2(t) < t_0$, we have data for an initial-value problem (which is analogous to the Cauchy problem for differential equations), and we must face the existence and uniqueness question. If we also have for the other argument $t - \tau_1(t) < t_0$, then $\dot{x}(t)$ must also be defined for values of t for which the argument is less than t_0. In addition $x(t)$ must be defined on the larger set of values of t for which $t - \tau_i(t) < t_0$, $i = 1, 2$.

Exercise 5-3 Using the foregoing description of an initial-value problem, describe the conditions which must be specified to solve an initial-value problem for an nth-order equation (with the nth-order derivative occurring on the left) of the form

$$x^{(n)}(t) = f(t, x(t), \ldots, x^{(n-1)}(t), x(t - \tau_1(t)), \ldots, x^{(n-1)}(t - \tau_n(t))) \qquad t \geq t_0$$

in which some of the argument deviations are less than t_0. Note that all these conditions can be given in terms of the derivatives of an initial function $\varphi(t)$.

If $t - \tau_i(t) \geq t_0$ is always satisfied, where $t \geq t_0$, one need only prescribe the values of $x(t)$ and its first $n - 1$ derivatives at t_0. A solution of an initial-value problem is a function which satisfies the initial conditions. It must also satisfy the equation, for all $t \geq t_0$. If t is restricted on the right, the solution must satisfy the equation on an interval of the form $t_0 \leq t \leq t_1$.

The initial-value problem is one among several interesting topics considered in the study of delay-differential equations. These topics can be arranged as in the chart for differential equations in chap. 4 of "NM." We shall touch on some of them, particularly those concerned with the construction of solutions for initial-value problems and with questions of stability. In general it is not known how to solve an initial-value delay-differential equation on both sides of t_0.

We mention that a boundary-value problem is another basic problem in the analysis of delay-differential equations. An example of such a problem is

$$\ddot{x}(t) + \lambda x(t) + M(t)x(t - \tau(t)) = 0 \qquad \text{on } [0, \pi]$$

(where λ is a real parameter and $M(t)$ and $\tau(t) \geq 0$ are continuous on $[0, \pi]$) subject to the boundary conditions

$$x(0) = x(\pi) = 0 \qquad x(t - \tau(t)) \equiv 0 \qquad \text{for } t - \tau(t) < 0$$

An existence theorem for this problem is available.[57] The problem has also been studied for eigenvalues and eigenfunctions and for oscillation theorems for the latter. The expansion of solutions of boundary-value problems in terms of eigenfunctions has also been studied for

$$\ddot{x}(t) + \lambda x(t) + \sum_{i=1}^{m} a_i(t)x(t - \tau_i(t)) + \int_0^t K(t,s)x(s)\, ds = 0$$

with λ real, $\tau_i(t) \geq 0$, $\tau_i(u) \leq u$, $0 \leq \tau_i(t) < 1$; $i = 1, \ldots, n$; $0 \leq t \leq u$; and

$$\int_0^u \int_0^u |K(t,s)|\, dt\, ds < \infty$$

with $a_i(t)$ real and continuous on $0 \leq t - \tau_i(t) \leq u$, $i = 1, \ldots, n$; and $a_i(t) = 0$ for $t - \tau_i(t) < 0$ subject to the boundary conditions

$$x(0) \cos \alpha + \dot{x}(0) \sin \alpha = 0$$
$$x(u) \cos \beta + \dot{x}(u) \sin \beta = 0$$

with α and β real.

The study of periodic and asymptotically periodic solutions of delay equations is another area of interest.

Yet another general pursuit related to delay-differential equations is the solution of variational problems which involve delay, e.g., to maximize the functional (which is essentially a generalization of the classical variational problem)[20,23]

$$\int_{t_0}^{t} f(t, x(t - \tau_1(t)), \ldots, x(t - \tau_n(t)), \dot{x}(t - \tau_1(t)), \ldots, \dot{x}(t - \tau_n(t))) \, dt$$

with $\tau_i(t) \geq 0$, $i = 1, \ldots, n$. We shall give a brief account of this type of problem in Sec. 5-9.

As in the theory of differential equations, dynamical systems with a delay in the argument are of substantial interest and are investigated for stationary points, etc.

If the arguments have the form $t \pm \tau_i$ with constant $\tau_i > 0$, the equation is called a differential-difference equation. We have already studied equations of this type, for example, $\dot{x}(t) = x(t + \alpha)$, in Chap. 3. We illustrate the idea of differential-difference equations with a system in which there is a constant time-lag parameter τ and a discrete variable k, arising in traffic dynamics. We have[10]

$$\frac{dy_k(t + \tau)}{dt} = \lambda[y_{k-1}(t) - y_k(t)] \qquad k = 1, \ldots, n$$

This system describes the motion of a line of vehicles obeying the traffic law, where y_k is the motion of the kth vehicle and y_{k-1} is the motion of the $(k - 1)$st vehicle, t is time, and λ is a positive constant of proportionality. Thus the acceleration of the kth vehicle is proportional to the difference of the velocities of the vehicle ahead and of its own and responds to this difference after a lapse of time $\tau > 0$. Here one determines the motion of the nth vehicle in terms of the motion of the leading one $y_0(t)$, subject to initial conditions. What are these conditions?

The next section is concerned with a simple equation of neutral type. Our primary interest is to expose the reader to a method of solution and its concomitant difficulties. It may be regarded as a scouting expedition which follows a well-established trail. Understanding of this example should be a helpful start.

The books by Pinney,[59] Bellman and Danskin,[4] Bellman and Cooke,[2] chap. 5 in the translation of the book by El'sgol'ts,[23] chap. 21

of Minorsky,[49] the superb presentation by Hale,[31] and the summarizing papers[51-54] by Myshkis (the pioneer in this field), Hahn,[29] and Zverkin et al.,[73] are particularly helpful. For stability theory, Krasovskii's book[43] is the outstanding reference.

Exercise 5-4 Verify that if

$$a\dot{x}(t) + b\dot{x}(t - \tau) + cx(t) + dx(t - \tau) = 0 \qquad t \geq 0$$

where a, b, c, d, τ are constants, $\tau > 0$, the equation is of:
1. Retarded type if $a \neq 0$, $b = 0$, and $d \neq 0$.
2. Neutral type if $a \neq 0$, $b \neq 0$, and either c or $d \neq 0$.
3. Advanced type if $a = 0$, $b \neq 0$, and $c \neq 0$.

The equation has a solution $x(t) = e^{\lambda t}$ if λ satisfies

$$a\lambda + b\lambda e^{-\tau\lambda} + c + de^{-\tau\lambda} = 0$$

If λ is a multiple root, then $x(t) = p(t)e^{\lambda t}$ [where $p(t)$ is a polynomial whose degree depends upon the multiplicity of λ] will also satisfy the equation.[2]

Exercise 5-5 Show that the equation

$$\ddot{x}(t) + x(-t) = 0$$

has the solution $x = e^t - e^{-t}$.
Also show that

$$\ddot{x}(t) + x(\pi - t) = 0$$

has the solutions $x = \sin t$, $x = e^t - e^{\pi-t}$.

5-2 A LINEAR EQUATION OF NEUTRAL TYPE

Even though we are generally interested in delay equations, the method of solution pursued in the following example is quite general and instructive for linear equations.

Consider the neutral equation[59]

$$\dot{x}(t) + a\dot{x}(t - 1) + bx(t - 1) = f(t) \tag{5-5}$$

with $x(t)$ and $\dot{x}(t)$ given on $-1 \leq t \leq 0$ and $\dot{x}(t)$ integrable and of bounded variation. A solution of this equation is an absolutely continuous function which satisfies the equation almost everywhere.

If we apply the Laplace transform with parameter s to (5-5), using the asterisk to indicate the transform, on simplifying we obtain

$$x^*(s) = [f^*(s) - (ase^{-s} + be^{-s}) \int_{-1}^{0} e^{-su}x(u) \, du \\ + x(0) + ax(-1)](s + ase^{-s} + be^{-s})^{-1} \tag{5-5a}$$

The last factor on the right is a meromorphic function. It has an infinite number of poles, as we shall presently show; we first give a special case of a theorem due to Borel.

Theorem 5-1 If $g(z)$ is an entire transcendental function and $p(z)$ is an arbitrary polynomial, then there is at most one expression of the form $g(z) - p(z)$ with a finite number of zeros; i.e., of two expressions $g(z) - p_1(z)$ and $g(z) - p_2(z)$, where $p_1(z)$ and $p_2(z)$ are given polynomials, at least one has an infinite number of zeros.

Definition: If $f(z)$ and $g(z)$ are entire functions, then $f(z)$ is said to be of slower growth than $g(z)$ if for any R we have on $|z| = R$ and for $\epsilon \geq 0$ and small

$$\log |f(R)| < [\log \max |g(R)|]^{1-\epsilon}$$

Theorem 5-2 (Borel) If the entire function $f(z)$ grows faster than the entire functions $g_1(z)$, $G_1(z)$ and $g_2(z)$, $G_2(z)$, then at most one of the equations $g_1 f = g_2$, $G_1 f = G_2$ has a finite number of zeros, where g_1 and G_1 are two specific instances of g_1; g_2 and G_2 are two specific instances of g_2.

We now apply Theorem 5-1 to our problem. The function

$$e^{-s}(as + b)$$

has exactly one zero, which in fact is at $s = -b/a$. Thus if we form the two equations

$$e^{-s}(as + b) = 0$$

and
$$e^{-s}(as + b) = -s \tag{5-6}$$

since the first has a finite number of roots, the second has an infinite number.

Remark: Note that the zeros of $s + e^{-s}(as + b)$ are isolated, since an analytic nonzero function $g(z)$ cannot have a zero as a limit point of zeros in the finite plane, i.e., cannot have an infinite number of zeros in any finite region of the plane.

We first write the solution for the case where all the zeros are simple. If we denote the zeros by $\ldots, s_{-2}, s_{-1}, s_0, s_1, s_2, \ldots$ (negative subscripts because in general they may exist in conjugates), then the Mittag-Leffler theorem says that we can write a meromorphic function, and, in particular, the one we are working with here, as

$$[s + e^{-s}(as + b)]^{-1} = \sum_{k=-\infty}^{\infty} \frac{c_k}{s - s_k} + h(s) \tag{5-7}$$

where $h(s)$ is an arbitrary entire function and must be zero for the solution of our equation (why?) and

$$c_k = \frac{1}{(d/ds)[s + e^{-s}(as + b)]_{s=s_k}}$$
$$= \frac{as_k + b}{as_k^2 + bs_k + b} \tag{5-8}$$

We now take the inverse transform

$$x(t) = \frac{1}{2\pi i} \int_{c-i\infty}^{c+i\infty} e^{st} x^*(s) \, ds$$

The inverse transform of $\displaystyle\sum_{k=-\infty}^{\infty} c_k \frac{f^*(s)}{s - s_k}$ yields a convolution, and we have

$$\sum_{k=-\infty}^{\infty} c_k e^{s_k t} \int_0^t e^{-s_k u} f(u) \, du \qquad (5\text{-}9)$$

Integrating by parts, we have

$$\int_{-1}^0 e^{-su} x(u) \, du = -\frac{x(0)}{s} + \frac{x(-1)e^s}{s} + \frac{1}{s} \int_{-1}^0 e^{-su} \dot{x}(u) \, du \qquad (5\text{-}10)$$

and we can write for the remaining expressions on the right of (5-5)

$$\left(-1 + \frac{s}{s + ase^{-s} + be^{-s}} \right) \left(-\frac{x(0)}{s} + \frac{x(-1)}{s} e^s + \frac{1}{s} \int_{-1}^0 e^{-su} \dot{x}(u) \, du \right)$$
$$+ \frac{x(0) + ax(-1)}{s + ase^{-s} + be^{-s}} \qquad (5\text{-}11)$$

Now we must compute the inverse transforms.   For example,

$$\frac{1}{2\pi i} \int_{c-i\infty}^{c+i\infty} e^{st} \frac{x(0)}{s} \, ds = x(0)$$

$$\frac{1}{2\pi i} \int_{c-i\infty}^{c+i\infty} e^{st} \frac{ds}{s} \int_{-1}^0 e^{-su} \dot{x}(u) \, du = \int_{-1}^0 \dot{x}(u) \, du = x(0) - x(-1)$$

$$\frac{1}{2\pi i} \int_{c-i\infty}^{c+i\infty} e^{st} \frac{x(-1)}{s} e^s \, ds = x(-1)$$

Also

$$\frac{1}{2\pi i} \int_{c-i\infty}^{c+i\infty} e^{st} \frac{x(-1)(a + e^s)}{s + ase^{-s} + be^{-s}} \, ds$$

$$= \frac{1}{2\pi i} \int_{c-i\infty}^{c+i\infty} e^{st} x(-1)(a + e^s) \sum_{k=-\infty}^{\infty} \frac{c_k}{s - s_k} \, ds$$

$$= \sum_{k=-\infty}^{\infty} c_k x(-1)(a + e^{s_k}) e^{s_k t}$$

$$= \sum_{k=-\infty}^{\infty} c_k x(-1) \left(-\frac{b}{s_k} \right) e^{s_k t} \qquad (5\text{-}12)$$

We have used the fact that s_k is a zero and thus replaced $a + e^{s_k}$ by its equal from the equation $s + ase^{-s} + be^{-s} = 0$. Finally

$$\frac{1}{2\pi i} \int_{c-i\infty}^{c+i\infty} e^{st} \left(\int_{-1}^{0} e^{-su} \dot{x}(u)\, du \right) \sum_{k=-\infty}^{\infty} \frac{c_k}{s - s_k}\, ds$$

$$= \sum_{k=-\infty}^{\infty} c_k \int_{-1}^{0} e^{-s_k(u-t)} \dot{x}(u)\, du \quad (5\text{-}13)$$

Our solution can now be written as

$$x(t) = \sum_{k=-\infty}^{\infty} \frac{as_k + b}{as_k{}^2 + bs_k + b}\, e^{s_k t} \left[\int_{0}^{t} e^{-s_k x} f(x)\, dx + \int_{-1}^{0} e^{s_k u} \dot{x}(u)\, du \right.$$

$$\left. - \frac{bx(-1)}{s_k} \right] \quad (5\text{-}14)$$

It turns out that (5-6) has, in addition, one double root and one triple root. Before discussing the solution for these cases we shall explain a method for determining all the roots for fixed real values of s. In the process of doing this we also determine the double and triple roots.

If we replace t by $t + 1$ in the homogeneous equation (5-5), take $f(t) \equiv 0$, and use operator notation, we have

$$EDx + aDx + bx = 0 \quad (5\text{-}15)$$

On ignoring x, substituting $E = e^D$, and replacing D by the complex variable $s = \alpha + i\beta$, we have, as in ordinary differential equations, the characteristic equation

$$s + e^{-s}(as + b) = 0 \quad (5\text{-}16)$$

which is (5-6). Usually the object is to determine the condition for stability, based on the nature of the roots of the equation.[47]

We can separate the equation into real and imaginary parts, obtaining

$$-\alpha e^{\alpha} = (a\alpha + b)\cos\beta + a\beta\sin\beta \quad (5\text{-}17)$$
$$-\beta e^{\alpha} = -(a\alpha + b)\sin\beta + a\beta\cos\beta \quad (5\text{-}18)$$

If we square these equations and add, we have

$$e^{2\alpha}(\alpha^2 + \beta^2) = (a\alpha + b)^2 + a^2\beta^2 \quad (5\text{-}19)$$

The problem now is to determine values of a and b which for a given value $\alpha = c$ satisfy the last equation for an infinite number of values of β; that is, the line $\alpha = c$ intersects the curve corresponding to the equation at infinitely many points.

Now for $\alpha = c$ and in order that the equation be satisfied for all β it is necessary to take $a^2 = e^{2c}$ so that the β terms drop out.   Thus

$$a = \pm e^c \qquad \text{and} \qquad b = \mp 2ce^c \tag{5-20}$$

As a check, if we substitute these values of a and b and $\alpha = c$ in (5-17) and (5-18), we have

$$-c = \mp c \cos \beta \pm \beta \sin \beta \tag{5-21}$$
$$-\beta = \pm c \sin \beta \pm \beta \cos \beta \tag{5-22}$$

Either of these two equations must be satisfied identically in β.   Eliminating β leads (either equation alone yields the same result) to

$$\beta = -c \left(\frac{\pm 1 - \cos \beta}{\sin \beta} \right) \equiv \frac{-c \sin \beta}{\pm 1 + \cos \beta} \tag{5-23}$$

establishing the sufficiency for the values of a and b.   From the relation

$$\frac{\sin \beta}{1 + \cos \beta} = \tan \frac{\beta}{2}$$

we have
$$\beta = \mp c \left(\tan \frac{\beta}{2} \right)^{\pm 1} \tag{5-24}$$

We now determine the multiple roots.   The equation

$$s + e^{-s}(as + b) = 0$$

together with the roots of the equations for the first and second derivatives determine a unique value $s = -2$ for a triple root, from which we easily have $a = e^{-2}$ and $b = 4e^{-2}$, and β can be computed for $c = -2$ from (5-24).   That roots of higher multiplicity cannot occur can be deduced from the fact that only three equations are needed to determine the unknowns a, b, and s.   For a double root we take only the first derivative and conclude in this case that $a = -(1 + c)e^c$ and $b = c^2e^c$. Substituting for a and b from (5-20), we get

$$-(1 + c) = \pm 1 \qquad c^2 = \mp 2c$$

These equations can be satisfied only if $c = -2$, which is the triple root value, or if $c = 0$, which gives $a = -1$, $b = 0$.   The values of β corresponding to $c = 0$ are $\beta = 0$, obtained from $\beta = -c \tan(\beta/2)$, and $\beta = \pm 2n\pi$ ($n = 1, 2, \ldots$), obtained from $\beta = c/\tan(\beta/2)$ as $c \to 0$ or directly from (5-21) and (5-22) on putting $c = 0$.   Thus we have obtained all the multiple and simple roots (for constant values of α in the latter case).

We now return to the solution.   The double root occurs at $s = c$, where c is real, $c \neq -2$, since $c = -2$ is a triple root.   For this case we exclude from the sum in (5-14) the expressions with the simple roots s_{-1} and s_1. In fact, $s_0 = 0$ is also excluded if $a < 0$ and is not excluded otherwise. With these exclusions the reader can show that we must add to the previous solution the factor

$$\frac{2}{c+2} e^{ct} \left\{ \int_0^t e^{-cu} f(u)(t-u) \, du + \int_{-1}^0 e^{-ct} \dot{x}(u)(t-u) \, du \right. $$
$$\left. - x(-1)\left[a + \frac{b(t+1)}{c} \right] \right\}$$

From the Mittag-Leffler theorem we have corresponding to the double root at $s = c$ a term of the form $A/(s-c)^2$, where

$$A = \frac{1}{(d^2/ds^2)[s + e^{-s}(as+b)]_{s=c}} = \frac{1}{c+2}$$

The inverse transform of $f^*(s)/(s-c)^2$ yields a convolution

$$\int_0^t e^{-c(u-t)} f(u)(t-u) \, du$$

Exercise 5-6 Attempt the solution for the case of the triple root.

A solution $x(t)$ is stable if $x(t) \to 0$ as $t \to \infty$. It is clear (from the exponential factor $e^{s_k t}$) that the solution will be stable if all the s_k have negative real parts. This condition is sufficient but not necessary.

Exercise 5-7 Show that the delay-differential equation

$$\dot{x}(t) = ax(t-1)$$

in which a is real, has the characteristic equation $s - ae^{-s} = 0$ and write its solution.

Exercise 5-8 Apply the Laplace-transform approach to

$$x(t+a) - x(t) = b\dot{x}(t) \qquad a > 0$$

with $x(t)$ given and integrable on $[0,a]$. This equation has an advanced argument and solutions will not, in general, go into the future.

Exercise 5-9 Supply the initial conditions and use Laplace-transform techniques to solve

$$\dot{x}(t) - bx(t-1) = \sum_{n=0}^{\infty} a_n t^n$$

Exercise 5-10 Show that a solution of the simple equation with retardation

$$\dot{x}(t) = a_1 x(t) + a_2 x(t-\tau)$$

where a_1, a_2, and τ are real and $x(t)$ is known in $0 \leq t \leq \tau$ can be expressed in the form $\sum_n c_n \exp(\lambda_n t)$, where c_n are constants and λ_n is a root of $\lambda = a_1 + a_2 e^{-\lambda t}$, provided the roots are simple.

Exercise 5-11   Show that the equation

$$\ddot{x}(t) + \tfrac{5}{2} x(t) + \tfrac{3}{2} x(\pi - t) = 0$$

has the solutions $x = \cos t$, $x = \cos 2t$.

5-3   WHAT IS A SOLUTION?

Recall that the solution $x(t_0, x_0; t)$ of an ordinary differential equation can be viewed as a continuous transformation of the space of initial points $x_0 \, \epsilon \, E^n$ defined on $t = t_0$ to points in E^n representing the state variables.

As we have seen in the material of earlier sections, for delay-differential equations (and even more generally for functional differential equations), we require an initial function φ defined over the lag period $\tau \geq 0$ or a translation of it.   (To simplify matters for discussion purposes one translates the lag interval to $[-\tau, 0]$.)   As the initial-value problem may require a prescription of n such functions or one function with its first $n - 1$ derivatives, all defined on $[-\tau, 0]$, the set of initial functions is a vector in E^n and is a continuous mapping of $[-\tau, 0]$ into E^n.   Thus, in general, we denote such vectors by φ and their space by $C([a,b], E^n)$, where we take $[a,b]$ to be $[-\tau, 0]$.   Here $\|\varphi\| = \sup_{a \leq \theta \leq b} |\varphi(\theta)|$.

Let x be any function with domain $[\sigma - \tau, \infty)$, $\tau \geq 0$, σ a real number, and range in E^n. Let x_t denote the restriction of x to $[t - \tau, t]$.   Thus for $t \geq \sigma$, x_t defined by $x_t(\theta) \equiv x(t + \theta)$, $-\tau \leq \theta \leq 0$ is an element of $C([-\tau, 0], E^n)$ or simply C; that is, its graph is the same as the graph of x on $[t - \tau, t]$ but shifted to $[-\tau, 0]$.   Note that[38] x and x_t, $t \geq \sigma$, respectively map $[\sigma - \tau, \infty)$ and $[-\tau, 0]$ to E^n, whereas $x(t)$ and $x_t(\theta)$ are their values at the indicated arguments.

Now if we write an equation of the form $\dot{x}(t) = X(x_t, t)$, we must define that portion of $x(t)$, given prior to any initial value of t, i.e., x_t. We require that $X(x_t, t)$, which is in E^n, have certain desirable properties, e.g., be continuous in the initial functions φ, with $\|\varphi\| < H$ for some $H > 0$, $t \, \epsilon \, [0, \infty]$.   All such φ define a space, which we denote by C_H. The function $x(\sigma, \varphi)$ with initial condition $\varphi \, \epsilon \, C([-\tau, 0], E^n)$, $\|\varphi\| < H$, that is, $\varphi \, \epsilon \, C_H$ at σ, is called a *solution* of $\dot{x}(t) = X(x_t, t)$ if there is a function from $[\sigma - \tau, \sigma + A]$, $A > 0$, into E^n such that:

1. $x_t(\sigma, \varphi) \, \epsilon \, C_H$, $\sigma \leq t \leq \sigma + A$.
2. $x_\sigma(\sigma, \varphi) = \varphi$.
3. $x(\sigma, \varphi)(t)$ satisfies the equation for $\sigma \leq t < \sigma + A$ (when $\sigma = 0$, it is sometimes completely omitted from the arguments).

It is convenient to examine the ideas of a solution of delay equations as special cases of more general equations known as *functional differential equations*. Thus, for example, if we consider the linear equation $\dot{x}(t) = \alpha x(t) + \beta x(t - \tau)$, where α and β are constants and x is a scalar, and define

$$\eta(\theta) = \begin{cases} 0 & \theta = -\tau \\ \beta & -\tau < \theta < 0 \\ \alpha + \beta & \theta = 0 \end{cases}$$

then the above equation is known to be equivalent to the linear functional differential equation

$$\dot{x}(t) = \int_{-\tau}^{0} x(t + \theta) \, d\eta(\theta) \equiv \int_{-\tau}^{0} x_t(\theta) \, d\eta(\theta) \equiv L(x_t)$$

Exercise 5-12   Demonstrate this equivalence.

In fact, this is a general form which applies to a vector equation $\dot{x}(t) = L(x_t)$, where L is linear in x_t and independent of t and is a continuous function on functions φ into E^n (where φ maps the interval $[-\tau, 0]$ into E^n).

On $[-\tau, 0]$, $x_t(\theta)$ is replaced by its equivalent $\varphi(\theta)$, and for the general linear case $\eta(\theta)$ is a matrix whose elements are functions of bounded variation, and the integral is in the sense of Stieltjes.[37,38]

Remark: The equation $\dot{x}(t) = L(x_t)$, of which the delay equation $\dot{x}(t) = \sum_{k=1}^{p} A_k x(t - \tau_k)$, $\tau_k \geq 0$, is a special case, is called an *autonomous linear functional differential equation* with constant coefficients and is itself a special case of the nonautonomous problem $\dot{x}(t) = X(x_t, t)$.

Remark: Hale has found it convenient in the study of qualitative properties of solutions to consider restriction of a function to an arbitrary interval $[t - \tau, t]$ and then translate it to $[-\tau, 0]$. Thus for each t one has a mapping from $C[-\tau, 0]$ to $C[-\tau, 0]$.

5-4   LINEAR DELAY - DIFFERENTIAL EQUATIONS

In this and in following sections we shall be concerned with the construction of solutions of delay equations, in particular, periodic solutions using perturbation techniques, and with questions of stability of solutions. It is important to emphasize that the theory of ordinary differential equations is helpful after it has been possible to reduce a problem in such a way that the theory is applicable. At times, completely new ideas must be introduced to deal with this new type of equation.

An nth-order nonhomogeneous delay-differential equation with variable coefficients can be given as follows:

$$x^{(n)}(t) = \sum_{i=0}^{n-1} \sum_{j=1}^{m} a_{ij}(t) x^{(i)}(t - \tau_j(t)) + f(t) \qquad (5\text{-}25)$$

where $a_{ij}(t)$, $f(t)$, and usually also $\tau_j(t) \geq 0$ are assumed continuous on $t_0 \leq t < t_1 \leq \infty$. The initial value function $\varphi(t)$ and its first $n - 1$ derivatives are continuous on the set $E = \bigcup_{j=1}^{m} E_j$, where $E_j : \{t - \tau_j(t) | t_0 \leq t \leq t_0 + \tau_j(t)\}$. For the analysis of stability, Eq. (5-25) may be replaced by a first-order system of the form (see Sec. 5-6 for discussion of some difficulties which may arise from such replacement.)

$$\dot{x}_k(t) = \sum_{i=0}^{n-1} \sum_{j=1}^{m} a_{ij} x_i(t - \tau_j(t)) + f_k(t) \qquad k = 0, \ldots, n - 1 \quad (5\text{-}26)$$

and involving n independent functions $x_0(t), \ldots, x_n(t)$.

Considering the homogeneous equation obtained from (5-25) by putting $f(t) \equiv 0$, we note that to each initial-value function $\varphi_p(t)$ and its $n - 1$ derivatives we have a solution $x_p(t)$, and a linear combination $\sum_{p=1}^{q} c_p x_p(t)$ is a solution corresponding to $\sum_{p=1}^{q} c_p \varphi_p(t)$.

The sum of a solution of the homogeneous problem with an initial function $\varphi(t)$ and of the nonhomogeneous problem with an initial function $\psi(t)$ is a solution of the nonhomogeneous problem with initial function $\varphi(t) + \psi(t)$. Myshkis has introduced the notion of a fundamental system of solutions for the homogeneous part of (5-25) by examining the metric spaces; one of these spaces is that of a set of initial conditions $\{\varphi(t)\}$ and the other of the corresponding solutions $\{x(t)\}$, which are required to satisfy in their respective metric the condition of the continuous dependence of solutions on the initial functions. Usually the fundamental system consists of an infinite number of solutions. Every solution of (5-25) can then be appropriately expressed by means of linear combinations of the elements of the system plus a particular solution. However, in general the method is not practical for obtaining solutions to equations. By making various assumptions on the coefficients, the deviations, etc., one can attempt to develop a solution assuming a particular appropriate form such as an integral or a series.

Exercise 5-13 For the following equation, due to El'sgol'ts,

$$x^{(n)}(t) = \sum_{j=0}^{n-1} a_j x^{(i)}(t - \tau)$$

where a_j and τ are constants, show that its solution $x(t)$ is defined by an n-times-differentiable initial function $\varphi(t)$ on $[0,\tau]$ and is given by

$$x(t) = \sum_{k=0}^{n-1} \frac{\varphi^{(k)}(0)}{k!} y(t) + \frac{1}{(n-1)!} \int_0^\tau x(t-s)\varphi^{(n)}(s)\, ds$$

where $y(t)$, $t > \tau$, is a solution of the equation for an initial function t^k with integer $k > 0$ on $[0,\tau]$ and $x(t) = 0$ for $t < 0$.

Solutions of such linear equations as (5-25), whose coefficients a_{ij} and deviations $\tau_j \geq 0$ are constants (the arguments may also take the form $t + \tau_j$), can often be obtained by means of Laplace transforms along lines similar to those used in the example of Sec. 5-2.

Applying the Laplace transform to such an equation gives rise to a so-called *quasi polynomial* in the parameter of the transform. An example of such a polynomial equation is

$$s^n = \sum_{i=0}^{n-1} \sum_{j=1}^{m} a_{ij} s^i e^{-\tau_j s}$$

One can write a similar equation for a system of equations. The inversion of Laplace transforms is itself a difficult task.

As we have seen in the example of Sec. 5-2, it usually is not possible to determine explicitly all the zeros of the quasi polynomial. However, knowledge of their location aids the analysis of stability. Many investigations have proceeded in this direction.

Often the treatment of linear difference equations with time lag τ gives rise to characteristic equations of the form

$$F(z) = z^n + az^{n-1} + \cdots - Ke^{i\theta}e^{-\tau z}(z^m + bz^{m-1} + \cdots) = 0$$

where $\tau > 0$, $K \geq 0$, and θ are real constants and a and b are complex constants. Krall has proved the following theorem:[42]

***Theorem* 5-3**   If $n > m$, the number of zeros of $F(z)$ with positive real part (or lying in any right half-plane) is finite; if $K \neq 0$, $F(z)$ has an infinite number of zeros with arbitrarily large negative real part.

If $n = m$ and $a = b$, when $K \neq 0$, $F(z)$ has an infinite number of zeros given by

$$\frac{1}{\tau} (\log K + i(\theta + 2k\pi)) + o(1)$$

as $k = 0, \pm 1, \pm 2, \ldots$, and only a finite number of other zeros. If $K < 1$, $F(z)$ has only a finite number of zeros with positive real part. If $K > 1$, $F(z)$ has only a finite number of zeros with negative real part.

If $n < m$, then the number of zeros of $F(z)$ with negative real part (or lying in any left half-plane) is finite; if $K \neq 0$, $F(z)$ has an infinite number of zeros with arbitrarily large positive real parts.

Exercise 5-14 Show by applying the Laplace transform with parameter s, integrating by parts, and equating to zero all outside factors that

$$\sum_{i=0}^{n} \sum_{k=1}^{m} a_{ik} x^{(i)} (t - \tau_{ik}) = 0$$

has the characteristic equation

$$\sum_{i=0}^{n} \sum_{k=1}^{m} a_{ik} s^i e^{-s\tau_{ik}} = 0$$

Exercise 5-15 Show that the solution of

$$y''(x) + ay'(x - \tau_1) + by(x - \tau_2) = w(x) \qquad \tau_1 > 0, \tau_2 > 0$$

with $y(x) = Y(x)$ given and integrable on $-\tau_2 \le x \le 0$ and $y'(x) = Y_1(x)$ given and integrable on $-\tau_1 \le x \le 0$ and where $w(x)$ is integrable is given by

$$y(x) = \int_0^x \int_0^u [w(v) - aY_1(v - \tau_1) - bY(v - \tau_2)] \, dv \, du + xY_1(0) + Y(0)$$

on the interval $[0, \tau]$, where $\tau = \min \{\tau_1, \tau_2\}$. If $\bar{\tau} = \max \{\tau_1, \tau_2\}$, then this solution can be extended to the interval $[\tau, \bar{\tau}]$ and then to $[\bar{\tau}, \bar{\tau} + \tau]$, etc.

Exercise 5-16 Write down the Laplace transform of the solution of the equation in the previous exercise.

Exercise 5-17 Show that

$$\dot{x}(t) = x(t + 1)$$

with $\qquad x(t) = t + 1 \qquad$ for $-1 \le t \le 0$

has the solution $x(t) = 1$ for $0 \le t \le 1$ and $x(t) = 0$ for $t > 1$.

Exercise 5-18 Show that a series solution of the homogeneous part of the equation

$$x^{(n)}(t) + a_1 x^{(n-1)}(t + c) + \cdots + a_{n-1} x^{(1)}[t + (n-1)c] + a_n x(t + nc) = f(t)$$

is given by $\qquad x(t) = A_0 + \sum_{j=1}^{\infty} A_j \frac{(t + jc)^i}{j!}$

where the coefficients are determined from the recursion formula

$$A_{n+j} + a_1 A_{n+j-1} + \cdots + a_n A_j = 0$$

and the solution of the entire equation is then given by

$$x(t) = \sum_{j=0}^{\infty} A_j \int_0^{t+cj} \frac{(t + cj - z)^{n+j-1}}{(n + j - 1)!} f(z) \, dz$$

where $|c| < e^{-1} \min 1/|\lambda_i|$, where λ_i is the ith root of the characteristic equation.

Exercise 5-19 An example of a general linear differential difference equation with variable coefficients is

$$a_{n0}(t)x^{(n)}(t) + \sum_{i=0}^{n-1}\sum_{k=1}^{m} a_{ik}(t)x^{(i)}(t - \tau_{ik}) = w(t)$$

Let $n = 2$ and $a_{ik}(t) = a_{ik}t + b_{ik}$, apply the Laplace transform, and discuss the solution.

Exercise 5-20 The equation

$$\dot{x}(t) = a(t)x[\tau(t)] + b(t) \qquad A \le t \le B$$

has variable lag parameter $\tau(t)$ and variable coefficients; it can be reduced to a Fredholm-type integral equation, and the existence of a solution can be deduced in that manner. Sketch this approach.

Remark on periodic solutions: A necessary and sufficient condition for the existence of periodic solutions of a linear homogeneous equation with constant coefficients and delays is the presence of purely imaginary roots of the characteristic quasi polynomial. The incommensurability of the absolute values of these roots gives rise to almost-periodic solutions.

If the forcing function of a nonhomogeneous problem is periodic, of period 2π, and, for example, is not resonant with the natural vibrations of the homogeneous equation, then one can expand $f(t)$ in a Fourier series. A periodic solution of period 2π is obtained by also expanding $x(t)$ in a Fourier series and determining the coefficients of the latter on substituting and equating coefficients.

For a linear dynamic system of n equations, which in vector notation has the form

$$\dot{x} = \sum_{p=1}^{n} A_p x(t - \tau_p) + f(t)$$

where all the components of f are periodic, of period 2π, and for which the quasi polynomial has k imaginary roots, we must have

$$\int_0^{2\pi} \sum_{i=1}^{n} y_{ij}(t)f_i(t)\,dt = 0 \qquad j = 1, \ldots, k \qquad i = 1, \ldots, n$$

where $\{y_{ij}\}$ are linearly independent periodic solutions of the adjoint system

$$\dot{y} = -\sum_{p=1}^{n} A'_p\, y(t + \tau_p)$$

where the matrix A'_p is the transpose of A.

As a particular example, a unique periodic solution of period ω of the linear system[30]

$$x(t) = A(t)x(t) + B(t)x(t - \tau) + f(t)$$

[in which the components of $f(t)$ have period ω] exists if and only if the only periodic solutions of period ω of the homogeneous system are the trivial ones. A periodic solution also exists in case the equation has a bounded solution for all t. The existence of periodic solutions of the equation $x(t) = \dot{x}(t + \alpha)$ is a special case.

STATIONARY POINTS

If the τ_{ij} are constants in the linear homogeneous system

$$\dot{x}_i = \sum_{k=1}^{m} \sum_{j=1}^{n} a_{ijk} x_j(t - \tau_{jk}) \qquad i = 1, \ldots, n$$

then its stationary, or singular, points $x_i \equiv 0$, $i = 1, \ldots, n$, can be classified according to the roots of its characteristic quasi polynomial. These are said to be of type (p,q,k) if p is the number (including multiplicity) of roots of the quasi polynomial having negative real parts, q is the number with positive real part, and k is the number with real part equal to zero. Thus $(\infty,0,0)$ and $(0,\infty,0)$ give, respectively, a general stable focus and a general unstable focus; $(p,q,0)$ is a saddle point. A center with $(0,0,\infty)$ cannot exist, since only a finite number of roots of the characteristic polynomial can lie on the imaginary axis; see Ref. 23.

5-5 EXISTENCE AND SOME METHODS OF SOLUTION

AN EXISTENCE THEOREM

Driver[16] gives an existence and uniqueness theorem for the delay-differential (or hysteretic-differential) system

$$\dot{x}_i(t) = F_i(t,x(s)) \qquad \alpha \leq s \leq t$$

for $t > t_0$, $i = 1, \ldots, n$, which can be represented in the form

$$\dot{x}(t) = F(t,x(.)) \qquad \text{for } t > t_0$$

In general $F(t,\psi(.))$ is an n-vector-valued functional determined by t and by the values of $\psi(s) = (\psi_1(s), \ldots, \psi_n(s))$ for $\alpha \leq s \leq t$.

Here $-\infty \leq \alpha \leq t_0 \leq \beta \leq \infty$, where α and β are given numbers. If $\alpha = -\infty$, then $-\infty < s \leq t$, etc.

Exercise 5-21 Which of the equations discussed in Secs. 5-1 and 5-2 can be considered as special cases of this system?

The norm of a vector $x \epsilon E^n$ will be $\|x\| = \max |x_i|$. If ψ maps $[a,b]$ into E^n, then we define

$$\|\psi\|^{[a,b]} = \sup_{a \leq s \leq b} \|\psi(s)\|$$

$C([a,b] \to R)$ denotes the class of continuous (vector-valued) functions which map the interval $[a,b]$ into the region R of E^n. If $\alpha = -\infty$ and D is a given domain of E^n, $\psi \epsilon C([\alpha,b] \to D)$ means that there exists a compact set $G_\psi \subset D$ such that $\psi \epsilon C((-\infty,t] \to G_\psi)$.

Thus in

$$\dot{x}(t) = F(t,x(.)) \qquad t_0 < t < \beta$$

the functional $F(t,\psi(.))$ takes values in E^n for $t_0 \leq t < \beta$ and

$$\psi \epsilon C([a,b] \to D)$$

F is said to be continuous in t if it is a continuous function of t for $t_0 \leq t < b$ whenever $\psi \epsilon C([\alpha,b] \to D)$. F is locally lipschitzian with respect to ψ if for every $\bar{\beta} \epsilon [t_0,b]$ and every compact set $G \subset D$ there is a constant L such that

$$\|F(t,\psi(.)) - F(t,\theta(.))\| \leq L\|\psi - \theta\|^{[\alpha,t]}$$

for $t \epsilon [t_0,\bar{\beta}]$ and for $\psi, \theta \epsilon C([\alpha,t] \to G)$.

We now state an existence theorem which combines theorems on local and extended existence (theorems 2 and 3 of Ref. 16).

Theorem 5-4 If $F(t,\psi(.))$ is continuous in t and locally lipschitzian with respect to ψ, and if φ (the initial function) belongs to $C([\alpha,t_0] \to D)$, then there exists an $h > 0$ such that a unique solution $x(t) = x(t,t_0,\varphi)$ exists for $\alpha \leq t < t_0 + h$, and this solution can be extended uniquely on $[\alpha,\gamma)$ where $t_0 < \gamma \leq \beta$, and if $\gamma < \beta$ and γ cannot be increased, then for any compact set $G \subset D$ there is a sequence $t_0 < t_1 < \cdots \to \gamma$ such that

$$x(t_k) \epsilon D - G \qquad \text{for } k = 1, 2, \ldots$$

that is, $x(t)$ comes arbitrarily close to the boundary of D, or else $x(t)$ is unbounded.

PICARD'S METHOD

Many delay-differential equations occur in the form

$$\dot{x}(t) = f[t,x(t),x(g(t))]$$
$$\equiv f(t,x,X)$$

where $g(t) \leq t$, and where f and g are given functions.

Suppose now that $x(t) \equiv \varphi(t)$ is continuous for $t \leq 0$, $\varphi(0) \equiv x_0 \equiv x(0)$, and consider the interval $-\infty < t < \tau$, $\tau > 0$. We shall study the solvability of this type of equation using Picard's method, already encountered in our study of ordinary differential equations and in Chap. 2. The approach can be generalized without difficulty to a vector equation.

We define

$$x^0(t) = \begin{cases} \varphi(t) & \text{on } -\infty < t \leq 0 \\ x_0 & \text{on } 0 \leq t \leq \tau \end{cases}$$

$$x^1(t) = \begin{cases} \varphi(t) & \text{on } -\infty < t \leq 0 \\ x^0(t) + \int_0^t f[s, x^0(s), x^0(g(s))] \, ds & \text{on } 0 \leq t \leq \tau \end{cases}$$

and similarly for $x^2(t)$, and so on. Suppose that $f(t,x,X)$ and $g(t)$ are continuous everywhere on $0 \leq t \leq \tau$. Let B be a bound on f in this interval. In addition, for any two points defined on this interval let f satisfy the Lipschitz condition

$$|f(t,x,X) - f(t,x^*,X^*)| \leq A|(x,X) - (x^*,X^*)|$$
$$\leq A \sqrt{(x-x^*)^2 + (X-X^*)^2}$$

We have the following estimates for the differences between successive iterations:

$$|x^1(t) - x^0(t)| = \left| \int_0^t f[s, x^0(s), X^0(g(s))] \, ds \right| \leq Bt \leq B\tau$$

$$|x^2(t) - x^1(t)| = \left| \int_0^t f[s, x^1(s), x^1(g(s))] - f[s, x^0(s), x^0(g(s))] \, ds \right|$$

$$\leq \int_0^t |.| \, ds \leq \int_0^t A \max \left[|x^1(s) - x^0(s)|, |x^1(g(s)) \right.$$
$$\left. - x^0(g(s))| \right] \, ds$$

$$\leq \int_0^t A \max (Bs, Bs) \, ds$$

$$\leq AB \frac{t^2}{2}$$

Exercise 5-22 Show that, in general, $|x(t) - x^{n-1}(t)| \leq BA^{n-1}(t^n/n!)$, and hence also show that $\lim_{n \to \infty} x^n(t) = x(t)$ exists by considering

$$\left| x(t) - x_0 - \int_0^t f[s, x(s), x(g(s))] \, ds \right|$$

$$\leq |x(t) - x^n(t)| - x_0 + x_0 + \left| \int_0^t f[s, x^{n-1}(s), x^{n-1}(g(s))] \right.$$
$$\left. - f[s, x(s), x(g(s))] \, ds \right| \to 0 \qquad \text{as } n \to \infty$$

Exercise 5-23 Show that $x(t)$ is unique. If you have difficulty, consult "NM," chap. 4.

Exercise 5-24   Apply the foregoing method to the solution of

$$\dot{x}(t) = ax(t) + bx(g(t)) \qquad \text{for } t < 0$$

with $x(0) = x_0$ and $-\infty < \alpha \le g(t) \le 0$, $\alpha \le t \le 0$. Show that

$$|x^n(t) - x^{n-1}(t)| \le |ax_0 + bx_0|(|a| + |b|)^{n-1}|\alpha|^n$$

and hence for convergence we must have

$$(|a| + |b|)|\alpha| < 1 \qquad |g(s)| > |s|$$

Exercise 5-25   Apply Picard's method to

$$\dot{x}(t) = a(t)x(t) + b(t)x(t - \tau) \qquad \text{on } t \le 0$$

with $x(0) = x_0$, and with a and b continuous in t. For example,

$$x^0(t) = x_0$$

and $\qquad |x^1(t) - x^0(t)| \le \int_t^0 (|a(s)| \, |x_0| + |b(s)| \, |x_0|) \, ds \le |x_0| \int_t^0 A(s) \, ds$

where $|a(t)| + |b(t)| \le A(t)$. For convergence one can assume that $\int_t^0 A(s) \, ds \le K < 1$ for all $t \le 0$.

Exercise 5-26   Prove the following lemma [due to W. T. Reid (*Trans. Am. Math. Soc.*, vol. 32, 1930)] useful for estimation purposes, often in conjunction with Picard's method.
If

$$v(t) \le C + \int_{t_0}^t v(s)u(s) \, ds \qquad t_0 \le t < \beta$$

where $C \ge 0$ and $u(t)$ and $v(t)$ are continuous and nonnegative functions, then

$$v(t) \le C \exp \left[\int_{t_0}^t u(s) \, ds \right]$$

Hint: Denote the integral by $R(t)$ and note that $R'(t) = u(t)v(t)$. Thus by multiplying by $u(t)$ obtain

$$R'(t) - u(t)R(t) \le Cu(t)$$

Then use $\exp \left[-\int_{t_0}^t u(s) \, ds \right]$ as an integrating factor. Integrate from t_0 to t, obtaining the desired result. With $u(t) \equiv A$, where A is constant, we have what is known as *Grünwald's inequality*.

Without a Lipschitz condition on f we can still prove local existence (but not uniqueness) using the Schauder-Leray theorem. Consider the rectangle $t_0 \le t \le t_0 + a$, $|x - \varphi(t_0)| \le b$, $|X - \varphi(t_0)| \le b$, $|f(t,x,X)| \le B$. Let $h = \min (a, b/B)$, and consider the Banach space of those functions $\psi(t)$ which are defined and continuous on $[t_0, t_0 + h]$, with

$$\|\psi\|^{[t_0, t_0+h]} = \sup_{t_0 \le t \le t_0+h} |\psi(t)|$$

Let S be the subset of functions ψ such that $\|\psi - \varphi(t)\|^{[t_0, t_0 + h]} \leq b$. To show that S is convex let ψ and $\theta \epsilon S$; then

$$\|\lambda\psi + (1 - \lambda)\theta - \varphi(t_0)\|^{[\cdot]} = \|\lambda\psi + (1 - \lambda)\theta - \lambda\varphi(t_0) - (1 - \lambda)\varphi(t_0)\|^{[\cdot]}$$
$$\leq \lambda\|\psi - \varphi(t_0)\|^{[\cdot]} + (1 - \lambda)\|\theta - \varphi(t_0)\|^{[\cdot]}$$
$$\leq \lambda b + (1 - \lambda)b = b$$

and hence S is convex.   For $\psi \epsilon S$ we define the transformation T mapping S into S as follows:

$$(T\psi)(t) = \varphi(t_0) + \int_{t_0}^t f(s, \psi(s), \psi(g(s)))\ ds$$

Then

$$\|T\psi - \varphi(t_0)\|^{[\cdot]} \leq \max_{t_0 \leq t \leq t_0 + h} \int_{t_0}^t |f[s, \psi(s), \psi(g(s))]|\ ds$$
$$\leq \max_{t_0 \leq t \leq t_0 + h} \int_{t_0}^t B\ ds = Bh \leq b$$

Thus T maps S into S.   To show that T is continuous we must show that if $\|\psi - \theta\|^{[\cdot]}$ is small for ψ and $\theta \epsilon S$, then $\|T\psi - T\theta\|^{[\cdot]}$ is also small. Now if $\|\varphi - \theta\|^{[\cdot]} < \delta$, then $|\psi(s) - \theta(s)|^{[\cdot]} < \delta$, and $|\psi(g(s)) - \theta(g(s))| < \delta$.   Also

$$|(T\psi)(t) - (T\theta)(t)| \leq \int_{t_0}^t |f[s, \psi(s), \psi(g(s))] - f[s, \theta(s), \theta(g(s))]|\ ds$$
$$\leq \int_{t_0}^t \frac{\epsilon}{h}\ ds \leq \frac{\epsilon}{h} h = \epsilon$$

Thus $\|T\psi - T\theta\|^{[\cdot]} < \epsilon$, and hence there exists an $x(t)$ such that

$$x(t) = \varphi(t_0) + \int_{t_0}^t f[s, x(s), x(g(s))]\ ds$$

EXPANSION IN POWERS OF τ

Approximate solution by expanding in powers of a small delay τ for the initial-value problem

$$\dot{x} = f(t, x(t), x(t - \tau)) \qquad t_0 \leq t \leq t_1 < \infty$$

with initial function $\varphi(t)$, is achieved by solving the ordinary differential equation obtained by replacing $x(t - \tau)$ above by its approximate expansion in powers of τ

$$x(t - \tau) = x(t) - \tau\dot{x}(t) + \cdots + \frac{(-1)^m \tau^m x^{(m)}(t)}{m!}$$

This approach may not lead to valid results.   El'sgol'ts points out from the theory of differential equations with a small parameter that the discarded term

$$\frac{\tau^{m+1}}{(m+1)!} x^{(m+1)}(t - \theta\tau)$$

for $m > 1$ is very large for small τ since it behaves like $1/\tau$ (Why?); it is small for $m = 1$, and the approximation is valid in that case. This approach is also valid for $m \leq k$ in an equation in which the kth is the highest derivative in which the argument $t - \tau$ occurs. The method is also useful in estimating the roots of the characteristic quasi polynomial.

Exercise 5-27 Apply this method to the equation

$$\dot{x}(t) = x(t - \tau) \qquad 0 \leq t < \infty, \varphi(t) \equiv 1$$

with $m = 0$ and again with $m = 1$, and point out which solution has better asymptotic properties.

Exercise 5-28 Consider

$$x(t - \tau) = f(t,x(t),\dot{x}(t))$$

and expand up to the $(m + 1)$st term (which involves θ). Solve and put on the left $x^{(m)}(t)$, and by successive integration over intervals obtain the same expression, with $x^{(m+1)}$ replaced by $\varphi^{(m+1)}$, where φ is the initial function. Note a behavior of order at least $1/\tau$ by differentiating this expression once to obtain $x^{(m+1)}(t)$ on the left and replacing $x^{(m)}(t)$ on the right by its expression in terms of lower derivatives. On the other hand, for $m = 1$ there is no small parameter.

EQUIVALENT LINEARIZATION[49]

By extending the method used in ordinary differential equations (for restrictions see Ref. 35) one assumes a solution of the form

$$x(t) = A(t) \sin (wt + b)$$

and develops the nonlinear terms and terms with lag in Fourier series, keeping only the first Fourier term and discarding the rest. Equating the coefficients of $\sin wt$ and $\cos wt$ leads to algebraic equations (or differential equations if A and w depend on t) from which A and w are determined. Hale has shown[35] that the foregoing procedure is valid. Other computational methods, e.g., Euler's method and parabolic methods, are also used to derive solutions of delay-differential equations.

5-6 PERTURBATION METHODS

Perturbation techniques have been recently studied by Hale and Perelló[38] and by Cooke[13] in the analysis of stability and in pursuit of periodic solutions. In the case of some ordinary differential equations, it was possible to study stability questions and the search for periodic solutions in terms of the eigenvalues of the characteristic equation. There the number of eigenvalues with their multiplicities was equal to the dimension of the space E^n in which the solution was to be found, and it was possible to analyze the properties of solutions by examining a related Jordan canonical form or an equivalent spectral representation. On page 225 of "NM" we studied the problem of finding periodic solutions to a nonlinear ordinary differential equation by means of perturbation techniques. The periodic solutions of the linear part played a significant role there, and knowledge of the structure of the eigenspace of the linear part was helpful in determining the existence of periodic solutions to the perturbed equation.

We have already seen that linear delay equations give rise to characteristic equations with an infinite number of roots. What is the structure of the space in this case, and how, for example, does one characterize the existence of periodic solutions or study the stability question?

One interesting approach is to start, for example, with a linear delay-differential equation with a perturbation parameter ϵ appearing as a multiplier of the terms involving the lag. By putting $\epsilon = 0$ one has an ordinary linear differential equation, for which the stability question can be studied. Then returning to the perturbed equation, by examining its characteristic roots for small values of ϵ one hopes to obtain information on the stability of the delay equation. This is precisely the approach taken by Cooke, in a paper which we now discuss briefly.

Although later we discuss stability in greater detail, we shall need the idea of stability for a few important concepts given below. A solution $x(t)$ of a delay-differential equation

$$\dot{x}(t) = f(t, x(t), x(t - \tau_1(t)), \ldots, x(t - \tau_m(t)))$$

is said to be stable with respect to perturbation of the initial set E_{t_0} if for every $\epsilon > 0$ there exists a $\delta(\epsilon) > 0$ such that whenever $|\psi(t) - \varphi(t)| < \delta(\epsilon)$, on E_{t_0}, we have $|x(t) - x(t)| < \epsilon$, for all $t > 0$, and any continuous function given on E_{t_0}. A stable solution is asymptotically stable if, in addition,

$$\lim_{t \to \infty} (x(t) - x(t)) = 0$$

ψ and φ denote initial functions. If we write $y(t) = x(t) - x(t)$, then we can investigate the stability of the trivial solution instead of $x(t)$. It is essential to note that the definition of stability depends on the choice of t_0; that is, there are examples of a solution of an equation stable on an initial set E_{t_0} but unstable on $E_{\bar{t}_0}$, with $\bar{t}_0 > t_0$.

Note that the initial-value problem for an nth-order delay equation can be posed in two ways. The first way is to give on the initial set $x^{(k)}(t) = \varphi^{(k)}(t)$, $k = 0, 1, \ldots, n - 1$; that is, the derivatives of $x(t)$ are defined in terms of the derivatives of $\varphi(t)$. The second way is to use arbitrary continuous functions $\varphi_k(t)$, $k = 0, 1, \ldots, n - 1$, each of which in general is not the kth derivative of $\varphi_0(t)$.

Exercise 5-29 Replace the equation

$$x^{(n)}(t) = f(t, x(t - \tau_1(t)), \ldots, x(t - \tau_m(t)), \ldots,$$
$$x^{(n-1)}(t - \tau_1(t)), \ldots, x^{(n-1)}(t - \tau_m(t)))$$

with $\tau_i(t) \geq 0$, $i = 1, \ldots, m$, by a system; for example, $\dot{x}_0(t) = x_1(t)$, $\dot{x}_1(t) = x_2(t)$, etc.

If in the problem of the last exercise we put $x_k(t) = \varphi_k(t)$, $k = 0, 1, \ldots, n - 1$, without the requirement that $\varphi_n(t) = \dot{\varphi}_{n-1}(t)$, that is, we pose the initial-value problem in the second way, then the problem of stability of the nth-order equation is not equivalent to that of the system in the sense of the first way of posing the initial-value problem. But in general this is an exceptional situation, as one often must pose the problems in the first way.

In a paper dealing with the problem of perturbation of the zeros of exponential polynomials Cooke[13] studies the linear delay equation with constant lag $\tau > 0$ and small parameter ϵ

$$\sum_{k=1}^{n} \epsilon^k [c_k(\epsilon) x^{(k+m)}(t) + d_k(\epsilon) x^{(k+m)}(t - \tau)] + \sum_{j=0}^{m} [a_j(\epsilon) x^{(j)}(t)$$
$$+ b_j(\epsilon) x^{(j)}(t - \tau)] = f(t, \epsilon) \qquad t > t_0 \quad (5\text{-}26a)$$

where the coefficients are continuous functions of ϵ and tend to constant values c_k, d_k, a_j, and b_j, respectively, as $\epsilon \to 0^+$ and

$$x^{(j)}(t) = \varphi^{(j)}(t, \epsilon) \qquad j = 0, 1, \ldots, m + n, t_0 - \tau \leq t \leq t_0$$

the superscripts indicating the order of derivatives. This type of perturbation is called *singular*. Note that when one sets $\epsilon = 0$, one loses terms in higher-order derivatives of the linear part and hence may also lose sight of initial data and even of stability.

The degenerate equation is obtained by setting $\epsilon = 0$. The object is to find a necessary and sufficient condition that as $\epsilon \to 0^+$, the solution of the perturbed equation approaches the solution of the unperturbed equation on any finite interval and that the exponential order, as $t \to \infty$, of the former approaches the exponential order of the latter. For ordinary differential equations obtained by considering the above equation in homogeneous form and without the terms involving the differences, the criterion is that all zeros of the polynomial $a_m + \sum\limits_{k=1}^{n} c_k z^k$ have negative real parts.

Exercise 5-30   Show that the above delay equation has the characteristic function

$$h(s,\epsilon) = \sum_{k=1}^{n} \epsilon^k [c_k(\epsilon) + d_k(\epsilon)e^{-\tau s}]s^{k+m} + \sum_{j=0}^{m} [a_j(\epsilon) + b_j(\epsilon)e^{-\tau s}]s^j$$

and for the degenerate case

$$h(s) = h(s,0) = \sum_{j=0}^{m} [a_j + b_j e^{-\tau s}]s^j$$

We define

$$\theta_R(z) = a_m + \sum_{k=1}^{n} c_k z^k$$

$$\theta_L(z) = b_m + \sum_{k=1}^{n} d_k z^k$$

and let
$$\theta(z,\epsilon) = \theta_R(z) + e^{-\tau z/\epsilon}\theta_L(z) \qquad \epsilon > 0$$
$$M(\epsilon) = \sup \{\text{Re } (\lambda(\epsilon))|h(\lambda(\epsilon),\epsilon) = 0\}$$
$$M(0) \equiv M \qquad M^* = \limsup_{\epsilon \to 0^+} M(\epsilon) \qquad \sigma^* = \inf \sigma_0$$

where σ_0 is completely regular; i.e., it satisfies the condition that there exist positive numbers ϵ_1, γ_1, and γ_2 such that:

1. $|\theta(z,\epsilon)| = |\theta_R(z) + e^{-\tau z/\epsilon}\theta_L(z)| \geq \gamma_1$ for z and ϵ satisfying $0 < \epsilon \leq \epsilon_1$, $\epsilon\sigma_0 \leq \text{Re } (z) < \infty$.
2. $|c_n + d_n e^{-\tau s}| \geq \gamma_2$, $\sigma_0 \leq \text{Re } (s) < \infty$.
3. $|a_m + b_m e^{-\tau s}| \geq \gamma_2$, $\sigma_0 \leq \text{Re } (s) < \infty$.

We shall refer to the fact that all zeros of $\theta_R(z)$ have negative real parts as condition 4. If this condition holds and $c_n + d_n e^{-\tau s}$ is negative for Re $(s) > M$, then one can show that $M^* = \max (M,\sigma^*)$.

Theorem 5-5   Let σ_0 be completely regular for Eq. (5-26a), and take σ_0 so that $h(s)$ has no zeros on the line Re $(s) = \sigma_0$. Let $P(\sigma_0)$ denote the half-plane Re $(s) \geq \sigma_0$. Then the set of zeros of $h(s,\epsilon)$ in $P(\sigma_0)$ approaches the set of zeros of $h(s)$ in $P(\sigma_0)$ in the sense that, given any $\rho > 0$, there exists an $\epsilon_1(\rho)$ such that for $0 \leq \epsilon \leq \epsilon_1$ all zeros of $h(s,\epsilon)$ in $P(\sigma_0)$ lie in circles of radius ρ centered at the zeros of $h(s)$ in $P(\sigma_0)$, and in each such circle the total multiplicity of zeros of $h(s,\epsilon)$ is equal to the total multiplicity of zeros of $h(\epsilon)$. Moreover, if all zeros of $\theta_R(z)$ have negative real parts, then condition 1 [in which we write $\epsilon\sigma_0 \leq \text{Re } (z) \leq \epsilon\sigma_1$ for σ_1 real, $\sigma_1 \geq \sigma_0$, instead of $\epsilon\sigma_0 \leq \text{Re } (z) < \infty$] is satisfied for every σ_0 and every $\sigma_1 \geq \sigma_0$ if and only if $b_m = 0$ and $d_k = 0$ ($k = 1, \ldots, n$). In this case

$\sigma^* = -\infty$, and the zeros of $h(s,\epsilon)$ in $P(\sigma_0)$ approach the zeros of $h(s)$ in $P(\sigma_0)$ for every σ_0.

The following is an example in which condition 4 holds and condition 2 holds for Re $(s) > M$ and $M^* = \sigma^* > M$. Suppose that

$$h(s,\epsilon) = \tfrac{3}{2}\epsilon^2 s^3 + \frac{\sqrt{3}}{2}\,\epsilon\sigma^2 + (1 - \sqrt{3}\,\epsilon)se^{-s} + e^{-s} + 2s + 2$$

Then also

$$h(s) = se^{-s} + e^{-s} + 2s + 2 = (s + 1)(e^{-s} + 2)$$

with zeros $\lambda = -1$, $\lambda = -\log(-2)$. Thus $M = -\log 2$. Since

$$a_m + b_m e^{-\tau s} = 2 + e^{-s} \qquad c_n + d_n e^{-\tau s} = \tfrac{3}{2}$$

conditions 2 and 3 [in which $\sigma_0 \le$ Re $(s) \le \sigma_1$] hold for Re $(s) > M$. Also $\theta_R(z) = 2 + (\sqrt{3}/2)z + \tfrac{3}{2}z^2$, $\theta_L(z) = 1$, and the zeros of $\theta_R(z)$ have negative real parts and $M^* = \max(M,\sigma^*)$. Now condition 1 (with ∞ replaced by $\epsilon\sigma_1$) is difficult to use in practice and is replaced by $1'$:

$$|\theta_R(iy)| = |\theta_L(iy)|e^{-\tau\sigma}$$

has no nonzero real root for any σ in $\sigma_0 \le \sigma \le \sigma_1$. Conditions $1'$ and 3 together are equivalent to

$$|\theta_R(iy)| \ne |\theta_L(iy)|e^{-\tau s} \qquad -\infty < y < \infty, \ \sigma_0 \le \sigma \le \sigma_1$$

We use this condition to find σ^*. We have

$$\left| 2 + \frac{\sqrt{3}}{2}\,iy - \tfrac{3}{2}y^2 \right| = e^{-\sigma}$$

There are real solutions y if and only if $e^{-2\sigma} \ge \tfrac{15}{16}$, and hence condition $1'$ is satisfied on $\sigma_0 \le \sigma \le \sigma_1$ whenever $2\sigma_0 > \log \tfrac{16}{15}$. We have

$$\sigma^* = \tfrac{1}{2}\log \tfrac{16}{15} > M$$

Hence the equation corresponding to $h(s,\epsilon)$ has for arbitrary ϵ solutions which are unbounded as $t \to \infty$, whereas all solutions of the equation corresponding to $h(s)$ approach zero as $t \to \infty$. The reason for this is that $h(s,\epsilon)$ has zeros of the form $i[(2\nu + \tfrac{2}{3})\pi]$, $\nu = 1, 2, \ldots$.

HALE'S METHOD[37,38,58]

Now let $\dot{x}(t)$ be the right-hand derivative of a function x at t, let F map $C \to E^n$, and consider the equation

$$\dot{x}(t) = F(x_t)$$

to which we assume there exists a unique solution depending continuously on $\varphi \in C_H$. Suppose that $F(0) = 0$ and F has a bounded Fréchet differential at zero. Thus we can write

$$\dot{x}(t) = L(x_t) + N(x_t)$$

where L is a bounded linear function, and the nonlinear term N satisfies $N(0) = 0$ and $N(\varphi) = o(\|\varphi\|)$ as $\|\varphi\| \to 0$. $L(\varphi)$ is the Fréchet differential of F at 0 evaluated at φ. We have already pointed out that

$$L(\varphi) = \int_{-\tau}^{0} [d\eta(\theta)]\varphi(\theta)$$

Consider the linear equation

$$\dot{u}(t) = L(u_t)$$

for which we know that a unique solution $u(\varphi)$ is defined for $t \in (0, \infty)$ and exists for any $\varphi \in C$. We define the semigroup of bounded linear operators $U(t):C \to C$ by $U(t)\varphi = u_t(\varphi)$. Thus $U(t + \tau) = U(t)U(\tau)$, $t > 0$, $\tau > 0$. The characteristic roots of $\dot{u}(t) = L(u_t)$ are the roots of the characteristic equation

$$\det \Delta(\lambda) \equiv \det [\lambda I - \int_{-\tau}^{0} [d\eta(\theta)]e^{\lambda\theta}] = 0$$

This equation has a finite number of roots with finite multiplicity in any half-plane Re $(z) \geq \gamma$. Suppose that none of these has a zero real part. Thus we have a finite number with positive real part, and the remaining infinity will have negative real parts. For a root λ of multiplicity k, there are at most k linearly independent solutions of the form

$$u(t) = p(t)e^{\lambda t}$$

$p(t)$ a polynomial of degree $\leq k - 1$ with coefficients in E^n. If X is the matrix of these k solutions, then there is a matrix B with λ as its only eigenvalue such that

$$Y(t) = Y(0)e^{Bt}$$

In this manner we define a subspace $P(\Lambda)$ corresponding to the finite roots and a basis for it. This space has a complementary space $Q(\Lambda)$ defined by considering the adjoint equation

$$\dot{v}(s) = - \int_{-\tau}^{0} [d\eta^T(\theta)]v(s - \theta) \qquad s \leq 0$$

and its characteristic equation.

Every element $\varphi \in C$ is a unique sum of its two projections, one belonging to each of these spaces. Note that P and Q are subspaces of C. The space P is an asymptotically stable subspace; i.e., it corresponds to initial values of solutions which tend to zero as $t \to -\infty$. The subspace Q has the same property as $t \to \infty$.

Returning to the equation in $N(x_t)$, suppose that $N(\varphi)$ is locally lipschitzian in the sense

$$|N(\varphi) - N(\varphi')| \leq \nu(\delta)\|\varphi - \varphi'\|$$

with $\nu(\delta)$ a continuous nondecreasing function and $\nu(0) = 0$. Then we have

$$x(t) = U(t)\varphi(0) + \int_0^t X(t - \tau)N(x_\tau) \, d\tau \qquad t \geq 0$$
$$x(\theta) = \varphi(\theta) \qquad\qquad\qquad \theta \, \epsilon \, [-\tau, 0]$$

where $X(t)$ is the $n \times n$ matrix whose columns are the solutions of the linear part with $X(t) \equiv 0$ for $t \, \epsilon \, [-\tau, 0]$ and $X(0) = I$. This representation can be decomposed into two components by projecting $X_0 \equiv X(0)$ on P and Q. Progress is being made by Hale and Perelló in studying periodic solutions in terms of this decomposition. The conditions on N, that is, $N(0) = 0$, and the local Lipshitz condition are used in the problem of the saddle point previously defined. The above decomposition of C can be made for any linear system whether the roots are on the imaginary axis or not. Generally when we seek periodic solutions for equations with a small parameter we do not have $N(0) = 0$ satisfied.

To illustrate Hale's basic theory with Perelló's recent addition to it we shall consider the problem of finding periodic solutions to the equation

$$\dot{u}(t) = -\alpha u(t - 1)(1 - \epsilon u^2(t)) \qquad \text{for } \alpha = \frac{\pi}{2} + \epsilon, \epsilon > 0$$

We first consider the linear part obtained by putting $\epsilon = 0$. We have

$$\dot{u}(t) = -\frac{\pi}{2}u(t - 1)$$

This has the characteristic equation

$$\lambda = -\frac{\pi}{2}e^{-\lambda}$$

which has the two imaginary roots $\pm i\pi/2$ with the corresponding families of periodic solutions $\sin(\pi/2)t$ and $\cos(\pi/2)t$. The remaining infinity of roots have negative real parts. Define

$$\varphi_1(\theta) = \sin\frac{\pi}{2}\theta$$
$$\qquad\qquad\qquad\qquad -1 \leq \theta \leq 0$$
$$\varphi_2(\theta) = \cos\frac{\pi}{2}\theta$$

Thus φ_1 and φ_2 generate a two-dimensional subspace $P(\Lambda)$

$$P = \{\varphi \epsilon C : \varphi = a\varphi_1 + b\varphi_2; a, b \text{ real}\}$$

To see how this space is transformed through the linear part of the differential equation we note that $\Phi = (\varphi_1, \varphi_2)$ is a basis of periodic solutions to the linear part. The derivative must be a linear combination of φ_1 and φ_2, or, in vector notation,

$$\dot{\Phi} = (\varphi_1, \varphi_2)B$$

Since φ_1 and φ_2 are known, so are $\dot{\varphi}_1$ and $\dot{\varphi}_2$, and hence B can be shown to be

$$B = \begin{pmatrix} 0 & -\dfrac{\pi}{2} \\ \dfrac{\pi}{2} & 0 \end{pmatrix}$$

Now the adjoint equation is

$$\dot{v}(t) = \frac{\pi}{2} v(t + 1)$$

which has the basic periodic solutions

$$\psi_1(\theta) = \sin \frac{\pi}{2} \theta$$
$$\psi_2(\theta) = \cos \frac{\pi}{2} \theta \qquad 0 \le \theta \le 1$$

The solutions of the two systems $\varphi \epsilon C$ and $\psi \epsilon C^*$ have the following bilinear form associated with them

$$(\psi, \varphi) = \psi(0)\varphi(0) - \frac{\pi}{2} \int_{-1}^{0} \psi(\xi + 1)\varphi(\xi) \, d\xi$$

Here also we define $\psi = (\psi_1, \psi_2)$. We compute

$$(\psi, \Phi) = (\omega_1, \psi_2) \begin{pmatrix} \varphi_1 \\ \varphi_2 \end{pmatrix} = \frac{1}{2} \begin{pmatrix} 1 & -\dfrac{\pi}{2} \\ \dfrac{\pi}{2} & 1 \end{pmatrix}$$

For convenience we change the basis in ψ so that the last product is the identity matrix; i.e., we use instead of ψ

$$\psi^* \equiv \psi(\psi, \Phi)^{-1} = \psi \frac{2}{1 + \pi^2/4} \begin{pmatrix} 1 & -\dfrac{\pi}{2} \\ \dfrac{\pi}{2} & 1 \end{pmatrix}$$

Thus $(\psi^*, \Phi) = I$. Now we have a coordinate system, so that any $\varphi \epsilon C$ can be written in the form

$$\varphi = \Phi c + \bar{\varphi}$$

Where $c = (\psi^*, \varphi)$ is the projection of φ on ψ^* and $\bar{\varphi} \epsilon C([-1,0] \to E^n)$. This representation is unique, since $(\psi^*, \bar{\varphi}) = 0$ permits one to study the equation by means of two parts, a finite- and an infinite-dimensional part.

We have

$$Q = \{\varphi \epsilon C : (\psi^*, \varphi) = 0\}$$

To obtain the solution of the perturbed equation we write

$$x_t = \Phi y(t) + \bar{x}_t$$
$$\varphi = \Phi c + \bar{\varphi}$$
with
$$y(t) = (\psi^*, x_t)$$

and our equation becomes

$$\dot{y}(t) = By(t) + \epsilon \psi^{*T}(0)f(\Phi y(t) + \bar{x}_t)$$

where f is the part of the original equation comprising the coefficient of ϵ after putting $\alpha = \pi/2 + \epsilon$ and simplifying. Also $\bar{x}_t$ is given by an integral equation since, as we have seen, x_t itself is. In general $\bar{x}_t$ is difficult to compute. However Perelló has shown that by putting $\bar{x}_t = 0$ we have an ordinary perturbed differential equation, to which ordinary perturbation procedures can be applied. Then one can examine the first approximation involving the first power of ϵ. If it is possible to determine periodic solutions in this manner, the problem of putting $\bar{x}_t = 0$ is legitimate, and one has the solution to the perturbed equation. Otherwise one must examine higher-order terms, i.e., terms in ϵ^2, ϵ^3, etc., a procedure that is under study.

Perelló has studied the construction of periodic solutions of the perturbed equation of the form

$$\dddot{z}(t) + a\ddot{z}(t) + b^2\dot{z}(t) + kz(t - \tau) + \epsilon\psi(z(t - \tau)) = 0$$

in which a, b^2, k, ϵ, and τ are positive constants and ψ is real in z and such that a unique solution exists for any $\varphi \epsilon C$ and $t \epsilon [0, \infty]$. By putting $\epsilon = 0$, periodic solutions of the resulting equation are obtained by imposing such conditions on ψ, a, b^2, k, and τ that its characteristic equation has two simple purely imaginary roots and all others have negative real parts.

Exercise 5-31 Put $\epsilon = 0$ and write the resulting equation in the vector form

$$\dot{u}(t) = Au(t) + B(t - \tau)$$

For the integral representing

$$\eta(\theta) = \begin{pmatrix} 0 & u(\theta) & 0 \\ 0 & 0 & u(\theta) \\ -ku(\theta + \tau) & -bu(\theta) & -au(\theta) \end{pmatrix}$$

where

$$u(t) = \begin{cases} 0 & t < 0 \\ 1 & t \geq 0 \end{cases}$$

show that

$$\Delta(\lambda) = \begin{pmatrix} \lambda & -1 & 0 \\ 0 & \lambda & -1 \\ ke^{-\lambda\tau} & b^2 & a + \lambda \end{pmatrix}$$

and show that for an eigenvector c of λ

$$\begin{pmatrix} \lambda c_1 - c_2 \\ \lambda c_2 - c_3 \\ ke^{-\lambda\tau}c_1 + b^2c_2 + ac_3 + \lambda c_3 \end{pmatrix} = 0$$

The remainder of the story is long and is better found in detail in the thesis by Perelló. He has constructed examples to show that more quantitative information on the periodic solution can be obtained by going to the higher-order (nonlinear) terms.

5-7 SOLUTION AND STABILITY OF A NONLINEAR DELAY EQUATION[41,70]

Consider the delay-differential equation

$$\frac{z'(u)}{z(u)} = \alpha - \beta z(u - \tau) \qquad \alpha > 0$$

The left side describes the net birthrate of a population $z(u)$ at time u, and the right side says that it is a constant minus a quantity proportional to the population at the previous time $u - \tau$.

If we put $u = \tau t$, multiply by $\tau^2\beta$, and define $x(t) \equiv \beta\tau z(\tau t)$, our equation becomes

$$x'(t) = [a - x(t - 1)]x(t) \tag{5-27}$$

with $a = \alpha\tau$.

The substitution

gives

$$ay(t) = x(t) - a$$
$$y'(t) = -ay(t - 1)[1 + y(t)] \qquad a > 0 \tag{5-28}$$

which is to be regarded as an equation in the unknown function $y(t)$ for $t > 0$. The values of $y(t)$ are assumed to be given by some $\psi(t)$ in $-1 < t \leq 0$, where it must also be bounded and Lebesgue-integrable. If $t > 0$, y is defined by

$$y(t) = y(0) + \int_0^t y'(s)\, ds \qquad (5\text{-}29)$$

Note that y' need not be the derivative of y at all points $0 \leq t \leq 1$ but must be where $t > 1$, since by (5-28) and (5-29) y' is absolutely continuous.

A function $y(x)$ which is such that $y(t) = \psi(t)$ for $-1 < t \leq 0$ and which satisfies the equation for x in some interval $(0,x)$, $x > 0$, is called a solution of the equation on $(0,x)$ corresponding to the initial function $\psi(t)$. The existence of a unique solution to our equation follows from the existence theorem given earlier, in which a more general system is considered.

Note that if $y(s) \neq -1$, $0 \leq s \leq t$, we can replace t in (5-28) by s, divide by $1 + y(s)$, integrate with respect to s from 0 to t, and take exponentials. We have

$$1 + y(t) = [1 + \psi(0)] \exp\left[-a \int_{-1}^{t-1} y(s)\, ds \right]$$

A step-by-step process leads to the solution for values of t in successive intervals. Thus one requires $y(t)$ to be initially given on an appropriate interval and then attempts to extend it continuously so that it satisfies the equation for all t to the right of the given interval.

Exercise 5-32 Write the solution for the interval $n \leq t \leq n + 1$.

Returning to (5-27), suppose that we are given the initial function

$$x(t) = \varphi(t) \qquad 0 \leq t \leq 1$$

The solution is then given by

$$x(t) = \varphi(1) \exp\left[a(t - 1) - \int_0^{t-1} \varphi(s)\, ds \right]$$

for $1 \leq t \leq 2$ and by

$$x(t) = x(n) \exp\left[a(t - n) - \int_{n-1}^{t-1} x(s)\, ds \right] \qquad (5\text{-}30)$$

for $n \leq t \leq n + 1$.

Remark: Two possible series solutions, due to Wright, are

$$x_1(t) = - \sum_{n=1}^{\infty} ac_n e^{nat}$$

$$x_2(t) = - \sum_{n=1}^{\infty} (-1)^n ac_n e^{nat}$$

where $c_1 = 1$ and $c_n = \dfrac{1}{n-1} \sum_{m=1}^{n-1} c_m c_{n-m} e^{-ma}$.   Both series converge for all t.

We now prove two theorems[41] illustrating the analysis of stability of Eq. (5-27).

Theorem 5-6   Let $a > 0$, $\varphi(1) > 0$, and let $x(t)$ be the solution of (5-27) corresponding to the initial function $\varphi(t)$.   Then one of the following holds:

1. $x(t)$ is asymptotic to $x = a$; that is, $x(t)$ and $x'(t)$ are monotone for large t and $\lim\limits_{t \to \infty} x(t) = a$, $\lim\limits_{t \to \infty} x'(t) = 0$.

2. $x(t)$ oscillates about $x = a$; that is, $x(t) = a$ assumes both positive and negative values for arbitrarily large t.

Proof: *Case 1:*   If $x(t) \geq a$ for all large t, then $a - x(t-1) \leq 0$, and $x'(t) = [a - x(t-1)]x(t) \leq 0$.   Furthermore,

$$x''(t) = [-x'(t-1)]x(t) + [a - x(t-1)]x'(t) \geq 0$$

Therefore, $x(t)$ increases monotonically to the limit a, and $x'(t)$ decreases monotonically to zero.

Case 2:   If $x(t) \leq a$ for all large t, then $a - x(t-1) \geq 0$, and because $x(t) > 0$ for $1 \leq t < \infty$, $x'(t) \geq 0$.   Furthermore,

$$x''(t) = x(t)[(a - x(t-1))^2 - x'(t-1)]$$

and because $x(t)$ increases monotonically to a,

$$x'(t) = [a - x(t-1)]x(t) \geq [a - x(t)]x(t) \geq [a - x(t)]^2 \geq 0$$

for large t.   Therefore, $x'(t)$ increases monotonically to 0.

Case 3:   If neither case applies, then $x(t)$ oscillates about $x = a$ for very large t.

Theorem 5-7   If $a \leq 0$ and $\varphi(1) < 0$, then $\lim\limits_{t \to \infty} x(t) = -\infty$.

Proof: If $\varphi(1) < 0$, then $x(t) < 0$.   Furthermore,

$$x'(t) = [a - x(t-1)]x(t) < 0$$

for all large t.   Therefore $x(t)$ decreases monotonically to $-\infty$.

We now catalogue the asymptotic behavior of $x(t)$ depending on different values of the parameter a and of $\varphi(1)$.

	$\varphi(1) < 0$		$\varphi(1) > 0$
$a < 0$	$\displaystyle\lim_{t\to\infty} x(t) =$ $\begin{cases} -\infty \\ 0 \\ a \end{cases}$	if $x(t) \leq a$ for a unit interval if $x(t) \geq a$ for a unit interval in all other cases (oscillates)	$\displaystyle\lim_{t\to\infty} x(t) = 0$
$a = 0$	$\displaystyle\lim_{t\to\infty} x(t) = -\infty$		$\displaystyle\lim_{t\to\infty} x(t) = 0$
$0 < a \leq 1/e$	$\displaystyle\lim_{t\to\infty} x(t) = -\infty$		$\displaystyle\lim_{t\to\infty} x(t) = a$ oscillates
$1/e < a < \pi/2$	$\displaystyle\lim_{t\to\infty} x(t) = -\infty$		$\displaystyle\lim_{t\to\infty} x(t) = a$ oscillates
$a > \pi/2$†	$\displaystyle\lim_{t\to\infty} x(t) = -\infty$		No characteristic pattern

† $\pi/2$ is *believed* to be the crucial value, according to Wright,[70-72] who tested several successive values approximating $\pi/2$.

5-8 A BRIEF DISCUSSION OF STABILITY, WITH EXAMPLES

Unlike ordinary differential equations, in which one can analyze the effect of disturbances at the initial point of a trajectory, for example, here one may be concerned with disturbing an initial function. Thus the concept of a Lyapunov function, which now depends not on an initial point but on an initial function, must be generalized to that of a Lyapunov functional.

We first introduce basic stability concepts and then proceed to give samples of basic theorems involving Lyapunov functionals.

Consider the system[16]

$$\dot{x}(t) = f(t, x(.)) \qquad \text{for } t > t_0 \qquad (5\text{-}31)$$

Let $f(t,0) \equiv 0$ for $t \geq t_0$ so that $x(t) \equiv 0$ will be a solution.

Definition: The trivial solution $x(t) \equiv 0$ of (5-31) will be called stable to the right of t_0 if for every $\bar{t}_0 \geq t_0$ and every $\epsilon > 0$ there exists a number $\delta = \delta(\epsilon, \bar{t}_0) > 0$ such that, for every continuous function φ from the interval $[\alpha, \bar{t}_0]$ to $\{x \in E^n : \|x\| < \delta\}$, an open ball of radius $\delta > 0$, that is,

$$\varphi \in C([\alpha, \bar{t}_0] \to B_\delta)$$

the solution $x(t,\bar{t}_0,\varphi)$ exists for all $t \geq \alpha$ and $\|x(t,\bar{t}_0,\varphi)\| < \epsilon$ for all $t \geq \alpha$. If δ is independent of $\bar{t}_0$, the trivial solution is called uniformly stable to the right of t_0.

We have asymptotic stability if in addition there is a $\delta_1 \equiv \delta_1(\bar{t}_0) > 0$ such that for every $\varphi \in C([\alpha,t_0] \rightarrow B_{\delta_1})$

$$\|\lim_{t \rightarrow \infty} x(t,t_0,\varphi)\| = 0$$

The trivial solution is called *uniformly asymptotically stable* to the right of t_0 if it is uniformly stable, δ_1 is independent of $\bar{t}_0$, and for every $\eta > 0$ there exists a number $T(\eta)$ such that, for every $\bar{t}_0 \geq t_0$, $\varphi \in C([\alpha,\bar{t}_0] \rightarrow B_{\delta_1})$ implies $\|x(t,t_0,\varphi)\| < \eta$ for all $t \geq \bar{t}_0 + T(\eta)$.

In "NM," sec. 4-11, we gave an introductory account of Lyapunov's direct, or second, method for analyzing the stability of dynamical systems of ordinary differential equations by means of an appropriately defined function $V(t,x)$, known as a *Lyapunov function*. The trivial solution $x = 0$ was shown to be stable if for solutions $x(t,t_0,x_0)$, with initial value $x(t_0) \equiv x_0$, $V[t,x(t,t_0,x_0)]$ is small whenever $V(t_0,x_0)$ is small. If

$$V[t,x(t,t_0,x_0)] \rightarrow 0$$

as $t \rightarrow \infty$, then the trivial solution is asymptotically stable. But the concept of a Lyapunov function is generally inadequate for studying the stability of delay-differential equations. Note that since the trivial solution to $\dot{x}(t) = -a(t)x(t)$ is stable, $a(t) \geq 0$ and continuous, the choice of $V(t,x) = x^2$ also works for the stability of

$$\dot{x}(t) = -x(t)x^2(g(t))$$

for $t > t_0$, where $\alpha \leq g(t) \leq t$, since $\dot{V} = -2x^2(t)x^2(g(t)) < 0$. Such a choice of a Lyapunov function fails for simple linear equations with constant delay and constant coefficients such as $\dot{x}(t) = -ax(t) - bx(t - \tau)$ for any $a, b \neq 0$ and $\tau > 0$. Since, for example, the initial condition for such an equation consists of more than a single point, we need the more general concept of a *Lyapunov functional*, introduced by Krasovskii.[43]

Note that in the definition of a Lyapunov function $V(t,x)$ the vector x is simply a point in euclidean n space. However, in the study of systems with deviating arguments we must indicate the dependence of x on the deviation, because different components of the equation may have different deviations in their arguments. This gives rise to the idea of a functional. In any case, we must emphasize that in a Lyapunov function, the x which appears in $V(t,x)$, need not be the solution vector but is any vector which satisfies appropriate conditions. To maintain this generality we use the notation $V(t,\psi(.))$ for a functional, where

$$\psi(t) = \{\psi_1(t), \ldots, \psi_n(t)\}$$

is a vector function defined for all t on a given interval $[a,b]$, which is often the interval of time for which a solution of the differential system or equation is desired. In particular, the introduction of this functional simplifies the task of developing a Lyapunov theory using prescribed initial conditions and coping with uniform asymptotic stability. One can give converse theorems (see "NM," theorem 4-8) for equations with deviating arguments in which, knowing asymptotic stability, it is possible to construct a functional in which $\psi(t)$ is replaced by the initial vector defined by the initial function $\varphi(t)$ and its derivatives. In this manner one can go on to conclude stability of differential equations under continuously acting disturbances. Such converse theorems are very difficult to establish with ordinary Lyapunov functions.

Often one assumes that the n components of $\psi(t)$ in E^n are continuous and writes $\psi(t) \in C([a,b] \to E^n)$.

If, in addition, we impose a boundedness condition on the norm of $\psi(t)$ in C, for example, $\|\psi\| < H$, then the mapping is considered into a subset B_H of E^n which satisfies this bound.

Let us illustrate with an example due to Krasovskii. The trivial solution $x \equiv 0$ of

$$\dot{x}(t) = -ax(t) - bx(t - \tau)$$

with $\alpha = t_0 - \tau$, and constant coefficients a, b, where b is real, with $|b| \leq a$, is stable for $t \geq t_0$. We define

$$V(t,\psi(.)) = \psi^2(t) + a \int_{t-\tau}^t \psi^2(s)\, ds \qquad t \geq t_0$$

where ψ is an arbitrary function, $\psi \in C([\alpha,t] \to E^1)$. We note that

$$\psi^2(t) \leq V(t,\psi(.)) \leq (1 + a\tau)\{\|\psi\|^{[\alpha,t]}\}^2$$

using the notation of Sec. 5-5.

Let φ be a given continuous initial function on $[\alpha,\bar{t}_0]$ for some $\bar{t}_0 \geq t_0$, and let $x(t,t_0,\varphi)$ be the corresponding solution. Then $V(t,\psi(.))$ is differentiable in t for $t > \bar{t}_0$, and

$$V(t,x(.)) = -ax^2(t) - 2bx(t)x(t - \tau) - ax^2(t - \tau)$$
$$\leq -(a - |b|)[x^2(t) + x^2(t - \tau)] \leq 0$$

It follows that for any $\epsilon > 0$, $|x(t,t_0,\varphi)| < \epsilon$ for all $t \geq \alpha$ if

$$\|\varphi\|^{[\alpha,t_0]} < \delta = \frac{\epsilon}{\sqrt{1 + a\tau}}$$

Let $B_H = \{x \in E^n : \|x\| < H\}$ be an open ball of radius $H > 0$ (possibly $H = \infty$), and let $f(t,\psi(.))$ be continuous in t. If $V(t,\psi(.))$ is a scalar-valued functional defined for $t \geq t_0$ and $\psi \in C([\alpha,t] \to B_H)$, then the derivative of V with respect to (5-31) is defined by[16]

$$\dot{V}(t,\psi(.)) = \lim_{\Delta t \to 0^+} \sup \frac{V(t + \Delta t, \psi^*(.)) - V(t,\psi(.))}{\Delta t}$$

where $\quad \psi^*(s) = \begin{cases} \psi(s) & \text{for } \alpha \leq s \leq t \\ \psi(t) + f(t,\psi(.))(s - t) & \text{for } t \leq s < t + \Delta t \end{cases}$

If the Lyapunov functional $V(t,\psi(.)) = V(t,\psi(t))$, a function, and if $V(t,x)$ has a continuous first derivative, then

$$\dot{V}(t,\psi(.)) = \frac{\partial V(t,x)}{\partial t}\bigg|_{x=\psi(t)} + \sum_{i=1}^n \frac{\partial V(t,x)}{\partial x_i}\bigg|_{x=\psi(t)} f_i(t,\psi(.))$$

We give here the following theorem on stability.

 Theorem 5-8 (*Stability conditions using a Lyapunov functional*) If there exists a functional $V(t,\psi(.))$, defined whenever $t \geq t_0$ and $\psi \in C([\alpha,t] \to B_H)$, such that:

1. $V(t,0) \equiv 0$,
2. $V(t,\psi(.))$ is continuous in t and locally lipschitzian with respect to ψ,
3. $V(t,\psi(.)) \geq w(\psi(t))$, where $w(y)$ is a positive continuous function for all $y \in B_H - \{0\}$ (positive definiteness), and
4. $\dot{V}(t,\psi(.)) \leq 0$,

then the solution $x(t) \equiv 0$ of (5-31) is stable to the right of t_0.

 Exercise 5-33 Prove this theorem by generalizing the procedure of the above example.

 Remark: If in condition 3 the inequality becomes $V(t,\psi(.)) \leq - w(\psi(t))$ with $w(y)$ still positive, then V is negative-definite.

 Krasovskii[43] gives the following theorem for the asymptotic stability of a simple case of (5-31). We first need the following definition.

 The functional $V[t,x(.)]$ is said to admit an infinitely small upper bound if there exists a continuous function $F(.)$ such that

$$V[t,x(.)] < F(\|x\|^{[t-\tau,t)}) \qquad \text{for } t > \tau, F(0) = 0$$

From here on we assume that $f(t,\psi(.)) = f(t,\psi(s))$, $t - \tau \leq s \leq t$.

 Theorem 5-9 If there corresponds to the system $\dot{x} = f(t,x(.))$ with $t \geq 0$ a functional $V(t,x(.))$ which is positive-definite, where $\|x(.)\|^{[t-\tau,t)} < H$, and which has an infinitely small upper bound, and if

$$\lim_{\Delta t \to t_0} \sup_t \frac{\Delta V}{\Delta t}$$

is negative-definite, then the null solution $x(t) = 0$ is asymptotically stable.

Example: This theorem is illustrated by taking the single equation

$$\dot{x}(t) = -ax(t) + b(t)x(t - \tau) \qquad a > 0$$

Then
$$V(x(.)) = \frac{1}{2a} x^2 + \alpha \int_{-\tau}^{0} x^2(s)\, ds$$

and dV/dt has the value

$$\lim_{\Delta t \to +0} \frac{\Delta V}{\Delta t} = -x^2(t) + \frac{b(t)}{a} x(t)x(t - \tau) + \alpha x^2(t) - \alpha x^2(t - \tau)$$
$$= -[\sqrt{1 - \alpha}\, x(t) - \sqrt{\alpha}\, x(t - \tau)]^2$$

by completing the square in $x(t - \tau)x(t)$, the right side of which is negative-definite if

$$q(4(1 - \alpha)\alpha) \geq \frac{b^2(t)}{a^2} \qquad \text{for some } q \,\epsilon\, (0,1)$$

which would imply the asymptotic stability of the origin.

We now give the converse theorem.

Theorem 5-10 If the null solution $x(t) \equiv 0$ of the system (5-31) is asymptotically stable uniformly to the right of t_0, then there is a number $H_0 > 0$ and a functional $V(t,\psi(.))$ defined on the continuous curves $\|\psi(.)\|^{(t-\tau,t)} < H_0$ for $t \geq 0$, positive-definite, having an infinitely small upper bound in this region and having, moreover, a right derivative $dV/dt = \overline{\lim}\, (\Delta V/\Delta t)$ for $\Delta t \to +0$ on the trajectory of the system. The derivative dV/dt is a negative-definite functional in some region $\|\psi(.)\|^{(t-\tau,t)} < H_0$, and the functional $V(t,\psi(.))$ satisfies a Lipschitz condition with respect to $\psi(.)$, that is,

$$|V(t,\psi''(.)) - V(t,\psi'(.))| \leq K\|\psi''(.) - \psi'(.)\|^{(t-\tau,t)}$$

The usefulness of this theorem is best demonstrated by the following powerful theorem, which is based on it. First consider the system[43]

$$\dot{x}_i = f_i(x_1(t), \ldots, x_n(t); x_1(t - \tau_{i1}(t)), \ldots, x_n(t - \tau_{in}(t)), t) \quad (5\text{-}32)$$

$0 \leq \tau_{ij}(t) \leq \tau;\ i,\ j = 1, \ldots, n$. Here f_i are defined for $|x_j| < H$, $|x_j(t - \tau_{ij}(t))| < H$ and satisfy a Lipschitz condition in these variables and $f_i(0, \ldots, 0; 0, \ldots, 0, t) = 0\ (i = 1, \ldots, n)$.

In addition to the above system, we consider the "perturbed" system of equations

$$\dot{x}_i = f_i(t, x_1(t), \ldots, x_n(t); x_1(t - \tau_{i1}^*(t)), \ldots, x_n(t - \tau_{in}^*(t)))$$
$$+ R_i(t, x_1(t), \ldots, x_n(t); x_1(t - g_{ij}(t)), \ldots, x_n(t - g_{ij}(t)))$$

with the conditions

$$i, j = 1, \ldots, n \qquad 0 \leq \tau_{ij}^*(t) \leq \tau \qquad 0 \leq g_{ij}(t) \leq \tau$$

where the continuous functions R_i are not required to reduce to zero when their arguments are zero and t is arbitrary.

Definition: The null solution $x = 0$ of the system (5-32) is called stable for persistent disturbances if for every $\epsilon > 0$ there exist positive numbers δ_0, η, Δ such that the solution $x(t,t_0,\varphi)$ of the second system satisfies the inequality

$$\|x(t,t_0,\varphi)\| < \epsilon$$

for all $t \geq t_0$, $t_0 \geq 0$, whenever the continuous initial curve φ satisfies the inequality and the perturbed time delay τ_{ij}^* and the functions R_i satisfy the inequalities

$$|R_i(t,x_1, \ldots ,x_n;y_1, \ldots ,y_n)| < \eta \qquad i = 1, \ldots , n$$

for $\|x\| < \epsilon$, $\|y\| < \epsilon$, and

$$|\tau_{ij}(t) - \tau_{ij}^*(t)| < \Delta \qquad i, j = 1, \ldots , n$$

Theorem 5-11 If the null solution $x = 0$ of the system (5-32) is uniformly asymptotically stable with respect to t_0 and the initial curve φ, then $x = 0$ is stable for persistent disturbances.

Example: Assume that the roots of the characteristic equation of the linear part of the system with constant coefficients and constant lags[43] given by

$$x_i(t) = \sum_{j=1}^{n} a_{ij}x_j(t) + f_i(t,x_1(t), \ldots ,x_n(t);x_1(t - \tau_{i1}), \ldots ,x_n(t - \tau_{in}))$$

$$i = 1, \ldots , n$$

with

$$|f_i| < \beta[\|x(t)\| + \|x(t - \tau_{ij})\|]$$

have negative real parts. Then a functional for this system is

$$V(x(.)) = \sum_{i,j=1}^{n} \left(b_{ij}x_ix_j + \epsilon_{ij} \int_{-\tau_{ij}}^{0} x_j^2(s) \, ds\right)$$

where we have from the condition on the linear part that

$$\sum_{j=1}^{n} (b_{ij}a_{jk} + b_{kj}a_{ji}) = -\delta_{ik}$$

The remainder of the analysis is left to the reader with the aid of the indicated references.

CONTOUR - MAPPING TECHNIQUES IN THE ANALYSIS OF STABILITY

Consider the complex s plane and the graph of $F(s)$ as s follows the contour shown in Fig. 5-1. It is not difficult to show that the graph of $F(s)$

in the complex $F(s)$ plane will encircle the origin of the $F(s)$ plane p times in a clockwise fashion as s follows the contour, where p is the difference between the number of zeros of $F(s)$ within the contour and the number of poles of $F(s)$ within the same contour. The particular contour shown will thus disclose the presence of any poles s_k with positive real part, which are the roots leading to unstable response. Nyquist has made use of such

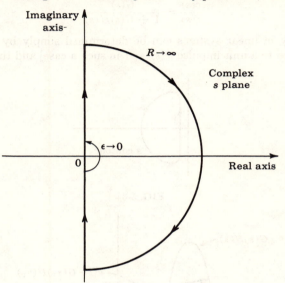

FIG. 5-1 s contour enclosing the right half of the s plane.

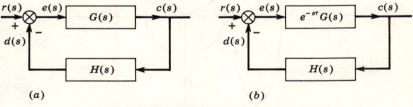

FIG. 5-2   Simple linear regulating systems.

contours, which have served for many years as an important technique in the analysis of stability of linear servomechanisms and regulating systems. As an example of its use consider the block diagram of the simple linear servomechanism shown in Fig. 5-2a. Here $c(s)$ is the output command of the system, $r(s)$ is its reference command, or input, $d(s)$ is the signal fed back to the comparator unit, $e(s)$ is the error, and $G(s)$ and $H(s)$ are the Laplace transforms of the responses of the forward and feedback elements to unit impulses. From the block diagram one can infer the following equations:

$$e(s) = r(s) - d(s)$$
$$c(s) = G(s)e(s)$$
$$d(s) = H(s)r(s)$$

from which we have

$$\frac{c(s)}{r(s)} = \frac{G(s)}{1 + G(s)H(s)}$$

Stability in linear systems can be determined simply by evaluating the response to a unit impulse [$r(s) = 1$ in such a case] and thus is deter-

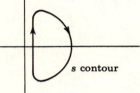

s contour

FIG. 5-3

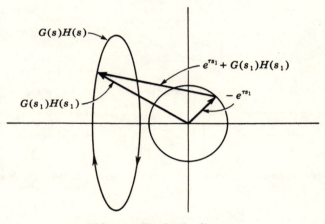

FIG. 5-4 The Satche diagram.

mined by the analysis of $G(s)(1 + G(s)H(s))^{-1} = c(s)$. Analysis shows that the poles of $c(s)$ are the zeros of $1 + G(s)H(s)$. To discover whether any of these have positive real parts, thus indicating instability, contour mapping can be used, with the contour of Fig. 5-3, graphing $1 + G(s)H(s)$ and observing the number of encirclements of the origin by the graph of $1 + G(s)H(s)$.

But it is often convenient to graph $G(s)H(s)$ as s follows the contour and observe the number of encirclements of the point $(-1 + i0)$ in the $G(s)H(s)$plane. In either case, the number of encirclements of the appropriate point equals the number of zeros of $1 + G(s)H(s)$ in the right half-plane (which is to be found) minus the number of poles of $1 + G(s)H(s)$ in the right half-plane [which is the number of poles of $G(s)$ in the right half-plane plus the number of poles of $H(s)$ in the right half-plane, which are generally known].

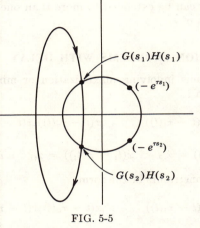

FIG. 5-5

Satche[62] extended the use of contour integration to systems with simple time delays. In such systems there is an element whose input is $f(t)$ and whose output is $f(t - \tau)$, where τ is a constant, and for this element the Laplace transform is $e^{-s\tau}$. Consider such a system in the simple form shown in Fig. 5-4 where, as before, $G(s)$ and $H(s)$ are simply fractions with numerator and denominator as polynomials in s.

Now the system transform is given by

$$\frac{c(s)}{r(s)} = \frac{e^{-s\tau}G(s)}{1 + e^{-s\tau}G(s)H(s)}$$

or

$$\frac{c(s)}{r(s)} = \frac{G(s)}{e^{s\tau} + G(s)H(s)}$$

Contour mapping to determine the number of encirclements of the origin by the graph of $e^{s\tau} + G(s)H(s)$ and foreknowledge of the poles of $e^{s\tau} + G(s)H(s)$ in the right half-plane [which equals the number of poles of $G(s)H(s)$ in the right half-plane] lead to the number of zeros of $e^{\tau s} + G(s)H(s)$ in the right half-plane and consequently to a determination of stability.

Satche suggests that this conclusion is reached more easily by graphing $-e^{\tau s}$ and $G(s)H(s)$ in the same plane, for $-e^{\tau s}$ is simply the unit circle as s follows the contour. Inspection of Fig. 5-5 shows that if the plot of $G(s)H(s)$ lies outside the unit circle, then there are no encirclements of the origin by $e^{\tau s} + G(s)H(s)$. Further thought shows that if $G(s)H(s)$ lies inside the unit circle in the interval $[s_1,s_2]$ (as shown in Fig. 5-5) while $-e^{\tau s}$ lies outside the $G(s)H(s)$ contour in the same interval, then the number of encirclements of the origin by $e^{\tau s} + G(s)H(s)$ will still be zero.

This approach can be extended to more than one delay and possibly to different values of τ.

5-9 OPTIMIZATION PROBLEMS WITH DELAY

Variational problems involving maximization or minimization of functionals of the form

$$F[x(t)] = \int_{t_0}^{t_1} f(t,x(t - \tau_1(t)), \ldots ,x(t - \tau_n(t));\dot{x}(t - \tau_1(t)),$$
$$\ldots ,\dot{x}(t - \tau_n(t))\, dt$$

with $x(t_0 - \tau_i(t_0)) = x_{i0} \qquad x(t_1 - \tau_i(t_1)) = x_{i1} \qquad i = 1, \ldots , n$

with or without constraints of the form

$$G[x(t)] = \int_{t_0}^{t_1} g(t,x(t - \tau_1(t)), \ldots ,x(t - \tau_n(t);\dot{x}(t - \tau_1(t)),$$
$$\ldots ,\dot{x}(t - \tau_n(t)))\, dt = L$$

where g and the constant L are given, have been investigated by El'-sgol'ts.[20,23] The treatment of the constraint problem leads to studying an expression of the form $F + \lambda G$, as in classical problems.

A generalization of the du Bois-Reymond lemma is used to conclude (after specifying conditions analogous to the usual ones) from

$$\int_{t_0}^{t_1} \sum_{i=1}^{n} y_i(t)\eta(t - \tau_i(t))\, dt = 0$$

with y_i continuous on $[t_0,t_1]$ and zero outside, $\tau_i(t) \geq 0$, and η arbitrary, etc., that

$$\sum_{i=1}^{n} \frac{y_i(\gamma_i(t))}{1 - \dot{\tau}'_i(\gamma_i(t))} = 0$$

where $\gamma_i(t)$ is the inverse function of $t - \tau_i(t)$. If η is replaced by its derivative in the integrand, the zero on the right side of the last condition is replaced by a constant C. The results can then be applied, for example, to the unconstrained problem to obtain as necessary conditions

$$\sum_{i=1}^{n} \frac{f'_{x(s-\tau_i(s))} - (d/ds)f'_{\dot{x}(s-\tau_i(s))}}{1 - \tau'_i(s)} \bigg|_{s=\gamma_i(t)} = 0$$

with the assumption that $\ddot{x}(t)$ exists on the external curve (primes indicate differentiation). Otherwise, i.e., without the assumption, we have

$$\sum_{i=1}^{n} \left[\int_{t_0}^{s} f'_{x(t-\tau_i(t))} \, dt - \frac{f'_{\dot{x}(s-\tau_i(s))}}{1 - \tau_i'(s)} \right]_{s=\gamma_i(t)} = C$$

Exercise 5-34 Develop a full statement of the lemmas mentioned above and obtain the necessary conditions.

These explorations have also been carried out for problems with variable end points.

Exercise 5-35 Show that the extremals of

$$\int_{0}^{t_1} [\dot{x}^2(t) + x^2(t - \tau)] \, dt$$

with $x(0) = x(t_1) = 0$ have the form

$$\ddot{x}(t) + x(t) = 0 \qquad 0 \le t \le t_1 - \tau$$
that is,
$$x(t) = c_1 \sin t$$
and
$$\ddot{x}(t) = 0 \qquad t_1 - \tau \le t \le t_1$$
that is,
$$x(t) = c_2(t - t_1)$$

The constants c_1 and c_2 are determined from continuity and smoothness conditions of $x(t)$ at $t_1 - \tau$.

Let $\bar{\tau} = \max \tau_j(t_1), j = 1, \ldots, n$, and let x_j and $\dot{x}_j$ be abbreviations for x and $\dot{x}$ with arguments involving τ_j; then another necessary condition for a minimum is that on the extremal curve the determinant of the matrix of partial derivatives $\{f_{\dot{x}_i \dot{x}_j}\}$ is nonnegative; for a maximum it must be nonpositive. The proof is obtained by applying the second variation to F and requiring that it be nonnegative.

REFERENCES

A large collection of references is included in the books and papers mentioned at the end of Sec. 5-1. We give here a small number, some of which we found helpful in writing this chapter.

1. Ålander, M.: Sur les solutions périodiques d'une certaine équation fonctionelle, *C. R. Acad. Sci. Paris*, vol. 244 pp. 2118–2120, 1957.

2. Bellman, R., and K. L. Cooke: "Differential-Difference Equations," vol. 6 of Mathematics in Science and Engineering, Academic Press Inc., New York, 1963.

3. ——— and J. M. Danskin: The Stability Theory of Differential-Difference Equations, in Proc. Sympos. Nonlinear Circuit Anal., pp. 107–123, New York, 1953.

4. ——— and ———: A Survey of the Mathematical Theory of Time-lag, Retarded Control, and Hereditary Processes, RAND Corporation, R-256, March 1, 1954.

5. Brownell, F. H., III: Nonlinear Delay Differential Equations, pp. 89–148 in "Contributions to the Theory of Nonlinear Oscillations," vol. 1, edited by S. Lefschetz, Annals of Mathematics Studies 20, Princeton University Press, Princeton, N.J., 1950.

6. Bruwier, L.: Sur les solutions périodiques d'une équation différentielle fonctionnelle I, II, *Bull. Soc. Roy. Sci. Liège*, vol. 4, pp. 336–342, 1935.

7. ———: Sur l'intégration, par une méthode symbolique, d'une équation aux derivées et aux différences melées, *Mathesis*, vol. 49, pp. 54–58, 1935.

8. ———: Sur les équations intégro-différentielles normales dont les termes intégraux contiennent la dérivée de la fonction inconnue I, II, III, *Bull. Soc. Roy. Sci. Liège*, vol. 4, pp. 119–124, 162–166, 217–219, 1935.

9. Chin, Yu Ming: On the Boundedness and Stability of Solutions of Difference-Differential Equations (in Chinese, English summary), *Shuxue Jinzhan [Peking] (Advancement in Math. Progress in Math.)*], vol. 4, pp. 288–296, 1958.

10. Chow, Tse-Sun: Operational Analysis of a Traffic-dynamics Problem, *Operations Res.*, vol. 6, November–December, 1958.

11. Cooke, K. L.: Forced Periodic Solutions of a Stable Non-linear Differential-Difference Equation, *Ann. of Math.*, vol. 61, pp. 381–387, 1955.

12. ———: The Asymptotic Behaviour of the Solutions of Linear and Nonlinear Differential-Difference Equations, *Trans. Amer. Math. Soc.*, vol. 75, pp. 80–105, 1953.

13. ———: The Condition of Regular Degeneration for Singularly Perturbed Linear Differential-Difference Equations, *J. Differential Equations*, vol. 1, no. 1, pp. 39–94, January, 1965.

14. Driver, R. D.: A Functional-Differential System of Neutral Type Arising in a Two-body Problem of Classical Electrodynamics, pp. 474–483 in "International Symposium on Nonlinear Differential Equations and Nonlinear Mechanics," Academic Press Inc., New York, 1963.

15. ———: Existence Theory for a Delay-Differential System, *Contrib. Differential Equations*, vol. 1, no. 3, pp. 317–336, 1963.

16. ———: Existence and Stability of Solutions of a Delay-Differential System, *Arch. Rational Mech. Anal.*, vol. 10, no. 5, pp. 401–426, 1962.

17. ———: A Two-body Problem of Classical Electrodynamics: the One-dimensional Case, *Ann. Physics*, vol. 21, no. 1, pp. 122–142, January, 1963.

18. ———: Note on a Paper of Halanay on Stability for Finite Difference Equations, *Arch. Rational Mech. Anal.*, vol. 18, no. 3, pp. 241–243, 1965.

19. ———: Existence and Continuous Dependence of Solutions of a Neutral Functional-Differential Equation, *Arch. Rational Mech. Anal.*, vol. 19, no. 2, pp. 149–166, 1965.

20. El'sgol'ts, L. E.: Variational Problems with a Delayed Argument, *Uspehi Mat. Nauk*, vol. 12, no. 1, pp. 257–258, 1957; *Amer. Math. Soc. Transl.*, vol. 16, pp. 468–469, 1960.

21. ———: Approximate Methods of Integration of Differential-Difference Equations (in Russian), *Uspehi Mat. Nauk*, vol. 8, no. 4, pp. 81–93, 1953.

22. ———: Stability of Solutions of Differential-Difference Equations (in Russian), *Uspehi Mat. Nauk*, vol. 9, no. 4, pp. 95–112, 1954.

23. ———: Qualitative Methods in Mathematical Analysis, *Amer. Math. Soc. Transl.*, vol. 12, 1964.

24. Frisch, R.: Propagation Problems and Impulse Problems in Dynamic Economics, in "Economic Essays in Honour of Gustav Cassel," pp. 171–205, George Allen & Unwin, Ltd., London, 1933.

25. ———— and H. Holme: The Characteristic Solutions of a Mixed Difference and Differential Equation Occurring in Economic Dynamics, *Econometrica*, vol. 3, pp. 225–239, 1935.

26. Gerasimov, S. G.: The Problem of Controlling Processes with Lag at Constant Speed (in Russian), Automation of Thermal Processes, *Trudy Moskov. Energet. Inst.*, 1948.

27. Germaidze, V. E.: Asymptotic Stability of Systems with Lagging Argument (in Russian), *Uspehi Mat. Nauk*, vol. 14, no. 4, pp. 149–156, 1959.

28. Goodwin, R.: The Non-linear Acceleration and the Persistence of Business Cycles, *Econometrica*, vol. 19, pp. 1–17, 1951.

29. Hahn, V. W.: Bericht über Differential-Differenzengleichungen mit festen und veränderlichen Spannen, *Jber. Deutsch. Math.-Verein.*, vol. 57, pp. 55–84, 1954.

30. Halanay, A.: Solutions périodiques des systèmes linéaires à arguments retardés, *C. R. Acad. Sci. Paris*, vol. 249, pp. 2708–2709, 1959.

30a. ————: Periodic solutions of linear systems with retardation, *Rev. Math. Pures Appl.*, vol. 6, pp. 141–158, 1961.

31. Hale, J. K.: Theory of Stability of Functional-Differential Equations, *RIAS Notes*, 1963.

32. ————: Functional-Differential Equations with Parameters, *Contrib. Differential Equations*, vol. 1, no. 4, March, 1963.

33. ————: Periodic and Almost Periodic Solutions of Functional-Differential Equations, *Arch. Rational Mech. Anal.*, vol. 15, no. 4, pp. 289–304, 1964.

34. ————: A Class of Functional-Differential Equations, *RIAS Notes*, 1963.

35. ————Averaging Methods in Functional Differential Equations, *J. Differential Equations* (to appear).

36. ————: A Stability Theorem for Functional-Differential Equations, *Proc. Nat. Acad. Sci. U.S.A.*, vol. 50, no. 5, pp. 942–946, November, 1963.

37. ————: Linear Functional-Differential Equations with Constant Coefficients, *Contrib. Differential Equations*, vol. 2, pp. 291–317, 1963.

38. ———— and C. Perelló: The Neighborhood of a Singular Point of Functional Differential Equations, *Contrib. Differential Equations*, vol. 3, no. 3, pp. 351–375, 1964.

39. Jones, G. S.: Periodic Motions in Banach Space and Applications to Functional-Differential Equations, *Contrib. Differential Equations*, vol. 3, no. 1, 1964.

40. ————: The Existence of Periodic Solutions of $f'(x) = -\alpha f(x - 1)\{1 + f(x)\}$. *J. Math. Anal. Appl.*, vol. 5, no. 3, pp. 435–450, December, 1962.

41. Kakutani, S., and L. Markus: On the Non-linear Difference-Differential Equation $y'(t) = [A - By(t - \tau)]y(t)$, in "Contributions to the Theory of Nonlinear Oscillations," vol. 4, edited by S. Lefschetz, Annals of Mathematics Studies, Princeton University Press, Princeton, N.J., 1958.

42. Krall, A. M.: On the Real Parts of Zeros of Exponential Polynomials, *Bull. Amer. Math. Soc.*, vol. 70, no. 2, p. 291, March, 1964.

43. Krasovskii, N. N.: "Stability of Motion: Applications of Lyapunov's Second Method to Differential Systems and Equations with Delay," Stanford University Press, Stanford, Calif., 1963.

44. ————: On the Application of Lyapunov's Second Method to Equations with Time Lags (in Russian), *Prikl. Mat. Meh.*, vol. 20, pp. 315–327, 1956.

45. Kushner, H. J.: On the Dynamical Equations of Conditional Probability Density Functions with Applications to Optimal Stochastic Control Theory, *J. Math. Anal. Appl.*, vol. 8, no. 2, pp. 332–344, April, 1964.

46. Lakshmikantham, V.: Functional Differential Systems and Extension of Lyapunov's Method, *J. Math. Anal. Appl.*, vol. 8, no. 3, 1964.

47. Larson, R. D.: Characteristic Roots of a Mixed Difference-Differential Equation, *Amer. Math. Monthly*, vol. 68, no. 9, November, 1961.

48. Levin, J. J.: The Asymptotic Behavior of the Solution of a Volterra Equation, *Proc. Amer. Math. Soc.*, vol. 14, no. 4, pp. 534–541, 1963.

49. Minorsky, N.: "Nonlinear Oscillations," chap. 21, D. Van Nostrand Company, Inc., Princeton, N.J., 1962.

50. ———: Sur les systèmes à l'action retardée, *C. R. Acad. Sci. Paris*, vol. 234, pp. 1945–1947, 1952.

51. Myshkis, A. D.: Hystero-differential Equations (in Russian), *Uspehi Mat. Nauk*, vol. 4, no. 1, pp. 190–193, 1949.

52. ———: General Theory of Differential Equations with a Retarded Argument (in Russian), *Uspehi Mat. Nauk*, vol. 4, no. 5, pp. 99–141, 1949; *Amer. Math. Soc. Transl.* 55, pp. 1–62, 1951.

53. ———: Supplementary Bibliographical Material to the Paper "General Theory of Differential Equations with a Retarded Argument" (in Russian), *Uspehi Mat. Nauk*, vol. 5, no. 2, pp. 148–154, 1950; *Project Rand Rep.* T-23, pp. 1–11, 1952.

54. ———: "Lineinye Differentsial'nye Uravneniya s Zapazdyvayushchim Argumentom" (Linear Differential Equations with Retarded Argument), Moscow and Leningrad, 1951; "Lineare Differentialgleichungen mit nacheilendem Argument," Deutscher Verlag der Wissenschaft, Berlin, 1955.

55. Newell, G. F.: Nonlinear Effects in the Dynamics of Car Following, *Operations Res.*, vol. 9, no. 2, March–April, 1961.

56. Nohel, J. A.: Some Problems in Nonlinear Volterra Integral Equations, *Bull. Amer. Math. Soc.*, vol. 68, no. 4, pp. 323–330, July, 1962.

57. Norkin, S. B.: Boundary Problem for a Second-order Differential Equation with a Retarded Argument, *Moskov. Gos. Univ. Uch. Zap.*, Mat. 8, vol. 181, pp. 59–72, 1956.

58. Perelló, Carlos: Periodic Solutions of Ordinary Differential Equations with and without Lag, Ph.D. dissertation, Brown University, Providence, R.I., 1965.

59. Pinney, E.: "Ordinary Difference-Differential Equations," University of California Press, Berkeley, Calif., 1958.

60. Robinson, L. B.: Sur une équation aux différences melées, *C. R. Acad. Sci. Paris*, vol. 201, p. 1319, 1935.

61. ———: Introduction to a Study of a Type of Functional Differential and Functional Integral Equations, *Math. Mag. Texas*, vol. 23, pp. 183–188, 1950.

62. Satche, M.: Discussion of a paper by Porsky: Stability of Linear Oscillating Systems with Constant Time Lag, *J. Appl. Mech. (ASME)*, vol. 16, pp. 419–420, 1949.

63. Shimanov, S. N.: Almost Periodic Solutions of Non-homogeneous Linear Differential Equations with Lag (in Russian), *Izv. Vyss. Ucebn. Zaved. Matematika*, vol. 4, pp. 270–274, 1958.

64. ———: Almost Periodic Oscillations in Nonlinear Systems with Lag (in Russian), *Dokl. Akad. Nauk SSSR*, vol. 125, pp. 1203–1206, 1959.

65. ———: On the Theory of Oscillations of Quasilinear Systems with Lag (in Russian), *Prikl. Mat. Meh.*, vol. 23, pp. 836–844, 1959.

66. ———: On Instability of Motion of Systems with Time Lag (in Russian), *Prikl. Mat. Meh.*, vol. 24, pp. 55–63, 1960.

67. Sokolov, A. A.: Criterion for Stability of Linear Control Systems with Distributed Parameters and Its Applications (in Russian), *Inzh. Sbornik*, vol. 2, pp. 4–26, 1946.

68. Tsien, H. S.: "Engineering Cybernetics," McGraw-Hill Book Company, New York, 1954.

69. Vorob'ev, Yu. V., and V. N. Drozdovich: Methods of Investigating the Stability of Control Systems with Distributed Parameters (in Russian), *Avtomat. i Telemeh.*, vol. 10, pp. 77–83, 1949.

70. Wright, E. M.: A Non-linear Difference-Differential Equation, *J. Reine Angew. Math.*, vol. 193–194, 1954–1955.

71. ———: On a Sequence Defined by a Non-linear Recurrence Formula, *J. London Math. Soc.*, vol. 20, pp. 68–73, 1945.

72. ———: The Non-linear Difference-Differential Equation, *Quart. J. Math. Oxford Ser.* vol. 17, pp. 245–252, 1946.

73. Zverkin, A. M., G. A. Kemenskii, S. B. Norkin, and L. E., El'sgol'ts: Differential Equations with a Deviating Argument, *Uspehi Mat. Nauk*, vol. 17, 1962; translation, *Math. Reviews*, May, 1963.

SIX

INTEGRAL
EQUATIONS†

6-1 INTRODUCTION

The subject of integral equations, as contrasted with that of differential equations, is of considerably more recent origin. Most of the development has been within the twentieth century, although many special cases were treated much earlier.

The first important distinction we wish to make is between the cases of variable limits and constant limits. An integral equation with a variable limit of integration such as

$$\varphi(x) + \int_0^x F(x,t,\varphi(t)) \, dt = g(x) \tag{6-1}$$

is often called a *Volterra equation*, after the Italian mathematician Vito Volterra, who studied such equations extensively. The case of constant limits would be typified by

$$\varphi(x) + \int_0^1 F(x,t,\varphi(t)) \, dt = g(x) \tag{6-2}$$

† This chapter was written by D. H. Hyers.

To illustrate how these two types may arise, we consider the nonlinear differential equation

$$\frac{d^2y}{dx^2} = f(x,y) \tag{6-3}$$

where f is continuous. First, let us prescribe the initial conditions

$$y(0) = a \qquad y'(0) = b \tag{6-4}$$

for the solution. Integrating (6-3) twice, we have

$$y = \int_0^x ds \int_0^s f(t,y(t)) \, dt + a + bx$$

By a change in the order of integration, the double integral becomes

$$\int_0^x f(t,y(t)) \, dt \int_t^x ds = \int_0^x (x - t)f(t,y(t)) \, dt$$

and the equation for y is

$$y(x) = \int_0^x (x - t)f(t,y(t)) \, dt + a + bx \tag{6-5}$$

It is easily seen that a continuous solution of (6-5) satisfies (6-3) and (6-4). Thus the initial-value problem for the differential equation is equivalent to an integral equation of the Volterra type.

On the other hand, let us consider the problem of solving the same differential equation (6-3) subject to the two-point boundary conditions

$$y(0) = 0 \qquad y(1) = 0 \tag{6-6}$$

Then (6-5) still holds with $a = 0$, providing that we select b so as to satisfy the second boundary condition. This gives

$$b = -\int_0^1 (1 - t)f(t,y(t)) \, dt$$

and substituting this result into (6-5), we obtain

$$y(x) + \int_0^1 x(1 - t)f(t,y(t)) \, dt + \int_0^x (t - x)f(t,y(t)) \, dt = 0$$

or

$$y(x) + \int_0^1 K(x,t)f(t,y(t)) \, dt = 0 \tag{6-7}$$

where

$$K(x,t) = \begin{cases} t(1 - x) & 0 \le t \le x \le 1 \\ x(1 - t) & 0 \le x \le t \le 1 \end{cases} \tag{6-8}$$

Again, a continuous solution of (6-7) can easily be shown to be a solution of the boundary-value problem given by Eq. (6-3) subject to Eq. (6-6). An equation of the form (6-7) is known as an *equation of Hammerstein*. It is a special case of Eq. (6-2).

The importance of the distinction between the two types of integral equations is illustrated when one uses the method of successive approximations and tries to find simple conditions for convergence.

In the case of (6-7), if f satisfies a Lipschitz condition with Lipschitz constant L and B is an upper bound for the absolute value of $K(x,t)$, then we should have to impose the stringent condition $BL < 1$ in order to obtain convergence of the successive approximations (see "NM," theorem 1-6). Also, the sequence of successive approximations would converge like a *geometric* series with common ratio BL.

On the other hand, in applying the Picard method of successive approximations to the Volterra equation, we can take advantage of the variable upper limit to obtain convergence like that of an *exponential* series, so that the stringent condition on the Lipschitz constant is not necessary in this case (see "NM," pages 185 to 186, for the proof of theorem 4-1).

6-2 SOME EXAMPLES OF PHYSICAL PROBLEMS LEADING TO INTEGRAL EQUATIONS

A NONLINEAR PROBLEM OF HEAT TRANSFER

The following boundary-value problem was investigated by Mann and Wolf[13] and by Roberts and Mann.[18]

$$\frac{\partial u}{\partial t} = \frac{\partial^2 u}{\partial x^2} \qquad x > 0, t > 0$$
$$u(x,0) = 0 \qquad x > 0$$
$$-u_x(0,t) = G[u(0,t)] \qquad t > 0$$

Here the function $G(u)$ is assumed to satisfy the following conditions:
1. G is continuous.
2. $G(1) = 0$.
3. G is monotonically decreasing.

The unknown function u is the temperature in the solid body occupying the half-space $x \geq 0$, the initial temperature of which is zero, and the temperature of the external medium is taken to be $u = 1$. The last boundary condition represents a sort of generalized radiation condition. Thus in the case of the Stefan-Boltzmann condition it would have the form $G = C((a + 1)^4 - (a + u)^4)$. The solution of the partial differential equation with the given initial condition and with the boundary condition $u_x(0,t) = -g(t)$ is

$$u(x,t) = \int_0^t \frac{g(s)}{\sqrt{\pi(t - s)}} e^{-x^2/4(t-s)} \, ds$$

This result can be established by means of Laplace transforms under the assumption that the temperature u remains bounded for all positive t (see, for example, page 136, prob. 6 of Ref. 3). Replacing $g(s)$ by $G[u(0,s)]$, we have

$$u(x,t) = \int_0^t \frac{G[u(0,s)]}{\sqrt{\pi(t-s)}} e^{-x^2/4(t-s)} ds$$

Thus the temperatures in the semi-infinite solid are determined by the boundary temperatures $u(0,t)$. Setting $x = 0$ in the last equation, we arrive at the integral equation of Volterra type

$$u(0,t) = \int_0^t \frac{G[u(0,s)]}{\sqrt{\pi(t-s)}} ds$$

for the unknown boundary temperatures. Other equations of a similar nature have been investigated by Padmavally[15] and by Levinson.[12]

DUFFING'S VIBRATION PROBLEM

The forced vibrations of finite amplitude of a pendulum are governed by the differential equation

$$\frac{d^2\varphi}{dt^2} + \alpha^2 \sin \varphi = F(t)$$

If we assume that driving function F is an odd-periodic function of period 2, then the problem of finding an odd-periodic solution with this same period is easily reduced to finding a solution on the interval $0 \le t \le 1$ which satisfies the boundary conditions $\varphi(0) = \varphi(1) = 0$. From the discussion in Sec. 6-1 we know that this boundary-value problem is equivalent to the integral equation

$$\varphi(t) = - \int_0^1 K(t,s)\{F(s) - \alpha^2 \sin \varphi(s)\} \, ds$$

where the kernel $K(t,s)$ is given by (6-8).

BENDING OF A ROD BY A LONGITUDINAL FORCE

When a thin uniform rod is hinged at one end and acted upon by a longitudinal compressive force P at the other, the equation for the bending moment is $M = -Py$, where y is the deflection of the rod from its original straight-line position. As is known by the theory of the bending of a beam, $M = EIk$, where E is Young's modulus, I is the moment of inertia of the cross section, and k is the curvature at the point under consideration. If we use the arc length s measured from the hinged end as the independent variable, then by elementary calculus the curvature is $k = y''(s)(1 - y'^2(s))^{-\frac{1}{2}}$, and the equation for the bending moment takes the form

$$y''(s) + \lambda y \sqrt{1 - y'^2(s)} = 0$$

where $\lambda = P/EI$ is a positive parameter. The boundary conditions appropriate to our problem are $y(0) = y(1) = 0$, if the length of the rod is taken to be unity. Next we put $x(s) = -y''(s)$. From the results of Sec. 6-1 we have

$$y(s) = \int_0^1 K(s,t)x(t) \, dt$$

where the kernel $K(s,t)$ is given by (6-8). By differentiation we get

$$y'(s) = \int_0^1 K_s(s,t)x(t) \, dt$$

where
$$K_s(s,t) = \begin{cases} 1 - t & s < t \\ -t & s > t \end{cases}$$

Thus the problem specified by the differential equation for y together with the boundary conditions can be replaced by the following integral equation

$$x(s) = \lambda \int_0^1 K(s,t)x(t) \, dt \sqrt{1 - \left[\int_0^1 K_s(s,u)x(u) \, du \right]^2}$$

for the unknown function x (the problem was formulated in this way by Krasnosel'skii,[9] pages 181 to 183).

WATER WAVES OF FINITE AMPLITUDE

A classical problem of Nekrasov and Levi-Civita is that of determining periodic waves of permanent type on a "heavy" liquid of infinite depth (see Stoker,[22] chap. 12). It can be reduced to the problem of finding a harmonic function θ and its (even) harmonic conjugate τ in the unit disk both of which vanish at the origin and satisfy the boundary condition

$$\frac{\partial \tau}{\partial \varphi} = \gamma e^{-3\tau} \sin \theta \qquad \text{for } r = 1$$

where r, φ are polar coordinates in the plane, and where

$$\gamma = \frac{g\lambda}{2\pi c^2}$$

in which g is the acceleration of gravity, λ is the wavelength, and c is the velocity of propagation of the wave. Let $\theta(\varphi)$ and $\tau(\varphi)$ denote the boundary values of the harmonic functions θ and τ, respectively. It is assumed that $\theta(\varphi)$ is an odd function while $\tau(\varphi)$ is even, which means that the solution is being restricted to wave profiles which are symmetric about a vertical line through a crest.

By using an appropriate Green's function the problem can be reduced to that of solving the integral equation

$$\theta(\varphi) = \gamma \int_0^\pi K(\varphi,s)e^{-3\tau(s)} \sin \theta(s) \, ds$$

where
$$K(\varphi,s) = \frac{1}{\pi} \log \left| \frac{\sin \frac{1}{2}(\varphi + s)}{\sin \frac{1}{2}(\varphi - s)} \right|$$

The above integral equation, or a slight variation of it, is called *Nekrasov's integral equation* since he was the first to formulate (and solve) the problem in this way. A discussion of this problem can be found in the article by Hyers in Ref. 1.

6-3 LINEAR INTEGRAL EQUATIONS OF FREDHOLM TYPE

The linear equation

$$\varphi(s) - \int_a^b K(s,t)\varphi(t) \, dt = f(s) \tag{6-9}$$

with fixed limits a and b, is called a *Fredholm equation of the second kind*, after the mathematician Ivar Fredholm, who wrote the definitive paper[6] on the theory of this equation. The function $K(s,t)$ is called the *kernel* of the equation. For simplicity we shall assume that $K(s,t)$ is continuous on the square $a \leq s \leq b$, $a \leq t \leq b$, although the theory can be generalized to certain classes of kernels which are not everywhere continuous. The function f is also assumed to be continuous, and we are interested in finding a continuous solution. There are two cases in which a Fredholm equation is easy to handle.

The first of these is the case of small kernels. When $|b - a|$ max $|K(s,t)| < 1$, the equation, rewritten as

$$\varphi(s) = \int_a^b K(s,t)\varphi(t) \, dt + f(s)$$

can be solved by iteration. The result can be expressed in the form

$$\varphi(s) = f(s) = \int_a^b R(s,t)f(t) \, dt \tag{6-10}$$

where $R(s,t)$, the so-called *resolvent kernel* of $K(s,t)$, is given by the Neumann series

$$-R(s,t) = K(s,t) + K^{(2)}(s,t) + \cdots + K^{(n)}(s,t) + \cdots \tag{6-11}$$

with
$$K^{(2)}(s,t) = \int_a^b K(s,\tau)K(\tau,t) \, d\tau$$

and in general

$$K^{(n)}(s,t) = \int_a^b K^{(n-1)}(s,\tau)K(\tau,t) \, d\tau$$

$n = 3, 4, 5, \ldots$. Under the assumption of small kernels made above, the series (6-11) converges uniformly, by comparison with a geometric series.

Exercise 6-1 Carry out the above iteration process to obtain (6-10).

The second case in which the Fredholm equation is easy to handle is the case of a *degenerate kernel*, i.e., when the kernel has the special form

$$K(s,t) = \sum_{j=1}^m \alpha_j(s)\beta_j(t) \tag{6-12}$$

where both sets $\alpha_1(s), \ldots, \alpha_m(s)$ and $\beta_1(t), \ldots, \beta_m(t)$ are linearly independent and continuous. In this case the integral equation (6-9) is equivalent to a system of m linear algebraic equations in m unknowns. In fact, if we substitute (6-12) into (6-9), we get

$$\varphi(s) - \sum_{j=1}^m \varphi_j\alpha_j(s) = f(s)$$

where $\varphi_j = \int_a^b \beta_j(t)\varphi(t) \, dt$. If we now multiply by $\beta_i(s)$ and integrate, we have

$$\varphi_i - \sum_{j=1}^m c_{ij}\varphi_j = f_i \qquad i = 1, \ldots, m \tag{6-13}$$

where $c_{ij} = \int_a^b \beta_i(s)\alpha_j(s) \, ds$ and $f_i = \int_a^b \beta_i(s)f(s) \, ds$. The system (6-13) is equivalent to (6-9). If φ_i is a solution of (6-13), the corresponding solution of (6-9) is given by

$$\varphi(s) = f(s) + \sum_{i=1}^m \varphi_i\alpha_i(s) \tag{6-14}$$

as is easily verified.

Exercise 6-2 Show that (6-14) provides a solution of (6-9) with the kernel given by (6-12), where the φ_i are any solution of the algebraic system (6-13).

There is an analogy between the theory of linear systems of algebraic equations and the theory of Fredholm integral equations, which has been exploited by Fredholm and others. We have seen that when the kernel is degenerate, the two theories are not merely analogous but are in fact identical. In order to pursue the analogy in the general case of an arbitrary continuous kernel, it is convenient to introduce a symbolic or operator notation

$$K\varphi = \int_a^b K(s,t)\varphi(t) \, dt \tag{6-15}$$

Thus when the kernel $K(s,t)$ is continuous, (6-15) defines a linear transformation or operator taking the space C of continuous functions on the closed interval $[a,b]$ into itself. The *transpose K'* of K is defined by

$$K'\varphi = \int_a^b K(t,s)\varphi(t) \, dt \tag{6-16}$$

If H and K are two such operators, the kernel of the "product" $L = HK$ of the two operators is given by

$$L(s,t) = \int_a^b H(s,\tau)K(\tau,t) \, d\tau \tag{6-17}$$

as is readily seen.

The following theorems are basic.

Theorem 6-1 (*First theorem of Fredholm*)   The integral equation $\varphi - K\varphi = f$ has a unique continuous solution for each continuous f if and only if the homogeneous equation $\varphi - K\varphi = 0$ has only the trivial solution $\varphi = 0$.

Theorem 6-2 (*Second theorem of Fredholm*)   The two integral equations $\varphi - K\varphi = 0$ and $\psi - K'\psi = 0$ have the same finite number of linearly independent solutions.

Theorem 6-3 (*Third theorem of Fredholm*)   In case the transposed homogeneous equation $\psi - K'\psi = 0$ has exactly n linearly independent solutions $\psi_1, \ldots, \psi_n$, then the integral equation $\varphi - K\varphi = f$ will have solutions if and only if the orthogonality conditions $\int_a^b f(t)$. $\psi_j(t) \, dt = 0$ $(j = 1, \ldots, n)$ are satisfied, and in this case there is an n-parameter family of solutions.

In the case when K corresponds to a degenerate kernel, these three theorems reduce to known results concerning linear systems of algebraic equations, in accordance with the above remarks [note that the transpose of a degenerate kernel corresponds to the transpose of the corresponding matrix c_{ij} in (6-13)]. Thus the Fredholm theorems hold when the kernel of K is degenerate.

The general case of a continuous kernel can be reduced to the preceding one by using a method due to Schmidt. The idea is to approximate the continuous kernel by degenerate kernels. This can always be done by virtue of the Weierstrass approximation theorem (cf. Ref. 4, pages 65 to 68) which states that a continuous function of two variables defined on a closed rectangle can be uniformly approximated by polynomials.

Thus any continuous kernel $K(s,t)$ can be expressed in the form

$$K(s,t) = P(s,t) + Q(s,t) \qquad (6\text{-}18)$$

where $P(s,t)$ has the form (6-12) of a degenerate kernel and the kernel $Q(s,t)$ is "small," in the sense that $|b - a| \max |Q(s,t)| < 1$. Since $Q(s,t)$ satisfies the condition for small kernels, it has a resolvent kernel $R(s,t)$. The integral equation $\varphi - K\varphi = f$ can then be written

$$\varphi - Q\varphi = P\varphi + f \qquad (6\text{-}19)$$

Putting $\varphi - Q\varphi = \omega$, we have $\varphi = \omega - R\omega$, in accordance with (6-10), and substituting this result into (6-19), we obtain

$$\omega = (P - PR)\omega + f \qquad (6\text{-}20)$$

for ω. Now it is not difficult to verify that the kernel of $P - PR$ is degenerate and that Eqs. (6-19) and (6-20) are equivalent, through the relations $\varphi = \omega - R\omega$ and $\omega = \varphi - Q\varphi$. Thus the general case has been reduced to the case of degenerate kernels. In this manner the validity of the three theorems of Fredholm can be demonstrated for arbitrary continuous kernels (for more details see Petrovskii[16]).

Exercise 6-3 Show (1) that Eqs. (6-19) and (6-20) are equivalent and that (6-20) has a degenerate kernel and (2) that the transposed equations of (6-19) and (6-20) are equivalent.

Remark: Although for purposes of simplicity of exposition we have assumed that the kernel was continuous in proving the three Fredholm theorems, this is not necessary. The important thing is that the functional transformation

$$g(s) = \int_a^b K(s,t)f(t) \, dt$$

be completely continuous in some appropriate function space. This will be true for the space $L_2(a,b)$ if the function $|K(s,t)|^2$ is integrable on the square $a \le s \le b$, $a \le t \le b$. For the proofs of the basic Fredholm theorems for such L_2 kernels, see Tricomi.[24]

Exercise 6-4 Given the kernel $K(s,t) = s - t$, find the iterated kernels $K^{(2)}(s,t)$, $K^{(3)}(s,t)$, etc., and then find an approximate solution to the integral equation $u(s) = s + \int_0^1 (s - t)u(t) \, dt$. Can you estimate the error in your approximate solution?

Exercise 6-5 Solve the following integral equations:

$$\varphi(s) - \int_0^1 st\varphi(t) \, dt = s^2$$

$$\varphi(s) - \int_0^1 (st - s^2 t^2)\varphi(t) \, dt = s^3$$

6-4   LINEAR EQUATIONS WITH SYMMETRIC KERNELS

The real kernel $K(s,t)$ is said to be *symmetric* when $K(t,s) = K(s,t)$.   The theory of linear integral equations with symmetric kernels is analogous to the theory of linear systems of algebraic equations whose coefficient matrix is symmetric.   As was indicated in "NM," chap. 1, the proper setting for the theory of linear integral equations is the Hilbert space $L_2(a,b)$, and the basic elements of the theory were developed there in secs. 1-6 to 1-8.   In the present section, we shall confine ourselves to the case of real functions, so that the inner product in the space is

$$(f,g) = \int_a^b f(s)g(s)\ ds$$

With a view to applications, we shall not demand that the kernel $K(s,t)$ itself be continuous, but rather we shall assume throughout this section that the following conditions, which we refer to as "standard assumptions," are satisfied:

1. $K(s,t)$ is integrable on the square $a \leq s \leq b,\ a \leq t \leq b$.
2.   The iterated kernel

$$K^{(2)}(s,t) = \int_a^b K(s,\sigma)K(\sigma,t)\ d\sigma$$

is continuous on this square.

3. $K(s,t) = K(t,s)$ for all (s,t) in the square.

Under these conditions $|K(s,t)|^2$ is evidently integrable on the square, since

$$\int_a^b \int_a^b K(s,t)K(s,t)\ ds\ dt = \int_a^b \int_a^b K(s,t)K(t,s)\ ds\ dt$$
$$= \int_a^b K^{(2)}(s,s)\ ds$$

and the latter integrand is continuous.   Also, if f is any function belonging to $L_2(a,b)$, we have by the Schwarz inequality

$$\left| \int_a^b K(s,t)f(t)\ dt \right| \leq \sqrt{\int_a^b [K(s,t)]^2\ dt \int_a^b f^2(t)\ dt}$$
$$\leq \sqrt{K^{(2)}(s,s) \int_a^b f^2(t)\ dt}$$

This implies that the function

$$g(s) = \int_a^b K(s,t)f(t)\ dt$$

is bounded for all s in the interval (a,b).   Moreover, *the transformation $f \to g$ thus defined is completely continuous.*   To demonstrate this, we again use the Schwartz inequality to obtain the following inequality:

$$|g(s_1) - g(s_2)| \leqq \sqrt{\int_a^b \{K(s_1,t) - K(s_2,t)\}^2 \, dt \int_a^b f^2(t) \, dt}$$

Now

$$\int_a^b \{K(s_1,t) - K(s_2,t)\}^2 \, dt = K^{(2)}(s_1,s_1) - 2K^{(2)}(s_1,s_2) + K^{(2)}(s_2,s_2)$$

$$\leqq |K^{(2)}(s_1,s_1) - K^{(2)}(s_1,s_2)| + |K^{(2)}(s_1,s_2) - K^{(2)}(s_2,s_2)|$$

and it follows from the continuity of the iterated kernel $K^{(2)}$ and the last two inequalities that the functions $g(s)$ form a uniformly bounded equicontinuous family when $f(t)$ ranges over a bounded set in $L_2(a,b)$. Hence by Ascoli's theorem (see "NM," page 47), the transformation $f \to g$ is a completely continuous transformation of $L_2(a,b)$ into the space $C[a,b]$ of continuous functions on $[a,b]$. But since uniform convergence implies convergence in the mean, the transformation is also completely continuous as a transformation into $L_2(a,b)$

It follows from the above remarks that the argument on page 31 of "NM" can be applied to prove that the transformation $f \to g$ can be represented by a symmetric Hilbert-Schmidt matrix. We denote the above linear operator $g(s) = \int_a^b K(s,t)f(t) \, dt$ by $g = Kf$, and it follows from theorem 1-4 of "NM" that the operator K has the following properties:

1. The operator K has real eigenvalues; i.e., there exist real numbers λ and functions φ in $L_2(a,b)$, not identically zero, such that $K\varphi = \lambda\varphi$.

2. Eigenfunctions corresponding to distinct eigenvalues are orthogonal.

3. If M denotes the space of all functions in $L_2(a,b)$ which are orthogonal to the null space of K, then there exists a complete orthonormal set of eigenfunctions φ_n in the subspace M, the corresponding eigenvalues λ_n are real and nonzero, and the sequence λ_n is either finite or else $\lambda_n \to 0$ as $n \to \infty$.

Remark: Since the function $g(s) = \int_a^b K(s,t)f(t) \, dt$ is continuous when f belongs to $L_2(a,b)$, it follows that all eigenfunctions $\varphi_n(s)$ are continuous.

On the basis of the above properties of the operator K, we can establish the following important result.

Theorem 6-4 (Hilbert-Schmidt expansion theorem) Every function of the form

$$f(s) = \int_a^b K(s,t)g(t) \, dt$$

where $K(s,t)$ satisfies our standard assumptions 1 to 3 and $g(t)$ belongs to the class $L_2(a,b)$, can be represented by its Fourier series with respect to the orthonormal system $\{\varphi_n\}$ of the eigenfunctions of $K(s,t)$ which correspond to nonzero eigenvalues

$$f(s) = \sum_{n=1}^{\infty} a_n \varphi_n(s)$$

where $a_n = \int_a^b f(s)\varphi_n(s) \, ds$, and this series converges absolutely and uniformly.

Proof: By hypothesis f is contained in the range of the operator K. It follows that f is orthogonal to the null space M of K, for if h belongs to M, then

$$(f,h) = (Kg,h) = (g,Kh) = 0$$

Hence, by property 3, the partial sums $\sum_{n=1}^{m} a_n \varphi_n$ converge in the mean to f. Put $g_n = (g,\varphi_n) = \int_a^b g(s)\varphi_n(s) \, ds$. Then by Bessel's inequality (see "NM," page 22) we have $\sum_{n=1}^{m} g_n{}^2 \leq (g,g)$ for all m, so that the series $\sum_{n=1}^{m} g_n{}^2$ converges. For each fixed s, the function $K(s,t)$, treated as a function of t, belongs to $L_2(a,b)$, since

$$\int_a^b \{K(s,t)\}^2 \, dt = \int_a^b K(s,t)K(t,s) \, dt = K^{(2)}(s,s)$$

Again for fixed s, the function $K(s,t)$ is clearly orthogonal to the null space N of the operator K. Hence by property 3 and the Parseval relation for a complete orthonormal set (see "NM," page 20) we have

$$K^{(2)}(s,s) = \int_a^b \{K(s,t)\}^2 \, dt = \sum_{n=1}^{\infty} \left\{ \int_a^b K(s,t)\varphi_n(t) \, dt \right\}^2 = \sum_{n=1}^{\infty} \lambda_n{}^2 \varphi_n{}^2(s)$$

Since $K^{(2)}(s,s)$ is continuous, it follows that the series $\Sigma \lambda_n{}^2 \varphi_n{}^2(s)$ converges uniformly on (a,b) by Dini's theorem (Ref. 4, page 57).

To prove that the series $\Sigma a_n \varphi_n(s)$ converges uniformly we consider the Cauchy remainder $\sum_{n=m}^{m+p} a_n \varphi_n(s)$. Since

$$a_n = (f,\varphi_n) = (Kg,\varphi_n) = (g,K\varphi_n) = (g,\lambda_n\varphi_n) = \lambda_n g_n$$

we can apply the Cauchy-Schwarz inequality for sums to obtain

$$\left[\sum_{n=m}^{m+p} a_n \varphi_n(s) \right]^2 = \left[\sum_{n=m}^{m+p} \lambda_n g_n \varphi_n(s) \right]^2 \leq \sum_{n=m}^{m+p} g_n{}^2 \sum_{n=m}^{m+p} \lambda_n{}^2 \varphi_n{}^2(s)$$

Since $\Sigma g_n{}^2$ converges and $\Sigma\lambda_n{}^2\varphi_n{}^2(s)$ converges uniformly, it follows that $\Sigma a_n\varphi_n(s)$ converges uniformly. But since the partial sums of this series converge in the mean to $f(s)$, it follows that the sum of the series must be $f(s)$. The theorem is proved.

We can use this expansion theorem to solve the Fredholm integral equation

$$\psi(s) - \mu \int_a^b K(s,t)\psi(t)\ dt = f(s) \tag{6-21}$$

where μ is a real parameter, f is continuous, and the f kernel satisfies our standard assumptions 1 to 3. Let us assume that μ is not equal to the reciprocal of any eigenvalue of K. In operator notation the equation is

$$\psi - \mu K\psi = f$$

Since $\psi - f$ is in the range of K, it can be expanded by the Hilbert-Schmidt theorem in the form $\psi - f = \Sigma c_n\varphi_n$, where, say,

$$c_n = (\psi - f,\ \varphi_n) = (\psi,\varphi_n) - (f,\varphi_n) = \psi_n - f_n$$

Taking inner products with φ_n, we have

$$c_n = (\psi - f,\varphi_n) = (\mu K\psi,\varphi_n) = \mu(\psi,K\varphi_n) = \lambda_n\mu(\psi,\varphi_n)$$
$$= \lambda_n\mu\psi_n$$

Thus we have

$$\psi_n - f_n = \lambda_n\mu\psi_n \qquad \text{or} \qquad \psi_n = \frac{f_n}{1 - \lambda_n\mu} \tag{6-22}$$

where we have used our assumption that $\lambda_n\mu \neq 0$ for all n. From (6-22) we immediately deduce that

$$c_n = \frac{\lambda_n\mu f_n}{1 - \lambda_n\mu} \tag{6-23}$$

Thus the solution of (6-21) is

$$\psi(s) = f(s) + \mu \sum_{n=1}^{\infty} \frac{\lambda_n f_n\varphi_n(s)}{1 - \lambda_n\mu} \tag{6-24}$$

and the series converges absolutely and uniformly.

So far we have discussed the nonsingular case in which the parameter μ is not a reciprocal eigenvalue. Suppose now that we are in the singular case in which $\mu = 1/\lambda$, where λ is an eigenvalue of K. By the Fredholm theory discussed in Sec. 6-3, the integral equation (6-21) will have a solution if and only if the function f is orthogonal to all those eigenfunctions which belong to the eigenvalue λ.   Thus if $\lambda = \lambda_{p+1} = \cdots = \lambda_{p+m}$, and if none of the other eigenvalues are equal to λ, we can solve Eq. (6-21) providing that the orthogonality conditions

$$\int_a^b f(s)\varphi_k(s)\ ds = 0 \qquad k = p + 1, \ldots, p + m$$

hold.   Under these conditions, Eq. (6-23) is still valid for $n \leq p$ and for $n > p + m$.   Since $\varphi_{p+1}(s), \ldots, \varphi_{p+m}(s)$ are solutions of the homogeneous equation corresponding to (6-21), the solution can still be written in the form

$$\psi(s) = f(s) + \frac{1}{\lambda_{m+1}} \sum_{n=1}^{\infty} c_n\varphi_n(s)$$

where $\qquad\qquad c_n = \dfrac{\lambda_n f_n}{\lambda_{m+1} - \lambda_n} \qquad n \leq p \text{ or } n > p + m$

and where the constants $c_{p+1}, \ldots, c_{m+p}$ are completely arbitrary.   The number m is the dimension of the null space of the operator $K - \lambda_{m+1}I$ and may be called the *geometric multiplicity* of the eigenvalue λ_{m+1}.   The Fredholm theory guarantees that m is a finite integer.

A symmetric kernel is called *positive* if all its nonzero eigenvalues are positive.   When $K(s,t)$ is positive, the quadratic integral form

$$\int_a^b \int_a^b K(s,t)f(s)f(t)\ ds\ dt$$

is nonnegative for all f in $L_2(a,b)$.   In case $K(s,t)$ is both positive and continuous, the expansion

$$K(s,t) = \sum_{n=1}^{\infty} \lambda_n\varphi_n(s)\varphi_n(t)$$

is valid, and the series converges uniformly.   This result is called the *bilinear formula* for $K(s,t)$, and its validity was proved by Mercer (see Ref. 4, page 138).

In the general case when the kernel is not positive, this series may not converge.   However, its partial sums will always converge in the mean to $K(s,t)$ under the standard assumptions 1 to 3.

Remark: Most books on integral equations use the reciprocals $\mu_n = 1/\lambda_n$ of our eigenvalues.   In the present section we follow the usage of matrix theory and linear-operator theory in order to be consistent with the terminology of the first chapter of "Nonlinear Mathematics."   In later sections we shall sometimes use the reciprocal eigenvalues μ_n, corresponding to the nonzero λ_n.

Exercise 6-6   Use the Hilbert-Schmidt expansion theorem to prove the bilinear formula

$$K^{(2)}(s,t) = \sum_{n=1}^{\infty} \lambda_n{}^2 \varphi_n(s)\varphi_n(t)$$

for the iterated kernel $K^{(2)}$.

Exercise 6-7   Show that the kernel $K(x,t)$ given by (6-8) has the eigenfunctions $\varphi_n(x) = \sqrt{2}\sin n\pi x$ corresponding to the eigenvalues $\lambda_n = 1/n^2\pi^2$. It follows that $K(x,t)$ is positive. *Hint:* Solve the boundary-value problem $d^2\varphi/dx^2 + \mu\varphi = 0$ with the boundary conditions $\varphi(0) = \varphi(1) = 0$.

Exercise 6-8   Show that the kernal

$$K(s,t) = \frac{1}{\pi}\log\left|\frac{\sin \tfrac{1}{2}(s+t)}{\sin \tfrac{1}{2}(s-t)}\right|$$

which occurs at the end of Sec. 6-2, satisfies standard assumptions 1 to 3, for $0 \le s \le \pi$, $0 \le t \le \pi$.

Exercise 6-9   Solve the integral equation

$$\psi(x) - 5\int_0^1 K(x,t)\psi(t)\,dt = x(1-x)$$

where $K(x,t)$ is given by (6-8).

Exercise 6-10   Prove that when the kernel is continuous and positive (in the sense that all its nonzero eigenvalues are positive), then the integral

$$J(f,f) = \int_a^b \int_a^b K(x,y)f(x)f(y)\,dx\,dy$$

is nonnegative for all continuous f.

6-5   NONLINEAR VOLTERRA EQUATIONS

The theory of Volterra integral equations is a direct generalization of the theory of ordinary differential equations. In particular, one can easily generalize the fundamental existence theorems for differential equations to Volterra integral equations. We shall be concerned with the integral equation

$$u(t) = g(t) + \int_0^t f(t,s,u(s))\,ds \tag{6-25}$$

When the integrand f is merely continuous, one can derive an existence theorem for this equation from the Schauder fixed-point theorem ("NM," page 45). The following form of the theorem was given by Sato.[20]

   ***Theorem* 6-5**   Let g be continuous on $0 \le t \le T$, and let D be the closed domain in (t,s,u) space defined by the inequalities $0 \le s \le t \le T$

and $|u - g(t)| \leq p$. Let f be continuous and have the bound $|f(t,s,u)| \leq M$ in D.

Then, if $r = \min \{T, p/M\}$, the integral equation (6-25) has a continuous solution on the interval $0 \leq t \leq r$.

Proof: Let $C[0,r]$ denote the space of continuous functions on the closed interval $0 \leq t \leq r$. From the hypotheses of the theorem it follows at once that the integral transformation $\bar{u} = Au$ which is defined by

$$Au = g(t) + \int_0^t f(t,s,u(s)) \, ds$$

takes the closed sphere S in the space $C[0,r]$ with center g and radius p into itself. Moreover, it is easily verified that the image AS is an equicontinuous set of functions. By Ascoli's theorem ("NM," page 47) the closure of the set AS is compact. If we denote by K the closed convex hull of AS, it follows that $K \subset S$, since S is convex and closed. Since AS has a compact closure, K is also compact (cf. prob. 5, page 134, of Ref. 23). Thus the transformation A takes the convex compact set K into itself, and the Leray-Schauder fixed-point theorem applies, to guarantee the existence of a continuous solution of (6-25).

The solution just demonstrated may not be unique. However, if the function f satisfies a Lipschitz condition, the uniqueness of the solution can be proved.

Corollary: Under the hypotheses of the theorem, let f in addition satisfy a Lipschitz condition

$$|f(t,s,u) - f(t,s,v)| \leq L|u - v|$$

for all (t,s,u) and (t,s,v) in D. Then there is a unique continuous solution $u(t)$ of the integral equation (6-25) on the interval $[0,r]$ which satisfies $|u(t) - g(t)| \leq p$ on this interval. Moreover, this solution is the uniform limit of the successive approximations $u_n(t)$ defined by

$$u_0(t) = g(t)$$
$$u_{n+1}(t) = g(t) + \int_0^t f(t,s,u_n(s)) \, ds$$

Exercise 6-11 Prove the above corollary. *Hint:* See "NM," proof of theorem 4-1, pages 185 to 186.

We turn next to some simple methods which can be used to obtain *estimates* for the solutions of Volterra equations. To start with, we shall assume that the integrand $f(t,s,u)$ is *strictly increasing* in u. For a thorough discussion of these ideas, the reader is referred to Walter;[27] we shall indicate only a few sample results.

***Theorem* 6-6**   Let $f(t,s,u)$ be defined for $0 \leq s \leq t \leq T$ and all real u, and let f be strictly increasing in u for each s and t. Let $f(t,s,\varphi(s))$ be absolutely integrable on the interval $0 \leq s \leq t$ for each t satisfying $0 < t \leq T$ and for each continuous function φ on the closed interval $[0,T]$. Let v, w be continuous functions defined on $[0,T]$ which satisfy the conditions:

1. $v(0) < w(0)$.
2. $v(t) - \int_0^t f(t,s,v(s))\,ds \leqq w(t) - \int_0^t f(t,s,w(s))\,ds$ when $0 < t \leq T$.

Then $v(t) < w(t)$ for $0 < t \leq T$.

Proof: If the conclusion of the theorem is false, there must be a point t_0 with $0 < t_0 \leq T$ such that $v(s) < w(s)$ for $0 < s < t_0$ and $v(t_0) = w(t_0)$. But in this case we have

$$
\begin{aligned}
v(t_0) &= v(t_0) - \int_0^{t_0} f(t_0,s,v(s))\,ds + \int_0^{t_0} f(t_0,s,v(s))\,ds \\
&\leqq w(t_0) - \int_0^{t_0} f(t_0,s,w(s))\,ds + \int_0^{t_0} f(t_0,s,v(s))\,ds \\
&< w(t_0) - \int_0^{t_0} f(t_0,s,w(s))\,ds + \int_0^{t_0} f(t_0,s,w(s))\,ds \\
&= w(t_0)
\end{aligned}
$$

which is a contradiction.

This result can be used in seeking bounds or approximations for solutions of Volterra equations. For example, if u is a solution of

$$u(t) = g(t) + \int_0^t f(t,s,u(s))\,ds \tag{6-25}$$

where f is strictly increasing in u, while the continuous functions v and w satisfy the inequalities

$$v(t) \leqq g(t) + \int_0^t f(t,s,v(s))\,ds \tag{6-26}$$

$$w(t) \geqq g(t) + \int_0^t f(t,s,w(s))\,ds \tag{6-27}$$

$$v(0) < u(0) < w(0) \tag{6-28}$$

when $0 < r \leq T$, we can apply the theorem in turn to v and u, and then to u and w, and conclude that

$$v(t) < u(t) < w(t) \tag{6-29}$$

for $0 < t \leq T$. Thus if inequality (6-28) *is true for all solutions of* (6-25), *we can conclude that the above relation between the three functions gives us an a priori bound on the solutions.*

Even if the function $f(t,s,u)$ is not monotonic in u, it may be possible to relate the given f to a positive strictly increasing function F by means of an inequality

$$|f(t,s,v) - f(t,s,u)| \leqq F(t,s,|v - u|) \tag{6-30}$$

The idea is to gain knowledge of our given integral equation (6-25) from the properties of the equation

$$\sigma(t) = d(t) + \int_0^t F(t,s,\sigma(s))\, ds$$

for which simple estimation theorems hold. The following result is based on this idea.

Theorem 6-7 Let $f(s,t,u)$ be defined for $0 \leq s \leq t \leq T$ and all real u, and let $f(t,s,\varphi(s))$ be absolutely integrable on the interval $0 \leq s \leq t$ for each t satisfying $0 < t \leq T$ and for each continuous function φ on the closed interval $[0,T]$. Let $F(t,s,u)$ satisfy the same conditions and in addition be strictly increasing in u for $u \geq 0$.

Let the inequality (6-30) hold for all real u and v, and let $u(t)$, $v(t)$, $p(t)$, $d(t)$, and $g(t)$ be continuous functions defined on the interval $[0,T]$. If for t in this interval the following conditions are satisfied:

1. $u(t)$ is a solution of (6-25),
2. $\left| v(t) - g(t) - \int_0^t f(t,s,v(s))\, ds \right| \leq d(t)$,
3. $p(t) > d(t) + \int_0^t F(t,s,p(s))\, ds$,

then the inequality $|v(t) - u(t)| < p(t)$ holds in the interval $[0,T]$.

Proof: We shall use the abbreviation $\int fw$ for the integral $\int_0^t f(t,s, w(s))\, ds$, etc. Then from the hypotheses of the theorem we have, for t in the interval $[0,T]$,

$$\begin{aligned} |v - u| &= |v - g - \int fu| = |v - g - \int fv + \int fv - \int fu| \\ &\leq d + |\int fv - \int fu| \leq d + \int |fv - fu| \leq d + \int F(|v - u|) \end{aligned}$$

Thus $|v - u| - \int F(|v - u|) \leq d \leq p - \int Fp$, so that condition 2 of Theorem 6-6 is satisfied, with $|v - u|$ replacing v and p replacing w. When $t = 0$, we have $|v(0) - u(0)| = |v(0) - g(0)| \leq d(0)$ and $d(0) < p(0)$ by conditions 2 and 3 of the present theorem. Thus $|v(0) - u(0)| < p(0)$, so that condition 1 of Theorem 6-6 is also satisfied for the functions $|v - u|$ and p. Then by Theorem 6-6 we have $|v - u| < p$ on the interval $[0,T]$, which completes the proof.

As a simple case illustrating this theorem, suppose that f satisfies a Lipschitz condition

$$|f(t,s,u) - f(t,s,v)| \leqq L|u - v| \tag{6-31}$$

Then we can take the function F to be simply $F(t,s,u) = Lu$ in order to satisfy (6-30). Let conditions 1 and 2 of Theorem 6-7 hold where d is a positive constant, and let $c > d$. The solution of the integral equation

$$p(t) = c + \int_0^t Lp(s)\, ds$$

is easily seen to be $p(t) = ce^{Lt}$.   Clearly we have $p(t) > d + \int_0^t Lp(s) \, ds$, so that condition 3 of the theorem is also satisfied for our chosen F.   The theorem gives the error bound $|v(t) - u(t)| < ce^{Lt}$ for our approximate solution $v(t)$, where c is any number greater than d and $0 \leq t \leq T$.

MONOTONICALLY DECREASING INTEGRANDS

Another case of interest is when the integrand $f(t,s,u)$ of (6-25) is monotonically decreasing in u

$$f(t,s,v) \leqq f(t,s,u) \qquad \text{for } v \geqq u$$

Theorem 6-8   Let $f(t,s,u)$ be defined for $0 \leq s \leq t \leq T$ and all real u, and let f be monotonically decreasing in u for fixed s, t.   Let $f(t,s,\varphi(s))$ be absolutely integrable on the interval $0 \leq s \leq t$ for each t satisfying $0 < t \leq T$ and for each continuous function φ on the closed interval $[0,T]$, and assume that $\lim\limits_{t \to 0} \int_0^t f(t,s,\varphi(s)) \, ds = 0$ for each such φ.

If $u(t)$ is any continuous solution of (6-25), and if $g(t)$, $v(t)$, and $w(t)$ are continuous functions on $[0,T]$ which satisfy there the inequalities

$$v(t) < g(t) + \int_0^t f(t,s,w(s)) \, ds$$
$$w(t) > g(t) + \int_0^t f(t,s,v(s)) \, ds \tag{6-32}$$

then the relations

$$v(t) < u(t) < w(t) \tag{6-33}$$

hold on $[0,T]$.

Proof: By setting $t = 0$ in (6-32) we see that the inequalities (6-33) hold for $t = 0$.   If they do not hold on $[0,T]$, there is a first point $t_0 > 0$ in this interval at which both do not hold.   Assuming that $v(t_0) = u(t_0)$, we have $v(t) < u(t) < w(t)$ for $0 \leq t < t_0$, and

$$v(t_0) = u(t_0) = g(t_0) + \int_0^{t_0} f(t_0,s,u(s)) \, ds$$
$$\geqq g(t_0) + \int_0^{t_0} f(t_0,s,w(s)) \, ds > v(t_0)$$

Thus we arrive at a contradiction.   Similarly it can be shown that $u(t_0) \neq w(t_0)$, so the theorem is proved.

For monotonically decreasing integrands the method of successive approximations is especially useful, even for numerical computations, as has been pointed out by Collatz and others (a brief general discussion is given in the article by Collatz in the book edited by Langer[11]).   We shall need the following corollary to Theorem 6-8.

Corollary: Let f satisfy the Lipschitz condition (6-31) in addition to the hypotheses of Theorem 6-8. Then equality signs are permitted in the inequalities (6-32) and (6-33).

Proof: Assume that $v \leq g + \int fw$ and $w \geq g + \int fv$. Put $p(t) = \alpha e^{Lt}$, where $\alpha > 0$. Then $\int_0^t Lp(\tau)\, d\tau = \alpha e^{Lt} - \alpha < p(t)$ for $0 \leq t \leq T$. Also, we may choose d as small as we please. Since f satisfies the Lipschitz condition, it follows that on $[0,T]$

$$\int f(v - p) - \int fv \leq L\int p < p$$

or

$$\int fv + p > \int f(v - p)$$

Hence

$$w + p \geq g + \int fv + p > g + \int f(v - p) \qquad (6\text{-}34)$$

Similarly, we have

$$\int fw - \int f(w + p) \leq L\int p < p$$

and

$$v - p \leq g + \int fw - p < g + \int f(w + p) \qquad (6\text{-}35)$$

Thus, we have from (6-34) and (6-35) that (6-32) and (6-33) are satisfied if we replace v by $v - p$ and w by $w + p$. It follows that $v - p < u < w + p$ on $[0,T]$. Since α may be taken arbitrarily small, it follows that $v \leq u \leq w$ on $[0,T]$, and the corollary is proved.

When $f(t,s,u)$ is monotonically decreasing in u and also satisfies a Lipschitz condition, let us once more consider the process of successive approximations. Beginning with a function v_0, let a sequence v_n be defined as usual by the recursion formula $v_{n+1} = g + \int fv_n$. Now if we assume that $v_0 \leq v_1$ and $v_0 \leq v_2$ (all equations and inequalities are to hold for all t on the interval $[0,T]$), then since $v_0 \leq g + \int fv_1$ and $v_1 = g + \int fv_0$, it follows from the corollary just proved that $v_0 \leq u \leq v_1$, where u is the unique solution of (6-25). Since $v_0 \leq v_2$, it follows that $\int fv_2 \leq \int fv_0$ by the monotonicity property of f. Hence $v_3 = g + \int fv_2 \leq g + \int fv_0 = v_1$. Also we have $v_2 = g + \int fv_1$ and $v_1 \geq v_3 = g + \int fv_2$. Again by the corollary it follows that $v_2 \leq u \leq v_1$. From $v_3 \leq v_1$ we get $\int fv_1 \leq \int fv_3$, and hence $v_2 = g + \int fv_1 \leq g + \int fv_3 = v_4$. Again since $v_2 \leq g + \int fv_3$ and $v_3 = g + \int fv_2$, we can apply the corollary once more to find that $v_2 \leq u \leq v_3$. So far we have shown that $v_0 \leq v_2 \leq u \leq v_3 \leq v_1$ and that $v_2 \leq v_4$. Proceeding in this way, we can prove by induction that

$$v_0 \leq v_2 \leq v_4 \leq \cdots \leq u \leq \cdots \leq v_5 \leq v_3 \leq v_1 \qquad (6\text{-}36)$$

Under our assumptions the sequence v_n is bound to converge to the unique solution u. Because of the alternating character of the sequence, it is obvious that at each step the error committed is less than the maximum of the absolute values of the difference between v_n and v_{n+1}. Thus we have a simple error estimate. Moreover, the same idea can often be used in problems where no Lipschitz condition holds when f is monotonically decreasing in u.

As an example, let us consider the following problem in differential equations which arises in the boundary-layer theory of fluid flow. The problem is to find the solution of the differential equation

$$y''' + 2yy'' = 0 \qquad 0 \leq t < \infty$$

which is to satisfy the initial conditions

$$y(0) = y'(0) = 0 \qquad y''(0) = 1$$

This is the simplest problem considered by Weyl in his paper[28] on the subject and occurs in connection with the boundary-layer flow past a half-plane. We divide the differential equation by y'' and then integrate the result, using the initial condition for y'' to get

$$\ln y'' = -2 \int_0^t y(s) \, ds$$

Two integrations by parts with the aid of the initial conditions for y and y' give

$$-\ln y'' = \int_0^t (t - s)^2 y''(s) \, ds$$

and if we put $\varphi(t) = -\ln y''(t)$, we have the integral equation

$$\varphi(t) = \int_0^t (t - s)^2 e^{-\varphi(s)} \, ds \tag{6-37}$$

for the determination of the function φ. In order to fit this equation into our theory, we may select T as large as we please but keep it fixed and put

$$f(t,s,u) = \begin{cases} (t - s)^2 e^{-u} & u \geq 0 \\ (t - s)^2 & u < 0 \end{cases}$$

Then it is easily seen that f is continuous and satisfies a Lipschitz condition with Lipschitz constant T^2. Also, f is clearly monotonically decreasing in u for each fixed t and s. If we set $v_0 \equiv 0$ and put

$$v_1(t) = \int_0^t (t - s)^2 e^{-v_0(s)} \, ds$$

$$v_2(t) = \int_0^t (t - s)^2 e^{-v_1(s)} \, ds$$

it is evident that $v_1(t) \geq v_0(t)$ and $v_2(t) \geq v_0(t)$ for $0 \leq t \leq T$. Thus all our hypotheses are satisfied, and we can conclude that the successive approximations v_n satisfy the inequalities (6-36) on the interval $[0,T]$, where u is the uniquely determined solution on this interval. It is not difficult to show that

$$|v_n(t) - v_{n-1}(t)| \leq \frac{2^n t^{3n}}{(3n)!}$$

and this inequality gives an upper bound for the error in the process of successive approximations.

Exercise 6-12   Prove the above inequality.

As another example, let us consider the heat-flow problem (the first example in Sec. 6-2). If we put $y(t) = u(0,t)$ for the unknown function, the integral equation to be solved is

$$y(t) = \int_0^t \frac{G[y(s)]\, ds}{\sqrt{\pi(t-s)}}$$

On physical grounds, we expect the temperature to vary between its initial value zero in the solid and the value one in the external medium. We shall assume, in addition to the hypotheses made in Sec. 6-2, that $G(u)$ satisfies a Lipschitz condition on the interval $0 \leq u \leq 1$, with the constant L. In order to be able to use our theory, which requires that the integrand $f(t,s,u)$ be defined for all real u, we may resort to the following artifice. We put

$$G^*(u) = \begin{cases} G(0) & \text{if } u < 0 \\ G(u) & \text{if } 0 \leq u \leq 1 \\ G(1) = 0 & \text{if } u > 1 \end{cases}$$

and consider the integral equation

$$y(t) = \int_0^t \frac{G^*[y(s)]\, ds}{\sqrt{\pi(t-s)}}$$

Now the integrand is defined for all real u and is monotonically decreasing in u, and it is easily shown that $G^*(u)$ satisfies a Lipschitz condition with the constant L

$$|G^*(u_1) - G^*(u_2)| \leq L|u_1 - u_2|$$

for all real values of u_1 and u_2. Let us start our successive-approximation process with $v_0(t) \equiv 0$. Then since the integrand is monotonically decreasing in u and nonnegative, it follows immediately that the next two iterates

$$v_1(t) = \int_0^t \frac{G^*[v_0(s)]\, ds}{\sqrt{\pi(t-s)}} \qquad v_2(t) = \int_0^t \frac{G^*[v_1(s)]\, ds}{\sqrt{\pi(t-s)}}$$

are each nonnegative, so that $v_0 \leq v_1$ and $v_0 \leq v_2$. Thus our "pincer" scheme for monotonically decreasing integrands works here. In order to obtain an error estimate for the successive approximation and at the same time establish the convergence of the sequence v_n, we shall need the formula

$$\int_0^t \frac{s^{n/2}\, ds}{\sqrt{\pi(t-s)}} = \frac{\Gamma(n/2+1)t^{(n+1)/2}}{\Gamma(n/2 + \frac{3}{2})}$$

which is easily established by using beta and gamma functions or a table or definite integrals. Applying this formula with $n = 0$, we get

$$|v_1(t) - v_0(t)| = v_1(t) = \int_0^t \frac{G^*(0)}{\sqrt{\pi(t - s)}} \, ds = \frac{G(0)t^{\frac{1}{2}}}{\Gamma(\frac{3}{2})} \leq \frac{Lt^{\frac{1}{2}}}{\Gamma(\frac{3}{2})}$$

since $G(0) = |G(0) - G(1)| \leq L|0 - 1| = L$. By mathematical induction with the aid of the Lipschitz condition for G^* and the above formula for the evaluation of the definite integral the inequality

$$|v_{n+1}(t) - v_n(t)| \leq \frac{L^{n+1}t^{(n+1)/2}}{\Gamma(n/2 + \frac{3}{2})}$$

can be established. This inequality is sufficient to prove that the sequence of successive approximations v_n will converge uniformly to a continuous limit function $y(t)$ on any given finite interval $[0,T]$. Moreover, the same inequality gives us a bound for the error for the approximate solution v_n in accordance with the theory for monotonically decreasing integrands.

Exercise 6-13 Carry out the details necessary to establish the last inequality.

We have established the existence of a solution $y(t)$ for the modified integral equation with G^* instead of G, as the limit of the above sequence of successive approximations $v_n(t)$. From the general theory for monotonically decreasing integrands we know that

$$0 \equiv v_0(t) \leq y(t) \leq v_1(t) \leq \frac{Lt^{\frac{1}{2}}}{\Gamma(\frac{3}{2})}$$

Hence, for sufficiently small values of t we have $0 \leq y(t) \leq 1$. We want to show that this same inequality is true for all t. If not, let t_1 be a point at which $y(t_1) > 1$, and put $t_0 = \sup [t : t < t_1, y(t) \leq 1]$. Then $y(t_0) = 1$ and $y(t) \geq 1$ when $t_0 < t \leq t_1$. Then we have

$$y(t_1) = \int_0^{t_1} \frac{G^*[y(s)] \, ds}{\sqrt{\pi(t_1 - s)}} = \int_0^{t_0} \frac{G^*[y(s)] \, ds}{\sqrt{\pi(t_1 - s)}}$$

$$\leq \int_0^{t_0} \frac{G^*[y(s)] \, ds}{\sqrt{\pi(t_0 - s)}} = y(t_0) = 1$$

which contradicts the assumption that $y(t_1) > 1$. It follows that $0 \leq y(t) \leq 1$ on any given interval $[0,T]$. Consequently,

$$G^*[y(s)] = G[y(s)]$$

and y satisfies the originally given integral equation.

Exercise 6-14 Referring to Theorem 6-6, find an "upper function" $w(t)$ and a "lower function" $v(t)$ for the integral equation $u(t) = \int_0^t [s + u^2(s)] \, ds$; that is, find v and w which satisfy the inequalities (6-26) to (6-28) when $g(t) \equiv 0$ and $f(t,s,u) = s + u^2$ on the interval $0 \le t \le 1$.

6-6 HAMMERSTEIN'S THEORY

In his basic paper[7] Hammerstein studied nonlinear integral equations of the form

$$\psi(x) + \int_B K(x,y)f(y,\psi(y)) \, dy = 0 \tag{6-38}$$

where B is a closed bounded simply connected region in the space of one or more dimensions, and where dy represents the appropriate element of integration. We shall give a brief account of some of Hammerstein's results, without giving the details of proof. The proofs of some of his basic theorems are given in chap. 4 of the book by Tricomi.[24]

In applications to boundary-value problems, the kernel $K(x,y)$ is often a Green's function for the domain B and for some differential operator (for a discussion of Green's functions see Courant and Hilbert[4]). In more than one dimension the Green's functions become infinite in some characteristic manner, depending on the dimension. Hammerstein's theory is general enough to include, for example, the Green's function for the Laplacian in two or three dimensions. The following assumptions are made throughout regarding the kernel, in addition to its integrability:

1. The *iterated kernel* $K^{(2)}(x,y) = \int_B K(x,s)K(s,y) \, ds$ is *continuous*.

2. $K(x,y)$ is *symmetric*.

3. $K(x,y)$ is *positive* in the sense that all its reciprocal eigenvalues are positive.

Let the reciprocal eigenvalues μ_n of the kernel $K(x,y)$ be arranged in their order of magnitude, and let the corresponding normal orthogonal eigenfunctions be denoted by $\varphi_n(x)$. Thus we have

$$0 < \mu_1 \le \mu_2 \le \mu_3 \le \cdots$$

where

$$\varphi_n(x) = \mu_n \int_B K(x,y)\varphi_n(y) \, dy$$

since the kernel is positive. Either the sequence μ_n is finite, or $\mu_n \to \infty$ with n.

Essentially, Hammerstein's method is to find a suitable variational problem and to relate the solution of Eq. (6-38) with a function which minimizes an appropriate functional. The functional to be minimized is

$$J(\omega) = \int_B \int_B K(x,y)\omega(x)\omega(y) \, dx \, dy + 2 \int_B F\left(x, \int_B K(x,y)\omega(y) \, dy\right) dx \tag{6-39}$$

where the function $F(x,u)$ is given by

$$F(x,u) = \int_0^u f(x,v)\, dv \qquad (6\text{-}40)$$

If $\omega(x)$ is a solution to the problem of making the functional $J(\omega)$ a minimum, then the function

$$\psi(x) = \int_B K(x,y)\omega(y)\, dy \qquad (6\text{-}41)$$

will satisfy the integral equation (6-38).   To see this, we take the Gateaux differential (or variation) of (6-39) and get

$$\begin{aligned}
\delta J &= 2\int_B \int_B K(x,y)\omega(x)\,\delta\omega(y)\, dy\, dx \\
&\quad + 2\int_B dx \int_B f\left(x, \int_B K(x,s)\omega(s)\, ds\right)K(x,y)\,\delta\omega(y)\, dy \\
&= 2\int_B \int_B K(x,y)\left\{\omega(x) + f\left(x, \int_B K(x,s)\omega(s)\, ds\right)\right\}\delta\omega(y)\, dy\, dx
\end{aligned}$$

For a minimum, $\delta J = 0$ for all continuous functions $\delta\omega$.   Thus we have, on using the symmetry of the kernel,

$$0 = \int_B K(y,x)\omega(x)\, dx + \int_B K(y,x)f\left(x, \int_B K(x,s)\omega(s)\, ds\right)dx$$

Therefore if we put $\psi(y) = \int_B K(y,x)\omega(x)\, dx$ in accordance with (6-41), we see that

$$\psi(y) + \int_B K(y,x)f(x,\psi(x))\, dx = 0$$

so that ψ satisfies (6-38).   The following theorems give conditions under which an absolute minimum of $J(\omega)$ exists and hence also a solution of (6-38) exists.

Theorem 6-9   If the kernel K satisfies the standard assumptions 1 to 3 and in addition is bounded on B, and if the function $f(x,u)$ is continuous and constants k and C_1 exist such that for all $x \,\epsilon\, B$ and all u, $F(x,u) = \int_0^u f(x,v)\, dv \geq -(k/2)u^2 - C_1$, where $0 < k < \mu_1$, then Eq. (6-38) has a continuous solution.

Theorem 6-10   If the kernel K satisfies only the standard assumptions and if the function f is continuous and satisfies the conditions

$$F(x,u) = \int_0^u f(x,v)\, dv \geq -\frac{k}{2}u^2 - C_1$$
$$|f(x,u)| \leq C_2|u| + C_3$$

for all x in B and all real u, where C_1, C_2, C_3, and k are positive constants with $k < \mu_1$, then Eq. (6-38) has a continuous solution.   In case $C_2 < \mu_1$ the first condition follows from the second.

Theorem 6-11 If the kernel K satisfies the standard assumptions and also its values $K(x,y)$ are never negative, then Eq. (6-38) will have a solution providing the function f is continuous and satisfies any one of the following conditions:

1. For $u \leq 0$, $0 \leq f(x,u) \leq C_4|u| + C_5$.
2. For $u \geq 0$, $0 \geq f(x,u) \geq -C_4|u| - C_5$.
3. For $u \leq 0$, $-C_6 \leq f(x,u) \leq C_4|u| + C_5$; for $u \geq 0, f(x,u) \geq -C_6$.
4. For $u \leq 0$, $f(x,u) \leq C_6$; for $u \geq 0$, $-C_4|u| - C_5 \leq f(x,u) \leq C_6$,

where C_4, C_5, C_6 are positive constants with $0 < C_4 < \mu_1$.

Theorems 6-9 and 6-10 are proved by means of a direct method of the calculus of variations with the aid of the Hilbert-Schmidt theory, which holds for kernels satisfying standard assumptions 1 to 3.

We may think of the function ω occurring in the functional (6-39) to be minimized being approximated by finite sums $\omega_m(x) = \sum_{j=1}^{m} b_j\varphi_j(x)$, and the solution approximated by a similar sum $\psi_m(x) = \sum_{j=1}^{m} a_j\varphi_j(x)$, where we are using the orthonormal system $\{\varphi_n\}$ of the eigenfunctions of the kernel $K(x,y)$ as our basis. In order to preserve the relation (6-41) between ψ_m and ω_m we must have $b_j = \mu_j a_j$, where μ_j is the reciprocal eigenvalue corresponding to φ_j.

If we replace ω in $J(\omega)$ by ω_m and use the relations (6-41) and $b_j = \mu_j a_j$, we obtain the following function of the parameters $a_1, \ldots, a_m$:

$$H_m(a_1, \ldots, a_m) = \sum_{j=1}^{m} \mu_j a_j{}^2 + 2 \int_B F\left(x, \sum_{j=1}^{m} a_i\varphi_i(x)\right) dx$$

As in the Rayleigh-Ritz method, we shall minimize this function of a finite number of parameters in the hope of obtaining an approximate solution of our problem. Under the hypothesis concerning $F(x,u)$ in Theorems 6-9 and 6-10 we have

$$H_m(a_1, \ldots, a_m) \geqq \sum_{j=1}^{m} \mu_j a_j{}^2 - 2 \int_B \left\{ \frac{k}{2}\left(\sum_{1}^{m} a_i\varphi_i(x)\right)^2 + C_1 \right\} dx$$

$$\geqq \sum_{j=1}^{m} (\mu_j - k)a_j{}^2 - 2C_1 \int_B dy$$

Since $k < \mu_1 \leq \mu_2 \leq \cdots \leq \mu_m$, it follows that the function H_m has an absolute minimum d_m when the variables $a_1, \ldots, a_m$ range over all real numbers. Hence there exists for each m a set of values of $a_1, \ldots, a_m$

for which $H_m = d_m$ and which satisfy the equations

$$\frac{\partial H_m}{\partial a_j} = 2\mu_j a_j + 2 \int_B f\left(x, \sum_{i=1}^{m} a_i \varphi_i(x)\right) \varphi_j(x) \, dx = 0 \qquad j = 1, 2, \ldots, m$$

or
$$a_j = -\frac{1}{\mu_j} \int_B f\left(x, \sum_{i=1}^{m} a_i \varphi_i(x)\right) \varphi_j(x) \, dx \qquad (6\text{-}42)$$

Thus for this set of a's we have for the approximation $\psi_m(x) = \sum_{j=1}^{m} a_j \varphi_j(x)$ the equation

$$\psi_m(x) = -\sum_{j=1}^{m} \frac{\varphi_j(x)}{\mu_j} \int_B f(y, \psi_m(y)) \varphi_j(y) \, dy \qquad (6\text{-}43)$$

From this result one can show that $\psi_m(x)$ is an approximate solution to Eq. (6-38) in the sense that

$$\lim_{m=\infty} \left[\psi_m(x) + \int_B K(x,y) f(y, \psi_m(y)) \, dy \right] = 0 \qquad (6\text{-}44)$$

providing that there is a constant A independent of m such that the inequality

$$\int_B [f(y, \psi_m(y))]^2 \, dy \leqq A \qquad (6\text{-}45)$$

holds for all m. This inequality can be shown to hold under either the hypotheses of Theorem 6-9 or those of Theorem 6-10.

The sequence of functions $\psi_m(x)$ may not converge. However, from the fact that the iterated kernel is continuous and from condition (6-45) it can be demonstrated by the same argument used in Sec. 6-4 in a similar connection that the functions

$$\omega_m(x) = \int_B K(x,y) f(y, \psi_m(y)) \, dy$$

form an equicontinuous family on the closed bounded domain B. Hence by Ascoli's theorem ("NM," page 47) it follows that there exists a uniformly convergent subsequence. If the limit of this subsequence is designated as $-\psi(x)$, it follows from (6-44) that $\psi(x)$ is a continuous solution of (6-38).

Theorem 6-11 can be proved easily on the basis of Theorem 6-10. In fact, let the kernel $K(x,y)$ satisfy standard assumptions 1 to 3, and in addition let $K(x,y) \geq 0$ for all x and y in B.

For condition 1, put

$$f^*(x,u) = \begin{cases} f(x,u) & \text{for } u < 0 \\ f(x,0) & \text{for } u \geq 0 \end{cases}$$

so that $f^*(x,u)$ satisfies the condition

$$|f^*(x,u)| \leq C_4|u| + C_5'$$

where $0 < C_4 < \mu_1$ for all x in B and all real u. Hence by Theorem 6-10 the integral equation

$$\psi^*(x) = -\int_B K(x,y)f^*(y,\psi^*(y))\, dy$$

has a continuous solution $\psi^*(x)$. But since the functions $K(x,y)$ and $f^*(x,u)$ are nonnegative, it follows that $\psi^*(x) \leq 0$. Hence

$$f^*(y,\psi^*(y)) = f(y,\psi^*(y))$$

so that ψ^* satisfies (6-38).

For condition 3, put

$$g(x) = C_6\int_B K(x,y)\, dy \qquad f^*(x,u) = f(x,\, u + g(x)) + C_6$$

It follows that $f^*(x,u)$ satisfies

$$0 \leq f^*(x,u) \leq C_4|u| + C_5'' \qquad \text{for } u \leq 0$$

where $C_5'' = C_5 + C_6$ and $f^*(x,u) \geq 0$ for $u \geq 0$. Hence by condition 1 it follows that the equation

$$\psi^*(x) + \int_B K(x,y)f^*(y,\psi^*(y))\, dy = 0$$

has a continuous solution $\psi^*(x) \leq 0$. By putting $\psi(x) = \psi^*(x) + g(x)$ the above equation reduces to (6-38), which therefore has a solution satisfying the inequality $\psi(x) \leq C_6\int_B K(x,y)\, dy$.

The remaining cases 2 and 4 follow immediately from 1 and 3, respectively, by using the "reflection"

$$f_1(x,u) = -f(x,-u)$$

UNIQUENESS

Theorem 6-12 Let the kernel satisfy standard assumptions 1 to 3, and let the function $f(x,u)$ be monotonically increasing in u for each fixed x in B and continuous for x in B and all real u. Then the integral equation (6-38) has at most one continuous solution.

Proof: Let $\psi(x)$ and $\psi_1(x)$ be two continuous solutions of (6-38), and put $\omega(x) = \psi_1(x) - \psi(x)$, $g(x,u) = f(x,\, u + \psi(x)) - f(x,\psi(x))$. Then (6-38) has at most one solution if and only if the integral equation

$$\omega(x) + \int_B K(x,y)g(y,\omega(y))\, dy = 0 \tag{6-46}$$

has only the identically vanishing solution. The hypotheses of the theorem imply that $g(x,0) = 0$ and that $g(x,u) \geqq 0$ for $u > 0$ and $g(x,u) \leq 0$ for $u < 0$, for all x in B. Let $\omega(x)$ be any solution of (6-46). Then if we multiply this equation by $g(x,\omega(x))$ and integrate over the domain B, we have

$$\int_B \omega(x)g(x,\omega(x))\ dx + \int_B \int_B K(x,y)g(y,\omega(y))g(x,\omega(x))\ dy\ dx = 0$$

Since the kernel K is positive, it follows that the second term is nonnegative, so that

$$\int_B \omega(x)g(x,\omega(x))\ dx \leqq 0$$

But, from the properties of the function g listed above, we know that the integrand of the last integral is nonnegative. Hence $\omega(x)g(x,\omega(x)) \equiv 0$ in B. Thus, if at some point x, $\omega(x) \neq 0$, then we must have

$$g(x,\omega(x)) = 0$$

at that point. Also, if $\omega(x) = 0$, then $g(x,\omega(x)) = g(x,0) = 0$. Consequently $g(x,\omega(x)) \equiv 0$ in B, and since $\omega(x)$ satisfies (6-46), it follows that $\omega(x) \equiv 0$ in B. This completes the proof of the theorem. Hammerstein proves another uniqueness theorem in case the function f satisfies a Lipschitz condition with the Lipschitz constant less than the smallest reciprocal eigenvalue μ_1. Moreover, under this condition it is true that the approximating sequence $\psi_m(x)$ given by (6-43) actually converges to the solution $\psi(x)$ of (6-38).

Exercise 6-15 On the basis of Hammerstein's theorems, prove that the integral equation

$$\psi(x) + \int_0^1 K(x,y)e^{\psi(y)}\ dy = 0$$

has a unique solution if $K(x,y)$ is the kernel given by (6-8).

Exercise 6-16 On the other hand, show that the integral equation

$$\psi(x) - \int_0^1 K(x,y)e^{\psi(y)}\ dy = 0$$

has no real solution if $K(x,y) = \alpha(x)\alpha(y)$, where $\alpha(x) \geq 1$ on $[0,1]$.

Exercise 6-17 Show that the integral equation

$$\psi(x) + \int_0^1 \alpha(x)\ \psi(y) \sin\left[\frac{\psi(y)}{\alpha(y)}\right] dy = 0$$

has infinitely many solutions if $\alpha(x) \geq 2$ on $[0,1]$.

6-7 NONLINEAR INTEGRAL EQUATIONS CONTAINING A PARAMETER—BRANCHING OF SOLUTIONS

We consider the integral equation containing the real parameter μ

$$\psi(s) - \mu \int_a^b K(s,t)f(t,\psi(t))\, dt = 0 \qquad (6\text{-}47)$$

For simplicity, we assume that the kernel $K(s,t)$ is continuous, while the function $f(s,y)$ is assumed to have a continuous partial derivative $f_y(s,y)$. Equation (6-47) may be thought of as a nonlinear operator equation

$$G(\psi,\mu) = 0 \qquad (6\text{-}48)$$

with ψ and the values of G in the space of continuous functions $C[a,b]$. Let $\psi = \psi_0$, $\mu = \mu_0$ be a solution of (6-48). The Fréchet derivative of $G(\psi,\mu)$ for fixed μ at the point $\psi = \psi_0$, $\mu = \mu_0$ is $I - \mu_0 H$, where H is the linear operator

$$H\omega = \int_a^b K(s,t)f_y(t,\psi_0(t))\omega(t)\, dt \qquad (6\text{-}49)$$

When μ_0 is not a reciprocal eigenvalue of the kernel

$$H(s,t) = K(s,t)f_y(t,\psi_0(t))$$

of this operator, the operator $I - \mu_0 H$ has a bounded inverse by Theorem 6-1. Hence by the implicit-function theorem for operator equations (see "MN," chap. 1) Eq. (6-48) has a solution $\psi = \psi(\mu)$ for all μ in the neighborhood of $\mu = \mu_0$. Moreover, this solution is *locally unique* in the sense that it is the only solution in the neighborhood of the point $\psi = \psi_0$, $\mu = \mu_0$ which approaches ψ_0 as $\mu \to \mu_0$.

Equation (6-47) has the obvious solution $\psi \equiv 0$ for $\mu = 0$. By the discussion just concluded, it will have locally unique solutions for small μ. However, if we let μ increase through a reciprocal eigenvalue μ_0 of the kernel $H(s,t)$, the number of solutions of (6-47) may change. When such a change occurs, the point $\psi = \psi_0$, $\mu = \mu_0$ is called a *branch point* or *bifurcation point* of the equation. From the above analysis it appears that in order for ψ_0, μ_0 to be a branch point it is necessary that μ_0 be a reciprocal eigenvalue of the kernel $H(s,t)$. In general, however, this condition is not sufficient.

Various methods have been devised to deal with the phenomenon of branching and in particular to discover whether a given reciprocal eigenvalue of the Fréchet derivative actually corresponds to a bifurcation point. The oldest of these methods is that due to Lyapunov and Schmidt and leads to the *bifurcation equation*.

We shall illustrate the Lyapunov-Schmidt method in the special case in which the function f has the form

$$f(t,y) = y + y^2 g(t,y) \tag{6-50}$$

where $g(t,y)$ is continuous in the pair t, y and analytic in y in some neighborhood of $y = 0$, and where the kernel $K(s,t)$ satisfies standard assumptions 1 to 3. We shall also assume that the reciprocal eigenvalue μ_1 of the kernel $K(s,t)$ is *simple*, in the sense that its eigenspace is of dimension one. Let $\varphi_1(s)$ be the normalized eigenfunction corresponding to μ_1, so that

$$\varphi_1(s) = \mu_1 \int_a^b K(s,t)\varphi_1(t)\, dt \qquad \int_a^b \varphi_1{}^2(t)\, dt = 1$$

Because of the special form (6-50) of the function f, it is evident that $\psi \equiv 0$ is a solution of (6-47) for all μ. Also, we have $H(s,t) = K(s,t)$ when we put $\psi_0(t) \equiv 0$. Following the approach of Schmidt, we form a new kernel

$$K_1(s,t) = K(s,t) - \mu_1{}^{-1}\varphi_1(s)\varphi_1(t) \tag{6-51}$$

which is constructed so that the equation

$$\psi(s) - \mu_1 \int_a^b K_1(s,t)\psi(t)\, dt = 0 \tag{6-52}$$

has only the trivial solution.

Exercise 6-18   Prove the last statement by using Theorem 6-3 together with the assumption that μ_1 is simple.

By Theorem 6-1, the equation

$$\psi(s) - \mu_1 \int_a^b K(s,t)\psi(t)\, dt = \omega(s) \tag{6-53}$$

has a continuous solution for all continuous functions $\omega(s)$. In order to take advantage of this fact, we first rewrite Eq. (6-47) in the form

$$\psi(s) - \mu \int_a^b K_1(s,t)\psi(t)\, dt = \mu \int_a^b K_1(s,t)\psi^2(t)g(t,\psi(t))\, dt$$

Next, we put

$$\mu = \mu_1 + \beta \qquad \alpha = \int_0^1 \varphi_1(t)\psi(t)\, dt \tag{6-54}$$

and use the relation (6-51) between the given kernel and the modified kernel to obtain

$$\psi(s) - \mu_1 \int_a^b K_1(s,t)\psi(t)\, dt = \alpha\varphi_1(s) + \beta \int_a^b K(s,t)\psi(t)\, dt$$
$$+ (\mu_1 + \beta) \int_a^b K(s,t)\psi^2(t)g(t,\psi^2(t))\, dt \tag{6-55}$$

The advantage of this form is that the linear operator $I - \mu_1 K_1$ appearing in the left member has an inverse, and moreover the right side is small in absolute value when the quantities α, β, and max $|\psi|$ are small. This remark leads to the following scheme. If we forget the definition (6-54) for the parameter α for a moment, Eq. (6-55) will have a solution $\psi(s;\alpha,\beta)$ depending on the two real parameters α and β, when $|\alpha|$ and $|\beta|$ are sufficiently small, by the implicit-function theorem for operator equations. To derive a solution of the given integral equation (6-47) from this two-parameter family of solutions we substitute the solution into Eq. (6-54) for α. The resulting equation is

$$\Lambda(\alpha,\beta) \equiv \int_a^b \varphi_1(t)\psi(t;\alpha,\beta) \, dt - \alpha = 0 \qquad (6\text{-}56)$$

and is known as the *bifurcation equation* of the problem. Since

$$\psi(t;0,0) \equiv 0$$

the bifurcation equation is satisfied by $\alpha = 0$, $\beta = 0$. If it has no other solutions, then the given integral equation acquires no new solutions as μ passes through the value μ_1. If it does have other solutions in the neighborhood of $\alpha = 0$, $\beta = 0$, then at least one new solution of the given integral equation will branch off from the zero solution at $\mu = \mu_1$. In the special case under consideration, we shall show that the bifurcation equation will have nontrivial solutions.

With obvious abbreviations Eq. (6-55) can be written as

$$\psi - \mu_1 K_1 \psi = \alpha\varphi_1 + \beta K\psi + (\mu_1 + \beta)K\psi^2 g(\psi) \qquad (6\text{-}57)$$

with the solution $\psi(\alpha,\beta)$, and the bifurcation equation is

$$\Lambda(\alpha,\beta) \equiv (\varphi_1,\psi(\alpha,\beta)) - \alpha = 0 \qquad (6\text{-}58)$$

Because of the analytic nature of the integral equation, it can be shown that $\psi(\alpha,\beta)$ is analytic in the variables α, β, and we can set

$$\psi = \alpha\psi_1 + \alpha^2\psi_{20} + \alpha\beta\psi_{11} + \alpha^3\psi_{30} + \alpha^2\beta\psi_{21} + \alpha\beta^2\psi_{12} + \cdots \qquad (6\text{-}59)$$

In this expansion we have omitted all terms which do not have α as a factor. This is permissible because if we set $\alpha = 0$ in our integral equation, the resulting equation obviously has $\psi = 0$ as a solution, and since the solution is locally unique, it follows that $\psi(0,\beta) = 0$. If we substitute the expansion (6-59) into Eq. (6-57) and equate coefficients, we find that the first three coefficients in the expansion (2-59) are determined by

$$\begin{aligned}
(I - \mu_1 K_1)\psi_1 &= \varphi_1 \\
(I - \mu_1 K_1)\psi_{20} &= \mu_1 K\varphi_1^2 g(0) \\
(I - \mu_1 K_1)\psi_{11} &= K\psi_1
\end{aligned} \qquad (6\text{-}60)$$

These equations can be solved by using the Hilbert-Schmidt expansion theorem, as indicated in Sec. 6-4, if we observe that the kernel $K_1(x,y)$ has the same eigenfunctions (and eigenvalues) as $K(x,y)$ except for the first one, φ_1. We find easily that $\psi_1 = \varphi_1$ and that $\psi_{11} = \mu_1^{-1}\varphi_1$. The bifurcation equation reduces to

$$\alpha^2(\varphi_1,\psi_{20}) + \alpha\beta(\varphi_1,\psi_{11}) + \alpha^3(\psi_{30},\varphi_1) + \alpha^2\beta(\psi_{21},\varphi_1) \\ + \alpha\beta^2(\psi_{12},\varphi_1) + \cdots = 0$$

Since every term has α as a factor, we have $\alpha = 0$ as one obvious root of the bifurcation equation. To look for other roots, we divide by α and obtain

$$\alpha(\psi_{20},\varphi_1) + \beta(\psi_{11},\varphi_1) + \alpha^2(\psi_{30},\varphi_1) + \alpha\beta(\psi_{21},\varphi_1) \\ + \beta^2(\psi_{12},\varphi_1) + \cdots = 0 \quad (6\text{-}61)$$

where the omitted terms are of order three or higher in α, β. The important thing is that the coefficient of β is $(\psi_{11},\varphi_1) = \mu_1^{-1}(\varphi_1,\varphi_1) = \mu_1^{-1} \neq 0$. It follows from the implicit-function theorem for analytic functions that the reduced bifurcation equation (6-61) is uniquely solvable for β as a function of α. Let us consider now the special case where $g(y) = -\frac{1}{6}y$. Then we find that $(\psi_{30},\varphi_1) = -\dfrac{1}{6}\displaystyle\int_0^1 \varphi_1^4(s)\,ds < 0$. When we solve Eq. (6-61) for β, we get $\beta = \left(\dfrac{\mu_1}{6}\displaystyle\int_0^1 \varphi_1^4(s)\,ds\right)\alpha^2 + O[\alpha^3]$. In this case, then, we see that the function $\beta = \beta(\alpha)$ does not vanish identically and that β is positive for small values of $|\alpha|$. We have at least one nonzero solution which tends to zero as $\alpha \to 0$. Actually, since $\psi_1 = \varphi_1 \not\equiv 0$, we see from (6-59) that ψ is not an even function of α, so that we get two nonzero solutions by taking α positive or negative. Thus as μ increases through the value μ_1, the solution $\psi \equiv 0$ branches into three solutions.

6-8 SOME RESULTS ON NONLINEAR OPERATOR EQUATIONS

The preceding sections have dealt with the classical theory of nonlinear integral equations. We shall now indicate very briefly some of the generalizations that are available in the contemporary mathematical literature. Most of these have to do with nonlinear operator equations in Banach spaces. One advantage of such a generalization is that systems of nonlinear integral equations can often be included in the theory. Another advantage is that the same theory may cover integral equations of various forms.

The idea of constructing a related variational problem which permits the use of direct variational methods was used by Hammerstein (see

Sec. 6-6). This idea has been modified and generalized by a number of mathematicians, and has resulted in the theory of potential operators (see Chap. 1), i.e., operators which can be obtained by taking the Gateaux or Fréchet differential of a "scalar" functional. This theory is described in some detail by Vainberg.[25] Briefer accounts are given in a survey article by Rall[17] and in chap. 6 of the book by Krasnosel'skii.[9]

Another more recent approach to the solution of nonlinear operator equations in Hilbert space (Zarantonello,[29] Minty,[14] Browder[2]) is based on the idea of a monotone operator, as defined in Chap. 1. We recall that a mapping F of a Hilbert space H into itself is called monotone if

$$\text{Re}\ (Fx - Fy, x - y) \geqq 0 \tag{6-62}$$

for all x and y in H. Zarantonello[29] used this concept together with the method of contractive averaging to prove the existence of solutions to nonlinear integral equations encountered in problems of cavity flows. To illustrate the method, let F be monotone and satisfy a Lipschitz condition

$$\|Fx - Fy\| \leqq \beta\|x - y\| \tag{6-63}$$

for all x and y in H. If $\beta < 1$, then of course we can solve the equation $x + Fx = z$ by successive approximations. However, if $\beta \geq 1$, we cannot apply the contraction-mapping method directly, and we proceed as follows. Let $0 < \alpha < 1$, and observe that the equation $x + Fx = z$ is equivalent to

$$x = (1 - \alpha)x - \alpha Fx + \alpha z \tag{6-64}$$

The idea is to choose the parameter α so that the operator

$$G = (1 - \alpha)I - \alpha F$$

has a Lipschitz constant less than one, permitting the solution of (6-64) by successive approximations. We have

$$Gx - Gy = (1 - \alpha)(x - y) - \alpha(Fx - Fy)$$
$$\alpha(Fx - Fy) = (1 - \alpha)(x - y) - (Gx - Gy)$$

Thus, from (6-63),

$$\begin{aligned}
\alpha^2\beta^2\|x - y\|^2 \geqq \alpha^2\|Fx - Fy\|^2 &= \|Gx - Gy - (1 - \alpha)(x - y)\|^2 \\
&= \|Gx - Gy\|^2 - 2\ \text{Re}\ (Gx - Gy, (1 - \alpha)(x - y)) \\
&\qquad\qquad\qquad\qquad + (1 - \alpha)^2\|x - y\|^2 \\
&= \|Gx - Gy\|^2 + 2\alpha(1 - \alpha)\ \text{Re}\ (Fx - Fy, x - y) \\
&\qquad\qquad\qquad\qquad - (1 - \alpha)^2\|x - y\|^2
\end{aligned}$$

Hence, by condition (6-62) we have

$$\|Gx - Gy\|^2 \leq [\alpha^2\beta^2 + (1 - \alpha)^2]\|x - y\|^2$$

so that G has the Lipschitz constant $\gamma = \sqrt{\alpha^2\beta^2 + (1 - \alpha)^2}$. If we choose $\alpha < 2/(1 + \beta^2)$, it follows that $\gamma < 1$, so that Eq. (6-64) is solvable by the contraction-mapping principle.

Exercise 6-19 Prove that the least value of γ obtainable by this method is $\gamma = \beta(1 + \beta^2)^{-\frac{1}{2}}$.

The approach of Zarantonello has been generalized by Kolodner,[8] who considers operator equations of the form $x = KFx + y$, where K is a linear bounded operator and F a lipschitzian operator on a Hilbert space. If λ^{-1} is in the resolvent set of K, we subtract λKx from both sides of the equation and operate on the result with $(I - \lambda K)^{-1}$ to get

$$x = (I - \lambda K)^{-1}K(F - \lambda I)x + (I - \lambda K)^{-1}y$$

Then even though KF does not have its Lipschitz constant less than one, it may be possible to choose the value of the parameter λ so that the operator appearing in the first term of the right member does have this desired property. Kolodner investigates conditions on the linear operator K and on the nonlinear F to ensure the success of the method. When $K = \pm I$, his results reduce to Zarantonello's.

The basic results of Minty are discussed in the survey by Dolph and Minty in Ref. 1 and in Chap. 1. Minty[14] showed that if F is defined on all of H and is monotone and continuous, then the equation $x + Fx = z$ has a unique continuous solution. This result is clearly more general than Zarantonello's; moreover, the methods of proof are completely different. A simplified version of the proof of the above theorem, together with an application to nonlinear boundary-value problems for elliptic partial differential equations, is given by Browder.[2]

The analytical theory of branching, which was illustrated in Sec. 6-7, has been developed by various authors, particularly by Russian mathematicians. An account of this is given in the survey article by Vainberg and Trenogin,[26] which includes an interesting case of singular nonlinear integral equations. Variational methods for problems of branching are given in the book by Vainberg[25] mentioned previously. Topological methods are taken up in the books by Cronin[5] and Krasnosel'skii.[9] One of the simpler of the geometrical or topological approaches to the solution of equations involving a parameter is the *method of cones*, which will now be indicated. Details of this theory may be found in chap. 5 of the book by Krasnosel'skii.[9] A brief account is also given in a survey article by Bueckner in Ref. 1.

Let E denote a real Banach space. Recall that a set K in E will be called a *cone* if the following conditions are satisfied:

1. K is closed.
2. If x and y are in K, so is $\lambda x + \mu y$ for all $\lambda \geq 0$ and $\mu \geq 0$.
3. If x and $-x$ both belong to K, then $x = 0$.

Examples are the set of all points in Euclidean space R^n whose coordinates are positive and the set of all nonnegative functions in the space $C[a,b]$ of continuous functions on $[a,b]$. Every such cone in a Banach space E gives rise to a *partial ordering* in E. We write $x \geq y$ if $x - y \, \epsilon \, K$. Elements of K are called *positive*, and an operator defined in E will be called *positive* (with respect to the cone K) if it takes K into itself. As an example, the nonlinear operator $Ty = \int_a^b g(s,t,y(t)) \, dt$, where $g(s,t,u)$ is continuous for $a \leq s, \, t \leq b$ and all real u, is a positive operator with respect to the cone of nonnegative functions in $C[a,b]$ providing that the function $g(s,t,u)$ is nonnegative everywhere.

Let B_r denote the closed ball with center at the origin and radius r. We put $K_r = K \cap B_r$.

Lemma: Let T be a transformation which maps K_r into K, and let T be completely continuous. Then if

$$\inf \, \{\|Tx\| : x \, \epsilon \, K_r\} > 0$$

there exists an $x_0 \, \epsilon \, K$ with $\|x_0\| = r$ and a positive number λ such that $Tx_0 = \lambda x_0$.

Proof: Define the operator S by $Sx = rTx/\|Tx\|$ for $x \, \epsilon \, K_r$. Then S takes the convex set K_r into itself and is completely continuous. Hence, by the Schauder fixed-point theorem, there exists an element $x_0 \, \epsilon \, K_r$ such that $Sx_0 = x_0$. Clearly $\|x_0\| = r$, and $Tx_0 = \lambda x_0$, where $\lambda = r^{-1}\|Tx_0\|$.

This lemma as it stands is of limited usefulness, and cannot be used when $T(0) = 0$, which case often occurs in applications. However, the lemma can be used to obtain a simple proof of the following theorem, which was proved by Rothe,[19] when E is a Hilbert space.

***Theorem* 6-13** Let F be a completely continuous operator which is positive with respect to the cone K, and let F satisfy the inequality

$$\inf \, \{\|Fx\| : x \, \epsilon \, K \qquad \|x\| = r\} > 0$$

for some positive number r.

Then there exists an element x_0 in K with $\|x_0\| = r$ and a positive number λ such that $Fx_0 = \lambda x_0$.

The proof is carried out by applying the lemma to the operator T which is defined on K_r by the formula

$$Tx = \|x\|F\left(\frac{r}{\|x\|}x\right) + (r - \|x\|)y$$

where y is a fixed nonzero element of the cone K.

This theorem could be applied, for example, to the nonlinear integral equation

$$\int_0^1 k(s,t)e^{x(t)} \, dt = \lambda x(t)$$

providing that the kernel $k(s,t)$ is continuous and has a *positive* lower bound on the square $0 \le s,\ t \le 1$, by taking $E = C[0,1]$. It is still too restrictive for many applications, however.

We shall indicate another result, which is a special case of theorem 2.4, page 268, of Krasnosel'skii.[9] In order to state this, it is necessary to introduce the concept of a *monotonic minorant*. An operator A on E into E is called *monotonic* (with respect to the cone K) if $x \le y$ implies that $Ax \le Ay$. An operator A is called a *minorant* of an operator T on K_r if $Tx \ge Ax$ for all x in K_r.

***Theorem* 6-14** For some fixed $r > 0$ let T be completely continuous and map K_r into K, let $y \in K$ be fixed, and let A be a monotonic minorant of T on K_r which has the property that $\beta A(ty) \ge ty$ for $0 < t < \gamma$, where β is some positive number and γ is such that $x - \gamma y$ is not in K for any x in K_r. Then for any given r_0, with $0 < r_0 < r$, there exists x_0 in K with $\|x_0\| = r_0$ and a positive number λ such that $Tx_0 = \lambda x_0$.

As an illustration of the last theorem, we consider the problem of the bending of a rod given as an example in Sec. 6-2. The integral equation of this example can be written as $Fx = \nu x$, where

$$Fx = \int_0^1 K(s,\sigma)x(\sigma) \, d\sigma \sqrt{1 - \left[\int_0^1 K_s(s,\tau)x(\tau) \, d\tau\right]^2}$$
$$\nu = \frac{EI}{P}$$

From the definition of $K_s(s,\sigma)$ it is obvious that $|K_s(s,\sigma)| \le 1$ for $0 < s,\ \sigma \le 1$. Hence we may take the Banach space E to be the space $C[0,1]$, and the operator F will be defined in the ball $\|x\| \le r$, where $0 < r < 1$. The cone Π of nonnegative functions in $C[0,1]$ is a natural one to use, since it seems intuitively evident that the curve of deflections y should be concave downward, so that $x(s) = -y''(s)$ should be positive. By the definition (6-8) of the kernel, we find that $K(s,\sigma)$ has nonnegative values, and since

$$\sqrt{1 - \left[\int_0^1 K_s(s,\tau)x(\tau) \, d\tau\right]^2} \ge \sqrt{1 - r^2}$$

when $\|x\| \le r < 1$, it follows that F maps the "sector" $\Pi_r = \Pi \cap S_r$ into the cone Π, where, as usual, S_r denotes the closed ball $\|x\| \le r$. In order to obtain a monotonic minorant of F, we define

$$Ax = \sqrt{1 - r^2} \int_0^1 K(s,\sigma)x(\sigma) \, d\sigma$$

Then $Fx \geqq Ax$ on Π_r, so that A is a minorant of F. Since A is a linear operator which is positive with respect to the cone Π, it follows that A is monotonic with respect to Π. To show that A satisfies the remaining conditions of Theorem 6-14, we put $z(s) = \sin \pi s$, which is the eigenfunction of the kernel $K(s,\sigma)$ corresponding to the smallest reciprocal eigenvalue $\mu_1 = \pi^2$. Clearly z belongs to Π. If we put $\beta = \pi^2/\sqrt{1 - r^2}$ and take $\gamma > r$, then for any real value of t we have

$$\beta A(tz) = \pi^2 t \int_0^1 K(s,\sigma) \sin \pi \sigma \, d\sigma = t \sin \pi s = tz$$

and if x is any element of Π_r, then

$$x(\tfrac{1}{2}) - \gamma z(\tfrac{1}{2}) = x(\tfrac{1}{2}) - \gamma \leq r - \gamma < 0$$

Thus $x - \gamma z$ does not lie in Π for any x in Π_r, and the hypotheses of Theorem 6-14 concerning the operator A are satisfied, with z replacing y. The only remaining hypothesis of the theorem to be verified is that the operator F is completely continuous, and this is left to the reader.

Exercise 6-20 Show that F is completely continuous in the ball S_r, where $0 < r < 1$.

We conclude that for each positive number $r_0 < r$ there exists a positive number ν and a nonnegative continuous function $x_0(s)$ such that $\|x_0\| = r$ and $Fx_0 = \nu x_0$.

Another problem which can also be handled by the method of cones is Nekrasov's equation, the last example in Sec. 6-2. This was done by Krasovskii,[10] who succeeded in obtaining new results for this old problem by this method.

REFERENCES

1. Anselone, P. M. (ed.): "Nonlinear Integral Equations," The University of Wisconsin Press, Madison, Wis., 1964.

2. Browder, F.: The Solvability of Non-linear Functional Equations, *Duke Math. J.*, vol. 30, pp. 557–566, 1963.

3. Churchill, R. V.: "Operational Mathematics," 2d ed., McGraw-Hill Book Company, New York, 1958.

4. Courant, R., and D. Hilbert: "Methods of Mathematical Physics," vol. 1, Interscience Publishers, Inc., New York, 1953.

5. Cronin, J.: "Fixed Points and Topological Degree in Nonlinear Analysis," American Mathematical Society, Providence, R.I., 1964.

6. Fredholm, I.: Sur une classe des équations fonctionelles, *Acta Math.*, vol. 27, pp. 365–390, 1903.

7. Hammerstein, A.: Nichtlineare Integralgleichungen nebst Anwendungen, *Acta Math.*, vol. 54, pp. 117–176, 1930.

8.   Kolodner, I.: Contractive Methods for Hammerstein Type Equations on Hilbert Spaces, *Dept. Math. Univ. New Mexico Tech. Rep.* 35, 1963; see also *Tech. Rep.* 47.

9.   Krasnosel'skii, M. A.: "Topological Methods in the Theory of Nonlinear Integral Equations," The Macmillan Company, New York, 1964.

9b.   ———: "Positive Solutions of Operator Equations," P. Noordhof, Ltd., 1964.

10.   Krasovskii, Yu. P.: The Theory of Steady-state Waves of Large Amplitude, *Soviet Physics Dokl.*, vol. 5, pp. 62–65, 1960.

11.   Langer, R. E. (ed.): "Boundary Problems in Differential Equations," The University of Wisconsin Press, Madison, Wis., 1960.

12.   Levinson, N.: A Nonlinear Volterra Equation Arising in the Theory of Superfluidity, *J. Math. Anal. Appl.*, vol. 1, pp. 1–11, 1960.

13.   Mann, W. R., and F. Wolf: Heat Transfer between Solids and Gases under Nonlinear Boundary Conditions, *Quart. J. Appl. Math.*, vol. 9, pp. 163–184, 1951.

14.   Minty, G. J.: Monotone (Non-linear) Operators in a Hilbert Space, *Duke Math. J.*, vol. 29, pp. 341–346, 1962.

15.   Padmaavally, K.: On a Nonlinear Integral Equation, *J. Math. Mech.*, vol. 7, pp. 533–555, 1958.

16.   Petrovskii, I. G.: "Lectures on the Theory of Integral Equations," Graylock Press, Albany, N.Y., 1957.

17.   Rall, L.: Numerical Integration and the Solution of Integral Equations by the Use of Riemann Sums, *SIAM Rev.*, vol. 7, pp. 55–64, 1965.

18.   Roberts, I. H., and W. R. Mann: A Non-linear Integral Equation of Volterra Type., *Pacific J. Math.*, vol. 1, pp. 431–445, 1951.

19.   Rothe, E. H.: *Amer. J. Math.*, vol. 66, pp. 245–254, 1944.

20.   Sato, T.: Sur l'équation intégrale non lineaire de Volterra, *Compositio Math.*, vol. 11, pp. 271–290, 1953.

21.   Schmidt, E.: Zur Theorie der linearen und nichtlinearen Integralgleichungen, III, *Math. Ann.*, vol. 65, pp. 370–399, 1908.

22.   Stoker, I. I.: "Water Waves," Interscience Publishers, Inc., New York, 1957.

23.   Taylor, A. E.: "Introduction to Functional Analysis," John Wiley & Sons, Inc., New York, 1958.

24.   Tricomi, F. G.: "Integral Equations," Interscience Publishers, Inc., New York, 1957.

25.   Vainberg, M. M.: "Variational Methods for the Study of Nonlinear Operators," Holden-Day, Inc., San Francisco, Calif., 1964.

26.   ——— and V. A. Trenogin: The Methods of Lyapunov and Schmidt in the Theory of Non-linear Equations and Their Further Development, *Russian Math. Surveys*, vol. 17, no. 2, pp. 1–60, 1962.

27.   Walter, W.: "Differential- und Integral-Ungleichungen," Springer-Verlag OHG, Berlin, 1964.

28.   Weyl, H.: On the Differential Equations of the Simplest Boundary-layer Problems, *Ann. of Math.*, vol. 43, pp. 381–407, 1942.

29.   Zarantonello, E. H.: Solving Functional Equations by Contractive Averaging, *Univ. Wisconsin Math. Res. Center Tech. Rep.* 160, 1960.

30.   ———: The Closure of the Numerical Range Contains the Spectrum, *Bull. Amer. Math. Soc.*, vol. 70, pp. 781–787, 1964.

SEVEN

INTEGRODIFFERENTIAL EQUATIONS

7-1 INTRODUCTION

The complexity of the questions encountered in the study of differential and integral equations is not mitigated in the analysis of integrodifferential equations, a combination of these subjects. In this chapter we give a condensed elementary discussion of examples and some theory of integrodifferential equations and an even briefer account of integrodifference and integrodifferential-difference equations, sometimes called recurrent integrodifferential equations. Although pertinent material of an abstract nature is difficult to present in a field as sparsely studied as this one, it is worthwhile to provide broad familiarity with some concepts occurring in the subject. As there is no adequate classification of these equations, we shall exemplify and list several which have proved to be of interest in research and in applications.

As an elementary illustration of how an integrodifferential equation arises we consider the problem of seeking a maximum or a minimum of a function defined by an integral where both the limits of integration and the integrand are functions of a parameter. The integro-differential equation arises by differentiating this function with respect to the parameter and equating to zero. Let

$$I(\alpha) = \int_{a(\alpha)}^{b(\alpha)} f(x,\alpha) \, dx$$

then the requirement that I be stationary with respect to α gives

$$\frac{dI}{d\alpha} = \int_{a(\alpha)}^{b(\alpha)} \frac{\partial f(x,\alpha)}{\partial \alpha} \, dx + f(b(\alpha),\alpha) \, \frac{db}{d\alpha} - f(a(\alpha),\alpha) \, \frac{da}{d\alpha} = 0$$

which is an integrodifferential equation.

A simple illustration of an integrodifferential equation is the linear equation

$$\frac{df(t)}{dt} = \int_0^t K(s)f(s) \, ds$$

in which $K(s)$, with $K(0) = 0$, is given and we wish to determine the unknown function f. Differentiation yields Hill's equation, which is a linear second-order differential equation with a variable coefficient. Thus we see how a simple integrodifferential equation is transformed into a differential equation with substantial stature. On the other hand, integration yields a Volterra-type integral equation. Thus, in addition to direct treatment of some equations, it is sometimes desirable to go from an integrodifferential equation to either a differential or an integral equation.

Exercise 7-1   Effect both foregoing reductions for the above example.

A relatively general form of a functional integrodifferential equation in a single variable x is

$$y^{(n)}(x) = f(x,y(x), \ldots ,y^{(n-1)}(x)\lambda \int_a^b K(x,t,y(t), \ldots ,y^{(m)}(t)) \, dt)$$

For $n \geq m$ this is known to have a unique solution $y(x)$, which satisfies

$$y^{(k)}(a) \equiv y_k \qquad k = 0, \ldots , n - 1$$

An example of a system of nonlinear integrodifferential equations is the Volterra system describing the growth of two conflicting populations, both subject to a factor of heredity

$$\frac{dx}{dt} = ax - bxy - K_1 x \int_0^t y(s) \, ds$$

$$\frac{dy}{dt} = -\alpha y + \beta xy + K_2 y \int_0^t x(s) \, ds$$

If we multiply the first equation by $1/x$ and the second by $1/y$ and differentiate with respect to t, we have the second-order system of nonlinear ordinary differential equations

$$xx'' = x'^2 - by'x^2 - K_1x^2y$$
$$yy'' = y'^2 + \beta x'y^2 + K_2xy^2$$

At $t = c$, $x = x_0$ and $y = y_0$, and we have

$$x' \Big|_{x=x_0} = ax_0 - bx_0y_0 \qquad y' \Big|_{y=y_0} = \alpha y_0 + \beta x_0y_0$$

that is, the derivatives are also given, and the system has a solution.

An example of a homogeneous system of linear integrodifferential equations is

$$L(y) + \int_a^x \sum_{j=1}^m P_j(x - t) \exp [\beta_j(t - x)]M[y(t)] \, dt = 0$$

where L and M are homogeneous linear differential operators with constant coefficients, $P_j(u)$ are polynomials in u, and β_j are constants.

The linear dynamical system is an example of a nonhomogeneous system

$$\sum_{k=1}^r \alpha_{ik}\ddot{x}_k(t) + \beta_{ik}x_k(t) + \gamma_{ik} \int_0^t x_k(\tau) \, d\tau + q_i = f_i(t) \qquad i = 1, \ldots, r$$

An equation is ordinary if all derivatives are taken with respect to the same variable; otherwise it is partial. An example of a partial integrodifferential equation is

$$u_{tt} + a^2u_{xxxx} + \int_0^t \varphi(t - \tau)u_{xxxx}(x,\tau) \, d\tau = F(x,t)$$

where u and u_{xx} vanish when $x = 0$, $x = L$, and u and u_t are prescribed when $x = 0$. Such equations may be parabolic, elliptic, or hyperbolic. The following is a system of partial integrodifferential equations

$$\frac{\partial u}{\partial x} = F(x,y, \int_0^y u(x,t) \, dt, u, v)$$
$$\frac{\partial v}{\partial y} = G(x,y, \int_0^x v(t,y) \, dt, u, v)$$

with the conditions

$$u(0,y) = 0 \qquad 0 \leq y \leq 1$$
$$u(x,0) = 0 \qquad 0 \leq x \leq 1$$

An illustration of an integrodifference equation is given by

$$f(t) - f(t + 1) = \int_{-1}^{1} K(s)f(t - s) \, ds = \int_{t-1}^{t+1} K(t - s)f(s) \, ds$$

$$0 \leq t < \infty$$

where, for example, $K(s)$ is defined for $-1 \leq s \leq 1$ and $f(t)$ is to be determined for $-1 \leq t \leq \infty$. This equation describes the efflux of gas from the open end of a tube.

Finally, an example of an integrodifferential-difference equation is given by the *integral equation of Palm*

$$f_n(t) = f_{n-1}(t) - \int_0^t (1 - e^{-x})f_n(t - x) \, df_{n-1}(x) \qquad n \geq 1, f_0(t) \text{ given}$$

It arises in queueing theory as follows. The infinite system L_1, $L_2, \ldots, L_n, \ldots$ of parallel channels is receiving traffic by a Poisson input. Each call enters at L_1, proceeding sequentially, if L_1 is occupied, along the line of channels until it arrives at an unoccupied channel. By assuming an exponential service time Palm derives an expression for $f_n(t)$, the probability that every call entering L_n during time $(t_0, t_0 + t)$ will be served, given that at moment t_0 a call was forwarded to L_{n+1}. We assume an exponential service time with unit expected value.

Exercise 7-2   Apply Laplace-transform methods to the solution of

$$f_n(t) = f_{n-1}(t) - \int_0^t f_n'(t - x) \, dx \qquad n \geq 1$$

with $f_n(0) = 0$ for all n and solve in terms of f_0.

Some equations known by specific names, many of which are in Ref. 33, are the following:

The integrodifferential equation of *Abel*:

$$\int_0^x \frac{F[x,y,\varphi(y),\varphi'(y)]}{(x - y)^\alpha} \, dy = f(x)$$

The *Darboux-Picard* equation:

$$z(x,y) = \sigma(x) + \tau(y) - z_0 + \int_0^x \int_0^y f(u,v,z(x,v),z_x(u,v),z_y(u,v)) \, du \, dv$$

Prandtl's circulation equation:

$$cy(x) + \frac{1}{\pi} \int_{-1}^{1} \frac{dy/d\xi}{\xi - x} \, d\xi = f(x)$$

where the Cauchy principal value of the integral is used.

The *neutron-transport* equation:

$$\mu \, \frac{\partial \varphi(x,\mu)}{\partial x} + \sigma\varphi(x,\mu) = \sigma \, \frac{c}{2} \int_{-1}^{1} \varphi(x,\mu) \, d\mu + s(x,\mu)$$

The equation of *radiative transfer:*

$$\mu \, \frac{\partial \varphi(r,\mu)}{\partial r} + \frac{1 - \mu^2}{r} \, \frac{\partial \varphi(r,\mu)}{\partial \mu} + \varphi(r,\mu) - \frac{1}{2} \int_{-1}^{1} \varphi(r,\mu) \, d\mu = 0$$

Czerny's integrodifferential equation for the temperature in melted glass under heat conduction and heat radiation:

$$k \int_{-a}^{a} T^4(\xi) E_i(-k|x - \xi|) \, d\xi + 2T^4(x) - \lambda T''(x) = T_i{}^4 K(-k(a + x))$$
$$+ \, T_r{}^4 K(-k(a - x))$$

The equation of *Boltzmann* (see Sec. 7-5 for details):

$$Df = J(ff_1)$$

where $J(ff_1)$ is the nonlinear operator of Boltzmann.

The *general retardation* system (in vector notation):

$$\dot{x}(t) = \int_{-\infty}^{0} x(t + s) d_s \eta(t,s) + f(t)$$

The nonlinear integrodifferential equation of *heat transfer* by conduction and radiation in the unknowns E_g and $T(x)$:

$$E_g \, \frac{1}{2\pi n^2} = \int_{0}^{\infty} \int_{0}^{h} k(\lambda) \, \text{sign} \, (x - t) K_1(k(\lambda)|x - t|) E(\lambda, T(t)) \, dt \, d\lambda$$
$$- \int_{0}^{\infty} [K_2(k(\lambda)x) E(\lambda, T_0) - K_2(k(\lambda)(h - x)) E(\lambda, T_h)] \, d\lambda - \frac{\kappa}{2\pi n^2} \, T'(x)$$

Among the early books on integrodifferential equations are those of Volterra[37] and Lichtenstein.[25]. Among recent books on integral equations chap. 17 of the book by Muskelishvili[27] is concerned with the solution of a well-known integrodifferential equation which arises in the theory of aircraft wings of finite span; Ref. 33, in which several papers attempt the numerical treatment of specific examples, has a useful list of references.

7-2   EXAMPLES

Integrodifferential equations are little known, even when they are linear. Therefore we illustrate the derivation and solution of some linear equations and go on to the study of a nonlinear problem in Sec. 7-5.

THE THEORY OF QUEUES

Assume that arrivals at a line in which customers wait to be served on a first-come first-served basis (called *ordered service*) at a counter, e.g., theater box office, occur at random by a Poisson distribution (see Chap. 4), and let the service-time distribution, assumed to be the same for all customers, be arbitrary. Note that $W(t)$, the waiting time in the queue at time t, jumps upward discontinuously at the arrival of a new customer (with service time greater than zero) by an amount equal to the service time of the new arrival; otherwise $W(t)$ approaches zero with a slope equal to minus unity. We have a Markoff process (a process in which the present state depends only on the one immediately preceding it) with a continuous parameter t. The example requires minimal knowledge of probability concepts and should not be bypassed by the diligent reader. See Ref. 34.

Let $P(w,t)$ be the probability of a customer's waiting in a queue an interval of time $W(t) \leq w$, given that the arrival occurs at time t.

The following argument pictures a large number N of single-channel queues with identical input and service distributions operating simultaneously. We wish to derive an expression for $P(w, t + \Delta t)$ in terms of $P(w,t)$ and $P(w + \Delta t, t)$. At time t one can divide these waiting lines into two sets:

1. Those for which the waiting time is $W(t) \leq w$. This number is $NP(w,t)$; that is, one multiplies N by the fraction which waits for a time $W(t) \leq w$.

2. Those for which the waiting time is $W(t) > w$. Their number is $N - NP(w,t)$.

At time $t + \Delta t$ the number in the first set becomes $NP(w + \Delta t, t)$ [there is no loss in generality in not writing $P(w + \Delta t, t + \Delta t)$] minus the number of queues whose waiting time has changed from $W(t) \leq w$ to $W(t) > w$ because of arrivals during Δt. We now obtain the latter quantity.

We first take the number of queues whose waiting time lies between x and $x + dx$ at time t. Since $P(x,t)$ is a cumulative distribution, we obtain the density function for $x > 0$ by differentiation. This number of queues is given by $N[\partial P(x,t)/\partial x] \, dx$ if $x > 0$ or by $NP(0,t)$ if $x = 0$.

Now the waiting time is made to exceed w during $(t, t + \Delta t)$ if there is an arrival during Δt and if the service time y of this arrival, when added to x, exceeds w, that is, $x + y > w$, from which $y > w - x$. We must therefore multiply this number of queues whose waiting time is x by the probability of an arrival during Δt, that is, $\lambda \, \Delta t$, and by the probability that the service time of the arrival exceeds $w - x$. If we assume that the density function of the service time y is $b(y)$, the probability of the last event is $\int_{w-x}^{\infty} b(y) \, dy$.

Therefore, for a fixed value $w > 0$ of waiting time, the number of waiting lines in the first case which change to the second set is given by

$$\left[N\lambda \, \Delta t \, \frac{\partial P(x,t)}{\partial x} \, dx \right] \left[\int_{w-x}^{\infty} b(y) \, dy \right] = N\lambda \, \Delta t \, \frac{\partial P(x,t)}{\partial x} \, dx \, B_c(w - x) \quad (7\text{-}1)$$

which must be summed over all $0 < x \leq w$. Here

$$B_c(y) \equiv 1 - \int_0^y b(u) \, du \equiv 1 - B(y)$$

In case $x = 0$, the number of waiting lines which change to the second set, i.e., wait $W(t) > w$, is given by

$$N\lambda \, \Delta t \, P(0,t) \int_w^{\infty} b(y) \, dy = N\lambda \, \Delta t \, P(0,t) B_c(w) \quad (7\text{-}2)$$

Hence

$$NP(w, t + \Delta t) = NP(w + \Delta t, t) - N\lambda \, \Delta t \int_{0+}^{w} \frac{\partial P(x,t)}{\partial x} B_c(w - x) \, dx$$
$$- N\lambda \, \Delta t \, P(0,t) B_c(w) \quad (7\text{-}3)$$

The fact that we have considered a large number of queues N has served its purpose of yielding an equation. We now divide through by N. Subtracting $P(w,t)$ from both sides, dividing by Δt, and passing to the limit yield the integrodifferential equation

$$\frac{\partial P(w,t)}{\partial t} = \frac{\partial P(w,t)}{\partial w} - \lambda \int_{0+}^{w} \frac{\partial P(x,t)}{\partial x} B_c(w - x) \, dx - \lambda P(0,t) B_c(w) \quad (7\text{-}4)$$

We have used the expansion for small $\Delta t > 0$

$$P(w + \Delta t, t) = P(w,t) + \frac{\partial P(w,t)}{\partial w} \Delta t + o(\Delta t)$$

The following conditions must also be satisfied: $P(w,0) = 1$ for all w, and $P(\infty, t) = 1$ for all t.

Integrating by parts gives

$$\frac{\partial P(w,t)}{\partial t} = \frac{\partial P(w,t)}{\partial w} - \lambda P(w,t) + \lambda \int_0^w P(w - v, t) \, dB_c(v) \quad (7\text{-}5)$$

Note that, if in (7-4) we write $B(t) = 1 - B_c(t)$ and take the lower limit from zero instead of zero plus (thus including the value of the integral at the origin), we have

$$\frac{\partial P(w,t)}{\partial t} = \frac{\partial P(w,t)}{\partial w} - \lambda P(w,t) + \lambda \int_0^w B(w - x) \, d_x P(x,t) \quad (7\text{-}6)$$

This integrodifferential equation, due to Takács, is also valid if λ is replaced by $\lambda(t)$.

On taking the Laplace transform of (7-4) with respect to w, when it exists (noting the presence of a convolution), we have

$$\frac{\partial P^*(s,t)}{\partial t} = sP^*(s,t) - P(0,t) - \lambda[sP^*(s,t) - P(0,t)]B_c^*(s) - \lambda P(0,t)B_c^*(s)$$

$$= sP^*(s,t)[1 - \lambda B_c^*(s)] - P(0,t)$$

where
$$P^*(s,t) = \int_0^\infty e^{-sw}P(w,t)\,dw$$

$$\int_0^\infty e^{-sw}\frac{\partial P(w,t)}{\partial w}\,dw = -P(0,t) + sP^*(s,t)$$

The steady-state, i.e., time-independent, result is obtained by setting the left side equal to zero, as the probabilities no longer change in time as $t \to \infty$, suppressing t, and solving for $P^*(s)$. This yields

$$sP^*(s) = \frac{P(0)}{1 - \lambda B_c^*(s)} = \frac{P(0)}{1 - \lambda[(1/s) - (b^*(s)/s)]} \tag{7-7}$$

But $\lim\limits_{w\to\infty} P(w) = 1$, and from Laplace-transform theory

$$\lim_{w\to\infty} P(w) = \lim_{s\to0} sP^*(s)$$

Since $\lim\limits_{s\to0} sP^*(s) = 1$, we now have

$$P(0) = \lim_{s\to0} \frac{s - \lambda[1 - b^*(s)]}{s}$$

This is an indeterminate form to which we must apply L'Hospital's rule. Recall that

$$\lim_{s\to0} b^*(s) = \lim_{s\to0} \int_0^\infty e^{-sy}b(y)\,dy = \int_0^\infty b(y)\,dy = 1$$

Thus, differentiating numerator and denominator separately with respect to s gives

$$P(0) = 1 - \frac{\lambda}{\mu} \equiv 1 - \rho$$

where
$$\frac{db^*(s)}{ds} = \int_0^\infty \frac{d}{ds}e^{-sy}b(y)\,dy = -\int_0^\infty ye^{-sy}b(y)\,dy$$

and, as $s \to 0$, this gives the mean of $b(y)$, which we denote by $1/\mu$. We finally have

$$sP^*(s) = \frac{1 - \rho}{1 - \lambda\{[1 - b^*(s)]/s\}} \tag{7-8}$$

On taking the inverse transform, this yields $P(w)$, giving the probability of not waiting in the queue longer than w in the steady state.

Example: Let $b(y) = \mu e^{-\mu y}$. Then

$$B_c(y) = 1 - \int_0^y \mu e^{-\mu t}\, dt = e^{-\mu y}$$

$$B_c^*(s) = \int_0^\infty e^{-sy} e^{-\mu y}\, dy = \frac{1}{s + \mu}$$

Therefore, from (7-7),

$$P^*(s) = P(0)\left[\frac{1}{s + \mu - \lambda} + \frac{\mu}{s(s + \mu - \lambda)}\right]$$

On observing that the second term is the Laplace transform of a convolution of an exponential and unity, we have

$$P(w) = P(0)\left[e^{-(\mu-\lambda)w} + \mu \int_0^w e^{-(\mu-\lambda)t}\, dt\right]$$

It is easy to verify that $P(0) = 1 - \lambda/\mu$, since $\lim_{w \to \infty} P(w) = 1$. Also,

$$W_q \equiv \int_0^\infty w\, \frac{dP(w)}{dw}\, dw = \frac{\lambda/\mu}{\mu(1 - \lambda/\mu)}$$

as can be seen by calculating dP/dw and integrating. This is a well-known result for the expected waiting time in the queue.

In case the density function $b(x)$ does not exist, we can use the Laplace-Stieltjes transform, that is,

$$\gamma(s,t) = \int_0^\infty e^{-sw}\, dP(w,t)$$

$$\beta(s) = \int_0^\infty e^{-sx}\, dB(x)$$

where $B(x)$ is the distribution of the cumulative service time.

It is clear from the definitions of this transform and the Laplace transform that, as $t \to \infty$, $sP^*(s) = \gamma(s)$, and $b^*(s) = \beta(s)$ when $b(t)$ exists. Thus we must have

$$\gamma(s) = \frac{1 - \rho}{1 - \lambda\{[1 - \beta(s)]/s\}} \tag{7-9}$$

For the time-dependent case we have without difficulty

$$\frac{\partial \gamma(s,t)}{\partial t} = \gamma(s,t)[s - \lambda + \lambda\beta(s)] - sP(0,t) \tag{7-10}$$

The solution of this linear differential equation with $\lambda(t)$ instead of λ and

$$\Lambda(t) = \int_0^t \lambda(u)\, du$$

is

$$\gamma(s,t) = \exp\{st - [1 - \beta(s)]\Lambda(t)\}$$
$$\cdot \left(1 - s\int_0^t \exp\{-sx + [1 - \beta(s)]\Lambda(x)\}P(0,x)\, dx\right) \tag{7-11}$$

which is actually the Laplace-Stieltjes transform of the original differential equation.

It can be shown that, if $1/\mu$ is the mean service time, if $\lim_{t \to \infty} \lambda(t) = \lambda$, and if $\lambda/\mu < 1$, then $\lim_{t \to \infty} P(w,t) = P(w)$ exists, is independent of the initial distribution $P(w,0)$, and is uniquely determined by its Laplace-Stieltjes transform (7-9), thus again providing a solution of the problem. If $\lambda/\mu \geq 1$, then $P(w)$ does not exist, and $\lim_{t \to \infty} P(w,t) = 0$ for all w.

Instead of using transforms, a numerical method of calculation has been suggested by Descamps. In the steady state, one can write

$$p(w) = \frac{dB(w)}{dw}$$

and obtain

$$p(w) - \lambda \int_0^w p(x)B_c(w - x)\, dx = \lambda \left(1 - \frac{\lambda}{\mu}\right) B_c(w) \qquad (7\text{-}12)$$

When $w \to 0$, we have

$$p(0) = \lambda \left(1 - \frac{\lambda}{\mu}\right)$$

By writing $q(x) = p(x)B_c(w - x) = p(x)B_c(n\Delta - x)$ one attempts to evaluate the above Volterra-type integral equation by breaking the interval of integration into n small subintervals, each of length Δ, and applying the trapezoidal method of approximation, that is,

$$\int_0^{n\Delta} q(x)\, dx = \Delta \left\{ \frac{q(0)}{2} + q(\Delta) + q(2\Delta) + \cdots + q[(n - 1)\Delta] + \frac{q(n\Delta)}{2} \right\}$$

Since $B_c(w)$ is known, one can use the equation in $p(w)$ to derive, step by step, $p(0)$, $p(\Delta)$, $p(2\Delta)$, etc.

Gnedenko's Generalization of Waiting Times—Another Integrodifferential Equation

Gnedenko has given a general formulation of the waiting-time problem. A customer arriving at a single-channel queue with ordered discipline may quit before joining, or he may leave after joining, or he may wait for service and then leave before its termination, or he may stay to finish service. Under this regime, let $D(w)$ be the distribution function for waiting in line an interval of time of length w, and let $G_x(w)$ be the conditional probability distribution that having waited in the line for a time x the customer will remain in service for a time of length w (that is, without finishing service, and thus he waits an additional amount w), $G_x(+0) = 0$. Let $B(y)$ be the service-time distribution function.

Let $W(t)$ be the stochastic variable indicating the waiting time to the beginning of service of a customer arriving at time t. Let

$$P(w,t) = \text{Prob } [W(t) \leq w]$$

Suppose that arrivals occur by a Poisson distribution with parameter λ. We consider the conditions for $W(t + \Delta t) \leq w$ to hold. They are:

1. $W(t) < w + \Delta t$, and no arrival occurs during $(t, t + \Delta t)$.

2. $W(t) = x < w + \Delta t$; there is an arrival at time $t + \epsilon$, $0 < \epsilon < \Delta t$, who leaves before the beginning of service; i.e., he waits for a time less than $x - \epsilon$. (This is governed by D.)

3. The same as in 2 except that the arrival stays for partial service and leaves before time in service $w - x + \Delta t$. Note that before service he waited longer than $x - \epsilon$. (This is governed by D and G.)

4. The same as in 3 except that the arrival remains to complete service, and hence he stays in service for longer than time $w - x + \Delta t$. Clearly he waits for service longer than $x - \epsilon$. (This is governed by D, G, and B.)

Using the laws of probability, we have

$$P(w, t + \Delta t) = (1 - \lambda \Delta t)P(w + \Delta t, t)$$
$$+ \int_0^t d\epsilon \int_0^{w + \Delta t} \{D(x - \epsilon) + [1 - D(x - \epsilon)]G_{x-\epsilon}(w - x + \Delta t)$$
$$+ [1 - D(x - \epsilon)][1 - G_{x-\epsilon}(w - x + \Delta t)]B(w - x + \Delta t)\} \, d_x P(x,t) + o(\Delta t) \quad (7\text{-}13)$$

In the usual way, by simplifying and dividing by Δt, and then allowing $\Delta t \to 0$, we have

$$\frac{\partial P(w,t)}{\partial t} = \frac{\partial P(w,t)}{\partial w} - \lambda P(w,t) + \lambda \int_0^w \{D(x) + [1 - D(x)]G_x(w - x)$$
$$+ [1 - D(x)][1 - G_x(w - x)]B(w - x)\} \, d_x P(x,t) \quad (7\text{-}14)$$

of which (7-6) is a special case.

As $t \to \infty$, this equation yields the corresponding equation for the stationary distribution $P(w)$, which can be shown to exist. It is obtained by equating the left side to zero and suppressing t as the argument on the right. The resulting equation can be simplified to yield

$$\frac{dP(w)}{dw} = \lambda \int_0^w [1 - D(x)][1 - G_x(w - x)][1 - B(w - x)] \, dP(x) \quad (7\text{-}15)$$

which has a unique solution with a discontinuity at $w = 0$ and is absolutely continuous for $w \neq 0$.

We now introduce some useful functions.

1. The distribution function of the effective waiting time is

$$P_1(w) = P(w) + [1 - P(w)]D(w) \quad (7\text{-}16)$$

obtained as the joint occurrence of the two events of waiting in line and leaving before service.

2. The distribution function of the total time in the system is

$$P_2(w) = P(w) - \frac{P'(w)}{\lambda} + [1 - P(w)] \, D(w) \quad (7\text{-}17)$$

3. The distribution function of the ratio of effective service time to the time necessary to complete service is

$$P_3(w) = 1 - \int_0^\infty \int_0^\infty [1 - G_x(wy)] \, dP(x) \, dB(y) \qquad 0 < w < 1 \quad (7\text{-}18)$$

4. The distribution function for the complete satisfaction of the demand for service, i.e., customer remains to complete service, is given by

$$P_4(w) = \int_0^\infty [1 - G_x(w)] \, dP(x) \tag{7-19}$$

5. The pure-loss probability is

$$\alpha_1 = \int_0^\infty D(x) \, dP(x) \tag{7-20}$$

6. The probability that service will be partially fulfilled is

$$\alpha_2 = \int_0^\infty [1 - D(x)] \int_0^\infty G_x(y) \, dB(y) \, dP(x) \tag{7-21}$$

Example: For lost calls the waiting density function is a Dirac delta function, and hence

$$D(w) = \begin{cases} 0 & w \le 0 \\ 1 & w > 0 \end{cases}$$

For a fixed wait in line of length τ, after which a customer reneges, i.e., leaves the line (one translates the origin in the lost-call case to τ), we have

$$D(w) = \begin{cases} 0 & w \le \tau \\ 1 & w > \tau \end{cases} \qquad G_x(u) \equiv 0$$

If τ is a random variable, then $G_x(w) = 0$; if the wait in the system is bounded by τ, then $G_x(w) = 1$ for $w + x > \tau$; and if τ is a stochastic variable, then

$$G_x(w) = \frac{D(w + x) - D(x)}{1 - D(x)} \tag{7-22}$$

For exponential service times with parameter μ for the case of leaving the waiting line (note that if a customer enters service, he remains to complete it) one can show that

$$P(w) = \begin{cases} \dfrac{P(0)}{1 - \rho} (1 - \rho\epsilon^{-(\mu-\lambda)w}) & \text{for } w < \tau \\ 1 - \rho P(0) e^{\lambda\tau - \mu w} & \text{for } w > \tau \end{cases} \tag{7-23}$$

where $\rho = \lambda/\mu$. Also

$$P_1(w) = \begin{cases} \dfrac{1 - \rho e^{-(\mu-\lambda)w}}{1 - \rho^2 e^{-(\mu-\lambda)\tau}} & \text{for } w < \tau \\ 1 & \text{for } w > \tau \end{cases} \tag{7-24}$$

$$\alpha_1 = \frac{\rho(1 - \rho)}{e^{(\mu-\lambda)\tau} - \rho^2} \tag{7-25}$$

which is the saltus of $P(w)$ at τ. In the service system there are no partial losses, i.e., once a unit enters service it completes service; hence, $\alpha_2 = 0$.

$$P_2(w) = \begin{cases} \dfrac{1 - e^{-(\mu-\lambda)w}}{1 - \rho^2 e^{-(\mu-\lambda)\tau}} & 0 < w > \tau \\ 1 - \rho \left[1 - \dfrac{P(0)}{1 - \rho} (1 - \rho e^{-(\mu-\lambda)\tau}) \right] e^{-\mu(w-\tau)} & w > \tau \end{cases} \tag{7-26}$$

FLUCTUATION OF STAR BRIGHTNESS[8]

We wish to find the probability distribution of the observed brightness I of stars assuming that the maximum distance to the limit, i.e., boundary, of the stellar system is L.   The stars are assumed to emit light of brightness η per unit distance along the line of sight (stars are not considered as point sources but are scattered, and hence η does not depend on the square of the distance).   The intensity of this light, however, is reduced by clouds of interstellar dust that are in front of the star along the line of sight.   The number of such clouds $n(s)$ in a line-of-sight distance interval $(0,s)$ is distributed according to the Poisson process

$$\frac{(\lambda s)^n e^{-\lambda s}}{n!} \qquad n = 0, 1, \ldots$$

where λ is the average number of clouds per unit distance.   Let q_i, called the *transparency factor*, $0 \leq q_i \leq 1$, $i = 1, \ldots, n(s)$, be the amount by which the ith cloud in this interval reduces the intensity of light of stars immediately behind it.   Each q_i has a frequency distribution $\psi_i(q_i)$.

We can express the observed brightness I as

$$I = \int_0^L \eta \prod_{i=1}^{n(s)} q_i \, ds \tag{7-27}$$

Thus the observed brightness from stars in the interval $(s, s + ds)$ is $\eta \, ds$ reduced by $\prod_{i=1}^{n(s)} q_i$ due to the clouds in $(0,s)$.   Integration of this quantity yields I.   If we multiply (7-27) by λ and divide by η, and if we define $u \equiv I\lambda/\eta$, $r \equiv \lambda s$, and $\xi \equiv \lambda L$, we have

$$u = \int_0^\xi \prod_{i=1}^{n(r)} q_i \, dr$$

Since $0 \leq q_i \leq 1$, $u \leq \xi$.

Our problem now is to determine $g(u,\xi)$, the frequency distribution of u for a given ξ for $u \leq \xi$.   Clearly, if $u > \xi$, $g(u,\xi) = 0$.   We shall be using the complement of the cumulative distribution; i.e., we define

$$f(u,\xi) \equiv \int_u^\xi g(u,\xi) \, du = \text{Prob} \left(\int_0^\xi \prod_{i=1}^{n(r)} q_i \, dr \geq u \right)$$

$$= \text{Prob} \left\{ \left[\int_0^\alpha \prod_{i=1}^{n(r)} q_i \, dr + \prod_{i=1}^{n(\alpha)} q_i \int_\alpha^\xi \prod_{i=1}^{n(r)-n(\alpha)} q_i \, dr \right] \geq u \right\} \tag{7-28}$$

where α satisfies $0 \le \alpha \le \xi$ and is otherwise arbitrary. If we replace $r - \alpha$ by r in the second integral, then since η has a Poisson distribution which is a stationary distribution i.e., time-homogeneous and hence independent of the position of an interval and dependent only on its length, we can write $n(r) = n(r - \alpha)$ and we have

$$f(u,\xi) = \text{Prob}\left\{\left[\int_0^\alpha \prod_{i=1}^{n(r)} q_i\, dr + \prod_{i=1}^{n(\alpha)} \int_0^{\xi-\alpha} \prod_{i=1}^{n(r)} q_i\, dr\right] \ge u\right\} \quad (7\text{-}29)$$

We derive an integrodifferential equation in f based on the following argument. Assume that $\alpha \ll 1$; that is, it is very small. Then up to terms in $O(\alpha^2)$ we find that in the interval $(0,\alpha)$ there is either no cloud, or there is one cloud. [Note that $r = \lambda s$ yields $r^n e^{-r}/n!$, the frequency of occurrence of n clouds in the interval $(0,r)$.] These two events occur with probabilities $1 - \alpha$ and α, respectively. Thus, if there is no cloud in $(0,\alpha)$, then $\prod_{i=1}^{n(\alpha)} q_i = 1$, with probability $1 - \alpha$. If there is one cloud, then $q \le \prod_{i=1}^{n(\alpha)} q_i \le q + dq$ with probability $\alpha \psi(q)\, dq$. Note that if there is one cloud, we can write $\prod_{i=1}^{n(\alpha)} q_i = q_1 \equiv q$. Thus $f(u,\xi)$ can be written as a sum of two expressions depending on the absence or presence of a cloud in $(0,\alpha)$. If it is present, we must include the fact that its factor q has a frequency distribution $\psi(q)$. We have

$$f(u,\xi) = (1 - \alpha)\, \text{Prob}\left\{\left[\alpha + \int_0^{\xi-\alpha} \prod_{i=1}^{n(r)} q_i\, dr\right] \ge u\right\}$$

$$+ \alpha \int_0^1 \psi(q)\, dq\, \text{Prob}\left\{\left[\theta\alpha + q \int_0^{\xi-\alpha} \prod_{i=1}^{n(r)} q_i\, dr\right] \ge u\right\} + O(\alpha^2) \quad (7\text{-}30)$$

where $0 \le \theta \le 1$. Now

$$\text{Prob}\left\{\left[\theta\alpha + q \int_0^{\xi-\alpha} \prod_{i=1}^{n(r)} q_i\, dr\right] \ge u\right\}$$

$$= \text{Prob}\left\{\left[\left(\int_0^\xi - \int_\xi^{\xi-\alpha}\right)\prod_{i=1}^{n(r)} q_i\, dr\right] \ge \frac{u}{q} - \frac{\theta\alpha}{q}\right\}$$

$$= \text{Prob}\left[\int_0^\xi \prod_{i=1}^{n(r)} q_i\, dr + \zeta\alpha \ge \frac{u}{q} - \frac{\theta\alpha}{q}\right] \quad 0 \le \zeta \le 1$$

$$= \text{Prob}\left[\int_0^\xi \cdots\, dr \ge \frac{u}{q} - \frac{\theta\alpha}{q} - \zeta\alpha\right]$$

Since we have an outside multiplier α, the expression $\theta\alpha/q + \xi\alpha$ is $O(\alpha^2)$. Consequently, the definition of f gives

$$f(u,\xi) = \alpha \int_0^1 \psi(q)f\left(\frac{u}{q}, \xi\right) dq + (1 - \alpha)f(u - \alpha, \xi - \alpha) + O(\alpha^2)$$

$$(7\text{-}30a)$$

Expansion about $\alpha = 0$ gives

$$f(u - \alpha, \xi - \alpha) = \left[f(u,\xi) + \frac{\partial f}{\partial \alpha}\bigg|_{\alpha=0}\right]\alpha + O(\alpha^2)$$

$$= f(u,\xi) + \left[\frac{\partial f}{\partial(u - \alpha)}\frac{d(u - \alpha)}{d\alpha}\bigg|_{\alpha=0}\right.$$

$$\left. + \frac{\partial f}{\partial(\xi - \alpha)}\frac{d(\xi - \alpha)}{d\alpha}\bigg|_{\alpha=0}\right]\alpha + O(\alpha^2)$$

$$= f(u,\xi) - \alpha\frac{\partial f}{\partial u} - \alpha\frac{\partial f}{\partial \xi} + O(\alpha^2)$$

Thus f satisfies the linear integrodifferential equation

$$f + \frac{\partial f}{\partial u} + \frac{\partial f}{\partial \xi} = \int_0^1 \psi(q)f\left(\frac{u}{q}, \xi\right) dq \qquad (7\text{-}31)$$

or, since $g = \partial f/\partial u$, $g(u,\xi)$ satisfies

$$g + \frac{\partial g}{\partial u} + \frac{\partial g}{\partial \xi} = \int_0^1 \psi(q)g\left(\frac{u}{q}, \xi\right)\frac{dq}{q} \qquad (7\text{-}32)$$

Remark: Now $u = \xi$ gives the probability that no cloud is in $(0,\xi)$, which occurs with probability $e^{-\xi}$. If we recall that $g(u,\xi)$ is a density function and hence its integral from 0 to ξ must be unity, we must define

$$g(u,\xi) \equiv \varphi(u,\xi) + e^{-\xi}\delta(u - \xi) \qquad (7\text{-}33)$$

where δ is the Dirac delta function, with an infinite spike at $u = \xi$ but with a unit value for its integral over u. Here φ gives the frequency of clouds up to ξ, and the second term gives it at ξ. Now for an arbitrary function $h(u - \xi)$ we have

$$\left(1 + \frac{\partial}{\partial u} + \frac{\partial}{\partial \xi}\right)h(u - \xi)e^{-\xi} = 0$$

Hence substituting for g from (7-33) into (7-32), we obtain

$$\varphi + \frac{\partial\varphi}{\partial u} + \frac{\partial\varphi}{\partial \xi} = \frac{u}{\xi}\frac{e^{-\xi}}{\xi} + \int_{u/\xi}^1 \psi(q)\varphi\left(\frac{u}{q}\right)\frac{dq}{q} \qquad (7\text{-}34)$$

We define the Mellin transform[32]

$$\mu(s,\xi) = \int_0^\infty g(u,\xi)u^s \, du \qquad (7\text{-}35)$$

with $g = 0$ for $u > \xi$,

$$q_s = \int_0^\infty \psi(q)q^s \, dq = \int_0^1 \psi(q)q^s \, dq \tag{7-36}$$

If we multiply (7-32) by u^s and integrate, we obtain

$$\frac{\partial \mu(s,\xi)}{\partial \xi} = \mu(s,\xi) + q_s\mu(s,\xi) + s\mu(s - 1, \xi) \tag{7-37}$$

with the initial condition

$$\begin{aligned} \mu(s,0) &= 0 \qquad s \neq 0 \\ \mu(0,0) &= 1 \end{aligned} \tag{7-38}$$

since $g(u,\xi) = \delta(u)$, $\xi = 0$; $\mu(0,\xi) = 1$, $\xi > 0$. Note that the inverse Mellin transform gives the desired result

$$g(u,\xi) = \frac{1}{2\pi i} \int_{\sigma-i\infty}^{\sigma+i\infty} \mu(s,\xi)u^{-s-1} \, ds \tag{7-39}$$

If we replace s by n, where $n \geq 0$ is an integer, then $\mu(n,\xi)$ is the nth moment of u. In that case (7-37) can be solved by iteration, yielding

$$\left[\prod_{j=1}^n \left(\frac{d}{d\xi} + \alpha_j \right) \right] \mu_n = n! \qquad \alpha_j = 1 - q_j$$

Thus

$$\mu_n = \sum_{k=1}^n A_k e^{-\alpha_k \xi} + \frac{n!}{\displaystyle\prod_{j=1}^n \alpha_j}$$

Since at $\xi = 0$, $\mu_n = 0$, $d^j\mu_n/d\xi^j = 0$, $j = 1, \ldots, n - 1$, we have

$$\sum_{k=1}^n A_k = - \frac{n!}{\displaystyle\prod_{j=1}^n \alpha_j}$$

$$\sum_{k=1}^n A_k\alpha_k{}^j = 0 \qquad j = 1, \ldots, n - 1$$

from which A_k can be determined, and hence

$$\mu_n = \frac{n! \displaystyle\sum_{k=0}^n \exp\left[-(1 - q_k)\xi\right]}{\displaystyle\prod_{\substack{j=0 \\ j \neq k}}^n (q_k - q_j)} \qquad n = 1, \ldots \tag{7-40}$$

Although, in general, it is not possible simply to write $\mu(s,\xi)$ from $\mu_n(\xi)$ and use the inverse transform to determine g, this can be done for special cases of $\psi(q)$. Note that if $\mu_n(\xi)$ involves terms of the form $\sum_{i=1}^{n} a_i$ (where a_i is constant), which can also be written as $\sum_{i=1}^{\infty} a_i - \sum_{i=1}^{\infty} a_{n+i}$, n can be replaced by s if $\sum_{i=1}^{\infty} a_i$ is absolutely convergent. If $\prod_{i=1}^{\infty} a_i$ is absolutely convergent, we can replace n by s in $\prod_{i=1}^{n} a_i \equiv \prod_{i=1}^{\infty} a_i \Big/ \prod_{i=1}^{\infty} a_{s+i}$. The term $n!$ can be replaced by the gamma function $\Gamma(s+1)$.

We shall illustrate with the situation $\xi \to \infty$ and hence $d\mu(\xi)/d\xi = 0$. In that case (7-32) becomes

$$g + \frac{dg}{du} = \int_0^1 \psi(q) g\left(\frac{u}{q}\right) \frac{dq}{q} \tag{7-41}$$

and

$$\mu_n = \frac{n!}{\prod\limits_{j=1}^{n} 1 - q_j} \tag{7-42}$$

If now we assume that all clouds are equally transparent with reducing factor q', then we can write $\psi(q') = \delta(q' - q)$, where δ is the Dirac delta function. Thus (7-41) becomes

$$g(u) + \frac{dg}{du} = \frac{1}{q} g\left(\frac{u}{q}\right)$$

Here we have $q_j = q^j$. We write

$$\frac{1}{\prod\limits_{j=1}^{n} 1 - q^j} = A_0 \prod_{j=1}^{\infty} 1 - q^{n+j}$$

where

$$\frac{1}{A_0} = \prod_{j=1}^{\infty} 1 - q^j$$

which converges for $q < 1$, because of the identity

$$\prod_{i=1}^{\infty} 1 - q^{i+n} = 1 + \sum_{j=1}^{\infty} (-1)^j q^{nj} \prod_{i=1}^{j} \frac{q^i}{1 - q^i} \tag{7-43}$$

This identity is derived in the following manner. Write

$$\varphi(x) = \prod_{j=1}^{p} 1 - xq^{n+j}$$

which satisfies the functional equation

$$(1 - xq^{n+p+1})\varphi(x) = (1 - xq^{n+1})\varphi(xq)$$

and write

$$\varphi(x) = \sum_{k=0}^{p} a_k x^k \qquad \varphi(xq) = \sum_{k=0}^{p} a_k (xq)^k$$

Substituting and equating coefficients of x^k yield

$$a_k(1 - q^k) = a_{k-1}(q^{n+p+1} - q^{n+p})$$

Dividing through by $1 - q^k$ and solving recursively give

$$a_k = q^{nk} \prod_{j=1}^{k} \frac{q^{p+1} - q^j}{1 - q^j} \qquad k = 1, 2, \ldots$$

from which

$$\varphi(x) = \prod_{j=1}^{p} 1 - xq^{n+j} = 1 + \sum_{k=1}^{p} x^k q^{nk} \prod_{j=1}^{k} \frac{q^{p+1} - q^j}{1 - q^j}$$

and the identity is obtained by putting $x = 1$, $p \to \infty$. We now have

$$\mu(s, \infty) = \mu(s) = A_0 s! \left[1 + \sum_{k=1}^{\infty} (-1)^k q^{sk} \prod_{j=1}^{k} \frac{q^j}{1 - q^j} \right] \qquad (7\text{-}44)$$

from which

$$g(u) = A_0 \left[e^{-u} + \sum_{k=1}^{\infty} (-1)^k e^{-u/q^k} \prod_{j=1}^{k} \frac{q^j}{1 - q^j} \right] \qquad (7\text{-}45)$$

The case $\xi \to \infty$, $\psi(q) = (m + 1)q^m$ yields

$$q_j = \frac{m + 1}{m + j + 1}$$

$$\mu_n = \frac{(m + n + 1)!}{(m + 1)!} \qquad (7\text{-}46)$$

$$\mu(s) = \frac{\Gamma(m + s + 2)}{(m + 1)!} \qquad (7\text{-}47)$$

$$g(u) = e^{-u} \frac{u^{m+1}}{(m + 1)!} \qquad (7\text{-}48)$$

In fact the equation can also be solved using the Laplace transform

$$G(s) = \int_0^\infty g(u) e^{-su} \, du$$

Remark: A solution for the case $\psi(q') = \delta(q' - q)$ is also available without the assumption $\xi \to \infty$ (Ref. 8).

AIRCRAFT WINGS OF FINITE SPAN

As a third example, consider the following singular integrodifferential equation, which occurs in aerodynamics

$$\frac{f(t_0)}{A(t_0)} - \frac{1}{\pi} \int_{-a}^{a} \frac{f'(t)}{t - t_0} \, dt = g(t_0) \tag{7-49}$$

with $f' = df/dt$, where f is unknown, $A(t)$ and $g(t)$ are given,

$$f(t) = f(-t) \qquad A(t) = A(-t) \qquad g(t) = g(-t)$$

and $\qquad f(a) = f(-a) = 0$

One can show that Eq. (7-49) can be written as[27]

$$A(t_0)f'(t_0) = \frac{-1}{\pi} \int_{-a}^{a} \frac{f(t)}{t - t_0} \, dt + A(t_0)g(t_0)$$

$$- \frac{A(t_0)}{\pi \sqrt{a^2 - t_0{}^2}} \int_{-a}^{a} R(t_0, t)f(t) \, dt \tag{7-50}$$

where $\qquad R(t_0, t) = \frac{1}{t - t_0} \left[\frac{\sqrt{a^2 - t^2}}{A(t)} - \frac{\sqrt{a^2 - t_0{}^2}}{A(t_0)} \right]$

Here we assume that

$$\frac{\sqrt{a^2 - t^2}}{A(t)} \qquad -a \leq t \leq a$$

has a continuous first derivative, and hence $R(t_0, t)$ is continuous in both arguments. Now integration by parts gives

$$\int_{-a}^{a} \frac{f(t)}{t - t_0} \, dt = \int_{-a}^{a} f(t) \, d \log |t - t_0| = - \int_{-a}^{a} f'(t) \log |t - t_0| \, dt$$

using $f(a) = f(-a) = 0$. Differentiation with respect to t_0 gives

$$\frac{d}{dt_0} \left[- \int_{-a}^{a} f'(t) \log |t - t_0| dt \right] = \int_{-a}^{a} \frac{f'(t)}{t - t_0} \, dt$$

since $\qquad \dfrac{d}{dt_0} \log |t - t_0| = \dfrac{1}{|t - t_0|} \dfrac{d}{dt_0} |t - t_0| = \dfrac{-1}{t - t_0}$

The interchange of differentiation and integration can be justified by breaking the integral from $-a$ to $t_0 - \epsilon$ and then from $t_0 + \epsilon$ to a and letting $\epsilon \to 0$. Thus (7-50) gives

$$\frac{d}{dt_0} [A(t_0)f'(t_0)] + \frac{1}{\pi} \int_{-a}^{a} \frac{f'(t)}{t - t_0} \, dt = h(t_0)$$

where $h(t_0)$ is d/dt_0 applied to the remaining terms of (7-50). Thus, using (7-49), we have

$$A(t_0) \frac{d}{dt_0} [A(t_0)f'(t_0)] + f(t_0) = A(t_0)[g(t_0) + h(t_0)]$$

Integrating this differential equation gives

$$f(t_0) = c_1 \cos \tau(t_0) + c_2 \sin \tau(t_0) + \int_0^{t_0} \{g(t) + h(t)\} \sin \{\tau(t_0) - \tau(t)\} \, dt$$

where $\tau(t_0) = \int_0^{t_0} \dfrac{dt}{A(t)}$

Note that $c_1 = f'(0)$ and $c_2 = 0$ from the condition defining the even functions involved. We finally have

$$f(t_0) + \frac{1}{\pi} \int_{-a}^{a} K(t_0,t) f(t) \, dt = J(t_0)$$

where $K(t_0,t) \equiv \displaystyle\int_0^{t_0} \dfrac{R(t_1,t)}{\sqrt{a^2 - t_1{}^2}} \cos [\tau(t_0) - \tau(t_1)] \, dt_1$

and $\qquad J(t_0) \equiv f(0) \cos \tau(t_0) + \displaystyle\int_0^{t_0} \{\sin [\tau(t_0) - \tau(t_1)] g(t_1)$

$$+ \frac{\cos [\tau(t_0) - \tau(t_1)]}{\pi \sqrt{a^2 - t_1{}^2}} \int_{-a}^{a} \frac{\sqrt{a^2 - t^2}}{t - t_1} g(t) \, dt\} \, dt_1$$

If $A(t) = A \sqrt{1 - t^2/a^2}\, p(t)$, where $p(t)$ is a rational function of t and A is constant, the foregoing equation can be easily solved. If $p(t)$ is constant, then $R(t_0,t) \equiv 0$, $K(t_0,t) \equiv 0$; and we have the elliptic wing.

7-3 AN EXAMPLE OF AN INTEGRODIFFERENCE EQUATION[1]

An illustration of an integrodifference equation related to the study of efflux of gas from the open end of a tube is

$$f(t) - f(t + 1) = \int_{-1}^{1} K(s) f(t - s) \, ds = \int_{t-1}^{t+1} K(t - s) f(s) \, ds$$
$$0 \leq t < \infty \qquad (7\text{-}51)$$

Here the kernel $K(s)$, $-1 \leq s \leq 1$, and the unknown function $f(t)$, $-1 \leq t < \infty$, must be continuous functions. Define

$$f_n(t) = f(t + n) \qquad 0 \leq t \leq 1, n \geq -1 \qquad (7\text{-}52)$$

then $f_n(t)$ are continuous. Because of the continuity of f we have

$$f_n(0) = f_{n-1}(1) \qquad n \geq 0 \qquad (7\text{-}53)$$

Now the integral on the right of (7-51) can be split for $n - 1 \leq t \leq n$, $t \geq 1$ into the sum of three integrals over $(t - 1, n - 1)$, $(n - 1, n)$, and $(n, t + 1)$. If we put $t = \tau + n$, in the three integrals let $s = \sigma + n - 1$, $s = \sigma + n$, and $s = \sigma + n + 1$, and substitute (7-52), we obtain equations such that whenever their solution is continuous we also have a solution of (7-51), but (7-53) must also be satisfied. Conversely a solution of (7-51) satisfies (7-52) and (7-53) and the system just described.

Exercise 7-3 Obtain this system and prove the last statement.

Exercise 7-4 Apply the Laplace transform to (7-51) and develop the result toward the solution as far as possible.

7-4 BRIEF ILLUSTRATION OF EXISTENCE

Beneš[4] studied the existence of solutions of the following nonlinear integrodifferential equation describing certain control mechanisms

$$\dot{x} = \omega - \int_0^t k(t - u)F[x(u)]\,du \qquad (7\text{-}54)$$

where $t \geq 0$, $\omega =$ constant, $x(0)$ are given, and F is nonlinear.

If we assume that:

1. k is bounded, $k \in L_1(0, \infty)$, $\|k\| = \int_0^\infty |k(u)|\,du$, and
2. F is bounded and satisfies the uniform Lipschitz condition

$$|F(x) - F(y)| \leq \beta|x - y|$$

then we can put (7-54) into the integral form

$$\begin{aligned}
x(t) &= x(0) + \omega t - \int_0^t \int_0^y k(y - u)F[x(u)]\,du\,dy \\
&= x(0) + \omega t - \int_0^t K(t - u)F[x(u)]\,du
\end{aligned} \qquad (7\text{-}55)$$

where $\qquad K(t) = \int_0^t k(z)\,dz$

It can be shown that there exists a unique solution of (7-55) satisfying (7-54) almost everywhere.

Another existence theorem concerning Eq. (7-54) can be proved using the Schauder-Tychonoff fixed-point theorem on a suitably restricted set of bounded continuous functions. The result states that for a given period T there exist ultimately periodic solutions of (7-54) of period T.

We shall not prove this theorem since its proof closely parallels that of the stability theorem given later for this same integrodifferential equation.

7-5 A NONLINEAR EQUATION[9,15]—BOLTZMANN'S EQUATION

In this section we discuss a well-known nonlinear integrodifferential equation known as *Boltzmann's transport equation*. It arises in the determination of the distribution of particles of an ideal gas in an enclosure on which there is an external force **F**. Each particle has six degrees of freedom, three for momentum and three for position. Let $f(\mathbf{r},\mathbf{v},t)$ be the desired distribution, where $\mathbf{r} = (r_1,r_2,r_3)$ is the position vector,

$\mathbf{v} = (v_1, v_2, v_3)$ is the velocity vector, and t is time.   Clearly

$$\int \cdots \int f(\mathbf{r}, \mathbf{v}, t) \, dr_1 \, dr_2 \, dr_3 \, dv_1 \, dv_2 \, dv_3 = N$$

the total number of particles at time t.   If we make the assumption that the particles are uniformly distributed in space, then f is independent of $\mathbf{r}$, and $\int\int\int f(\mathbf{v}, t) \, dv_1 \, dv_2 \, dv_3 = N/V$, where V is the volume of the enclosure.   Let us now examine the behavior of the system over a small time interval $t + \Delta t$.   Recall that distance is velocity multiplied by the time Δt and velocity is acceleration times time, that is, $\mathbf{F}/m$, where m is the mass of a particle, multiplied by Δt.   If force, such as the gravitational force, depends on position only, then the six-dimensional volume at t, $dr_1 \, dr_2 \, dr_3 \, dv_1 \, dv_2 \, dv_3$, is equal to the volume at time $t + \Delta t$, $dr_1' \, dr_2' \, dr_3' \, dv_1' \, dv_2' \, dv_3'$; that is, the jacobian is unity.   This is a special case of a theorem due to Liouville, but it can be established independently.   We have for the distribution in the new volume element

$$f\left(\mathbf{r} + \mathbf{v}\,\Delta t, \, \mathbf{v} + \frac{\mathbf{F}}{m}\,\Delta t, \, t + \Delta t\right) d^3 r' \, d^3 v' = f(\mathbf{r}, \mathbf{v}, t) \, d^3 r \, d^3 v \quad (7\text{-}56)$$

and because of the last remark we can cancel the volume elements on both sides.   Thus the distribution does not change in a small time interval provided that there are no collisions which would result in the loss of particles outside the six-dimensional incremental volume element.   If there is a collision, then to a first-order approximation

$$f(\mathbf{r} + \mathbf{v}\,\Delta t, \, \mathbf{v} + \frac{\mathbf{F}}{m}\,\Delta t, \, t + \Delta t) = f(\mathbf{r}, \mathbf{v}, t) + \left(\frac{\partial f}{\partial t}\right)_{\text{coll}} \Delta t \quad (7\text{-}57)$$

We have added a term indicating the rate of change of f in a small time interval to what we obtained above when no collisions occurred.   If we expand the left side to first-order terms and let $\Delta t \to 0$, we have

$$\left(\frac{\partial}{\partial t} + \mathbf{v} \cdot \boldsymbol{\nabla}_r + \frac{\mathbf{F}}{m} \cdot \boldsymbol{\nabla}_v\right) f(\mathbf{r}, \mathbf{v}, t) = \left(\frac{\partial f}{\partial t}\right)_{\text{coll}} \quad (7\text{-}58)$$

where, for example,

$$\boldsymbol{\nabla}_r = \left(\frac{\partial}{\partial r_1}, \frac{\partial}{\partial r_2}, \frac{\partial}{\partial r_3}\right)$$

Note that $(\partial f/\partial t)_{\text{coll}}$ is the difference during Δt between the number of molecules entering the volume element because of collisions outside it causing molecules to be thrown in and the number leaving the volume element because of collisions in the volume element.

To estimate this difference we assume that no more than two particles collide at any moment and ignore the walls of the container and the effect of the external force on the collision probability.   We also assume that the velocity of a molecule is independent of its position.

Within the spatial volume $d^3\mathbf{r}$ at $\mathbf{r}$, we examine R, the rate of change of $f(\mathbf{r},\mathbf{v},t)$ due to collisions of molecules lying within a "volume" element $d^3\mathbf{v}_1$ about $\mathbf{v}_1$ with molecules having any velocity $\mathbf{v}_2$. Because of the assumption that the distribution of velocities is independent of position, the same expression will be used to represent the distribution of molecules with velocity $\mathbf{v}_2$. Since $\mathbf{v}_2$, the velocity of the incident molecules, can take on any value, we must integrate over all possible values of $\mathbf{v}_2$. These incident molecules in their collision with target molecules having velocity $\mathbf{v}$ encounter only a portion of the area of each molecule which stands in their path. This area is denoted by $\sigma(\Omega)$, the differential cross section, where Ω is the angle between the initial relative-velocity vector for the two colliding molecules and the resultant relative-velocity vector after the collision. The incident molecules with any velocity $\mathbf{v}_2$ can collide with molecules of velocity $\mathbf{v}_1$ from any direction within the volume $d^3\mathbf{r}$ about $\mathbf{r}$; hence we must integrate $\sigma(\Omega)$ over all possible values of Ω. Performing the indicated integrations and multiplying the result by the spatial density of $\mathbf{v}_1$ molecules in $d^3\mathbf{v}_1$, we have

$$R = \int d^3\mathbf{v}_2 \int d\Omega \, \sigma(\Omega)|\mathbf{v}_1 - \mathbf{v}_2|f(\mathbf{r},\mathbf{v}_1,t)f(\mathbf{r},\mathbf{v}_2,t) \qquad (7\text{-}59)$$

The collision between molecules with velocities $\mathbf{v}_1$ (initially within the spatial volume $d^3\mathbf{r}$) and $\mathbf{v}_2$ produces molecules with final velocities $\mathbf{v}_1$ and $\mathbf{v}_2$, respectively. Similarly for $\bar{R}$, the rate of change in f due to collisions of molecules outside the volume element. Our concern is the collision between molecules with velocities $\mathbf{v}_1'$ (assumed fixed) and $\mathbf{v}_2'$ (assumed arbitrary), after which one of the molecules is found in the spatial volume with final velocity $\mathbf{v}_1$, and the other has final velocity $\mathbf{v}_2$. The differential cross section $\sigma(\Omega')$ for this collision can easily be shown to be equal to $\sigma(\Omega)$.

The same considerations also establish the following equalities

$$\begin{aligned} |\mathbf{v}_2 - \mathbf{v}_1| &= |\mathbf{v}_2' - \mathbf{v}_1'| \\ d^3\mathbf{v}_1 \, d^3\mathbf{v}_2 &= d^3\mathbf{v}_1' \, d^3\mathbf{v}_2' \\ \sigma(\Omega) &= \sigma(\Omega') \end{aligned} \qquad (7\text{-}59a)$$

Hence

$$\bar{R}' = \int d^3\mathbf{v}_2 \int d\Omega \, \sigma(\Omega) \, |\mathbf{v}_1 - \mathbf{v}_2|f(\mathbf{r},\mathbf{v}_1',t)f(\mathbf{r},\mathbf{v}_2',t)$$

$$\left(\frac{\partial}{\partial t} + \mathbf{v}_1 \cdot \nabla_r + \frac{\mathbf{F}}{m} \cdot \nabla_{\mathbf{v}_1} \right) f(\mathbf{r},\mathbf{v}_1,t) = \left[\frac{\partial f}{\partial t} (\mathbf{r},\mathbf{v}_1,t) \right]_{\text{coll}}$$

$$\bar{R}' - \bar{R} = \int d^3\mathbf{v}_2 \int d\Omega \, \sigma(\Omega)|\mathbf{v}_1 - \mathbf{v}_2|[f(\mathbf{r},\mathbf{v}_1',t)f(\mathbf{r},\mathbf{v}_2',t) - f(\mathbf{r},\mathbf{v}_1,t)f(\mathbf{r},\mathbf{v}_2,t)]$$

$$(7\text{-}60)$$

We solve (7-60) at equilibrium, i.e., when

$$\left[\frac{\partial f}{\partial t} (\mathbf{r},\mathbf{v}_1,t) \right]_{\text{coll}} = 0$$

This condition does not imply that collisions do not occur but that the net effect of the collisions is zero. (Every molecule entering the volume element with a given velocity by a collision is matched by a molecule leaving the volume element with the same velocity.) Therefore for the case at equilibrium we must solve

$$0 = \int d^3 \mathbf{v}_2 \int d\Omega \, \sigma(\Omega) |\mathbf{v}_1 - \mathbf{v}_2| \{f_1' f_2' - f_1 f_2\} \tag{7-61}$$

where the subscripts on f correspond to those of its $\mathbf{v}$ argument.

We now show that a necessary condition for the vanishing of (7-61) is that[15]

$$f_1' f_2' = f_1 f_2 \tag{7-61a}$$

To see this, let f satisfy (7-60), and define

$$H(t) = \int d^3 \mathbf{v} \, f(\mathbf{r},\mathbf{v},t)_{\text{coll}} \log f(\mathbf{r},\mathbf{v},t)_{\text{coll}} = \int d^3 \mathbf{v} \, f(\mathbf{v},t) \log f(\mathbf{v},t)$$

only if f is independent of $\mathbf{r}$ and there are no external forces. Differentiating with respect to t yields the integrand $(\partial f/\partial t)(1 + \log f)$ from which $\partial f/\partial t = 0$; that is Eq. (7-61) implies $dH/dt = 0$. We first note that $dH/dt \leq 0$ for all t. To see this we substitute for $\partial f/\partial t$ from (7-60) having replaced $\mathbf{v}$ by $\mathbf{v}_1$. We take half of this quantity and half of the result obtained by interchanging $\mathbf{v}_1$ and $\mathbf{v}_2$. [Note that since $\sigma(\Omega)$ is invariant when $\mathbf{v}_1$ and $\mathbf{v}_2$ are interchanged, so is the integrand of the new result.] We obtain

$$\frac{dH}{dt} = \tfrac{1}{2} \int d^3 \mathbf{v}_1 \int d^3 \mathbf{v}_2 \int d\Omega \, \sigma(\Omega) |\mathbf{v}_2 - \mathbf{v}_1| (f_1' f_2' - f_1 f_2)(2 + \log f_1 f_2) \tag{7-61b}$$

Now because of (7-59a) and the fact that the argument used to develop (7-60) is symmetric if the elements of each pair $\mathbf{v}_1$, $\mathbf{v}_2$ and $\mathbf{v}_1'$, $\mathbf{v}_2'$ are interchanged, i.e., we can think of inverse collisions instead, using the same cross section, we can rewrite the above integral replacing the unprimed elements with primed ones and vice versa. If we take half of (7-61b) and half of this same equation switching primes, we have

$$\frac{dH}{dt} = \tfrac{1}{4} \int d^3 \mathbf{v}_1 \int d^3 \mathbf{v}_2 \int d\Omega \, \sigma(\Omega) |\mathbf{v}_2 - \mathbf{v}_1| (f_1' f_2' - f_1 f_2)(\log f_1 f_2 - \log f_1' f_2')$$

whose integrand cannot be positive. Thus $dH/dt \leq 0$. Now since $\partial f/\partial t = 0$ implies $dH/dt = 0$ we have the result $f_1 f_2 = f_1' f_2'$.

Exercise 7-5 Work out the foregoing in detail.

Taking the logarithm of both sides of (7-61a), we obtain an expression which resembles a conservation law, i.e., the energy- and momentum-conservation laws, and we have

$$\log f_1' + \log f_2' = \log f_1 + \log f_2$$

Hence it is appropriate to seek a solution involving energy and momentum terms, i.e., $mv^2/2$ and mv, respectively. In the one-dimensional such a solution is given by

$$f(\mathbf{r},\mathbf{v}) = A e^{B|v-v_0|^2}$$

Note that in equilibrium f is independent of t. The constant A, when determined, is known as *Boltzmann's constant*. The above solution has the proper form with a v^2 term for energy and a v term for momentum. A, B, and v_0 are constants. This expression is known as the *Maxwell-Boltzmann*, or maxwellian, *distribution* and is a solution of Boltzmann's equation at equilibrium.

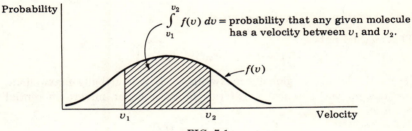

FIG. 7-1

Figure 7-1 represents the maxwellian distribution of molecular velocities in a gas. The solution given above, with constants A and B determined from physical considerations, was actually discovered by Maxwell to be the distribution of velocities of molecules in a gas before Boltzmann's equation was set down.

Let us now write Boltzmann's equation (7-60) in the abbreviated form

$$Df = J(ff_1) \tag{7-62}$$

where $J(ff_1)$ is the nonlinear operator of collision, and where f corresponds to f_2 in (7-60) and f_1 to f_1. Several methods for deriving a solution have been considered, and here we give a glimpse of some. (See Ref. 16.)

THE ENSKOG - CHAPMAN AND HILBERT METHOD OF NORMAL SOLUTIONS

Here f is developed in a series of powers of a small parameter ϵ:

$$f = \sum_{n=0}^{\infty} \epsilon^n f^{(n)} \tag{7-63}$$

This yields an infinite system of linear integral equations of the form

$$n^2 I(\varphi^{-(n)}) = \psi^{(n)} \equiv D^{(n)} - \sum_{m=1}^{n-1} J(f^{(m)} f_1^{(n-m)}) \qquad (7\text{-}64)$$

where one puts $f^{(n)} = f^{(0)} \varphi^{(n)}(\mathbf{v},\mathbf{r},t)$ and where $\psi^{(n)}$ (defined by the process of decomposition of Enskog) is a known function of $(\mathbf{v},\mathbf{r},t)$ which depends only on the approximations of lower order in n. The linear integral operator I is common to all the equations of the system, and it has the form

$$n^2 I(\varphi) = \iint f^{(0)} f_1^{(0)} \partial(\varphi) g\sigma \, d\Omega \, d\mathbf{v}_1 \equiv n^2 f^{(0)} L(\varphi) \qquad (7\text{-}65)$$

with $\partial(\varphi) = \varphi' + \varphi_1' - \varphi - \varphi_1$. The passage from the nonlinear operator J to the linear operator L thus defined depends on the smallness of ϵ.

THE METHOD OF LINEARIZATION

This method puts

$$f = f^{(0)}(1 + \varphi) \qquad |\varphi| \ll 1$$

$f^{(0)}$ can be locally or globally maxwellian. If it is globally maxwellian, one has, in the absence of exterior force, a linear integrodifferential equation

$$\frac{\partial \varphi}{\partial t} + \mathbf{v} \cdot \boldsymbol{\nabla}_r \varphi = n_0^2 L(\varphi)$$

the operator $L(\varphi)$ being defined by (7-65) to within a constant factor. To solve such an equation one seeks proper functions of the operator L defined by $L(\varphi) = \lambda \varphi$.

THE METHOD OF GRAD

This consists of putting for f the following development in tensor polynomials of Hermite in the form

$$f = f^{(0)} \sum_{n=0}^{\infty} \frac{1}{n!} \alpha_\nu^{(n)}(\mathbf{r},t) H_\nu(\mathbf{w}) \qquad (7\text{-}66)$$

where the first $H(\mathbf{w})$ (with $\mathbf{w} = \sqrt{m/kT} \, \mathbf{v}$) are

$$H^{(0)} = 1 \qquad H_i^{(1)} = w_i \qquad H_{ij}^{(2)} = w_i w_j - \delta_{ij}$$

By substituting (7-16) in (7-62) one has an infinite system of coupled differential equations, generally nonlinear, in the coefficients $a_\nu^{(n)}(\mathbf{r},t)$. These, on the other hand, can be expressed linearly by means of the successive moments of f, using orthogonal relations satisfied by the Hermite polynomials. Thus the differential system satisfied by $a_\nu^{(n)}(\mathbf{r},t)$ is equivalent to a system of equations satisfied by the successive moments of f. In this method the development (7-66) stops at terms of order n, thus truncating the infinite system of differential equations at the nth equation.

THE METHOD OF ITERATION

This method is concerned with the solution according to Maxwell, Ikenberry, and Truesdell, etc., of the infinite differential system of the equations of moments by successive iterations with respect to a maxwellian distribution. The symmetry of the collision operator with respect to solutions in the space of velocities naturally leads to writing the equations of moments for the spherical moments; these are related to the coefficients $a_\nu^{(n)}$. In this manner the method of iteration appears as another process of solution (for a maxwellian interaction) of the system of equations of moments and is equivalent to Grad's system.

7-6 SOME METHODS OF SOLUTION

In this section we shall be concerned with various (not yet systematized) methods of solving integrodifferential equations. Although some equations are reduced to integral or differential equations, the solutions of some equations are developed directly.

SOLUTION OF A FUNCTIONAL INTEGRODIFFERENTIAL EQUATION BY THE METHOD OF UPPER AND LOWER FUNCTIONS.[39]

Consider the nonlinear functional integrodifferential equation

$$y''(x) = F\left[x,\lambda \int_a^x K(x,t,y(t),y'(t))\,dt\right] \qquad (7\text{-}67)$$

We seek a solution $y(x)$ which, together with its first- and second-order derivatives, is continuous in $[a,b]$ and satisfies initial conditions

$$y(a) = y_0 \qquad y'(a) = y_0' \qquad (7\text{-}68)$$

where a, b, y_0, y_0' are given numbers.

With the aid of definite conditions on F and K, we construct, as in the last chapter, a sequence of upper and lower functions which converge to the desired solution in an arbitrary closed interval $[a,b]$. We assume that:

1. $F(x,u)$ is continuous and bounded together with its derivative F_u and satisfies $\lambda F_u > 0$ in the region $D = \{a \leq x \leq b, \; -\infty < u < \infty\}$.

2. $K(x,t,y,y')$ is continuous and bounded together with its derivatives K_y and K'_y, which satisfy $K_y > 0$, $K'_y \geq 0$ in $B = \{a \leq t \leq x \leq b, \; -\infty < y, y' < \infty\}$.

***Theorem* 7-1**   If $z(x)$ is continuous and has first- and second-order continuous derivatives in $[a,b]$, and if it satisfies (7-68) and the inequality

$$z''(x) > F\left[x, \lambda \int_a^x K(x,t,z(t),z'(t))\, dt\right] \qquad \text{for } a \leq x \leq b \quad (7\text{-}69)$$

then $z(x) > y(x)$, $x \,\epsilon\, (a,b)$, where $y(x)$ is continuous together with its first- and second-order derivatives in $[a,b]$ and satisfies (7-67) and (7-68).

Proof: From (7-67) and (7-69) we have $z''(a) > y''(a)$.   Because of the continuity of $z''(x)$ and $y''(x)$, $z''(x) > y''(x)$ holds in $[a, a + \epsilon]$, where $\epsilon > 0$, that is, in a neighborhood of a.   Now we show that $z''(x) > y''(x)$ for all $x \,\epsilon\, [a,b]$.   This is shown by first assuming the opposite.   Let $\bar{x} > a$ be the first point of $[a,b]$ at which $z''(\bar{x}) = y''(\bar{x})$.   Hence $z''(x) > y''(x)$ for $a \leq x < \bar{x}$.   Then by (7-68) $z'(x) > y'(x)$ and $z(x) > y(x)$ for $x \,\epsilon\, (a,\bar{x})$.

Note that because

$$z''(x) - y''(x) > 0 \qquad \text{in } (a,\bar{x}) \qquad\qquad (7\text{-}70)$$

$z'(x) - y'(x)$ is monotone increasing, and thus

$$z'(x) - y'(x) > z'(a) - y'(a) = 0 \qquad \text{for } x \,\epsilon\, (a,\bar{x})$$

Likewise $z(x) - y(x) > z(a) - y(a) = 0$.   From (7-67) and (7-69), using the mean-value theorem for derivatives first with respect to F then with respect to K, we have

$$z''(x) - y''(x) > F[.] - F[.]$$
$$= \lambda F_u^* \int_a^x [K(x,t,z(t),z'(t)) - K(x,t,y(t),y'(t))]\, dt$$
$$= \lambda F_u^* \int_a^x \{K_y^*[z(t) - y(t)] + K_{y'}^*[z'(t) - y'(t)]\}\, dt \quad (7\text{-}71)$$

where F_u^*, K_y^*, and $K_{y'}^*$ indicate the corresponding mean values of the derivatives.   From (7-71), with $x = \bar{x}$, we have

$$z''(\bar{x}) - y''(\bar{x}) > \lambda(F_u^*)_{\bar{x}} \int_a^{\bar{x}} \{(K_y^*)_{\bar{x}}[z(t) - y(t)]$$
$$+ (K_{y'}^*)_{\bar{x}}[z'(t) - y'(t)]\}\, dt > 0 \quad (7\text{-}72)$$

where, for example, $(K_u^*)_{\bar{x}}$ is K_u^* evaluated at $\bar{x}$.

But this contradicts the assumption that $z''(\bar{x}) = y''(\bar{x})$.   Hence $z''(x) > y''(x)$, $x \,\epsilon\, [a,b]$.

Analogously it can be proved that if $v(x)$ is continuous, together with its first- and second-order derivatives in $[a,b]$, and satisfies (7-68) and an inequality in $v''(x)$ whose sense is opposite to that given in (7-69) for $x \in [a,b]$, then

$$v(x) < y(x) \qquad \text{for } x \in (a,b) \tag{7-73}$$

The functions $z(x)$ and $v(x)$ thus defined will be called *upper* and *lower functions*, respectively, for the solution of (7-67).

Clearly $z(x) > v(x)$, $x \in (a,b)$. We now show that the sets of upper and lower functions are nonempty

Exercise 7-6   Show that the following is true:

$$\overbrace{\int_a^x \int_a^{\xi_1} \cdots \int_a^{\xi_{n-1}}}^{n\text{-fold}} f(\xi_n) \, d\xi_n \, d\xi_{n-1} \cdots d\xi_1 = \frac{1}{(n-1)!} \int_a^x (x-\xi)^{n-1} f(\xi) \, d\xi$$

$$\equiv I_n[f(x)]$$

With $a = 0$, the integral on the right is known as the *Riemann-Liouville integral*. Show that $I_{n+m} = I_n I_m$.

Suppose that $\sup |K(x,t,y,y')| = M$ for $(x,t,y,y') \in B$. We define

$$\omega(x) = y_0 + y_0'(x-a) + \int_a^x (x-\xi)F[\xi,\lambda(M+1)(\xi-a)] \, d\xi \tag{7-74}$$

and show that it is an upper function. If (7-74) is true, $\omega(a) = y_0$, and $\omega'(a) = y_0'$. By the mean-value theorem, we have

$$\omega''(x) - F[x,\lambda \int_a^x K(x,t,\omega(t),\omega'(t)) \, dt]$$

$$= F[x,\lambda \int_a^x (M+1) \, dt] - F[x,\lambda \int_a^x K(x,t,\omega(t),\omega'(t)) \, dt]$$

$$= \lambda F_u^* \int_a^x [M+1 - K(x,t,\omega(t),\omega'(t))] \, dt > 0 \tag{7-75}$$

Thus $\omega(x)$ is an upper function. A lower function is analogously constructed using the infimum of K.

Let us now construct by successive approximations, sequences of upper and lower function $\{z_n\}$ and $\{v_n\}$. Suppose that $z_0(x)$ and $v_0(x)$ are the initial choices. They satisfy inequalities similar to (7-69) and its analogue in $v(x)$. They also satisfy $z_0(x) > y(x) > v_0(x)$, $x \in (a,b)$. Using these, we have for the next iteration

$$z_1(x) = y_0 + y_0'(x-a) + \int_0^x (x-\xi)F[\xi,\lambda \int_a^\xi K(\xi,t,z_0(t),z_0'(t)) \, dt] \, d\xi \tag{7-76}$$

$$v_1(x) = y_0 + y_0'(x-a) + \int_0^x (x-\xi)F[\xi,\lambda \int_a^\xi K(\xi,t,v_0(t),v_0'(t)) \, dt] \, d\xi \tag{7-77}$$

Now (7-76) implies

$$z_1(a) = y_0 \qquad z_1'(a) = y_0' \tag{7-78}$$

Also

$$z_1''(x) = F[x, \lambda \int_a^x K(x,t,z_0,z_0') \, dt] \tag{7-79}$$

$$z_0''(x) - z_1''(x) > 0 \qquad x \, \epsilon \, [a,b] \tag{7-80}$$

But because

$$z_0(a) - z_1(a) = 0 \qquad \text{and} \qquad z_0'(a) - z_1(a) = 0$$

we obtain

$$z_0'(x) > z_1'(x) \qquad \text{and} \qquad z_0(x) > z_1(x) \qquad x \, \epsilon \, (a,b) \tag{7-81}$$

Furthermore from (7-79), (7-81), and the mean-value theorem, we have [using (7-79) for z_1'']

$$z_1''(x) - F[x, \lambda \int_a^x K(x,t,z_1(t),z_1'(t)) \, dt]$$
$$= \lambda F_u^* \int_a^x \{K_y^*[z_0(t) - z_1(t)] + K_{y'}^*[z_0'(t) - z_1'(t)]\} \, dt \tag{7-82}$$

From (7-78), (7-81), and (7-82) it follows that z_1 is an upper function and $z_0(x) > z_1(x)$.

Analogously we can show that $v_1(x)$ is a lower function and that $v_1(x) > v_0(x)$ for $x \, \epsilon \, (a,b)$.

In this manner we construct inductively sequences of upper and lower functions which satisfy

$$z_0(x) > z_1(x) > \cdots > z_n(x) > \cdots > y(x) \tag{7-83}$$
$$v_0(x) < v_1(x) < \cdots v_n(x) < \cdots < y(x) \tag{7-84}$$

and are given by relations similar to (7-76) and (7-77) with the appropriate subscripts on z and v.

It is always true that $z(x) > v(x)$ and $z'(x) > v'(x)$ for any two upper and lower functions $z(x)$ and $v(x)$. Thus we have a monotone decreasing sequence of upper functions and a monotone increasing sequence of lower functions.

Since for every n, $v_n(x) < \omega(x)$, the sequences $\{v_n(x)\}$ and $\{v_n'(x)\}$ converge. Similarly for the sequences $\{z_n(x)\}$ and $\{z_n'(x)\}$.

We shall show that these four sequences converge uniformly for $x \, \epsilon \, (a,b)$.

For this purpose let δ_0 be the larger of

$$\sup [z_0(x) - v_0(x)] \qquad \sup [z_0'(x) - v_0'(x)] \qquad x \, \epsilon \, [a,b]$$

and N the larger of $\sup K_y$ and $\sup K_{y'}$ in

$$B_0 = \{a \le t \le x \le b, v_0(x) \le y(x) \le z_0(x), v_0'(x) \le y'(x) \le z_0'(x)\}$$

and define sup $\lambda F_u(x,u) \equiv P$ in D.   Then we obtain from (7-76) and (7-77)

$$z_1(x) - v_1(x) = \int_0^x (x - \xi)\lambda F_u^*(\xi)\, d\xi \int_a^\xi \{K_y^*[z_0(t) - v_0(t)]$$
$$+ K_{y'}^*[z_0'(t) - v_0'(t)]\}\, dt$$
$$< 2\delta_0 PN \frac{(x - a)^3}{3!} \tag{7-85}$$

$$z_1'(x) - v_1'(x) = \int_0^x \lambda F_u^*(\xi)\, d\xi \int_a^\xi \{K_y^*[z_0(t) - v_0(t)]$$
$$+ K_{y'}^*[z_0'(t) - v_0'(t)]\}\, dt$$
$$< 2\delta_0 PN \frac{(x - a)^2}{2!} \tag{7-86}$$

Analogously

$$z_2(x) - v_2(x) = \int_a^x \lambda F_u^*(\xi)\, d\xi \int_a^\xi \{K_y^*[z_1(t) - v_1(t)] + K_{y'}^*[z_1'(t) - v_1'(t)]\}\, dt$$
$$< 2\delta_0 P^2 N^2 \frac{(x - a)^5}{5!}\left(\frac{x - a}{6} + 1\right)$$

$$z_2'(x) - v_2'(x) = \int_a^x \lambda F_u^*(\xi)\, d\xi \int_a^\xi \{K_y^*[z_1(t) - v_1(t)] + K_{y'}^*[z_1'(t) - v_1'(t)]\}\, dt$$
$$< 2\delta_0 P^2 N^2 \frac{(x - a)^4}{4!}\left(\frac{x - a}{5} + 1\right) \equiv 2\delta_0 L P^2 N^2 \frac{(x - a)^4}{4!}$$

and hence

$$z_2(x) - v_2(x) < 2\delta_0 L P^2 N^2 \frac{(x - a)^5}{5!}$$

By induction we obtain

$$z_n(x) - v_n(x) < \frac{2\delta_0}{L\sqrt{LPN}} \frac{[\sqrt{LPN}\,(x - a)]^{2n+1}}{(2n + 1)!}$$

and $$z_n'(x) - v_n'(x) < \frac{2\delta_0}{L} \frac{[\sqrt{LPN}\,(x - a)]^{2n}}{(2n)!}$$

Therefore as $n \to \infty$, $z_n(x) - v_n(x) \to 0$ and $z_n'(x) - v_n'(x) \to 0$ uniformly in $[a,b]$.   Thus

$$z_n(x) \to \bar{y}(x) \qquad z_n'(x) \to w(x)$$
$$v_n(x) \to \bar{y}(x) \qquad v_n'(x) \to w(x)$$

But since the functions are continuous and uniformly convergent, the limit functions are continuous.

If we integrate

$$z_n'(x) = y_0' + \int_a^x F\left[\xi, \lambda \int_a^\xi K(\xi, t, z_{n-1}(t), z_{n-1}'(t))\, dt\right] d\xi$$

and pass to the limit, we obtain

$$y(x) = y_0 + y_0'(x - a) + \int_a^x (x - \xi) F\left[\xi, \lambda \int_a^\xi K(\xi, t, \bar{y}(t), w(t))\, dt\right] d\xi$$

where

$$w(x) = y_0' + \int_a^x F\left[\xi, \lambda \int_a^\xi K(\xi, t, \bar{y}(t), w(t))\, dt\right] d\xi$$

From this it follows that $w(x) = \bar{y}'(x)$, and the following holds:

$$\bar{y}''(x) = F\left[x, \lambda \int_a^x K(x, t, \bar{y}(t), \bar{y}'(t))\, dt\right]$$

We now show that the solution is unique. Suppose we have two solutions $\bar{y}_1(x)$ and $\bar{y}_2(x)$, and let δ be the larger of

$$\sup |\bar{y}_1 - \bar{y}_2| \qquad \sup |\bar{y}_1' - \bar{y}_2'| \qquad x \,\epsilon\, [a, b]$$

Then in a manner analogous to the above we have

$$|\bar{y}_1(x) - \bar{y}_2(x)| < \frac{2\delta}{L\sqrt{LPN}} \frac{[\sqrt{LPN}\,(x - a)]^{2m+1}}{(2m + 1)!} \qquad m = 1, 2, \ldots$$

and taking the limit as $m \to \infty$, we have

$$|\bar{y}_1(x) - \bar{y}_2(x)| = 0 \qquad x \,\epsilon\, [a, b]$$

Thus $\bar{y}_1(x) \equiv \bar{y}_2(x)$, $x \,\epsilon\, [a, b]$, and the solution is unique.

MIXED INTEGRODIFFERENTIAL - DIFFERENCE EQUATIONS

From Sec. 7-2 the reader has a better idea regarding the solution of the mixed linear integrodifferential-difference equation with constant coefficients

$$\sum_{i=0}^m a_i y^{(i)}(t) + \sum_{i=0}^m \sum_{j=0}^n b_{ij} y^{(i)}(t - \tau_j) + \sum_{j=0}^n \int_a^b g_j(s) y^{(i)}(t - s)\, ds = w(t)$$

$$(7\text{-}87)$$

or a system of such equations in which appropriate initial conditions are given. Here the coefficients and the g's are given.

Exercise 7-7 Write down the equations of a system generalizing from the above equation by using additional subscripts, for example, $y_h{}^{(j)}$, a_{khi}, b_{khij}, g_{khi}, and $w_k(t)$, introducing appropriate summations.

Exercise 7-8 Write down the solution of the above equation in the case of simple characteristic roots.

ATOMIC SCATTERING[26]

A well-known system of linear (partial) integrodifferential equations which arises in the theory of atomic scattering is

$$\left(\frac{d^2}{dr^2} + k_m{}^2\right)f_m(r) = \sum_{j=1}^{n} \left(V_{mj}(r)f_j(r) + \int_0^\infty K_{mj}(r,r')f_j(r')\,dr'\right)$$

$$m = 1, \ldots, n$$

with $f_m(0) = 0$, $m = 1, \ldots, n$;

$$f_1(r) \sim \sin k_1 r + a \cos k_1 r$$
$$f_m(r) \sim \beta_m \exp(ik_m r) \qquad m > 1$$

as $r \to \infty$. The constants a and β_m are to be determined. The above system expressed in vector notation is

$$(\nabla_r{}^2 + k^2)f(r) = V(r)f(r) + \int_0^\infty K(r,r')f(r')\,dr'$$

It represents the relative motion of two nuclei in a collision. Here $f(r)$ is the relative radius vector of the two nuclei, k^2 is proportional to the relative kinetic energy of the colliding nuclei, $V(r)$ is an averaged potential energy arising from integration, and the integral involves intersections in the exchange of particles between the nuclei. Since the system is linear, if $f_{m1}, \ldots, f_{mn}$ is a system of n linearly independent solutions, we can determine the constants $\alpha_1, \ldots, \alpha_n$ such that $\sum_{j=1}^{n} \alpha_j f_{mj}$ has the correct asymptotic form. To obtain such a set we first solve for $f_m{}^{(0)}(r)$ in

$$\left(\frac{d^2}{dr^2} + k_m{}^2\right)f_m{}^{(0)}(r) = \sum_{j=1}^{n} V_{mj}(r)f_j{}^{(0)}(r)$$

and use it to evaluate

$$I_0 = \sum_{j=1}^{n} \int_0^\infty K_{mj}(r,r')f_j{}^{(0)}(r')\,dr'$$

Then we solve for $f_m{}^{(1)}(r)$ in

$$\left(\frac{d^2}{dr^2} + k_m{}^2\right)f_m{}^{(1)}(r) = \sum_{j=1}^{n} V_{mj}(r)f_j{}^{(1)} + I_0$$

and use it to obtain I_1, etc. The method can be shown to converge.

FROM AN INTEGRODIFFERENTIAL TO A DIFFERENTIAL EQUATION

Sometimes an integrodifferential equation can be made to correspond to a differential equation by specializing its kernel.   Thus

$$f(x,y) = g(x,y) + \int_0^y K(y,u) \frac{\partial f(x,u)}{\partial x} \, du \qquad (7\text{-}88)$$

with g and K given, g continuous in $0 \le x \le 1$, $0 \le y$ and k continuous for $y > u > 0$, with

$$|K(y,u)| < M|y - u|^{k-1} \qquad k > 0$$

for some M and k, corresponds to the Cauchy problem

$$\frac{\partial^k f}{\partial y^k} = \frac{\partial f}{\partial x} + \frac{\partial^k g}{\partial y^k} \qquad (7\text{-}89)$$

$$\frac{\partial^i f}{\partial y^i}\bigg|_{y=0} = \frac{\partial^i g}{\partial y^i}\bigg|_{y=0} \qquad i = 0, 1, \ldots, k - 1$$

if we put $\quad K(y,u) = \dfrac{(y - u)^{k-1}}{(k - 1)!} \qquad k$ an integer

To see this we integrate (7-89) or differentiate (7-88) k times with respect to y.   Now we can state that if g has all order derivatives with respect to x continuous in the region $0 \le x \le 1$, $0 \le y$, and in any bounded subset of this region we have uniformly for any constant A

$$\lim_{n \to \infty} \frac{\partial^n g}{\partial x^n} \frac{A^n}{\Gamma(kn)} = 0$$

(7-88) has a unique solution with the same properties as g.   It is given by[31]

$$f(x,y) = g(x,y) + \sum_{n=1}^{\infty} T^n \left[\frac{\partial^n g(x,y)}{\partial x^n} \right]$$

where T is the integral operator

$$T[h(y)] = \int_0^y K(y,u)h(u) \, du$$

Now
$$|T[h(y)]| \le M\Gamma(k)I_k|h(y)|$$
$$|T^n[h(y)]| \le M^n\Gamma^n(k)I_{nk}|h(y)|$$

where I_k and I_{nk} are the Riemann-Liouville integrals.

Exercise 7-9 By using the second inequality to estimate $|T^n \left[\partial^{n+m} g(x,y)/\partial x^{n+m} \right]|$ show that the series

$$\sum_{n=1}^{\infty} T^n \left[\partial^{n+m} g(x,y)/\partial x^{n+m} \right] \qquad m = 0, 1, \ldots$$

is uniformly convergent in any bounded subset of $0 \leq x \leq 1$, $0 \leq y$. Thus conclude that f has continuous derivatives of all orders in this region. The mth derivative is obtained by differentiating the above expression m times for f with respect to x. Show that the solution is unique by writing $F(x,y) = f_1(x,y) - f_2(x,y)$ where f_1 and f_2 are two solutions, estimating $|F| = |T^n F|$ and showing that $T^n F \to 0$ as $n \to \infty$, that is, that $F = 0$.

SUCCESSIVE APPROXIMATIONS

Consider the equation

$$\frac{\partial \varphi}{\partial x} + \frac{\partial \varphi}{\partial y} = \int_x^y [\varphi(x,\xi)F(\xi,y) - \varphi(\xi,y)F(x,\xi)] \, d\xi$$

Letting $u = (y - x)/2$ and $v = (y + x)/2$, we have

$$\varphi(x,y) = \int_u^v f(\zeta - u, \zeta + u) \, d\zeta + \theta(y - x)$$

where θ is an arbitrary differentiable function and vanishes like φ for $x = y$. If we put

$$\theta(y - x) = \int_0^{y-x} g(s) \, ds$$

then the solution can be obtained by successive approximations using[37]

$$\varphi(x,y) = \int_0^{2u} g(s)\psi(s;x,y) \, ds$$

where $\quad \psi(s;x,y) = \sum_{n=1}^{\infty} \psi_n(s;x,y)$

$$\psi_1 = 1$$

$$\psi_n = - \int_u^v d\zeta \int_0^{2u} d\xi [\psi_{n-1}(s; \zeta - u + \xi, \zeta + u)$$
$$\cdot F(\zeta - u, \zeta - u + \xi)$$
$$- \psi_{n-1}(s; \zeta - u, \xi + \zeta - u)F(\xi + \zeta - u, \zeta + u)]$$

Remark: In passing, we mention that one can apply the numerical method of Runge-Kutta to the solution of an integrodifferential equation of the form[30]

$$f'(x) = F\{x,f(x), \int_{x_0}^x g[x,f(x),s,f(s),f'(s)] \, ds\}$$

with $f(x_0) = y_0$, by reduction to the canonical form

$$f'(x) = F[x,f(x),G(x)]$$
$$\varphi(x_0) = y_0$$

where $\quad\quad\quad G(x) = \int_{x_0}^x H[x,f(x),s,f(s),G(s)] \, ds$

$$H(x,y,s,z,w) = g[x,y,s,z,F(s,z,w)]$$

The discrete system[37]

$$f_i(y_1, \ldots, y_n) = z_i \qquad i = 1, \ldots, n$$

can be replaced by the continuous functional equation

$$F[y(t),s] = z(s)$$

where t, with $a \le t \le b$, replaces the indices $1, \ldots, n$, and the parameter s replaces the index i. Similarly the system of ordinary differential equations

$$\frac{dy_i}{dx} = f_i(y_1, \ldots, y_n, x) \qquad i = 1, \ldots, n$$

can be replaced by

$$\frac{\partial y(x,s)}{\partial x} = F[y(x,t);x,s] \qquad a \le t \le b \tag{7-90}$$

which is a first-order ordinary functional-differential equation and also an integrodifferential equation for the reason that we can expand F in a series of integrals using an extension of Taylor's theorem. We write

$$f(\epsilon) \equiv F[y(t) + \epsilon\varphi(t)]$$

for arbitrary $\varphi(t)$ with specified end-point condition and from

$$f(1) = f(0) + \frac{\partial f}{\partial \epsilon}\Big|_{\epsilon=0} + \cdots + \frac{1}{(n-1)!}\frac{\partial^{n-1}f}{\partial\epsilon^{n-1}}\Big|_{\epsilon=0}$$
$$+ \frac{1}{n!}\frac{\partial^n f}{\partial\epsilon^n}\Big|_{\epsilon=\theta} \qquad 0 < \theta < 1$$

we have

$$F[y(t) + \varphi(t)] = F[y(t)] + \sum_{i=1}^{n-1}\frac{1}{i!}\int_a^b \cdots \int_a^b F^{(i)}[y(t);\zeta_1, \ldots, \zeta_i]$$
$$\cdot \varphi(\zeta_1) \cdots \varphi(\zeta_i)\, d\zeta_1 \cdots d\zeta_i + \frac{1}{n!}\int_a^b \cdots \int_a^b$$
$$F^{(n)}[y(t) + \theta\varphi(t), \zeta_1, \ldots, \zeta_n]\varphi(\zeta_1) \cdots \varphi(\zeta_n)\, d\zeta_1 \cdots d\zeta_n$$

If the functional F has derivatives of all orders, then

$$F[y(t) + \varphi(t)] = F[y(t)] + \sum_{n=1}^{\infty}\frac{1}{n!}\int_a^b \cdots \int_a^b F^{(n)}[y(t);\zeta_1 \cdots \zeta_n]$$
$$\cdot \varphi(\zeta_1) \cdots \varphi(\zeta_n)\, d\zeta_1 \cdots d\zeta_n$$

Note that by integrating from x_0 to x, (7-90) can be reduced to an integral equation

$$y(x,s) = y(x_0,s) + \int_{x_0}^x F[y(x,t);x,s]\, dx \qquad a \le t \le b$$

and then one can use Picard's method for its solution.

PALM'S EQUATION

To solve Palm's recurrent integrodifferential equation, given in Sec. 7-1, we write[19]

$$f_n(t) = \sum_{k=0}^{n} C_{nk} e^{-a_{nk}t}$$

where $\quad C_{nk} = \dfrac{B_{n-1}(1 - a_{nk})}{B_n'(-a_{nk})}$

and $\quad B_n(t) = \lambda^{n+1} + \sum_{j=0}^{n} \binom{n+1}{j} t(t+1) \cdots (t+n-j)\lambda^j$

with $\quad B_n'(t) = \dfrac{dB_n(t)}{dt}$

and where a_{nk}, $k = 1, \ldots, n$, are the roots of $B_n(t) = 0$.

We have $\quad\quad\quad\quad a_{ni} > a_{n,i-1} + 1 \quad\quad 1 \le i \le n$

$$-a_{n0} < -a_{n+1,0} < 0$$

$$\frac{B_{n-1}(1 - a_{nk})}{B_n'(-a_{nk})} \ge 0 \quad\quad\quad \text{for all } n, k$$

Since $f_n(0) = 1$ for all n,

$$\sum_{k=0}^{n} C_{nk} = 1 \quad\quad 0 \le C_{nk} \le 1, 0 \le k \le n$$

Exercise 7-10 Show that

$$f'(x) = e^{-x^2}(1 + x^2) + x - f(x) + \int_0^x x^2 e^{-sx} f(s) f'(s)\, ds$$

with $f(0) = 0$ has the solution $f(x) = x$.

Exercise 7-11 Prove that

$$f'(x) = 1 + 2x - f(x) + \int_0^x x(1 + 2x)e^{s(x-s)}f(s)\, ds$$

with $f(0) = 1$ has the solution $f(x) = e^{x^2}$.

Exercise 7-12 Consider the equation

$$\dot{x}(t) = x(t) + \mu f(t) + \int_0^t K(t,s)x(s)\, ds$$

Write down its solution by iteration. *Hint:* Solve $\dot{x} = x(t) + \mu f(t)$ for the first trial. Assume that $K(t,s) \equiv \tilde{K}(t - s)$ and develop the solution. Finally let $K(t,s) = \sum_{i=1}^{n} a_i(t)b_i(s)$ and give a solution method.

Exercise 7-13   Consider the integrodifferential equation

$$\sum_{k=0}^{n} \int_{-\infty}^{\infty} f^{(k)}(x - y) \, dg_k(y) = 0$$

where $f(x)$ is the unknown function.   One method of solution is to assume an expansion for $f(x)$ which yields a solution in the form $\sum_{m=-\infty}^{\infty} A_m(x) \exp(\omega_m x)$, where $A_m(x)$ are polynomials and ω_m are zeros of

$$K(\omega) \equiv \sum_{k=0}^{n} \omega^k \int_{-\infty}^{\infty} e^{-\omega y} \, dg_k(y)$$

and the degree of each $A_m(x)$ is less than the order of the zero of $K(\omega)$ at $\omega = \omega_m$. Justify the method and the expansion.

7-7   STABILITY

In a preceding section we considered the nonlinear integrodifferential equation[4]

$$\dot{x} = \omega - \int_0^t k(t - u) F[x(u)] \, du \tag{7-91}$$

Now we determine conditions on ω, k, and F such that a solution x of (7-91) is stable in the sense that $\dot{x}(t) \to 0$ as $t \to \infty$.

To do this Beneš uses the Schauder-Tychonoff fixed-point theorem. Equation (7-91) can be reduced to the integral equation

$$x(t) = x(0) + \omega t - \int_0^t K(t - u) F[x(u)] \, du \tag{7-92}$$

where $K(t) = \int_0^t k(z) \, dz$.

We rewrite (7-92) as the functional equation

$$g(t) = F\left(x(0) + \omega t - \int_0^t K(t - u) g(u) \, du\right) \qquad t \geq 0 \tag{7-93}$$

and seek a solution g of this equation in the space of bounded continuous functions $B \cap C$.

If we define a transformation T by

$$Tg(t) = \begin{cases} F\left(x(0) + tK(\infty) \limsup_{t \to \infty} g(t) - \int_0^t K(t - u) g(u) \, du\right) & t \geq 0 \\ F(x(0)) & t < 0 \end{cases}$$

then we shall look for fixed points of T. We have replaced ω by

$K(\infty) \lim_{t \to \infty} \sup g(t)$ since for stability we must have $g(t) \to \omega/K(\infty)$ in (7-93).

The following metric will be used in $B \cap C$. For $g_1, \, g_2 \, \epsilon \, B \cap C$, define

$$\rho(g_1, g_2) = \sum_{m=1}^{\infty} 2^{-m} \max_{-m \leq u \leq m} |g_1 - g_2| + \lim_{t \to \infty} \sup |g_1 - g_2| \qquad (7\text{-}94)$$

This is a modification of the topology of uniform convergence on compact sets.

It can be shown that the above metric makes $B \cap C$ a locally convex linear topological space. We leave as an exercise to show that if $g_n \to g$ uniformly then $g_n \to g$ in the above metric.

The following two additional assumptions are made:

1. There is a positive integrable function ψ such that

$$|K(\infty) - K(t)| \leq \psi(t) \qquad t \geq 0$$

and $\psi(t) \to 0$ as $t \to \infty$.

2. There is a positive integrable solution h of the integral inequality

$$\sup_{u} F(u) \int_{t}^{\infty} \psi(u) \, du + |K(\infty)| \int_{t}^{\infty} h(u) \, du + \int_{0}^{t} \psi(t - u) h(u) \, du$$
$$\leq \beta^{-1} h(t)$$

for $t \geq 0$, such that $h(t) \to 0$ as $t \to \infty$. β is the constant given in a Lipschitz condition on F.

For convenience of notation below we set

$$Lg(t) = \begin{cases} x(0) + \bar{g} \int_{0}^{t} [K(\infty) - K(u)] \, du \\ x(0) + \int_{0}^{t} K(t - u)[\bar{g} - g(u)] \, du & \begin{array}{l} t \geq 0 \\ t < 0 \end{array} \end{cases}$$

[where $\bar{g} = \lim_{t \to \infty} \sup g(t)$] so that $Tg(t) = F(Lg(t))$.

Define S to be functions g in $B \cap C$ such that:

1. $g(\infty)$ exists.
2. $g(t) = F(x(0))$ for $t \leq 0$.
3. $|g| \leq \sup_{u} |F(u)|$.
4. $|g(\infty) - g(t)| \leq h(t)$ for $t \geq 0$.
5. g has modulus of continuity m given by

$$m(\epsilon) = \beta \left[4 \sup_{u} |F(u)| k\epsilon + m_0(\epsilon) \int_{0}^{\infty} h(u) \, du \right] \qquad \epsilon > 0$$

where m_0 is modulus of continuity of K.

Definition: If f is given on (a,b) and I is a subinterval of (a,b), the oscillation of f in I is defined by[14]

$$\omega(f,I) = \sup_{x \in I} f(x) - \inf_{x \in I} f(x)$$

The modulus of continuity of a continuous function f on $[a,b]$ is defined by

$$m(\delta,f) = \sup \omega(f,I)$$

for all intervals I in $[a,b]$ of length $\leq \delta$. Note that m is continuous and nondecreasing in δ and tends to zero as $\delta \to 0$.

We show that the set S and the mapping T satisfy the hypotheses of the foregoing fixed-point theorem. Thus we must show that S is compact and T is a continuous map of S into itself.

The reader can verify that S is closed. To show that S is compact it is sufficient to show that it is sequentially compact.

Let $\{g_n\}$ be any sequence of S. For each u, $g_n(\infty)$ exists, and $|g_n(\infty)| \leq \sup_u |F(u)|$. Select a subsequence g_{n_i} such that $g_{n_i}(\infty)$ converges as $i \to \infty$. From this, select a subsequence $g_{n_{i_j}}$ for which $g_{n_{i_j}}(t)$ converges uniformly in t for every compact set. This can be done by the compactness of the real numbers. Since the metric we have defined is essentially that of uniform convergence on compact sets, the sequence $g_{n_{i_j}}$ converges in the metric. Thus S is compact.

Now we show that T is continuous on S and leave as an exercise the fact that T maps S into S.

Let $\{g_n\}$ be a sequence of S converging to g. Then $g \in S$ and for $t \geq 0$

$$Tg_n(t) - Tg(t) = F(\alpha(t) + \delta(t)) - F(\alpha(t)) \tag{7-95}$$

where

$$\alpha(t) = x(0) + g(\infty)K(\infty)t - \int_0^t K(t - u)g(u)\, du$$

and

$$\delta(t) = \int_0^t K(t - u)[g_n(u) - g(u)]\, du - tK(\infty)[g_n(\infty) - g(\infty)]$$

Using (7-95) and the fact that F satisfies the Lipschitz condition, we need to show that $\delta(t)$ can be made small uniformly in t by choosing n large enough.

From the above definition of $\delta(t)$ it can be shown that for all t and y

$$|\delta(t)| \leq \sup_{0 \leq u \leq y} |g(u) - g_n(u)| \int_0^y |K(\infty) - K(u)|\, du + 2|K(\infty)|h(y)$$
$$+ 2 \sup_u |F(u)| \int_y^\infty |K(\infty) - K(u)|\, du + |K(\infty)| \int_0^y |g(\infty)$$
$$- g_n(\infty) - g(u) + g_n(u)|\, du$$

Choose y large enough to make the second and third terms small, and then choose n large. Then if $g_n \to g$ in the given metric, $Tg_n \to Tg$ uniformly and thus in the given metric.

We now define the functional M by

$$Mg = \int_0^\infty [g(\infty) - g(u)]\, du$$

whenever this integral converges absolutely. If $g \,\epsilon\, S$,

$$|Mg| \le \int_0^\infty h(u)\, du \equiv \|h\|$$

Theorem 7-2   Let $MK(.) > 0$ and $K(\infty) > 0$, and let x and y be real numbers, with $0 < x < y \le \sup_u |F(u)|$.

Define intervals

$$I = [xK(\infty), yK(\infty)]$$
$$J = [x(0) + xMK - K(\infty)\|h\|,\ x(0) + yMK + K(\infty)\|h\|]$$

and suppose the set $Z = \{u \,|\, x \le F(u) \le y\}$ has the property that $J \subseteq Z$.

Then there exists an $\omega \,\epsilon\, I$ for which the solution of (7-92) has the properties:

1. $\dot{x}(t) \to 0$ as $t \to \infty$.
2. $|F(x(\infty)) - F(x(t))| \le h(t)$.
3. $x(\infty) \,\epsilon\, J$.

Proof: Let $A = \{g \,\epsilon\, S \,|\, x \le g(\infty) \le y\}$. A is convex, and the same argument as for S shows that A is closed. For $g \,\epsilon\, A$ we have

$$|Mg| \le \|h\| \qquad Tg(\infty) = F(x(0) + g(\infty)MK + K(\infty)Mg)$$

Since $J \subseteq Z$, $Lg(\infty) \,\epsilon\, Z$, and so $Tg \,\epsilon\, A$. That is, $TA \subseteq A$. Since A is a closed subset of a compact set, it is also compact. Thus T is a continuous map of a compact convex set into itself.

From the Schauder-Tychonoff theorem there is a $g^* \,\epsilon\, A$ such that $Tg^* = g^*$. Thus g^* satisfies the functional equation (7-93). Properties 2 and 3 are a result of $g^* \,\epsilon\, A$, when $x(t)$ is defined as $Lg^*(t)$. An abelian argument may now be use to show that $\dot{x}(t) \to 0$ as $t \to \infty$.

Exercise 7-14   Show that if $g_n \to g$ uniformly, then $g_n \to g$ in the given metric.

Exercise 7-15   Show that S is closed.

Exercise 7-16   Show that $TS \subseteq S$.

Levin and Nohel[19] have established a stability theorem for the equation

$$\dot{x}(t) = -\int_0^t a(t - \tau) g(x(\tau))\, d\tau$$

Note that the linear case is obtained if $g(x) \equiv x$. Also note that if $a(t) \equiv a(0)$, the equation reduces to the nonlinear oscillator

$$\ddot{x} + a(0)g(x) = 0$$

Theorem 7-3   Let $a(t)$ and $g(x)$ satisfy

$$a(t) \in C[0,\infty), (-1)^k a^{(k)}(t) \geq 0$$

$$\text{(completely monotone)} \quad 0 < t < \infty \,; k = 0, 1, \ldots$$

$$g(x) \in C(-\infty,\infty), xg(x) > 0 \quad x \neq 0, \ G(x) \equiv \int_0^x g(\xi)\, d\xi \to \infty, |x| \to \infty$$

If $a(t) \not\equiv a(0)$ and $y(t)$ is any solution on $0 \leq t < \infty$, then

$$\lim_{t \to \infty} y^{(j)}(t) = 0 \quad j = 0, 1, 2, \ldots$$

It is interesting to note that the use of Lyapunov theory to prove the theorem (the proof is long and will not be given here) leads to requirements on $a(t)$ which coincide with those given in the hypotheses.   Let

$$V(t) = G(y(t)) + \frac{1}{2} \int_0^t \int_0^t a(\tau + s)g(u(t - \tau))g(u(t - s))\, d\tau\, ds$$

be the Lyapunov function.

If also $a'(t) \in L_1(0,T)$ for each $0 < T < \infty$, $V(t)$ is nonnegative if the second term is nonnegative, which is the case if $a(\tau + s)$ is of positive type on the square $0 < \tau, s < t$ for each $0 < t < \infty$, that is, if and only if we have

$$a(t) = \int_{-\infty}^{\infty} \exp\left[-\xi t\right] d\alpha(\xi) \quad 0 < t < \infty$$

where $\alpha(\xi)$ is nondecreasing on $-\infty < \xi < \infty$.

Now

$$V'(t) = \int_0^t \int_0^t a'(\tau + s)g(u(t - \tau))g(u(t - s)\, d\tau\, ds$$

is nonpositive if $-a'(\tau + s)$ is of positive type on the square $0 < \tau$, $s < t$ for each $0 < t < \infty$.   This is compatible with the condition on $a(t)$ if and only if $\alpha(-\infty) = \alpha(0-)$.   Indeed the conditions on $a(t)$ given in the theorem and the requirement that $a(t)$ be representable in the above form are equivalent as asserted by a theorem of Bernstein (see Widder, "The Laplace Transform" p. 160).

Levin and Nohel[20] have also considered the delay equation

$$x(t) = -\frac{1}{L} \int_{t-L}^{t} (L - (t - \tau))g(x(\tau))\, d\tau \quad 0 < t < \infty$$

where $L > 0$ is a given constant and $g(x)$ is a possibly nonlinear restoring force of a given spring.   On differentiation, this equation becomes

$$\ddot{x}(t) + g(x(t)) = \frac{1}{L} \int_{t-L}^{t} g(x(\tau))\, d\tau \quad 0 \leq t < \infty$$

For large t the behavior of this equation and of the following equation

$$\ddot{x} + g(x) = 0$$

whose equivalent system is $\dot{x} = y$, $\dot{y} = -g(x)$, are shown to be related.

Exercise 7-17 Give a brief justification for the derivation of the linear equation due to Volterra

$$\ddot{x}(t) + cx(t) = \int_{t-L}^{t} F(t - \tau)x(\tau) \, d\tau$$

where c and L are given positive constants.

An existence and uniqueness theorem of the initial-value problem

$$\frac{\partial \varphi}{\partial t} = \int_a^b K(x,s,t,\varphi(s,t)) \, ds + F(x,t,\varphi(x,t)) \tag{7-96}$$

with $\varphi(x,t_0) = f(x)$, $a \leq x \leq b$ for some $t_0 \geq 0$, has been proved by Liberman,[22] with a Lipschitz condition with respect to φ imposed on the given function K and F in the domain $a \leq x \leq b$, $a \leq s \leq b$, $|\varphi| \leq r$, $0 \leq t < \infty$.

The trivial solution of (7-56) is called *uniformly asymptotically stable* with respect to x if (1) given $\epsilon > 0$, one can find $\delta > 0$ and $T \geq t_0$ such that every solution φ with $|\varphi(x,t_0)| < \delta$ satisfies $|\varphi(x,t)| < \epsilon$ for all $t > T$, $a \leq x \leq b$; (2) corresponding to every $\epsilon > 0$ there exists a $\delta > 0$ such that $|\varphi(x,t_0)| < \delta$ implies $|\varphi(x,t)| < \epsilon$ for all $t > t_0$, $a \leq x \leq b$. If we replace K and F in (7-96) by $\bar{K}$ and $\bar{F}$ which satisfy the same assumptions as K and F, and if there exist $\gamma_1(t)$ and $r_1(t)$ such that in the given domain

$$|\bar{K} - K| < \gamma_1(t)|\varphi| \tag{7-97}$$
$$|\bar{F} - F| < r_1(t)|\varphi| \tag{7-98}$$

we have the following theorem.

***Theorem* 7-4** If (1) the trivial solution of (7-96) is uniformly asymptotically stable and (2) $\alpha > 0$, $B \geq 1$ are constants such that every solution $\varphi(x,t)$ of (7-96) with $|\varphi(x,t_0)| < \delta < r$ satisfies

$$|\varphi(x,t)| < B\delta e^{-\alpha(t-t_0)} \qquad \text{for } t > t_0 \tag{7-99}$$

and (3) the relations (7-97) and (7-98) are satisfied, and (4)

$$T^{-1} \int_t^{t+T} \gamma_1(s) \, ds < \eta_1 \qquad T^{-1} \int_t^{t+T} r_1(s) \, ds < \eta_2 T = \alpha^{-1} \log 4B$$

and (5) η_1 and η_2 are sufficiently small and γ_1 and γ_2 are also sufficiently small in the mean, then the solution $\bar{\varphi} \equiv 0$ of the equation in $\bar{K}$ and $\bar{F}$ is also uniformly asymptotically stable, and each solution $\bar{\varphi}$ in the domain satisfies (7-99).

A similar theorem for persistent disturbances is also available.

The foregoing results are applicable to the integrodifferential equation with retardation obtained from (7-96) by replacing $\varphi(s,t)$ by $\varphi(s, t - \tau_1(t))$, $\varphi(s, t - \tau_2(t))$, . . . , $\varphi(s, t - \tau_n(t))$ in both K and F. Here $\tau_i \geq 0$, $i = 1, \ldots, n$. If $\tau_i(t) \leq \tau$, the initial function $f(x,t)$ is defined for $a \leq x \leq b$, $t_0 - \tau \leq t \leq t_0$ (Ref. 23).

REFERENCES

1.   Anselone, P. M., and H. F. Bueckner: On a Difference-Integral Equation, *J. Math. Mech.*, vol. 11, no. 1, pp. 81–100, 1962.

2.   Baumann, V.: Eine nichtlineare Integrodifferentialgleichung der Wärmeübertragung bei Wärmeleitung und -strahlung, *Math. Z.*, vol. 64, pp. 353–384, 1956.

3.   Bellman R., R. Kalaba, and G. M. Wing: Invariant Imbedding and Neutron Transport in a Rod of Changing Length, *Proc. Nat. Acad. Sci. U.S.A.*, vol. 46, pp. 128–130, 1960.

4.   Beneš, V. E.: A Fixed Point Method for Studying the Stability of a Class of Integro-differential Equations, *J. Math. and Phys.*, vol. 40, no. 1, April, 1961.

5.   ———: Ultimately Periodic Solutions to a Non-linear Integro-differential Equation, *Bell System Tech. J.*, vol. 41, no. 1, pp. 257–268, January, 1962.

6.   Buckingham, R. A., and P. G. Burke: The Solution of Integro-differential Equations Occurring in Nuclear Collision Problems, in Ref. 33.

7.   Bueckner, H. F.: An Iterative Method for Solving Nonlinear Integral Equations, in Ref. 33.

8.   Chandrasekhar, S., and G. Münch: The Theory of the Fluctuations in Brightness of the Milky Way. I, II, III, *Astrophys. J.*, vol. 112, p. 380, 1950; vol. 114, p. 110, 1951; vol. 115, p. 94, 1952.

9.   Chapman, S., and T. G. Cowling: "The Mathematical Theory of Nonuniform Gases," 2d ed., Cambridge University Press, London, 1952.

10.   Dascalopoulos, D.: Existenz und Eindeutigkeit der Lösung von linearen Volterraschen Integrodifferentialgleichungen und Systemen vom Faltungstyp mit konstanten Koeffizienten, *Math. Ann.*, vol. 148, pp. 67–82, 1962.

11.   Davis, H. T.: "Introduction to Nonlinear Differential and Integral Equations," Dover Publications, Inc., New York, 1961.

12.   Douglas, J., Jr., and B. F. Jones, Jr.: Numerical Methods for Integro-differential Equations of Parabolic and Hyperbolic Types, *Numer. Math.*, vol. 4, pp. 96–102, 1962.

13.   Hellman, O.: Asymptotic Solution of Palm's Integral Equation, *Operations Res.*, vol. 2, no. 4, p. 553, July–August, 1963.

14.   Hille, E.: "Analysis," vol. 1, The Blaisdell Publishing Company, 1964.

15.   Huang, K.: "Statistical Mechanics," John Wiley & Sons, Inc., New York, 1963.

16.   Jancel, R., and T. Kahan: Relations entre les diverses solutions de l'équation intégro-différentielle de Boltzmann, *C. R. Acad. Sci. Paris*, vol. 254, pp. 1929–1931, 1962.

17.   Keller, H. B.: Convergence of the Discrete Ordinate Method for the Anisotropic Scattering Transport Equation, in Ref. 33.

18.   Levin, J. J.: The Asymptotic Behavior of the Solution of a Volterra Equation, *Proc. Amer. Math. Soc.*, vol. 14, pp. 534–541, 1963.

19.   ——— and J. A. Nohel: Note on a Linear Volterra Equation, *Proc. Amer. Math. Soc.*, vol. 14, no. 6, pp. 924–929, 1963.

20. —— and ——: On a Nonlinear Delay Equation, *J. Math. Anal. Appl.*, vol. 8, no. 1, pp. 31–44, 1964.

21. —— and ——: On a System of Integro-differential Equations Occurring in Reactor Dynamics, II, *Arch. Rational Mech. Anal.*, vol. 11, no. 3, pp. 210–243, 1962.

22. Liberman, L. H.: On the Stability of Solutions of Integro-differential Equations, *Izv. Vyss. Ucebn. Zaved. Matematika*, nos. 3 and 4, pp. 142–151, 1958; *Math. Reviews*, August, 1964.

23. ——: Integro-differential Equations with Lagging Argument and the Stability of Their Solutions, *Izv. Vyss. Ucebn. Zaved. Matematika*, nos. 6 and 7, pp. 161–175, 1958; *Math. Reviews*, August, 1964.

24. ——: The Problem of Stability of the Solutions of Integro-difference Equations, *Izv. Vyss. Ucebn. Zaved. Matematika*, no. 1(8), pp. 91–104, 1959; *Math. Reviews*, August, 1964.

25. Lichtenstein, L.: "Vorlesungen über einige Klassen nichtlinearer Integralgleichungen und Integro-Gleichungen nebst Anwendungen," Springer-Verlag OHG, Berlin, 1931.

26. McCarroll, R.: Solution of Integro-differential Equations Arising in the Theory of Atomic Scattering, in Ref. 33.

27. Muskhelishvili, N. I.: "Singular Integral Equations," Erven P. Noordhoff, NV, Groningen, Netherlands, 1953.

28. Pomey, L.: Sur les équations intégro-différentielles linéaires à plusieurs variables, *C. R. Acad. Sci. Paris*, vol. 177, p. 1094, 1923.

29. Pouzet, P.: Méthode d'intégration numérique des équations intégrales et intégro-différentielles du type Volterra de seconde espèce formulés de Runge-Kutta, in Ref. 33.

30. ——: Algorithme de traitement des équations intégro-différentielles du 1ᵉʳ ordre de type Volterra, *Rev. Française Traitment Information*, vol. 7, no. 3, pp. 251–254, 1964.

31. Pucci, C.: An Integro-differential Equation, in Ref. 33.

32. Ramakrishnan, A., and P. M. Matthews: On the Solution of an Integral Equation of Chandrasekhar and Münch, *Astrophys. J.*, January, 1954.

33. Rome Symposium on the Numerical Treatment of Ordinary Differential Equations, Integral and Integro-differential Equations, "Proceedings of the Rome Symposium (PICC), September 20–24, 1960," Birkhäuser Verlag, Basel, 1961.

34. Saaty, T. L.: "Elements of Queueing Theory," McGraw-Hill Book Company, New York, 1961.

35. Slugin, S. N.: Approximate Solution of Integro-differential Equations by Chaplygin's Method, *Izv. Vyss. Ucebn. Zaved. Mathematika*, no. 1, pp. 211–221, 1957.

36. Underhill, L. H.: A Self-consistent One-group Method for the Boltzmann Transport Equation in Neutronics, in Ref. 33.

37. Volterra V.: "Theory of Functionals and of Integral and Integro-differential Equations," Dover Publications, Inc., New York, 1959.

38. Wing, G. M.: Analysis of a Problem of Neutron Transport in a Changing Medium, *J. Math. Mech.*, vol. 2, no. 1, 1962.

39. Zeragija, P. K.: Solution of a Non-linear Integro-differential Equation by the Method of Upper and Lower Functions, *Soobsc. Akad. Nauk Gruzin. SSR*, vol. 34, pp. 3–10, 1964; *Math. Reviews*, vol. 29, p. 302, 1965.

EIGHT

STOCHASTIC DIFFERENTIAL EQUATIONS†

8-1 INTRODUCTION

The behavior of natural phenomena is governed by chance and does not follow strict deterministic laws. For example, electric current exhibits visible chance fluctuations; the trajectory of a projectile is influenced by the uncertainty of the initial speed; and the demand for goods and services varies in some random manner.

Consequently, mathematical descriptions of natural systems in terms of suitable equations for appropriate functions, as presented earlier in this book, must be modified to account for chance effects. This can be achieved by regarding these system functions not as deterministic variables but rather as *random variables* with associated probabilistic, i.e., stochastic, properties. In this way, equations governing the behavior of systems become equations for random variables, i.e., *stochastic equations*. This change of interpretation corresponds to the change of the point of view from the deterministic to the stochastic mathematical model.

† This chapter was written by R. Syski.

This chapter is devoted to ordinary stochastic differential equations, say

$$\dot{x}(t) = K[t,x(t)] \tag{8-1}$$

where $x(t)$ is the unknown random variable, $\dot{x}(t) = dx(t)/dt$, and $K(.,.)$ is a given operator; the variable t is usually interpreted as time. The solution $x(t)$ of the equation is also referred to as the random (or stochastic) integral of the equation.

The main questions concerning stochastic differential equations are essentially the same as those for nonrandom equations, viz., the conditions for existence and uniqueness of solutions and determination of explicit solutions and their properties. The introduction of probabilistic aspects, however, leads immediately to several new problems, as well as to specific mathematical difficulties. Consequently, the study of stochastic equations emphasizes probabilistic problems and in this respect differs from its nonrandom counterpart.

First of all, a random variable $x(t)$, for fixed t, is a measurable function on a given probability space, and when t varies, one obtains a family of random variables (indexed by t), i.e., by definition, a *stochastic process*, or, in different words, a *random function* $x(.)$, or briefly x. Thus x is in fact a mapping from the index space T, where $t \in T$, into a space $\mathfrak{IC}$ of random variables, and $x(t) \in \mathfrak{IC}$ is a value of x at a point t. Consequently, Eq. (8-1) is a differential equation for functions taking values in an abstract function space $\mathfrak{IC}$. This immediately requires that proper meaning should be attached to concepts like the continuity, differentiability, and integration of such abstract-valued functions. Equations of this type with $\mathfrak{IC}$ a Banach space are now studied intensively, and in probabilistic situations the best-known special case is that of $\mathfrak{IC}$ being the space L_2 (of functions with integrable absolute square) which corresponds to the space of random variables having second moments. In this connection it must be added that the concept of a stochastic integral, which arises in a natural way, has received much attention in its own right; such integrals exhibit peculiar properties in some cases.

Furthermore, the fact that $x(t)$ is a function on a probability space indicates that x is actually a function of two variables t and ω (a point in probability space). Consequently, Eq. (8-1) is formally a partial differential equation for a real-valued function $x = x(.,.)$ of two variables, with variable ω suppressed. These two interpretations of the random function x are evidently linked together, and their interrelationship reflects the relationship between the *stochastic integrals* and the *sample-function integrals*, the latter being simply the integrals of Eq. (8-1) regarded as the usual nonrandom equation.

The mathematical difficulties mentioned above are responsible for the fact that the development of the theory of stochastic differential equations did not follow the same path as the theory of nonrandom equations; the concept of stability, for example, is still not in its final form. Instead, the theory is more concerned with probabilistic aspects, i.e., with the study of stochastic processes defined by the equations. To solve the stochastic equation means essentially to find the stochastic properties of the random variables $x(t)$. The explicit form of the solution is useful, but it is not the main task. More precisely, one is interested in the determination of the distribution of $x(t)$ or joint distributions at several instants t. One also seeks characteristic functions, moments, and mean values.

Most commonly one tries to solve the equation as if it were deterministic and then to investigate stochastic properties of $x(t)$ in terms of given stochastic properties of random elements entering into the form of $x(t)$. This procedure (essentially the sample-function approach) does not always lead to correct results unless the exact conditions of its validity are established. Moreover, in some important cases the explicit form of $x(t)$ in terms of random elements cannot be obtained.

It is convenient to distinguish three basic types of stochastic differential equations according to the form in which random elements enter into equations:

1. Random initial conditions.
2. Random forcing functions.
3. Random coefficients.

In the first case, the given probabilistic properties of the initial conditions (regarded as random variables) determine those of the solution $x(t)$ through the explicit form of $x(t)$. The typical example is when $x(t)$ represents the motion of a particle subject to uncertainty only in its initial position and governed by deterministic forces.

In the second case, the stochastic process representing random forcing functions is given, and the properties of the solution $x(t)$ are developed through probabilistic argument. The typical example of this sort is the study of the output of a physical system when the input is a brownian-motion process. (Here, the nonexistence of the derivative leads to additional difficulties.)

Finally, in the third case, the parameters of the system are regarded as the random variables, and the stochastic properties of $x(t)$ reflect the influence of randomness imposed by the structure of the system. For example, the output of an electric circuit with randomly varying capacity is described by equations of this type.

These three types are not mutually exclusive, and most of the discussion is concerned with equations of mixed types. As mentioned earlier, all mixed cases are special forms of Eq. (8-1) with proper interpretation of K.

This chapter is divided in three main parts, according to the above classification. The presentation starts with equations with random initial conditions (Sec. 8-2) because they require a minimum of concepts from probability theory and thus form a natural bridge between deterministic and stochastic equations.

This section is followed by equations with a random forcing function (Sec. 8-3), stochastic operations being taken in the sense of the mean square. The emphasis is on the brownian-motion process and the Poisson process, and the section closes with the discussion of the Ito equation and the Markoff process. The treatment is entirely probabilistic, and the required tools are explained in detail.

Equations with random coefficients are discussed in Sec. 8-4, but the treatment is limited to some particular cases because the subject is less developed than the previous two. Section 8-5 is devoted to general properties (existence theorems, the sample-function approach, etc.), and the appendix in Sec. 8-6 contains some basic concepts from probability theory used in the main text.

The whole chapter is written from the point of view of applied mathematics, but familiarity with probability theory is not assumed. Hence, the probabilistic concepts are introduced gradually; Sec. 8-2 uses only very elementary concepts, and the presentation progresses at a reasonable rate to highly advanced concepts from the theory of stochastic processes (in Sec. 8-5). For this reason proofs using concepts that are too technical are given only in outline or are omitted. The emphasis is on ideas, and the interested reader can find details of proofs in the literature.

However, the chapter is written as an introductory text and not as a survey for specialists. Illustrative examples are frequently given to help in understanding, and several exercises are included.

The list of references contains those works which were helpful in the preparation of the text, but a few additional references are also included. However, no claim to completeness is made, and many specialized papers are deliberately omitted.

Much work has been done in the theory of stochastic equations, but even more remains to be done. Many challenging and difficult problems still await solution. In general, the study of stochastic processes defined by stochastic differential equations is a large open field. Among the topics not discussed in this chapter are partial differential equations and integral equations and equations for abstract-valued random variables.

As usual, the question of good notation is the crucial one, especially in a text on probability theory. Thus, the important distinction between a random variable and its value is strictly observed here, and different letters are used for them. Random variables are denoted by italic letters, $x(t)$, $y(t)$, $z(t)$, etc., but for their values Greek letters, λ, μ, ζ, etc., will be used. Thus, the value of $x(t)$ at a point ω is denoted by $x(t,\omega) = \lambda$, say.

As is usual, ω will be suppressed when no confusion can arise. The stochastic process (random function) will be denoted by x or occasionally by $x(.)$, or more specifically $\{x(t), 0 \leq t < \infty\}$.

8-2 RANDOM INITIAL CONDITIONS

INTRODUCTION

The passage from deterministic, i.e., nonrandom, differential equations to stochastic equations is perhaps the easiest when random elements enter only through initial conditions. In such situations the probabilistic argument is the simplest, and the stochastic properties of the solutions are deterministically related to the stochastic properties of the initial conditions. The practical problem is the determination of the distribution function of new random variables related functionally to the random variables representing the initial conditions, in terms of the distribution function of these initial conditions. In principle, this task is reduced to an ordinary integration. Nevertheless, the case of random initial conditions provides a good introduction to more complicated probabilistic aspects and at the same time presents many interesting problems in its own right.

Consider, for the purpose of illustration, a linear differential equation of the second order

$$\frac{d^2x(t)}{dt^2} + a(t)\,\frac{dx(t)}{dt} + b(t)x(t) = 0$$

Let $x(t)$ be a solution satisfying the given initial conditions x^0 and x_1^0 imposed on $x(t)$ and on its derivative $dx(t)/dt$ at the initial point $t = 0$

$$x(t) = g(x^0, x_1^0, t)$$

In order to pass to a stochastic equation, suppose that the initial conditions, instead of having fixed definite values, are subjected to random fluctuations. More precisely, the initial conditions x^0 and x_1^0 are regarded as random variables with a given joint distribution. In other words, instead of specifying the initial conditions exactly, one can assign only probabilities to the collections of values the initial conditions may assume.

Consequently the solution $x(t)$ given by a function g of random variables x^0 and x_1^0, with g assumed to be sufficiently regular, is itself a random variable depending on t. A similar meaning can be attached to the derivatives $dx(t)/dt$ and $d^2x(t)/dt^2$. With this interpretation, the above differential equation becomes a stochastic differential equation.

The problem can now be formulated as follows. Given stochastic properties of initial conditions (say, in the form of their joint distribution function), find the stochastic properties of the random variable $x(t)$ which is the solution of the differential equation. One can ask here for the distribution function of $x(t)$ at individual values of t (to determine the mean, the variance, etc.), or for the joint distribution at two values t and t_1 (to determine autocorrelation), or for the joint distribution of $x(t)$ and its derivative, and so on.

In this form, the problem is simple in principle. For the purpose of presentation, it is convenient to regard $x(t)$ as describing a motion with t as time. The function $x(t)$ is a random variable, but random elements are introduced only through initial conditions. Once started, the random variable $x(t)$ develops with time according to the deterministic law described by the function g. If the initial conditions x^0 and $x_1{}^0$ lie initially within a given set S (in a two-dimensional space) with a specified probability attached to this set, then, as time progresses, the values of $x(t)$ and $dx(t)/dt$ will lie at each t in the set S_t (in this space) which has been obtained by transferring all points of S by the motion. As the motion itself is deterministic, it is evident that the probability of finding $x(t)$ and $dx(t)/dt$ in S_t should be the same as that of finding x^0 and $x_1{}^0$ in S. In other words, both sets S_t and S have the same probability. This means that the joint distribution function of $x(t)$ and its derivative can be found from the joint distribution function of x^0 and $x_1{}^0$ by standard methods of change of variables in double integrals; the distribution function of $x(t)$ is then obtained as the marginal distribution.

The same principle applies to the more complicated problems and to the system of equations. Before theoretical considerations are entered upon, several simple examples are given to illustrate basic ideas.

Example 1 (*Linear motion*): The equation of motion is

$$\frac{dx(t)}{dt} = c$$

where c is a nonnegative constant and $0 \leq t < \infty$. The solution is then $x(t) = ct + x^0$, with x^0 the initial condition. Suppose now that x^0 is a random variable with a given distribution. The mean $E\{x(t)\}$ and variance var $\{x(t)\}$ of the random variable $x(t)$ can be calculated directly, even without the explicit form of this distribution:

$$E\{x(t)\} = ct + E\{x^0\}$$
$$\text{var}\ \{x(t)\} = \text{var}\ (ct + x^0) = \text{var}\ x^0$$

Suppose now that x^0 has a gaussian distribution with the mean zero and variance σ^2, that is, the probability that x^0 assumes values between

λ and $\lambda + d\lambda$ is given by $f_0(\lambda)\, d\lambda$, where the density $f_0(\lambda)$ has the form

$$f_0(\lambda) = \frac{1}{\sqrt{2\pi}\,\sigma} \exp\left(-\frac{\lambda^2}{2\sigma^2}\right) \qquad -\infty < \lambda < \infty$$

In order to find the distribution of $x(t)$ proceed as follows:

$$\boldsymbol{P}\{x(t) < \mu\} = \boldsymbol{P}\{ct + x^0 < \mu\} = \boldsymbol{P}\{x^0 < \mu - ct\}$$
$$= \int_{-\infty}^{\mu - ct} f_0(\lambda)\, d\lambda = \int_{-\infty}^{\mu} f_0(\alpha - ct)\, d\alpha = \int_{-\infty}^{\mu} f(\alpha, t)\, d\alpha$$

Thus $x(t)$ again has gaussian density $f(\alpha, t)$

$$f(\alpha, t) = \frac{1}{\sqrt{2\pi}\,\sigma} \exp\left[-\frac{(\alpha - ct)^2}{2\sigma^2}\right]$$

with the same variance σ^2 but with the mean ct.

Example 2 (Harmonic oscillator): The equation of motion is

$$\frac{d^2x(t)}{dt^2} + \omega^2 x(t) = 0$$

and the solution is

$$x(t) = a \cos(\omega t + \varphi)$$

where a, ω, and φ are, respectively, the amplitude, frequency, and initial phase of the oscillations. Consider now a different problem, viz., keep a, ω, and φ constant but assume that time t is the random variable with given distribution. In other words, find the distribution of the positions $x(t)$ when the motion is observed at random times.

Consider only the half-period $(-\varphi/\omega, \pi/\omega - \varphi/\omega)$, and suppose that time t has uniform distribution with the density

$$f(t) = \frac{\omega}{\pi} \qquad -\frac{\varphi}{\omega} \le t \le -\frac{\varphi}{\omega} + \frac{\pi}{\omega}$$

Proceeding as in the previous example, for $-a \le \lambda \le a$, one has

$$\boldsymbol{P}\{x(t) < \lambda\} = \boldsymbol{P}\left\{t < \frac{1}{\pi} \arctan \frac{\lambda}{a} - \frac{\varphi}{a}\right\} = \int_{\alpha}^{\lambda} f_1(\alpha)\, d\alpha$$

and the density of $x(t)$ is found to be

$$f_1(\alpha) = \begin{cases} \dfrac{1}{\pi} \dfrac{1}{\sqrt{a^2 - \alpha^2}} & -a \le \alpha \le a \\ 0 & \alpha < -a \text{ and } \alpha > a \end{cases}$$

The mean of $x(t)$ is zero, and the variance is $\frac{1}{2}a^2$.

Example 3 (Kinetic energy): Let the solution $x(t)$ of the equation of motion represent the velocity of a gas molecule (in one dimension), and suppose that in equilibrium the density of the random variable $x(t)$ is given by the following law (independent of time t):

$$f(\lambda) = \begin{cases} k\lambda^2 e^{-\beta\lambda^2} & \lambda > 0 \\ 0 & \lambda \leq 0 \end{cases}$$

where β is a physical (positive) constant and $k = 4\beta^{3/2}\pi^{-1/2}$.

It is required to find the distribution of a functional of the motion, say the kinetic energy E of the molecule of mass m:

$$E = \tfrac{1}{2}mx^2(t)$$

The new random variable E also assumes only nonnegative values. The same argument as before shows that the probability that E assumes values between μ and $\mu + d\mu$ is $f_1(\mu)\,d\mu$, where the density $f_1(\mu)$ of E has the form of a gamma distribution:

$$f_1(\mu) = \begin{cases} \dfrac{2}{\pi^{1/2}}\left(\dfrac{2/\beta}{m}\right)^{3/2}\mu^{1/2}\exp\left(-\dfrac{2\beta}{m}\,\mu\right) & \text{for } \mu > 0 \\ 0 & \text{for } \mu \leq 0 \end{cases}$$

The mean and the variance of the energy E are $3m/4\beta$ and $3m^2/8\beta^2$, respectively.

Exercise 8-1   Verify Examples 1 to 3.

CHANGE OF RANDOM VARIABLES

Transformation of random variables is the basic tool for determining distribution functions for solution of equations with random initial conditions. Since this technique will be used repeatedly in this section, it is convenient to describe it separately in somewhat general terms.

Let $x = (x_1, \ldots, x_n)$ be an n-dimensional random vector of n random variables x_i $(i = 1, \ldots, n)$ defined on some probability space. Let g be a transformation of the random vector x into a new (also n-dimensional) random vector $y = (y_1, \ldots, y_n)$, defined on the same probability space. The mapping g is deterministic, i.e., nonrandom, and is assumed to be Borel-measurable (see Sec. 8-6). The problem is to find the distribution of the transformed vector $y = g(x)$ in terms of the distribution of the original vector x.

The basic principle is that an event in probability space should have the same probability irrespective of whether it is described in terms of the random vector x or the random vector y, both vectors x and y being related by a nonrandom transformation g. In symbols (referring to sets in probability space)

$$\boldsymbol{P}(y \,\epsilon\, gS) = \boldsymbol{P}(x \,\epsilon\, S)$$

where gS is the image (in the range space of y) of a set S in the n-dimensional range space of x. Now let P_x and P_y be the distributions of x and y, referred to their respective range spaces. The above basic identity becomes

$$P_y(gS) = P_x(S)$$

Formally, this principle expresses the property that the transformation g, mapping the original probability P_x onto the induced probability P_y, is *probability-preserving*. It follows that the distribution functions of vectors x and y will differ in form.

In the following only a very special type of transformation g will be required. In fact, let g be defined by the set of n relations

$$y_i = g_i(x_1, \ldots ,x_n) \qquad i = 1, \ldots , n$$

where the functions g_i are continuous, have continuous partial derivatives, and define one-to-one correspondence between x and y. Hence, the inverse functions

$$x_i = h_i(y_1, \ldots ,y_n) \qquad i = 1, \ldots , n$$

characterizing the inverse transformation $g^{-1} = h$ exist, are also one-to-one, and have continuous partial derivatives.

Assume further that the n-dimensional random vector $(x_1, \ldots ,x_n)$ possesses the joint density $f_x(\lambda_1, \ldots ,\lambda_n)$. It is required to find the joint density f_y for y. The answer is provided by the following lemma.

Lemma 1: Under the conditions stated above, the joint density of a transformed random vector $(y_1, \ldots ,y_n)$ is

$$f_y(\mu_1, \ldots ,\mu_n) = f_x(\lambda_1, \ldots ,\lambda_n) \left| \frac{\partial(\lambda_1, \ldots ,\lambda_n)}{\partial(\mu_1, \ldots ,\mu_n)} \right|$$

where $\lambda_i = h_i(\mu_1, \ldots ,\mu_n)$, $i = 1, \ldots , n$, and the term on the extreme right is the *absolute* value of the jacobian of the transformation (expressed as a function of $\mu_1, \ldots , \mu_n$):

$$\frac{\partial(\lambda_1, \ldots ,\lambda_n)}{\partial(\mu_1, \ldots ,\mu_n)} = \begin{vmatrix} \dfrac{\partial h_1}{\partial \mu_1} & \cdots & \dfrac{\partial h_1}{\partial \mu_n} \\ \cdots & \cdots & \cdots \\ \dfrac{\partial h_n}{\partial \mu_1} & \cdots & \dfrac{\partial h_n}{\partial \mu_n} \end{vmatrix}$$

Proof of Lemma 1 follows from the familiar rules of the change of variables in definite integrals because

$$P(y \,\epsilon\, gS) = \int_{gS} f_y(\mu_1, \ldots ,\mu_n)\, d\mu_1 \cdots d\mu_n$$
$$= \int_S f_x(\lambda_1, \ldots ,\lambda_n)\, d\lambda_1 \cdots d\lambda_n = P(x \,\epsilon\, S)$$

In particular, for $n = 1$, Lemma 1 yields the case discussed in Examples 1 to 3

$$f_y(\mu) = f(h(\mu))|h'(\mu)|$$

The restriction to a one-to-one transformation has been made here only for simplicity. The case when the g_i are many-to-one can be treated similarly by separate consideration of the regions on which the one-to-one condition holds.

Consider now a different problem, which is very common in applications. Suppose that there are given random variables $x_1, \ldots, x_n$ with the joint density $f_x(\lambda_1, \ldots, \lambda_n)$, and it is required to find the density of a new random variable y defined by a single equation

$$y = g(x_1, \ldots, x_n)$$

where g is a continuous function with continuous partial derivatives.

Lemma 2: The density of $y = g(x_1, \ldots, x_n)$ is given by

$$f(\mu) = \int_{-\infty}^{\infty} \cdots \int_{-\infty}^{\infty} f_y(\mu, \mu_2, \ldots, \mu_n)\, d\mu_2 \cdots d\mu_n$$

For proof, observe that Lemma 2 is a special case of Lemma 1. That is, it suffices to put $y_1 = y$ and introduce $n - 1$ auxiliary random variables y_i, $i = 2, \ldots, n$. Lemma 1 then gives the joint density $f_y(\mu, \mu_1, \ldots, \mu_n)$, and the marginal density $f(\mu)$ is obtained by integrating out variables $\mu_2, \ldots, \mu_n$. Q.E.D.

Recall also that the expectation of any transformed random variables $y = g(x)$ can be computed either with the distribution of y or of x, and the result is the same

$$E\{y\} = \int_{-\infty}^{\infty} \mu f(\mu)\, d\mu = \int_{-\infty}^{\infty} g(\lambda) f_x(\lambda)\, d\lambda = E\{gx\}$$

SINGLE DIFFERENTIAL EQUATION

The basic theory of stochastic differential equations with random initial conditions is most conveniently illustrated for the case of a single equation. Although the results are a formal specialization of those for systems of equations, the single equation nevertheless deserves separate discussion.

The single differential equation of nth order can be written in the form

$$x^{(n)}(t) = K[x(t), x^{(1)}(t), \ldots, x^{(n-1)}(t), t] \tag{8-2}$$

where $x(t)$ is an unknown random variable and its derivatives with respect to the argument t are denoted by

$$x^{(i)}(t) = \frac{d^i}{dt^i} x(t) \qquad i = 0, 1, \ldots, n$$

Conditions of existence and properties of solutions of such stochastic differential equations follow from the theory of a larger class of equations, presented in Sec. 8-5; see also remarks on systems. However, one can expect from the parallelism between stochastic and ordinary functions of t that in the present case the theory will be similar to that of ordinary differential equations. In other words, it suffices for present purposes to regard Eq. (8-2) as an equation for an ordinary function of t, i.e., for sample functions, the randomness being introduced only through the initial conditions. The derivatives $x^{(i)}(t)$ and integrals of $x(t)$ should be interpreted in the same manner. This point of view is adopted in this section in order to concentrate on the probabilistic properties of the solution $x(t)$; the validity of such approach will be justified in Sec. 8-5.

It is known from the general theory of equations (cf. "NM," sec. 4-3) that whenever a function K is continuous and satisfies a Lipschitz condition on its first n arguments, then there exists the unique function $x(t)$ satisfying the equation and the given initial conditions imposed on $x(t)$ and its first $n - 1$ derivatives at the initial point $t = 0$:

$$x^{(i)}(0) = x_i{}^0 \qquad i = 0, 1, \ldots, n - 1$$

Hence, the solution $x(t)$ and its derivatives can be written as ordinary functions of random initial conditions

$$x^{(i)}(t) = g_i(x^0, x_1{}^0, \ldots, x_{n-1}^0, t) \qquad i = 0, 1, \ldots, n - 1 \qquad (8\text{-}3)$$

assumed to hold for $0 \leq t < \infty$.

As it stands, Eq. (8-2) is satisfied by the random variable $x(t)$, with randomness introduced through initial conditions only. Hence, the system of n functions g_i, $i = 0, 1, \ldots, n - 1$, represents the transformation of a random vector $(x^0, x_1{}^0, \ldots, x_{n-1}^0)$ into random vector $(x(t), x^{(1)}(t), \ldots, x^{(n-1)}(t))$ prevailing at t. This point of view allows one to determine the stochastic properties of the solution of Eq. (8-2).

***Theorem* 8-1**   Let $x(t)$ be the unique solution of a differential equation (8-2) and suppose that the initial conditions $x^0, x_1{}^0, \ldots, x_{n-1}^0$ have the joint density $f_0(\lambda_0, \lambda_1, \ldots, \lambda_{n-1})$. Suppose that the function K possesses partial derivatives $\partial K(\mu_0, \mu_1, \ldots, \mu_{n-1}, t)/\partial \mu_j$, which are continuous functions of $(\mu_0, \mu_1, \ldots, \mu_{n-1}, t)$. Then the joint density of $x(t)$ and its first $n - 1$ derivatives at t is given by

$$f(\mu_0, \mu_1, \ldots, \mu_{n-1}, t) = f_0(\lambda_0, \ldots, \lambda_{n-1}) \left| \frac{\partial(h_0, \ldots, h_{n-1})}{\partial(\mu_0, \ldots, \mu_{n-1})} \right| \qquad (8\text{-}4)$$

where $h_i = g_i{}^{-1}$ and $\lambda_i = h_i(\mu_0, \mu_1, \ldots, \mu_{n-1}, t)$, $i = 0, 1, \ldots, n - 1$. The jacobian is explicitly given as a function of $\mu_0, \mu_1, \ldots, \mu_{n-1}, t$ by

$$\frac{\partial(h_0, \ldots, h_{n-1})}{\partial(\mu_0, \ldots, \mu_{n-1})} = \exp\left(- \int_0^t \frac{\partial K}{\partial \mu_{n-1}} \, d\tau \right)$$

where in $\partial K/\partial \mu_{n-1}$ all μ_i should be expressed in terms of λ_i prior to integration.

Proof: This theorem is the corollary of Theorems 8-3 and 8-5 proved below. It is, however, of independent interest to outline the basic idea of the proof. One treats here the equation $d^n\mu_0/dt^n = K(\mu_0, \mu_1, \ldots, \mu_{n-1}, t)$, whose solution is

$$\mu_i = g_i(\lambda_0, \lambda_1, \ldots, \lambda_{n-1}, t)$$

It is known from general theory that the functions g_i are continuous in all their arguments simultaneously and that existence of continuous partial derivatives $\partial K/\partial \mu_j$ implies existence of partial derivatives $\partial g_i/\partial \lambda_j$ continuous in all arguments simultaneously. Hence, Lemma 1 of Sec. 8-2 is applicable. Q.E.D.

As a corollary, the marginal density $f(\mu, t)$ of $x(t)$ can be obtained from Lemma 2 on page 355. Furthermore, the assignment of the initial density f_0 must be compatible with the values assumed by the initial conditions. The actual evaluation of formulas in Theorem 8-1 is usually complicated, and frequently it is possible to determine only the appropriate transforms or some parameter (like the mean) of the required densities. Examples are given below to illustrate the procedure. However, in the frequently occurring case of a linear homogeneous equation of nth order, the above theorem simplifies considerably, as follows.

***Theorem* 8-2** Let the random variable $x(t)$ be the unique solution of the linear differential equation

$$x^{(n)}(t) + a_1(t)x^{(n-1)}(t) + \cdots + a_{n-1}(t)x^{(1)}(t) + a_n(t)x(t) = 0$$

where the functions $a_i(t)$, $i = 1, \ldots, n$, are continuous over some interval $0 \leq t \leq L < \infty$. Suppose that initial conditions x^0, x_1^0, $\ldots$, x_{n-1}^0 have the joint density $f_0(\lambda_0, \lambda_1, \ldots, \lambda_{n-1})$. Then the joint distribution of $x(t)$ and its first $n-1$ derivatives at t has a density of the form

$$f(\mu_0, \mu_1, \ldots, \mu_{n-1}, t) = f_0(\lambda_0, \ldots, \lambda_{n-1}) \exp\left(\int_0^t a_1(\tau)\, d\tau\right)$$

where λ_i are linear combinations of $\mu_0, \ldots, \mu_{n-1}$ (with coefficients depending on t).

If one expresses $x(t)$ as a linear combination of initial conditions

$$x(t) = c_0 x^0 + c_1 x_1^0 + \cdots + c_{n-1} x_{n-1}^0 \qquad c_0 \neq 0$$

the density of the random variable $x(t)$ is given by

$$f(\mu, t) = \int_{-\infty}^{\infty} \cdots \int_{-\infty}^{\infty} f_0(\lambda_0, \lambda_1, \ldots, \lambda_{n-1})\, d\lambda_1 \cdots d\lambda_{n-1}$$

where $\quad \lambda_0 = \dfrac{\mu}{c_0} - \displaystyle\sum_{k=1}^{n-1} \dfrac{c_k}{c_0}\, \lambda_k$

Proof: The first assertion is the immediate corollary of Theorem 8-1, and the second follows from Lemma 2. The reader is invited to provide the direct proof based on properties of the wronskian of the fundamental set of linearly independent solutions. Q.E.D.

Exercise 8-2 Prove Theorem 8-2 using the wronskian.

Example 4 (*The simplest harmonic function*): Consider

$$\frac{d^2x(t)}{dt^2} = 0 \qquad 0 \leq t < \infty$$

whose solution is

$$x(t) = x^0 + u^0t$$
$$u(t) = u^0$$

where $u(t) = dx(t)/dt$, and the initial conditions are $x(0) = x^0$ and $u(0) = u_0$. Assume now that x^0 and u^0 are *independent* random variables, having gaussian distribution, with

$$E\{x^0\} = E\{u^0\} = 0 \qquad \text{var}\ (x^0) = \sigma_0{}^2 \qquad \text{var}\ (u^0) = s_0{}^2$$

Hence their joint density is the product of their marginal densities

$$f_0(\lambda,\lambda_1) = \frac{1}{2\pi\sigma_0 s_0} \exp\left[-\frac{1}{2}\left(\frac{\lambda^2}{\sigma_0{}^2} + \frac{\lambda_1{}^2}{s_0{}^2}\right)\right]$$

where λ and λ_1 are values of x^0 and u^0, respectively.

Directly, from the linearity of $x(t)$ and $u(t)$, we get

$$E\{x(t)\} = E\{u(t)\} = 0$$
$$\text{var}\ \{x(t)\} = s_0{}^2t^2 + \sigma_0{}^2 \equiv \sigma^2$$
$$\text{var}\ \{u(t)\} = s_0{}^2 \equiv s^2$$
$$E\{x(t)u(t)\} = E\{(u^0)^2t + x^0u^0\} = s^2t$$

In the present case, the functions g and g_1 have the form

$$\mu = \lambda + \lambda_1 t \qquad \mu_1 = \lambda_1$$

where μ and μ_1 are values of $x(t)$ and $u(t)$, respectively. The inverse functions h and h_1 are

$$\lambda = \mu - \mu_1 t \qquad \lambda_1 = \mu_1$$

and the jacobian takes the value

$$\begin{vmatrix} \dfrac{\partial h}{\partial \mu} & \dfrac{\partial h}{\partial \mu_1} \\ \dfrac{\partial h_1}{\partial \mu} & \dfrac{\partial h_1}{\partial \mu_1} \end{vmatrix} = \begin{vmatrix} 1 & -t \\ 0 & 1 \end{vmatrix} = 1$$

The joint density of $x(t)$ and $u(t)$, according to Theorem 8-2, is, after some algebra,

$$f(\mu,\mu_1,t) = \frac{1}{2\pi s\sigma \sqrt{1-\rho^2}} \exp\left[-\frac{1}{2(1-\rho^2)}\left(\frac{\mu^2}{\sigma^2} - \frac{2\rho\mu\mu_1}{\sigma s} + \frac{\mu_1{}^2}{s^2}\right)\right]$$

where ρ is the correlation coefficient

$$\rho = \frac{st}{\sigma} = \frac{s_0 t}{\sqrt{s_0{}^2 t^2 + \sigma_0{}^2}}$$

Thus $f(\mu,\mu_1,t)$ is gaussian, and consequently the marginal densities are also gaussian; the density for $x(t)$ has zero mean and variance $\sigma^2 = \sigma^2(t)$, whereas the density for $u(t)$ is the same as that for u^0.

It should be observed, however, that although initial values of $x(t)$ and $u(t)$ at $t = 0$ were independent, the random variables $x(t)$ and $u(t)$ for $t > 0$ are correlated, hence dependent, with the correlation coefficient ρ depending on time.

Example 5 (*Harmonic oscillator*): The equation of motion is

$$\frac{d^2x(t)}{dt^2} + \omega^2 x(t) = 0 \qquad 0 \le t < \infty$$

where the frequency ω is constant; the solution is

$$x(t) = x^0 \cos \omega t + u^0 \omega^{-1} \sin \omega t$$
$$u(t) = -\omega x^0 \sin \omega t + u^0 \cos \omega t$$

where $u(t) = dx(t)/dt$, and the initial conditions are $x(0) = x^0$, $u(0) = u^0$. Assume, as before, that x^0 and u^0 are independent random variables with gaussian distribution and zero means and variances $\sigma_0{}^2$ and $s_0{}^2$, respectively.

The discussion in Example 4 is applicable here, and the joint density $f(\mu,\mu_1,t)$ for $x(t)$ and $u(t)$ has the same form as that in Example 4, with

$$\sigma^2 = \sigma_0{}^2 \cos^2 \omega t + s_0{}^2 \omega^{-2} \sin^2 \omega t$$
$$s^2 = s_0{}^2 \cos^2 \omega t + \omega^2 \sigma_0{}^2 \sin^2 \omega t$$
$$\rho\sigma s = (s_0{}^2 - \omega^2\sigma_0{}^2)\omega^{-1} \sin \omega t \cos \omega t$$

Similarly, the marginal distributions of $x(t)$ and $u(t)$ are gaussian, with zero mean and variances σ^2 and s^2, respectively. Observe that when initially the potential energy equals the kinetic energy, then $s_0{}^2 = \omega^2\sigma_0{}^2$, and therefore $f(\mu,\mu_1,t) = f_0(\lambda,\lambda_1)$ for all t. In this case, the random variables $x(t)$ and $u(t)$ are uncorrelated, and, being gaussian, they remain independent. In addition, $\sigma^2 = \sigma_0{}^2$ and $s^2 = s_0{}^2$, so that marginal distributions of $x(t)$ and $u(t)$ remain constant independent of time (stationary process). It may be noted that the density for the second derivative is also gaussian, with mean zero and variance $\omega^4\sigma^2$.

Example 6 (*Nonlinear motion*):   The equation of motion for a falling body in a medium with resistance proportional to the square of the velocity is

$$m \frac{d^2x(t)}{dt^2} + p \left[\frac{dx(t)}{dt} \right]^2 = mg$$

where m, p, and g are positive constants.   When one writes $dx(t)/dt$ for the velocity $u(t)$, the above equation becomes

$$\frac{du(t)}{dt} + \frac{p}{m} u^2(t) = g \qquad t \geq 0$$

which is a special case of the Riccati equation, with the initial condition $u(0) = u^0$.

Direct solution yields

$$u(t) = a \frac{e^{\alpha t} - ke^{-\alpha t}}{e^{\alpha t} + ke^{-\alpha t}}$$

with $\qquad \alpha = \sqrt{\frac{gp}{m}} \qquad a = \sqrt{\frac{gm}{p}} \qquad k = \frac{a - u^0}{a + u^0}$.

The solution can be put in more convenient form

$$u(t) = a \frac{a \tanh \alpha t + u^0}{a + u^0 \tanh \alpha t} \qquad 0 \leq t < \infty$$

with $u(0) = u^0$ and $u(\infty) = a$.   Furthermore, $u(t)$, as the function of t, is monotonic (increasing if $a > u^0$ and decreasing if $a < u^0$).   The inverse function is therefore

$$u^0 = a \frac{a \tanh \alpha t - u}{u \tanh \alpha t - a} \qquad 0 \leq t < \infty$$

Hence $\qquad \dfrac{du^0}{du} = a^2 \dfrac{1 - \tanh^2 \alpha t}{(a - u \tanh \alpha t)^2} \geq 0$

Let the random variable u^0, with values ζ, have the density $f_0(\zeta)$ for $\zeta > 0$ and $f_0(\zeta) = 0$ for $\zeta \leq 0$.   By Theorem 8-2 the density $f(\mu,t)$ of the random variable $u(t)$, with values μ, has the form

$$f(\mu,t) = f_0(\zeta) \frac{d\zeta}{d\mu}$$

where the above expressions should be substituted for ζ and $d\zeta/d\mu$.

For example, suppose that u^0 has the uniform distribution

$$f(\zeta) = \begin{cases} \dfrac{1}{K} & \text{for } 0 < \zeta \leq K \\ 0 & \text{otherwise} \end{cases}$$

The mean of $u(t)$ is found to be

$$E\{u(t)\} = \frac{a}{K} \int_0^K \frac{a \tanh \alpha t + \zeta}{a + \zeta \tanh \alpha t} \, d\zeta$$
$$= \left(\frac{a}{\sinh \alpha t}\right)^2 \left[\frac{\sinh 2\alpha t}{2a} - \log \frac{a + K \tanh \alpha t}{a}\right]$$

with $E\{u(0)\} = \frac{1}{2}K$ and $E\{u(\infty)\} = a$.

Exercise 8-3 Verify Examples 4 to 6.

SYSTEM OF DIFFERENTIAL EQUATIONS

Consider now a system of n first-order stochastic differential equations with n unknowns $x_1(t), \ldots , x_n(t)$:

$$\frac{dx_i(t)}{dt} = K_i[x_1(t), \ldots ,x_n(t),t] \qquad i = 1, \ldots , n \qquad (8\text{-}5)$$

with $0 \le t \le a < \infty$. It will be convenient to write

$$\frac{dx_i(t)}{dt} = \dot{x}_i \qquad i = 1, \ldots , n$$

where $\qquad x_i(t) = x_i \qquad x_i(0) = x_i{}^0$
and $\qquad x = (x_1, \ldots ,x_n) \qquad x^0 = (x_1{}^0, \ldots ,x_n{}^0)$

Hence, symbolically,

$$\dot{x}_i = K_i(x,t) \qquad i = 1, \ldots , n$$

or $\qquad\qquad \dfrac{dx_1}{K_1} = \cdots = \dfrac{dx_n}{K_n} = dt \qquad\qquad (8\text{-}6)$

The functions x_i are random variables, the operators K_i are functions of $n + 1$ variables $x_1, \ldots , x_n$ and t, and the values of K_i are also random variables. However, randomness is introduced only through initial conditions x^0, which are assumed to possess a given joint distribution. The development with t of a solution vector x is governed by differential equations in a deterministic way. In other words, Eqs. (8-5) can be solved as if they were ordinary nonrandom equations, and to make the solution vector x a random vector, it suffices to treat the initial conditions as random variables. Thus one can write the solution in the form

$$x_i(t) = g_i(x_1{}^0, \ldots ,x_n{}^0,t) \qquad i = 1, \ldots , n \qquad (8\text{-}7)$$

or simply $x_i = g_i(x^0,t)$, which reduces to x^0 for $t = 0$.

The complete existence and uniqueness proof involves too many preparatory concepts, and is deferred to Sec. 8-5. For present purposes it suffices to treat system (8-6) as an ordinary nonrandom system. It then follows from general theory (cf. "NM," theorem 4-1) that (8-6) possesses a unique solution (in some interval) satisfying the initial conditions provided that the functions K_i are continuous in their arguments and satisfy the Lipschitz condition (if there is a common Lipschitz constant, the solution can be uniquely extended to indefinitely large values of t). Formally, Eqs. (8-6) and (8-7) refer to the random variables x_i and $x_i{}^0$, but in view of the above remarks, they are also valid for values (depending on t) of these random variables, i.e., for sample functions; it will be always clear from the context which meaning is intended.

Equations (8-5) admit the geometric interpretation; i.e., they describe the motion of a point representing a dynamical system. The solution vector describes the trajectory (with time t as a parameter) in the n-dimensional *phase space E_n.* The motion begins from an initial position subjected to indeterminacy, but once started, the motion follows deterministic law given by the functions g (*cryptodeterministic process* in Whittaker's terminology). This point of view provides a link between probabilistic and analytic aspects of the system of equations, as shown below.

Let P_0 be the initial point of the motion, with coordinates

$$\lambda = (\lambda_1, \ldots, \lambda_n)$$

where λ is the value of the random vector x^0 of initial conditions. The probability that P_0 lies in some set S_0 in the n-dimensional space can be written as

$$P\{P_0 \epsilon S_0\} = \int_{S_0} \cdots \int f_0(\lambda_1, \ldots, \lambda_n) \, d\lambda_1 \cdots d\lambda_n \qquad (8\text{-}8a)$$

where $f_0(\lambda)$ is the joint density of the initial conditions, assumed to be given. At the time t, the dynamical system arrives at the point P_t, with coordinates $\mu = (\mu_1, \ldots, \mu_n)$, where μ is the value of the random vector x at t. The probability that P_t lies in some set S_t is, similarly,

$$P\{P_t \epsilon S_t\} = \int_{S_t} \cdots \int f(\mu_1, \ldots, \mu_n, t) \, d\mu_1 \cdots d\mu_n \qquad (8\text{-}8b)$$

where $f(\mu,t)$ is the joint density of the random variables x_i, $i = 1, \ldots, n$, at time t.

Since the motion of the dynamical system is governed by deterministic differential equations (apart from random initial conditions), the probability of finding P_t in S_t at time t must be the same as the probability of finding P_0 in S_0 initially. Thus, (8-8a) and (8-8b) should coincide; hence, in symbols,

$$\int_{S_0} f_0(\lambda) \, d\lambda = \int_{S_t} f(\mu,t) \, d\mu \qquad (8\text{-}9)$$

This is, however, in agreement with the principle of preservation of probability, described earlier. Hence, the following theorem holds.

Theorem 8-3 Let $x_1(t), \ldots, x_n(t)$ be the unique solution of the system (8-5), and suppose that the initial conditions $x_1^0, \ldots, x_n^0$ have joint density $f_0(\lambda_1, \ldots, \lambda_n)$. Suppose that the functions K_i possess partial derivatives $\partial K_i(\mu_1, \ldots, \mu_n, t)/\partial \mu_j$, which are continuous functions of (μ, t). Then the joint density $f(\mu, t)$ of $x_1(t), \ldots, x_n(t)$ is given by

$$f(\mu_1, \ldots, \mu_n, t) = f_0(\lambda_1, \ldots, \lambda_n) \left| \frac{\partial(h_1, \ldots, h_n)}{\partial(\mu_1, \ldots, \mu_n)} \right| \qquad (8\text{-}10)$$

where $\quad h_i = g_i^{-1} \quad$ and $\quad \lambda_i = h_i(\mu, t) \qquad i = 1, \ldots, n$

Proof (Outline): As before, one treats here the system $\dot{\mu}_i = K_i(\mu, t)$, whose solution is $\mu_i = g_i(\lambda, t)$. It is known from general theory that the functions g_i are continuous in λ and t simultaneously and that existence of the continuous partial derivatives $\partial K_i(\mu, t)/\partial \mu_j$ (which implies Lipschitz conditions) implies the existence of the partial derivatives (with respect to initial values) $\partial g_i(\lambda, t)/\partial \lambda_j$ continuous in λ and t simultaneously. Hence Lemma 1 on page 354 is applicable. Q.E.D.

This theorem will be improved below by giving the explicit expression for the jacobian of the transformation. Prior to that (and at the risk of some repetition) it is worthwhile to give an independent dynamical justification of Theorem 8-3 and to prove the most general form of the fundamental Liouville theorem for dynamical systems.

Theorem 8-4 (*Liouville-Gibbs*) The density $f(\mu, t)$ of the solution vector x of system (8-5) is an integral invariant of the motion and satisfies the *Liouville equation*

$$\frac{\partial f}{\partial t} + \sum_{i=1}^{n} \frac{\partial(fK_i)}{\partial \mu_i} = 0 \qquad (8\text{-}11)$$

Proof: A function $f(\mu, t)$, a *dynamical variable*, is an integral invariant of motion if for any region S_t (depending on t) the integral

$$I(t) = \int_{S_t} f(\mu, t) \, d\mu$$

is independent of t. According to Poincaré's theorem, for f to be such an invariant it should be a multiplier of the system (8-6); i.e., it should satisfy the first-order partial differential equation (8-11).

Consider $I(t + h)$, and let $\zeta = (\zeta_1, \ldots, \zeta_n)$ be coordinates at $t + h$, obtained from $\lambda = (\lambda_1, \ldots, \lambda_n)$ at t; similarly, let S_{t+h} be the region transformed from S_t. Then

$$I(t + h) = \int_{S_{t+h}} f(\zeta, t + h) \, d\zeta = \int_{S_t} f(\zeta, t + h) \frac{\partial \zeta}{\partial \mu} \, d\mu$$

Now $\qquad \zeta_i = \mu_i + h\dfrac{d\mu_i}{dt} + \cdots = \mu_i + hK_i + \cdots$

so that for the jacobian, $\dfrac{\partial \zeta}{\partial \mu} = 1 + h\left(\dfrac{\partial K_1}{\partial \mu_1} + \cdots + \dfrac{\partial K_n}{\partial \mu_n}\right) + \cdots$

However

$$f(\zeta, t + h) = f(\mu,t) + h\left(\frac{\partial f}{\partial \mu_1}\frac{d\mu_1}{dt} + \cdots + \frac{\partial f}{\partial \mu_n}\frac{d\mu_n}{dt} + \frac{\partial f}{\partial t}\right) + \cdots$$

Thus

$$f(\zeta, t + h)\frac{\partial \zeta}{\partial \mu} = f(\mu,t) + h\left(\frac{\partial f}{\partial \mu_1} K_1 + \cdots + \frac{\partial f}{\partial \mu_n} K_n + \frac{\partial f}{\partial t}\right.$$
$$\left. + f\frac{\partial K_1}{\partial \mu_1} + \cdots + f\frac{\partial K_n}{\partial \mu_n}\right) + \cdots$$

Hence $\qquad \dfrac{dI(t)}{dt} = \displaystyle\int_{S_t}\left(\frac{\partial f K_1}{\partial \mu_1} + \cdots + \frac{\partial f K_n}{\partial \mu_n} + \frac{\partial f}{\partial t}\right) d\mu$

Hence, $dI(t)/dt = 0$ whatever the region S_t may be if and only if (8-11) holds. Furthermore $I(0) = I(t)$ is just (8-9). Q.E.D.

As a corollary of Theorem 8-4 one obtains immediately the expression for the jacobian in Theorem 8-3. Recall that the divergence of the n-dimensional vector $K = (K_1, \ldots ,K_n)$ is defined by

$$\text{div } K = \sum_{i=1}^{n}\frac{\partial K_i}{\partial \mu_i} \equiv \Phi(\mu,t) \tag{8-12}$$

and that the total derivative of $f(\mu,t)$ is

$$\frac{df}{dt} = \frac{\partial f}{\partial t} + \sum_{i=1}^{n} K_i \frac{\partial f}{\partial \mu_i}$$

When the terms of the total derivative of f are collected, Eq. (8-11) assumes the equivalent form

$$\frac{df}{dt} + f \text{ div } K = 0 \tag{8-13}$$

and its solution is

$$f(\mu,t) = f_0(\lambda) \exp\left(- \int_0^t \Phi(\mu,\tau)\, d\tau\right) \tag{8-14}$$

where in $\Phi(\mu,\tau)$ all μ_i should be expressed prior to integration, in terms of the initial values λ_i, by $\mu_i = g_i(\lambda,t)$.

It can be verified that solving (8-11) by the Lagrange method amounts to solving the system

$$\frac{dt}{1} = \frac{d\mu_1}{K_1} = \cdots = \frac{d\mu_n}{K_n} = -\frac{df}{f\Phi}$$

expressed in terms of λ, and the result is precisely (8-14).

Formula (8-14) gives the explicit time solution of Eq. (8-11) with arbitrary initial $f_0(\lambda)$ assumed given. Comparison of (8-10) with (8-13) yields the following theorem.

Theorem 8-5   The Jacobian occurring in Theorem 8-3 is given by

$$J = \left| \frac{\partial(h_1, \ldots ,h_n)}{\partial(\mu_1, \ldots ,\mu_n)} \right| = \exp\left[-\int_0^t \Phi(\mu,\tau)\, d\tau \right] \qquad (8\text{-}15)$$

in terms of t and initial values λ.   *Note:* After integration convert back to μ by using $\lambda_i = h_i(\mu,t)$.

Formula (8-15) is of intrinsic interest, and it is worthwhile to give the independent proof; then Theorem 8-5 and Eq. (8-10) can be used to prove Theorem 8-4.

Alternative proof of Theorem 8-5:   Let λ be a vector of initial coordinates, and let μ be a vector of coordinates at t.   Differentiation of the determinant (written as function of λ)

$$J^{-1} = \frac{\partial(\mu_1, \ldots ,\mu_2)}{\partial(\lambda_1, \ldots ,\lambda_n)} \equiv Y$$

yields

$$\frac{dY}{dt} = \sum_{k=1}^n \frac{\partial(\mu_1, \ldots ,\mu_{k-1},\rho,\mu_{k+1}, \ldots ,\mu_n)}{\partial(\lambda_1, \ldots ,\lambda_n)} = \sum_{k=1}^n Y_k$$

where Y_k has been defined by the above expression and $\rho = d\mu_k/dt$.   But

$$\frac{\partial\rho}{\partial\lambda_k} = \sum_{i=1}^n \frac{\partial\rho}{\partial\mu_i} \frac{\partial\mu_i}{\partial\lambda_k}$$

Hence substitution gives

$$Y_k = \sum_{i=1}^n \frac{\partial\rho}{\partial\mu_i} \frac{\partial(\mu_1, \ldots ,\mu_{k-1},\mu_i,\mu_{k+1}, \ldots ,\mu_n)}{\partial(\lambda_1, \ldots ,\lambda_n)}$$

$$= \frac{\partial\rho}{\partial\mu_k} \frac{\partial(\mu_1, \ldots ,\mu_k, \ldots ,\mu_n)}{\partial(\lambda_1, \ldots ,\lambda_n)} = \frac{\partial\rho}{\partial\mu_k} Y$$

because all other determinants have two equal rows and thus vanish. Now, in view of Eqs. (8-5),

$$\sum_{k=1}^n Y_k = Y \sum_{k=1}^n \frac{\partial\rho}{\partial\mu_k} = Y\Phi(\mu,t)$$

or

$$\frac{dJ}{dt} + J\Phi(\mu,t) = 0$$

with $\mu_i = g_i(\lambda,t)$.   Q.E.D.

Remarks: By analogy with hydrodynamics, the Liouville equation (8-11) written in the form

$$\frac{\partial f}{\partial t} + \text{div } (fK) = 0 \qquad (8\text{-}16)$$

is called the *equation of continuity* of an ideal fluid of density f and expresses conservation of probability in the flow (no sources or sinks). The flow is called *stationary* if f does not depend explicitly on t; that is, $\partial f/\partial t = 0$. The fluid is called *incompressible* if $df/dt = 0$; by (8-14) this is equivalent to the condition div $K = 0$. Hence, by (8-15), $J = 1$, and $f = f_0$; in particular, when f_0 is constant, the flow preserves volume in E_n space. Since μ represents the position vector, K becomes the velocity vector. For irrotational flow, $K = \mathrm{grad}\ V$, where $V = V(\mu)$ is the velocity potential; hence

$$\frac{df}{dt} + f\,\nabla^2 V = 0$$

which reduces to $\nabla^2 V = 0$ for incompressible flow.

The continuity equation (8-16) or (8-13) is of basic importance in physics, especially in statistical mechanics, hydrodynamics, and potential theory. Its derivation (essentially equivalent to Theorem 8-4) and extensions to nonconservative flows are discussed in most texts of theoretical physics; for probabilistic applications see, for example, Moyal,[74] Kac,[52] Kurth,[63] and also Sec. 8-3 below.

APPLICATIONS

System (8-5) of first-order differential equations with random initial conditions possesses many interesting theoretical properties and has numerous applications to analytical dynamics and statistical mechanics. In this subsection, a brief discussion of these applications is given, with emphasis on probabilistic aspects; the reader should consult the references for a complete treatment.

Transformations

Equations (8-7), written briefly in vector form, as

$$x = g(x^0, t)$$

where g is a vector with components g_i, describe the motion of the dynamical system in the phase space E_n and induce a group of transformations T_{ts}, which map a random vector x into a random vector $T_{ts}x$. The transformation T_{ts} is defined by the corresponding transformation of coordinates in E_n, that is, by mapping a point P_s into a point P_t

$$P_t = T_{ts}P_s$$

where P_t is the value assumed by the random vector x. In particular, the initial point P_0 is mapped into a point $P_t = T_{t0}P_0$, and this corresponds to the mapping

$$x = T_{t0}x^0 = g(x^0,t)$$

For three instants of time, say τ, s, and t, one has successively

$$T_{t\tau}P_\tau = P_t = T_{ts}P_s = T_{ts}T_{s\tau}P_\tau$$

and the group property takes the form

$$T_{t\tau} = T_{ts}T_{s\tau}$$

with $T_{tt} = I$ (identity). The inverse transformation represents reversible motion (a semigroup is obtained if no reversal of time is allowed). Transformations T_{ts} are therefore determined by the differential equations (8-5)

$$\dot{x} = K(x,t)$$

and the form of T_{ts} depends on the vector K. The introduction of a continuous transformation T_{ts} on E_n makes available the whole theory of continuous groups of transformations (Lie groups). The connection between such groups of transformations and differential equations is well known and has many applications in analytic dynamics; see, for example, Goursat,[41] Whittaker,[87] Pontryagin,[77] and Eisenhart.[32] Extension from point transformations in E_n to transformations on space of random vectors, as needed here, retains classical properties.

For an autonomous system (8-5), i.e., when K_i does not involve t explicitly, the transformation T_{ts} depends on t and s in the form of the difference $t - s$ (homogeneity), and Eq. (8-7) induces a one-parameter group of transformations T_t, with the group property

$$T_{t+s} = T_t T_s \qquad T_0 = I$$

This type of transformation has received a considerable amount of attention (especially in the case of semigroups). Problems of interest in theoretical dynamics include the classification of motions according to properties of trajectories ($P_t = T_t P_0$ when t varies) and study of integral invariants (invariant measures) determined by equations of motion, culminating in the Poincaré recurrence theorem and the Birkhoff ergodic theorem. Let

$$x = T_t x^0$$

Direct differentiation yields

$$\frac{dx}{dt} = \lim_{h \to 0} \frac{T_{t+h}x^0 - T_t x^0}{h} = Dx$$

where D is the *infinitesimal transformation* of the homogeneous group, defined symbolically by

$$\lim_{h \to 0} \frac{T_h - I}{h} = D$$

Hence the vector equation $\dot{x} = K(x)$ can be written in operator form

$$\dot{x} = Dx$$

with $Dx = K(x)$. The solution is, symbolically,

$$x = e^{Dt}x^0$$

so, formally,

$$T_t = e^{Dt}$$

For a full discussion of the validity of these operations see Hille and Phillips,[46] where applications to Lie semigroups can also be found.

Let $V(x)$ be any functional of x (*dynamical variable*) not involving t explicitly and defined by its values $V(\mu) = V(\mu_1, \ldots, \mu_n)$, where μ is the value of x. Define

$$V(x) = V(T_t x^0) = T_t V(x^0)$$

Then direct differentiation yields, as above,

$$\frac{dV(x)}{dt} = DV(x)$$

or, symbolically, $\dot{V}(x) = V(\dot{x})$. On the other hand,

$$\frac{dV(\mu)}{dt} = \sum_{i=1}^{n} \frac{\partial V}{\partial \mu_i} K_i(\mu) \equiv DV(\mu)$$

One obtains thus the explicit expression for the infinitesimal transformation D on E_n; this transformation has been the principal tool in the study of Lie groups. In particular, letting $V(\mu) = f(\mu)$, one has $df/dt = Df$.

The functional V is said to be an *invariant* of the one-parameter group (T_t) if $V = T_t V$ for all t; it can be shown that the only invariants in such a group are the integrals of the equation $DV = 0$.

It must be added that transformations T_t form a special case of a group of *contact transformations* defined by $2n$ equations

$$\bar{\lambda}_i = g_i(\lambda_1, \ldots, \lambda_n, p_1, \ldots, p_n, t)$$
$$\bar{p}_i = \gamma_i(\lambda_1, \ldots, \lambda_n, p_1, \ldots, p_n, t)$$

with the property that

$$\sum_i p_i \, d\lambda_i - \sum_k \bar{p}_k \, d\bar{\lambda}_k = dG$$

The difference should be a total differential of some function G of vectors λ and p. These transformations map a point λ with additional parameters $(p_1, \ldots, p_n)$ into a point $\bar{\lambda}$ at t with parameters $(\bar{p}_1, \ldots, \bar{p}_n)$ in such a way that the new coordinates depend on λ and p but not on higher derivatives; thus, in dynamical applications a point of contact of two curves is preserved, hence the name. See Whittaker[87] and Eisenhart.[32]

Typical Cases

Several special cases of the system (8-5) occur frequently in applications and deserve special mention.

Case 1: Consider the *linear autonomous system*

$$\dot{x}_i = a_{i1}x_1 + \cdots + a_{in}x_n \qquad i = 1, \ldots, n \qquad (8\text{-}17)$$

which can be written in matrix notation, with $\mathfrak{A} = (a_{ij})$, as

$$\dot{x} = \mathfrak{A}x$$

The solution is

$$x = e^{\mathfrak{A}t}x^0 \qquad t \geq 0$$

The inverse relation

$$x^0 = e^{-\mathfrak{A}t}x$$

gives for the jacobian

$$J = e^{-(\text{Tr }\mathfrak{A})t}$$

where Tr $\mathfrak{A}$ is the trace of the matrix $\mathfrak{A}$. The Liouville equation (8-13) is now

$$\frac{df}{dt} + f(\text{Tr }\mathfrak{A}) = 0$$

with solution $f = f_0 J$.

By linearity, the vector of the mean values satisfies the equation of the same form as (8-17)

$$E\{x\} = e^{\mathfrak{A}t}E\{x^0\}$$

Example 7: Take $n = 2$ and $a_{11} = a_{22} = a$, $a_{12} = a_{21} = b$, with $a > b$. The joint density is found to be

$$f(\mu_1,\mu_2,t)$$
$$= f_0[e^{-at}(\mu_1 \cosh bt - \mu_2 \sinh bt), e^{-at}(-\mu_1 \sinh bt + \mu_2 \cosh bt)]e^{-2at}$$

Case 2 (Autonomous system): Suppose that functions K_i do not involve t explicitly:

$$\dot{x}_i = K_i(x_1, \ldots, x_n) \qquad i = 1, \ldots, n \qquad (8\text{-}18)$$
or
$$\dot{x} = K(x)$$

The solution is

$$x = g(x^0, t) \qquad x^0 = g(x^0, 0)$$

where g is a vector with components g_i; similarly, the inverse relation is $x^0 = h(x, t)$. By the property of jacobians, differentiation of g yields

$$J^{-1} = \frac{\partial g}{\partial \lambda}$$

as a function of (λ, t), where λ is the value of x^0. This is in agreement with Theorem 8-5. Although the function

$$\Phi(\mu) = \sum_{i=1}^{n} \frac{\partial}{\partial \mu_i} K_i(\mu)$$

does not involve t explicitly, the jacobian J is time-dependent because μ in $\Phi(\mu)$ must be expressed in terms of λ and t prior to integration in (8-15).

Case 3: Return now to the single differential equation (8-2) of nth order. As is well known, this equation can be written in the form (8-5) by letting

$$x^{(i)} = x_{i+1} \qquad \text{for } i = 0, 1, \ldots, n-1$$

Hence
$$\dot{x}_i = x_{i+1} \qquad i = 1, 2, \ldots, n-1 \qquad (8\text{-}19)$$
$$\dot{x}_n = K(x_1, \ldots, x_n, t)$$

Evidently,
$$\Phi(\mu, t) = \frac{\partial K}{\partial \mu_n}$$

and the solution for density is $f = f_0 J$ with jacobian

$$J = \exp\left(-\int_0^t \frac{\partial K}{\partial \mu_n} \, d\tau\right)$$

This yields the proof of Theorem 8-1. Furthermore, for the linear equation, $\partial K / \partial \mu_n = -a_1(t)$, and the statement of Theorem 8-2 is obtained.

Example 8: Consider the Riccati equation (cf. Example 6):

$$\dot{x} + ax^2 = 0 \qquad a > 0$$

whose solution is

$$x = \frac{x^0}{1 + ax^0 t}$$

Let μ and λ be values of x and x^0, respectively. Direct evaluation of the jacobian yields

$$J = \frac{dh}{d\mu} = (1 - at\mu)^{-2}$$

On the other hand, $K = -a\mu^2$, and

$$\frac{\partial K}{\partial \mu} = -\frac{2ax^0}{1 + ax^0 t} = -\frac{2a\lambda}{1 + a\lambda t}$$

and the above formula yields the equivalent value

$$J = (1 + a\lambda t)$$

Case 4 (*Hamiltonian equations*): One of the most important systems of equations in theoretical dynamics is provided by canonical equations of motion. A dynamical system with n degrees of freedom can be regarded as a point in $2n$-dimensional phase space E_{2n} with coordinates $q_1, \ldots, q_n, p_1, \ldots, p_n$, where q_i and p_i are generalized positions and momenta, respectively. The equations of motion (derived from Newton's laws) assume canonical form and are known as *hamiltonian equations:*

$$\frac{dq_i}{dt} = \frac{\partial H}{\partial p_i} \qquad \frac{dp_i}{dt} = -\frac{\partial H}{\partial q_i} \qquad i = 1, \ldots, n \qquad (8\text{-}20)$$

where H is the *hamiltonian*

$$H = H(q_1, \ldots, q_n, p_1, \ldots, p_n, t)$$

H represents the total energy in the dynamical system, excluding dissipative forces, e.g., friction.

The solution of the equations can be written as a function of the initial conditions:

$$\begin{aligned} q_i(t) &= g_i(q_1{}^0, \ldots, q_n{}^0, p_1{}^0, \ldots, p_n{}^0, t) \\ p_i(t) &= \gamma_i(q_1{}^0, \ldots, q_n{}^0, p_1{}^0, \ldots, p_n{}^0, t) \end{aligned} \qquad i = 1, \ldots, n$$

Formally, q_i and p_i are random variables, but as they depend only on random initial conditions, the interpretation already used in this section should eliminate any confusion.

The above equations determine a group of transformations T_{ts} (analytic automorphisms of phase space) and describe hamiltonian flow; for conservative dynamical systems, a one-parameter group of transformations T_t is obtained. It is well known that T_{ts} and T_t are contact transformations, and it can be proved that canonical equations are invariant under any contact transformation.

For the hamiltonian system, the Liouville equation (8-11) assumes its usual form

$$\frac{\partial f}{\partial t} + \{f, H\} = 0 \qquad (8\text{-}21)$$

where the term in braces is called a *Poisson bracket:*

$$\{f,H\} = \sum_{i=1}^{n} \left(\frac{\partial f}{\partial q_i} \frac{\partial H}{\partial p_i} - \frac{\partial f}{\partial p_i} \frac{\partial H}{\partial q_i} \right) \qquad (8\text{-}22)$$

These formulas are usually proved directly by noting that, in view of the hamiltonian equations, the rate of change of f is

$$\frac{df}{dt} = \frac{\partial f}{\partial t} + \sum_{i=1}^{n} \left(\frac{\partial f}{\partial q_i} \frac{dq_i}{dt} + \frac{\partial f}{\partial p_i} \frac{dp_i}{dt} \right) = \frac{\partial f}{\partial t} + \{f,H\}$$

and the invariance of f means $df/dt = 0$. In particular $\{H,H\} = 0$; hence, unless the hamiltonian is explicitly time-dependent, $dH/dt = 0$, so that the total energy H remains constant along every trajectory of the conservative motion. Furthermore,

$$\Phi(q,p,t) = \sum_{i=1}^{n} \left[\frac{\partial}{\partial q_i} \left(\frac{\partial H}{\partial p_i} \right) + \frac{\partial}{\partial p_i} \left(- \frac{\partial H}{\partial q_i} \right) \right] = 0$$

so that the jacobian of transformation is $J = 1$.

Thus, by virtue of the Liouville equation (also known as the *principle of conservation of density in phase,* a name due to Gibbs), the representative points move in phase space as if they constituted an incompressible fluid of varying density. The Liouville theorem asserts then that regions in phase space obtained by transformation of some initial region will continue to occupy the same volume but with altered shape during the motion (the *principle of conservation of extension in phase,* in Gibbs's terminology); in other words, the integral

$$v(S_t) = \int \underbrace{\cdots \int}_{S_t} dq_1 \cdots dq_n \, dp_1 \cdots dp_n$$

is the invariant of the motion. Formally this can be expressed as follows.

Theorem 8-6 (*Liouville theorem*) The phase volume, i.e., Lebesgue measure, is invariant under hamiltonian flow.

The idea of introducing probability density f in the phase space is due to Gibbs, who considered the *ensemble* of replicas of a given dynamical system and defined f as the limit of frequency on the assumption of uniform distribution of the system in the ensemble (equal a priori probabilities); cf. Gibbs.[40] For further discussion of hamiltonian equations in dynamics and statistical mechanics see, for example, Whittaker[87] and Kurth.[63]

Exercise 8-4 Show that for a stationary hamiltonian system, $\{f,H\} = 0$ and that this equation is satisfied when the density f is a function of the total energy H.

Verify that Gibbs's *canonical ensemble*

$$f = e^{\alpha(\psi - H)}$$

where α and ψ are physical constants, gives the equilibrium distribution in phase space.

Exercise 8-5   Restrict the motion of the hamiltonian system to the energy surface E defined by the requirement that the total energy $H(q_1, \ldots, p_n)$ be a constant on E. Assume the initial distribution to be given by the surface integral (Gibbs's *microcanonical ensemble*)

$$P(S) = k \int_S \|\text{grad } H\|^{-1} \, ds$$

where $S \subset E$, k is the normalization constant for $P(E) = 1$ and

$$\|\text{grad } H\| = \Big[\sum_{i=1}^{n} \Big(\frac{\partial H}{\partial p_i} \Big)^2 + \Big(\frac{\partial H}{\partial q_i} \Big)^2 \Big]^{1/2}$$

Show that the Liouville theorem implies that

$$P(T_t S) = P(S)$$

where T_t is the transformation defined by the hamiltonian motion.

Exercise 8-6 (Poincaré recurrence theorem)   In the notation of Exercise 8-5 suppose that $P(S) > 0$ for $S \subset E$, with $P(E) = 1$.   Then all points of S, except those in a subset of S of probability zero, will return to S under the motion T_t infinitely often.   For proof see Kac,[52] Kurth,[63] and Halmos.[42]

Exercise 8-7   The famous Bolzmann equation for the density f has the form

$$\frac{\partial f}{\partial t} + \{f, H\} = \int dw \int d\Omega \, gI \, (g, \theta)(f'f_1' - ff_1)$$

where the left side is the Liouville operator for Hamiltonian systems [cf. (8-21)] with the hamiltonian

$$H = U(q) + \frac{1}{2m} \sum_{i=1}^{n} p_i^2$$

where vectors q and p are replaced by a position vector μ and a velocity vector v. The right side describes the effect of collisions; the prime and the index 1 of f's refer to the velocities only, so, for example, $f_1 = f(\mu, w, t)$, and the four velocities refer to the velocities of the binary collisions $(v, w) \rightleftharpoons (v', w')$; $g = |v - w| = |w' - v'|$ is the relative velocity turning over the angle θ in a collision, and $I(g, \theta)$ represents the differential collision cross section for a collision into the solid angle $d\Omega$.

1.   The derivation of the Bolzmann equation, as an extension of the Liouville equation, still is an open question; see however, Kurth,[63] Kac,[52] and Khintchine [58]

2.   The Bolzmann H theorem.   Let (*negative entropy*)

$$H(t) = \iint f \log f \, d\mu \, dv$$

Show that

$$\frac{d}{dt} H(t) \leq 0$$

and the equality occurs only when

$$f = Ae^{-\beta H}$$

where H is the hamiltonian and A and β are constants; see Sec. 7-5.

Remark: The Bolzmann equation has often been criticized. Moreover, its consequence, the Bolzmann H theorem, appears to contradict Poincaré's recurrence theorem; cf. Kac.[52]

DEGREE OF RANDOMNESS

By way of introduction to the more advanced considerations in Sec. 8-5, it is advisable to comment on the general properties of the random variables $x(t)$ obtained as solutions of stochastic differential equations with random initial conditions. These random variables are of the general form

$$x(t) = g(x^0, t)$$

where g is an ordinary nonrandom function and x^0 is a random vector of initial conditions. The random variable $x(t)$ is one-dimensional and may be either a solution of a single equation or a component of a solution vector in the case of the system of equations; the form of g depends on the equations in question.

Obviously, $x(t)$ are elements of some function space $\mathcal{K}$ of random variables, assumed to be given. Varying t over a parameter set T, one obtains a family of random variables

$$x = \{x(t), \ t \in T\}$$

that is a *stochastic process* determined by stochastic equations. The name *random function* is used for x because it defines a mapping from T into $\mathcal{K}$, with value of x at t being $x(t)$. Thus stochastic differential equations are, in fact, differential equations for a random function x whose values are random variables. This point of view is of considerable importance in connection with existence proofs for stochastic differential equations.

Still another interpretation is possible. Random variables $x(t)$ and x^0 are defined on some probability space $(\Omega, \mathfrak{F}, P)$ (see Sec. 8-6), and their values at point ω in Ω are $x(t, \omega)$ and $x^0(\omega)$, respectively. Fixing the point ω and allowing t to vary, one obtains a function $x(., \omega)$ of t, known as a *sample function* (or realization) of the process. The collection of all sample functions constitutes a sample space of the process. Evidently

$$x(t, \omega) = g(x^0(\omega), t)$$

and it follows from the properties of g that sample functions are continuous and differentiable in t.

Fix now the value of the initial conditions, say $x^0(\omega) = \lambda$. Since g is an ordinary function, one finds, in agreement with earlier considerations, the following equality of sets:

$$\{\omega : x^0(\omega) = \lambda\} = \{\omega : x(t,\omega) = g(\lambda,t)\}$$

This means that all ω in this set correspond to the same sample function

$$x(.,\omega) = g(\lambda,.)$$

The function $g(\lambda,.)$ describes a trajectory of the motion in phase space E_n associated with fixed initial conditions λ. It is more convenient to disregard space Ω, refer instead to phase space E_n, and call $g(\lambda,.)$ a sample function in sample space E_n. The stochastic process x is accordingly defined by an ordinary function g of t and x^0, with x^0 being a (multidimensional) random variable with a given probability distribution.

Stochastic properties of the random function x are expressed by the joint distributions of the random variable $x(t)$ at a finite number of distinct points. In particular, the distribution of $x(t)$ at fixed t is of great interest. It is evident that the flexibility of the process will depend on the number of random variables in the random vector of initial conditions x^0.

It will now be shown that the order of a differential equation determines the flexibility of the stochastic process obtained from it, in the sense that the higher the order, the less deterministic the process. For this purpose the concept of degree of randomness is needed; cf. Moyal.[74] It is said that the process $\{x(t),\ t \epsilon T\}$ has n *degrees of randomness* if n is the smallest integer such that if the values $\zeta_1,\ \ldots\ ,\ \zeta_n$ of $x(t)$ are known exactly at n distinct points $t_1 < \cdots < t_n$, then there exists (with probability one) a functional relationship between the values of $x(t)$ at $t > t_n$ and these ζ_i and t; that is,

$$P\{x(t,\omega) = \varphi(\zeta_1,t_1;\ \ldots\ ;\zeta_n,t_n)\} = 1 \qquad (8\text{-}23)$$

where φ is an ordinary function of the ζ_i and t_i. The process is said to be completely random if no such relationship exists for any finite n (this includes situations when the process is specified either by denumerable or nondenumerable selections of values of t).

The following theorems show that the stochastic processes considered here are not completely random.

***Theorem* 8-7** A stochastic process $\{x(t),\ t \epsilon T\}$ determined by a differential equation of the nth order has exactly n degrees of randomness.

Proof: The solution $x(t)$ is determined analytically in terms of n functionally independent random initial conditions $x_1^0,\ \ldots\ ,\ x_n^0$. Thus it is possible to define $x(t)$ at n distinct points by

$$x(t_i) = g(x_1{}^0, \ldots ,x_n{}^0,t_i) \qquad i = 1, \ldots , n$$

Since the function g has continuous partial derivatives with respect to the first n arguments, the joint density f of $x(t)$ at n points $t_1, \ldots , t_n$ is (by the procedure explained in Sec. 8-2) $f = f_0 J$, where f_0 is the density of initial conditions, and

$$J^{-1} = J^{-1}(t_1, \ldots ,t_n) = \frac{\partial(g, \ldots ,g)}{\partial(\lambda_1, \ldots ,\lambda_n)}$$

(with λ_i the value of $x_i{}^0$), provided the jacobian does not vanish identically for all values of the t_i.

If values of $x(t)$ are known at n or more points, say $k \geq n$, then the values of $x(t_i)$ for $i = n, \ldots , k$ can be also expressed as functions of values of $x_1{}^0, \ldots , x_n{}^0$. If, in turn, $x_1{}^0, \ldots , x_n{}^0$ are expressed in terms of $x(t_i)$ for $t_1, \ldots , t_n$ and the corresponding values substituted into expressions for $x(t_i)$, $i = n, \ldots , k$, then $x(t_i)$, $i = n, \ldots , k$, become functions of $x(t_i)$, $i = 1, \ldots , n$. On the other hand, if the values of $x(t)$ are known at fewer points than n, say $k < n$, then the remaining $n - k$ values are subject to chance mechanism of the process. Consequently, the joint distributions are all singular at $k \geq n$ points, with the probability concentrated on an n-dimensional hypersurface, but are not necessarily singular at $k < n$ points. And that is just condition (8-23). Q.E.D.

For example, in the case of the first-order differential equation only one initial condition appears, and the one-dimensional distribution of $x(t)$ is sufficient to describe the stochastic properties of the whole process; in fact, for any two instants t and t', $x(t) = \varphi(x(t'))$, and the two-dimensional distribution has already become degenerate. By Theorem 8-7 the nth-order differential equation leads to the n-dimensional distribution, which is sufficient to describe all stochastic properties of the process.

Example 9: Let

$$\frac{d}{dt} x(t) + ax(t) = 0 \qquad t \geq 0$$

The solution is
$$x(t) = x_0 e^{-at}$$
Hence,
$$x(t) = x(t')e^{-a(t-t')}$$

for any pair t and t'. Evidently

$$E\{x(t)x(t')\} = e^{-a(t+t')}E\{x_0{}^2\}$$
$$E\{x(t)\} = e^{-at}E\{x_0\} \qquad \mathrm{var}\, x(t) = e^{-2t}\,\mathrm{var}\, x_0$$
and
$$\mathrm{cov}\,(x(t),x(t')) = e^{-a(t+t')}\,\mathrm{var}\, x_0$$

so that the correlation coefficient between $x(t)$ and $x(t')$ is one.

Example 10:   Let

$$\frac{d^2x(t)}{dt^2} + \omega^2 x(t) = u(t) \qquad t \geq 0$$

where $u(t)$ is a nonrandom function.   The solution is

$$x(t) = x_0 \cos \omega t + \dot{x}_0 \omega^{-1} \sin \omega t + \frac{1}{\omega} \int_0^t \sin \omega(t - \tau) u(\tau) \, d\tau$$

Denote the above integral by $i(t)$, and write μ and i for values of $x(t)$ and $i(t)$ at time t and μ' and i' at time t', respectively.   Then the values λ and $\dot\lambda$ of initial conditions x_0 and $\dot{x}_0$ are

$$\lambda = \frac{(\mu - i) \sin \omega t' - (\mu' - i') \sin \omega t}{\sin \omega(t' - t)}$$

$$\dot\lambda = \frac{-(\mu - i) \cos \omega t' + (\mu' - i') \cos \omega t}{\sin \omega(t' - t)} \omega$$

Hence

$$\frac{\partial(\lambda,\dot\lambda)}{\partial(\mu,\mu')} = \frac{\omega}{\sin \omega(t' - t)}$$

Consequently the joint density of $x(t)$ and $x(t')$ is

$$f(\mu,t;\mu',t') = f_0(\lambda,\dot\lambda) \frac{\omega}{|\sin \omega(t' - t)|}$$

where λ and $\dot\lambda$ in the initial density f_0 should be expressed in terms of μ, μ', t and t', by the above-quoted formulas.

In the case of a system of m differential equations of the nth order, the solution $x(t) = (x_1(t), \ldots, x_m(t))$ now requires $n \times m$ initial conditions (n for each of m coordinates), so by the same argument as in Theorem 8-7, the number of degrees of randomness is nm.   However, the vector stochastic process $\{x(t), t \in T\}$, whose random variables $x(t)$ are m-dimensional vectors, will be specified completely by its mn-dimensional distributions at n points only (cf. the case discussed in Theorem 8-3).

Return now to the single differential equation of the nth order, and suppose that the joint distribution of $x(t)$ and its first $n - 1$ derivatives has been found.   Then the joint distribution of any group of these derivatives can be obtained as the marginal distribution.   However, the nth derivative is functionally related by the equation itself to $x(t)$ and its first $n - 1$ derivatives, and thus its distribution is determined by n conditions (n degrees of randomness).   This is in agreement with the procedure by which the distribution of a first derivative is determined from the joint distribution of $x(t)$ and $x(t + h)$.

So far, the definition of a derivative of random variable $x(t)$ has been taken directly by analogy from the derivative of the nonrandom function $x(t)$; the same remark applies to the integrals of $x(t)$ with respect to t.

In essence, this means that derivatives and integrals have been obtained from derivatives and integrals of sample functions and then regarded as random variables. It is of great importance to examine the conditions for the validity of such a procedure; this will be discussed later in this chapter in connection with more general concepts of stochastic differentiation and integration.

As a final remark, observe that the averaging and differentiation commute, so that for the linear differential equation the mean $E\{x(t)\}$ satisfies the same equation as $x(t)$, with averaged initial conditions. However, this is not true for nonlinear equations.

8-3   RANDOM FORCING FUNCTION

INTRODUCTION

The theory of stochastic differential equations to be presented in this section leans heavily on the theory of stochastic processes; in fact, differential equations are used to investigate properties of stochastic processes. Nevertheless, this field has not been sufficiently explored, and most of the results now available refer to individual equations or special stochastic processes.

As a primary datum, it is assumed that there is a given stochastic process (or a random function) $z = \{z(t), t \geq 0\}$ with specified stochastic properties. The differential equation to be considered here relates the unknown random function $x = \{x(t), t \geq 0\}$ and its derivatives to the given random function z. The solution of the equation defines then a new stochastic process $\{x(t), t \geq 0\}$ in terms of the given z process. In particular, linear equations of the nth order occur very frequently:

$$x^{(n)}(t) + a_1(t)x^{(n-1)}(t) + \cdots + a_n(t)x(t) = z(t)$$

From physical applications, the function $z(t)$ is known as a *forcing function;* hence the title of this section.

In contrast with the equations studied in Sec. 8-2, where $z(t)$ was an ordinary nonrandom function, the stochastic properties of the solution $x(t)$ depend significantly on the stochastic properties of $z(t)$ and less on the initial conditions. It is this dependence on $z(t)$ which complicates the analysis and actually leads to a new, difficult, and interesting theory, one essentially different from that developed in Sec. 8-2.

In order to illustrate the problem, without entering into probabilistic details, consider the simple case of the first-order equation

$$\frac{dx(t)}{dt} + ax(t) = z(t) \qquad t \geq 0$$

whose solution is formally

$$x(t) = x(0)e^{-at} + \int_0^t e^{-a(t-\tau)}z(\tau)\, d\tau \qquad t \geq 0$$

It can be seen that $x(t)$ depends on the values of $z(\tau)$ over the interval $[0,t]$, and it is not immediately clear whether the integral is a random variable at all. This brings up the question of what is meant by a *stochastic integral*, i.e., an integral of random variables, in what sense is it defined, and what are its probabilistic properties. In particular, it is important to know whether the integral in question can be defined by the corresponding integral of sample functions (or realizations) of the random function z, that is, with $z(t)$ treated as if it were an ordinary nonrandom function of t.

A related problem is the meaning attached to the *stochastic* differentiability of $x(t)$, and then the question of existence and uniqueness of stochastic differential equations should be examined. A point of view helpful in these considerations regards a random function, say x, as a function on real line to the space of random variables. Thus, the value of x at t is a random variable $x(t)$, a point in a function space of random variables. Problems of continuity, differentiability, and integrability of random functions now become the same problems for the ordinary functions taking values in abstract space; the same comment applies to differential equations.

Furthermore, explicit expressions for distribution (or its transform) of $x(t)$ are required in terms of those for $z(t)$; this task, important for applications, is seldom easy. However, the integration and averaging commute, so that in the above example

$$E\{x(t)\} = e^{-at}E\{x(0)\} + \int_0^t e^{-a(t-\tau)}E\{z(\tau)\}\, d\tau$$

This section begins with a discussion of random functions continuous in the mean. Next, the best-known case of brownian motion is examined in detail, and the section closes with discussion of a diffusion process.

MEAN - SQUARE PROPERTIES

The best-known case of stochastic equations is that in which random functions are of the second order; that is, the random variables possess finite variances. This subsection is devoted to a description of the properties of such random functions, with emphasis on applications to stochastic equations. In particular, the stochastic integral will be introduced. Since the material is classical and is available in standard texts, e.g., Loève,[69] only a brief summary is given here. The reader may omit this subsection on the first reading and return to it as the need arises.

Let $(\Omega, \mathfrak{F}, P)$ be a given probability space (see Sec. 8-6), and consider the space $L_2 = L_2(\Omega, \mathfrak{F}, P)$ of all (complex-valued) random variables on Ω with finite second absolute moment, the equivalent random variables being identified. With the inner product defined by the second mixed moment, the space L_2 is a Hilbert space, and the norm convergence is known as the *mean-square convergence*.

Let T be a real line or its subset, and let $z = z(.)$ denote a random function defined on T with values $z(t)$ in L_2. The function R on $T \times T$ defined by the inner product

$$R_z(t,t') = (z(t),z(t')) = E\{z(t)\overline{z(t')}\}$$

where the bar denotes the complex conjugate and E stands for expectation, is called a *covariance* (or autocovariance) *function* of the random function z. Evidently, the norm of $z(t)$ is

$$\|z(t)\|^2 = E\{|z(t)|^2\} = R_z(t,t)$$

and the distance is

$$E\{|z(t) - z(t')|^2\} = \|z(t) - z(t')\|^2$$
$$= R_z(t,t) - R_z(t,t') - R_z(t',t) + R_z(t',t')$$

The mean-square properties of the random function z follow from the corresponding ordinary properties of its covariance function. By the Cauchy condition, the necessary and sufficient condition for a random variable $z(t)$ to converge in mean square to $z(t_0)$ when $t \to t_0$ is that

$$\|z(t) - z(t')\|^2 = E\{|z(t) - z(t')|^2\} \to 0$$

when $t \to t_0$ and $t' \to t_0$ or alternatively that the covariance $R_z(t,t')$ converges in the ordinary sense to $R_z(t_0,t_0)$. More generally, the random functions z_h converge in mean square to a random function s as $h \to h_0$ if and only if the functions $(z_h(t),z_h'(t))$ converge to a finite function as $h \to h_0$, $h' \to h_0$; then $R_{z_h}(t,t') \to R_z(t,t')$.

A random function z is *continuous* in mean square at t if

$$\lim_{h \to 0} \|z(t + h) - z(t)\|^2 = 0$$

It follows that the necessary and sufficient condition for this is that $R_z(t,t')$ be continuous at $t = t'$; if this condition holds on the whole diagonal line $t = t'$, then z is mean-square-continuous for all t.

A random function z is *mean-square-differentiable* at t if there is a random variable $\dot{z}(t)$ in L_2, called a mean-square derivative of z at t, such that

$$\lim_{h \to 0} \left\| \frac{z(t + h) - z(t)}{h} - \dot{z}(t) \right\|^2 = 0$$

The necessary and sufficient condition for the existence of $\dot{z}(t)$ at t is that the partial derivatives $\partial R/\partial t$, $\partial R/\partial t'$, and $\partial^2 R/(\partial t\, \partial t')$ should exist at $t = t'$. Again, if this condition holds on the whole diagonal line $t = t'$, then the random function z is differentiable in mean square for all t.

A random function z is of *bounded variation* on the interval $[a,b]$ if

$$\sup \sum_i \| [z(t_i) - z(t_i')] \| < \infty$$

over all finite partitions of the interval; the above supremum is known as the *total variation*. A random function z is of bounded variation if and only if $R_z(t,t')$ is of the bounded variation on the rectangle $[a,b]^2$. Furthermore, if $R_z(t,t')$ is of bounded variation, then the limits $R_z(t \pm 0, t' \pm 0)$ exist and differ from $R_z(t,t')$ on a countable set of (t,t') only. Hence, one-sided mean-square limits $z(t \pm 0)$ exist, and the random function z is mean-square-continuous outside a countable set of values of t.

The Riemann-Stieltjes *stochastic integral in mean square* is defined as follows. Let $y(.)$ and $z(.)$ be two second-order random functions with covariances R_y and R_z, respectively. It is assumed that the products $y(t)z(t')$ also belong to L_2. Consider a partition of the interval $[a,b]$

$$a = t_0 < t_1 < \cdots < t_n = b$$

with $\partial = \max (t_i - t_{i-1})$, $i = 1, \ldots, n$. Let

$$S_\partial = \sum_{i=1}^{n} y(s_i)[z(t_i) - z(t_{i-1})]$$

where $t_{i-1} \le s_i \le t_i$, $i = 1, \ldots, n$; note that $S_\partial \epsilon L_2$. If the $\lim_{\partial \to 0} S_\partial$ exists in the mean-square sense for a net of partitions (ordered by inclusion) and is independent of the choices of the corresponding s_i, then this limit, denoted by

$$\varphi = \int_a^b y(t) \, dz(t)$$

defines the mean-square integral of the random function y with respect to the random function z. The integral is a random variable in L_2, defined up to an equivalence. The extension to an integral over an infinite interval is defined in the usual way, by going to the limit in the mean-square sense.

It is important to observe that the above definition depends on the increments $\Delta z(t)$ of the random function z and not on its values $z(t)$; thus the increments of the covariance $R_z(t,t')$ are of importance below:

$$\Delta\Delta' R_z(t,t') = E\{\Delta z(t) \, \overline{\Delta' z(t')}\}$$

When $y(t)$ and $\Delta z(t')$ are independent, e.g., when $y(t)$ and $z(t)$ are independent, then the *existence criterion* asserts that the integral φ exists if the repeated integral

$$\|\varphi\|^2 = \int_a^b \int_a^b R_y(t,t') \, dd'R_z(t,t')$$

exists. This will be the case when, in addition, R_y is bounded and R_z is of bounded variation on rectangle $[a,b]^2$.

Specializations to the mean-square integrals commonly enoountered are obtained when the random function z possesses additional properties. The random function $z = z(.)$ is said to have *independent increments* if for $t_1 < \cdot\,\cdot\,\cdot < t_n$ the differences $z(t_2) - z(t_1), \ldots, z(t_n) - z(t_{n-1})$ are mutually independent. The random function z is said to have *uncorrelated increments* if

$$E\{\Delta z(t) \, \overline{\Delta'z(t')}\} = E(\Delta z(t))E(\overline{\Delta'z(t')})$$

for nonoverlapping intervals Δt and $\Delta' t'$. The random function z is said to have *orthogonal increments* if the above product vanishes and if

$$E\{|\Delta z(t)|^2\} = \Delta R_z(t)$$

or, in symbolic form,

$$E\{|dz(t)|^2\} = dR_z(t)$$

Let the random function z with orthogonal increments be normalized to vanish at $t = 0$, that is, $\mathbf{P}\{z(0) = 0\} = 1$. Then the function $R_z(t) = R_z(t,t)$ is monotone nondecreasing, and $R_z(0) = 0$. Evidently,

$$R_z(t,t') = \begin{cases} R_z(t) & \text{for } t < t' \\ R_z(t') & \text{for } t > t' \end{cases}$$

and it follows that the random function z is mean-square-continuous if $R_z(t)$ is a continuous function of t, but z cannot be mean-square-differentiable even if $R_z(t)$ is differentiable, because the first partial derivatives of $R_z(t,t')$ are discontinuous at $t = t'$.

If the distribution of $\Delta z(t)$ depends only on Δt, the random function with independent increments is said to have stationary increments. It follows that the random function with orthogonal increments will have stationary increments if and only if $R_z(t) = c^2|t|$, where c is a constant.

When the random function z has orthogonal increments $\Delta z(t)$, independent of the random variables $y(t)$, the double integral in the existence criterion reduces to the single integral

$$\|\varphi\|^2 = \int_a^b R_y(t) \, dR_z(t)$$

Consider two (second-order) random functions y_1 and y_2, and let φ_1 and φ_2 be their mean-square integrals with respect to the random function z with orthogonal increments independent of $y_1(t)$ and $y_2(t)$. Then the relation

$$(\varphi_1, \varphi_2) = \int_T E\{y_1(t)\overline{y_2(t)}\} \, dR_z(t)$$

establishes a norm-preserving correspondence between integrals $\varphi \in L_2$ and the corresponding random functions y with the norm defined by

$$\|y\|_0^2 = \int_T \|y(t)\|^2 \, dR_z(t) = \|\varphi\|^2$$

This point of view is especially useful when $y(t)$ is replaced by a nonrandom function of t and the stochastic integral is defined first on a linearly dense set and then extended; cf. Doob.[29]

It may be added that a stochastic integral of a complex-valued square-integrable bounded function g on a real line taken with respect to a second-order random function $z(.)$ with orthogonal increments can also be defined as a spectral integral taken with respect to a projection-valued measure. This construction is well known in Hilbert-space theory; see, for example, Halmos[43] or Moyal[74] for the stochastic point of view.

Let $\mathfrak{z}$ be a fixed random variable in L_z, and consider the random function z whose random variables are projections

$$z(t) = P_t\mathfrak{z} \qquad -\infty < t < \infty$$

where P_t is a resolution of identity. Then

$$R_z(t) = (P_t\mathfrak{z}, \mathfrak{z}) = \|P_t\mathfrak{z}\|^2$$

with $R_z(-\infty) = 0$ and $R_z(+\infty) = \|\mathfrak{z}\|^2$. The scalar product

$$(P_t\mathfrak{z}, \mathfrak{s}) = E(z(t)\overline{\mathfrak{s}})$$

is, for any random variable $\mathfrak{s}$ in L_2, a function of t of bounded variation. Then there exists a linear operator A on L_2 such that

$$(A\mathfrak{z}, \mathfrak{s}) = \int_{-\infty}^{\infty} g(t) \, d(P_t\mathfrak{z}, \mathfrak{s})$$

with

$$\|A\mathfrak{z}\|^2 = \int_{-\infty}^{\infty} |g(t)|^2 \, dR_z(t)$$

The dependence of A on function g and projection P_t is denoted by writing

$$A = \int_{-\infty}^{\infty} g(t) \, dP_t$$

This representation of the operator A is known as the *spectral integral*, P_t being the spectral measure of interval $[0, t]$. The stochastic integral is obtained by referring to the range of A, that is,

$$A\mathfrak{z} = \int_{-\infty}^{\infty} g(t) \, dz(t) = \int_{-\infty}^{\infty} g(t) \, dP_t\mathfrak{z}$$

The indefinite integral is defined in the usual manner

$$\varphi(t) \ = \ \int_a^t y(s) \, dz(s) \qquad a \le t \le b$$

This integral defines a new stochastic process $\{\varphi(t), t \, \epsilon \, [a,b]\}$ whose random variables have the norm

$$E\{|\varphi(t)|^2\} \ = \ \int_a^t \int_a^t R_y(s,s') \, dd' R_z(s,s')$$

In applications, one often encounters a generalization of the above mean-square integral to the case when the random variables $y(t)$ depend on some parameter, say τ. The existence criterion remains the same but with covariance involving τ. In particular, when $y(t,\tau)$ becomes an ordinary function of two variables $g(t,\tau)$, then

$$\varphi(t) \ = \ \int_a^t g(t,\tau) \, dz(\tau)$$

Such integrals define a *linear stochastic process* $\{\varphi(t), t \ge 0\}$. For example, linear differential equations with constant coefficients lead to stochastic integrals of the form

$$\varphi(t) \ = \ \int_{-\infty}^t g(t - \tau) \, dz(\tau) \qquad -\infty < t < \infty$$

with z having orthogonal increments; such integrals occur in stationary stochastic processes.

The analysis of sample properties of second-order random functions involves concepts of the separability and measurability of random functions (see Sec. 8-6); because of limitations of space this discussion must be limited to the brief remarks made below. Suppose that the second-order random function $z(.)$ is defined on the closed interval $[a,b]$ and possesses a continuous covariance function $R_z(t,t')$. It can be shown that the random function z is mean-square-continuous, hence measurable, and is also sample integrable and sample square-integrable. Furthermore, the continuity of R_z implies (by the existence criterion) that the indefinite mean-square integral

$$x(t) \ = \ \int_a^t z(s) \, ds \qquad a \le t \le b$$

exists, and is a random variable (up to an equivalence). It can easily be shown that its mean-square derivative is

$$\frac{dx(t)}{dt} \ = \ z(t)$$

The following important theorem holds in the present case.   If the random function is mean-square-continuous, then its stochastic integral $x(t)$, defined above, and its sample integral defined as a random variable $x(t,.)$ with values (for almost all individual sample functions)

$$x(t,\omega) = \int_a^t z(s,\omega)\, ds \qquad a \le t \le b$$

coincide.

Observe, however, that for the random function z with orthogonal increments, the integral

$$\varphi(t) = \int_a^t g(s)\, dz(s) \qquad a \le t \le b$$

cannot, in general, be defined as an ordinary Stieltjes integral in the individual sample functions because the sample functions of z are not of bounded variation, except in special cases.

LINEAR EQUATIONS

Existence Theorem

In this subsection a linear differential equation will be examined in detail when the forcing function is taken from the second-order random function.   To simplify the writing the discussion will be confined to the second-order equation of the form

$$d\dot{x}(t) + a(t)\dot{x}(t)\, dt + b(t)x(t)\, dt = dz(t) \qquad 0 \le t \le T \qquad (8\text{-}24)$$

where $a(t)$ and $b(t)$ are complex-valued functions defined and continuous on $[0,T]$, with $0 < T < \infty$.   Here, $x(t)$ is the unknown random variable, and $z(t)$ is the random variable of the second-order random function $z = z(.)$, defined on $[0,T]$.   The equation has been written as a difference equation because in many interesting cases the mean-square derivatives of $z(t)$ and $\dot{x}(t)$ do not exist, and special meaning must be attached to the differentials $dz(t)$ and $d\dot{x}(t)$.

The formal solution, by analogy with the classical case, can be written as

$$x(t) = x_0 u(t) + x_1 v(t) + \int_0^t g(t,\tau)\, dz(\tau) \qquad 0 \le t \le T \qquad (8\text{-}25)$$

where
$$x_0 = x(0) \qquad x_1 = \dot{x}(0) \qquad\qquad\qquad (8\text{-}26)$$

the functions $u(t)$ and $v(t)$ form the fundamental system of solutions

$$u(0) = 1 \qquad \dot{u}(0) = 0 \qquad v(0) = 0 \qquad \dot{v}(0) = 1 \qquad (8\text{-}27)$$

and $g(t,\tau)$ is a Green's function for a one-point boundary condition, with the properties

$$g(t,\tau), \frac{\partial g}{\partial t}, \text{ and } \frac{\partial^2 g}{\partial t^2} \text{ are continuous on the rectangle } [0,T]^2 \quad (8\text{-}28a)$$

$$g(t,t) = 0 \qquad \frac{\partial g}{\partial t} = 1 \text{ at } \tau = t \qquad \text{for } 0 \leq t \leq T \quad (8\text{-}28b)$$

$$\frac{\partial^2 g}{\partial t^2} + a(t)\frac{\partial g}{\partial t} + b(t)g = 0 \qquad \text{in } [0,T]^2 \quad (8\text{-}28c)$$

The following theorem confirms that the above formal solution is indeed a general solution of the stochastic differential equation.

Theorem 8-8 If the second-order random function $z(.)$ defined on $[0,T]$, $0 < T < \infty$, has the covariance function $R_z(t,t')$ continuous and of bounded variation on rectangle $[0,T]^2$, then the differential equation (8-24) has a unique solution on $[0,T]$ satisfying the initial conditions (8-26), with x_0, x_1 in L_2. The general solution is given explicitly by (8-25), where $g(t,\tau)$ is a Green's function with properties (8-28).

Proof (Outline): Evidently, the complementary function $x_0 u(t) + x_1 v(t)$ belongs to L_2, is mean-square-continuous, twice mean-square-differentiable, and satisfies the homogeneous equation. Consider now the integrals

$$y_i(t) = \int_0^t g^{(i)}(t,\tau)\, dz(\tau) \qquad \frac{\partial^i}{\partial t^i} g(t,\tau) = g^{(i)}(t,\tau) \qquad i = 0, 1, 2$$

By continuity of $R_z(t,t')$, the random function z is mean-square-continuous, and the bounded variation of $R_z(t,t')$ implies that integrals $y_i(t)$ exist. For $i = 0$, the integral $y(t)$ defines a new random function y with the covariance function on $[0,T]^2$

$$R_y(t,t') = \int_0^t \int_0^{t'} g(t,\tau)\overline{g(t',\sigma)}\, ddR_z(\tau,\sigma) \quad (8\text{-}29)$$

Since $R_y(t,t')$ is continuous, the random function y is mean-square-continuous. Also, in view of the properties of $g(t,\tau)$, the partial derivatives $\partial R_y/\partial t$, $\partial R_y/\partial t'$, and

$$\frac{\partial^2 R}{\partial t\, \partial t'} = \int_0^t \int_0^{t'} g^{(1)}(t,\tau)\overline{g^{(1)}(t',\sigma)}\, ddR_z(\tau,\sigma)$$

exist, so that the random function y is mean-square-differentiable. Since $\partial^2 R_y/(\partial t\, \partial t') = R_{\dot{y}}(t,t')$, it follows that the mean-square derivative of y for all $0 \leq t \leq T$ is

$$\dot{y}(t) = y_1(t)$$

Now, let Δ be a difference operator of step h, and consider

$$\Delta \dot{y}(t) - h y_2(t) - \Delta z(t) = \int_0^t [g^{(1)}(t+h, \tau) - g^{(1)}(t,\tau) - h g^{(2)}(t,\tau)]\, dz(\tau)$$
$$+ \int_t^{t+h} [g^{(1)}(t+h, \tau) - 1]\, dz(\tau)$$

The first integral converges in mean square to zero when $h \to 0$; the same is true for the second integral because (by the mean-value theorem)

$$g^{(1)}(t+h, \tau) = g^{(1)}(t+h, \tau) - g^{(1)}(\tau, \tau) + 1 = h g^{(2)}(t + \theta h, \tau) + 1$$

where $0 < \theta < 1$ whenever $t \leq \tau \leq t + h$. Thus

$$\Delta \dot{y}(t) - h y_2(t) - \Delta z(t) \xrightarrow[h \to 0]{\text{m.s.}} 0$$

Using (8-28c), one has

$$y_2(t) = -a(t) y_1(t) - b(t) y(t) = -a(t) \dot{y}(t) - b(t) y(t)$$

Hence
$$\Delta \dot{y}(t) + [a(t) \dot{y}(t) + b(t) y(t)] h - \Delta_z(t) \xrightarrow[h \to 0]{\text{m.s.}} 0$$

so that $y(t)$ is a particular integral of (8-24). The proof of uniqueness is analogous to that in the classical case. Q.E.D.

The above proof indicates that $x(t)$ is mean-square-continuous on $[0,T]$ and that $\dot{x}(t)$ exists as a mean-square derivative and is mean-square-continuous on $[0,T]$ and also that the mean-square derivative

$$\frac{d}{dt}[\dot{x}(t) - z(t)]$$

exists.

Cumulant Generating Function

The next problem is the determination of the distribution function of the random variable $x(t)$, which is the solution (8-25) of the stochastic equation; a more general problem requires the determination of a joint distribution function at several points $t_1, \ldots, t_n$, which in turn will characterize stochastic properties of the random function $x = x(.)$. In principle the distribution function $F(\lambda, t)$ of $x(t)$ can be derived from the distribution functions of random variables forming $x(t)$, in view of the theorem which asserts that if $x(\tau)$ converges in mean square to $x(t)$ when $\tau \to t$, then the distribution function $F(\lambda, \tau)$ of $x(\tau)$ converges in ordinary sense to the distribution function $F(\lambda, t)$ of $x(t)$ at every point of continuity of $F(\lambda, t)$. The explicit results, however, can be obtained only in special cases, some of which are mentioned below.

Suppose first that the initial conditions vanish identically, so that $x(t)$ reduces to the stochastic integral

$$y(t) = \int_0^t g(t, \tau)\, dz(\tau) \qquad 0 \leq t \leq T < \infty \qquad (8\text{-}30)$$

with $y(0) = \dot{y}(0) = \ddot{y}(0) = 0$ and Green's function $g(t,\tau)$. Since $y(t)$ is the mean-square limit of Stieltjes sums, the distribution function of $y(t)$ is, by the above rule, the limit of distribution functions of these sums (in turn, the distribution function of a sum of random variables can be expressed in principle in terms of the joint distribution function of its summands; see Sec. 8-2). It is more convenient to work with the characteristic function instead, as shown below.

In order to get workable explicit formulas the following assumption is imposed in this subsection, unless explicitly stated otherwise.

Assumption: The random function $z = z(.)$ has independent increments, and

$$P\{z(0,\omega) = 0\} = 1$$

Thus $R_z(t,t') = \begin{cases} E\{z(t)\}E\{\overline{z(t') - z(t)}\} + E\{|z(t)|^2\} & \text{for } t < t' \\ E\{z(t) - z(t')\}E\{\overline{z(t')}\} + E\{|z(t')|^2\} & \text{for } t > t' \end{cases}$

and $\text{cov}\,\{z(t),z(t')\} = \text{var}\,\{z(s)\}$ with $s = \min\,(t,t')$

The characteristic function of $y(t)$ is given by

$$E(e^{i\theta y(t)}) = \lim_{n \to \infty} E\left\{\exp\left[i\theta \sum_{h=1}^{n} g(t,\tau_h)\,\Delta_h z(\tau)\right]\right\}$$

Since $\Delta z(\tau)$ are independent random variables for nonoverlapping intervals, the above characteristic function factorizes. Taking logarithms of both sides, one has

$$\log E(e^{i\theta y(t)}) = \lim_{n \to \infty} \sum_{k=1}^{n} \log E(e^{i\theta g(t,\tau_k)\Delta_k z(\tau)})$$

The logarithm of a characteristic function is called a *cumulant generating function*, and is denoted by $C(\theta,t)$. In this notation the above relation becomes

$$C_y(\theta,t) = \lim_{n \to \infty} \sum_{k=1}^{n} C_z(\theta g(t,\tau_k),\Delta\tau_k) = \int_0^t C_z(\theta g(t,\tau),d\tau) \tag{8-31}$$

where the integral on the right is an ordinary Stieltjes integral.

Differentiation of $C_y(\theta,t)$ yields immediately

$$E\{y(t)\} = \int_0^t g(t,\tau)\,dm(\tau) \tag{8-32}$$

$$\text{var}\,\{y(t)\} = \int_0^t |g(t,\tau)|^2\,d\sigma^2(\tau)$$

where $m(t) = E\{z(t)\}$

$\sigma^2(t) = \text{var}\,\{z(t)\} = R_z(t) - m^2(t)$

Similarly, the second mixed derivative of the joint cumulant generating function $C_y(\theta_1, \theta_2, t)$, given by (8-39) below, yields

$$\text{cov}\{y(t), y(t')\} = \int_0^s g(t, \tau)\overline{g(t', \tau)}\, d\sigma^2(\tau) \tag{8-33}$$

with $s = \min(t, t')$. Thus

$$R_y(t, t') = \text{cov}\{y(t), y(t')\} + E\{y(t)\}\overline{E\{y(t')\}}$$

cf. also (8-29).

Formula (8-31) provides the required answer, but it is possible to put it into another form. Since the distribution function of increments $\Delta z(t)$ is infinitely divisible, the cumulant generating function for $z(t)$, with $z(0) = 0$, is given by the Levy formula

$$C_z(\theta, t) = i\theta M(t) - \frac{\theta^2}{2} S^2(t) + \fint_{-\infty}^{\infty} \left(e^{i\theta\zeta} - 1 - \frac{i\theta\zeta}{1 + \zeta^2} \right) dL(t, \zeta)$$

where $M(t)$, $S(t)$, and $L(t, \zeta)$ are real functions vanishing with t, $S^2(t)$ is nondecreasing, and the barred integral sign means that the origin is excluded from integration. For the proof, see, for example, Loève.[69] $L(t, .)$ is defined for all real ζ except $\zeta = 0$ and is a nondecreasing function of ζ in each of the intervals $(-\infty, 0-)$ and $(0+, +\infty)$, with $L(t, \pm\infty) = 0$, continuous on the right for $\zeta \neq 0$, and for some $\zeta > 0$

$$\fint_{-\delta}^{\delta} \zeta^2\, dL(t, \zeta) < \infty$$

Substitution of the above $C_z(\theta, t)$ into (8-31) yields the general expression for the cumulant generating function of the random function y

$$C_y(\theta, t) = i\theta \int_0^t g(t, \tau)\, dM(\tau) - \frac{\theta^2}{2} \int_0^t |g(t, \tau)|^2\, dS^2(\tau)$$

$$+ \int_0^t \fint_{-\infty}^{\infty} \left(e^{i\theta g(t, \tau)\zeta} - 1 - \frac{i\theta g(t, \tau)\zeta}{1 + \zeta^2} \right) dd L(\tau, \zeta) \tag{8-34}$$

The following theorem summarizes these results.

Theorem 8-9 The distribution function $F(\lambda, t)$ of the stochastic integral $y(t)$, defined by (8-30) with respect to the random function z with independent increments and with cumulant generating function $C_z(\theta, t)$, can be obtained by inversion from the cumulant generating function $C_y(\theta, t)$ given by (8-31) or, alternatively, by (8-34). The moments are given by (8-32) and (8-33).

Expression (8-34) indicates that the random function y can be considered as the sum of a gaussian random function and a set of Poisson random functions. In fact, for these two special processes the above expressions simplify considerably. When the increments are gaussian, with mean $m(t)$ and variance $\sigma^2(t)$, the cumulant generating function takes the form

$$C_z(\theta,t) = i\theta m(t) - \frac{\theta^2}{2}\sigma^2(t) \tag{8-35}$$

so that only the first two integrals remain in (8-34). Thus, the $y(t)$ process is also gaussian, with mean and variance given by (8-32).

On the other hand, when increments are poissonian, with mean ct, the cumulant generating function is

$$C_z(\theta,t) = ct(e^{i\theta} - 1) \tag{8-36}$$

That is, $M(t) = ct/2$, $S^2(t) = 0$, and $L(t,.)$ has a jump of height ct at $\zeta = 1$ and is constant otherwise. Now, (8-31) gives immediately

$$C_y(\theta,t) = c \int_0^t (e^{i\theta g(t,\tau)} - 1)\, d\tau \tag{8-37}$$

so that $y(t)$ is actually a sum of Poisson random variables, as was to be expected. Evidently

$$E\{y(t)\} = c \int_0^t g(t,\tau)\, d\tau \qquad \text{var } \{y(t)\} = c \int_0^t |g(t,\tau)|^2\, d\tau \tag{8-38}$$

Theorem 8-9 can be extended to two or more points. The following two formulas are of importance for the $y(t)$ process. The cumulant generating function at two points $t_1 < t_2$ takes the form

$$C_y(\theta,t_1,\theta_2,t_2) = \int_0^{t_1} C_z(\theta_1 g(t_1,\tau) + \theta_2 g(t_2,\tau),d\tau) + \int_{t_1}^{t_2} C_z(\theta_2 g(t_2,\tau),d\tau) \tag{8-39}$$

Similarly, the cumulant generating function for $y(t)$ and its derivative $\dot{y}(t)$ is

$$C_{y,\dot{y}}(\theta_1,\theta_2,t) = \int_0^t C_z(\theta_1 g(t,\tau) + \theta_2 g^{(1)}(t,\tau),d\tau) \tag{8-40}$$

Density Equation

Although Theorem 8-9 yields an explicit expression for the unknown distribution function $F(\lambda,t)$, it is possible to obtain equations for $F(\lambda,t)$ which often are easier to handle. The relationship between stochastic differential equations and equations for distribution functions is of intrinsic interest, and in the present case it can be established in the following manner.

Consider again the stochastic integral (8-30) and suppose, for simplicity, that its distribution function $F(\lambda,t)$ possesses the density $f(\lambda,t)$, with integrable characteristic function $K(\theta,t)$. Since

$$C_y(\theta,t) = \log K(\theta,t)$$

differentiation yields the equation for $K(\theta,t)$:

$$\frac{\partial K(\theta,t)}{\partial t} = \left[\frac{\partial}{\partial t} C_y(\theta,t)\right] K(\theta,t) \tag{8-41}$$

When the form of $C_y(\theta,t)$ is given, (8-41) can be inverted to give $f(\lambda,t)$.
Now, from (8-34)

$$\frac{\partial}{\partial t} C_y(\theta,t) = i\theta\alpha(t) - \frac{\theta^2}{2}\beta(t)$$

$$+ i\theta \int_0^t g^{(1)}(t,\tau) \int\!\!\!\!\!\!-_{-\infty}^{\infty} \left(\zeta e^{i\theta g(t,\tau)\zeta} - \frac{\zeta}{1+\zeta^2}\right) ddL(\tau,\zeta)$$

where $\quad \alpha(t) = \displaystyle\int_0^t g^{(1)}(t,\tau)\, dM(\tau) \qquad \beta(t) = \displaystyle\int_0^t \frac{\partial}{\partial t}|g(t,\tau)|^2\, dS^2(\tau) \geq 0$

Hence, by inversion

$$\frac{\partial}{\partial t} f(\lambda,t) = -\alpha(t)\frac{\partial}{\partial\lambda} f(\lambda,t) + \tfrac{1}{2}\beta(t)\frac{\partial^2}{\partial\lambda^2} f(\lambda,t)$$

$$+ \int_0^t g^{(1)}(t,\tau) \int\!\!\!\!\!\!-_{-\infty}^{\infty} \frac{\partial}{\partial\lambda}\left[-\zeta f(\lambda - \zeta g(t,\tau),t) + \frac{\zeta}{1+\zeta^2} f(\lambda,t)\right] ddL(\tau,\zeta)$$

$$\tag{8-42}$$

with the initial condition $f(\lambda,0) = \partial(\lambda)$. This leads to the following theorem.

***Theorem* 8-10** Under the conditions of Theorem 8-9 the density $f(\lambda,t)$, assumed to exist, satisfies the difference-differential equation (8-42).

Equation (8-42) can be regarded as the generalization of the Liouville equation from Sec. 8-2, and it has, in fact, the form of the Feller-Ito equation for a Markoff process. Special forms of (8-42) are well known; e.g., for a gaussian process one obtains the familiar diffusion equation

$$\frac{\partial f}{\partial t} = -\alpha(t)\frac{\partial f}{\partial\lambda} + \tfrac{1}{2}\beta(t)\frac{\partial^2 f}{\partial\lambda^2}$$

The analysis can be extended to multinormal densities; cf. examples below. However, when in the expansion of $C_z(\theta,t)$ only the first two terms are retained, then (8-31) implies that $M(t) = m(t)$ and $S^2(t) = \sigma^2(t)$; hence the above diffusion equation can serve as the first approximation.

Remarks: Return now to the general solution (8-25) with the arbitrary initial conditions

$$x(t) = w(t) + y(t) \tag{8-43a}$$

where $w(t) = x_0 u(t) + \dot{x}_0 v(t)$ and $y(t)$ is the stochastic integral (8-30). It is reasonable to assume that the random variables $w(t)$ and $y(t)$ are independent for each t. Since the random variable $w(t)$ of the complementary solution has two degrees of randomness, it is apparent that one should consider jointly $x(t)$ and its derivative $\dot{x}(t)$, the second derivative being functionally related

$$\dot{x}(t) = \dot{w}(t) + \dot{y}(t) \qquad (8\text{-}43b)$$

It is seen from these equations that the vector random variable $\{x(t),\dot{x}(t)\}$ depends only on $(x_0,\dot{x}_0)$ and on (a linear combination of) the independent increments $\Delta z(\tau)$, which are also independent of $(x_0,\dot{x}_0)$. Similarly for any fixed s, $0 \le s \le t$, the vector $\{x(t),\dot{x}(t)\}$ depends only on the vector $\{x(s),\dot{x}(s)\}$ and independent increments $\Delta z(\tau)$ for $s \le \tau \le t$. Thus, the conditional distribution of $\{x(t),\dot{x}(t)\}$ given $\{x(\tau),\dot{x}(\tau)\}$ for $0 \le \tau \le s$ is the same as the conditional distribution given $\{x(s),\dot{x}(s)\}$ only; in other words, the vector process $\{[x(t),\dot{x}(t)], 0 \le t < \infty\}$ is markovian. By the same argument, the marginal processes $\{x(t), 0 \le t < \infty\}$ and $\{\dot{x}(t), 0 \le t < \infty\}$ are second-order markovian. More generally, under the same assumptions a linear differential equation of the nth order will lead to a markovian process for the n-dimensional vector of $x(t)$ and its first $n - 1$ derivatives, or nth-order markovian process for one-dimensional marginal random variables. Direct proofs of these statements are cumbersome, and are omitted.

In principle, the conditional distribution for the vector random variable $\{x(t),\dot{x}(t)\}$ can be obtained by standard methods. For simplicity, let us suppose that all random variables in (8-43a) possess densities. Let $f_1(\mu,\dot{\mu})$ be the joint density of x_0 and $\dot{x}_0$, assumed to be given. Let $f_2(\lambda,\dot{\lambda},t)$ be the joint density of random variables $y(t)$ and $\dot{y}(t)$, which can be obtained, say, from (8-40). The argument from Sec. 8-2 shows that the conditional density of $\{x(t),\dot{x}(t)\}$, given $\{x_0,\dot{x}_0\}$, provided $w(t)$ and $y(t)$ are independent, equals precisely the joint density of $y(t)$ and $\dot{y}(t)$, these random variables being expressed by (8-43a) and (8-43b), respectively. In symbols, the transition probability density for the vector process $\{x(t),\dot{x}(t)\}$ is

$$f(\eta,\dot{\eta},t|\mu,\dot{\mu}) = f_2[\eta - \mu u(t) - \dot{\mu}v(t), \, \dot{\eta} - \mu\dot{u}(t) - \dot{\mu}\dot{v}(t), \, t]$$

where η and $\dot{\eta}$ are values of $x(t)$ and $\dot{x}(t)$, respectively.

Partial differential equations for f can be obtained the same way as explained above. It is also convenient to work with the conditional cumulant generating function, relative to the initial conditions. The above form of f implies that

$$\begin{aligned}
C_{x,\dot{x}}(\theta_1,\theta_2,t|x_0 = \mu, \, \dot{x}_0 = \dot{\mu}) &= i\theta_1[\mu u(t) + \dot{\mu}v(t)] \\
&\quad + i\theta_2[\mu\dot{u}(t) + \dot{\mu}\dot{v}(t)] + C_{y,\dot{y}}(\theta_1,\theta_2,t) \quad (8\text{-}44)
\end{aligned}$$

from which the marginal cumulant generating function for $x(t)$ or $\dot{x}(t)$ can be obtained by letting $\theta_2 = 0$ or $\theta_1 = 0$. Similarly, one can evaluate the cumulant generating function of x at two or more points; cf. (8-39). Expressions of this kind are too involved when written explicitly; below the method is illustrated for some well-known examples.

However, it should be kept in mind that the primary datum is the random function $z(.)$ and that the distribution of $x(t)$ is not determined solely by the initial conditions but principally by the values of $z(.)$ over the whole interval $[0,t]$. In contrast with the situation treated in Sec. 8-2, the distribution of initial conditions plays only a minor role, and one is principally interested in conditional distributions of $x(t)$ relative to the initial conditions (whose distribution is then chosen arbitrarily). Furthermore, for stable motion $w(t)$ tends to zero when t tends to infinity, and the influence of the initial conditions disappears (ergodicity).

Stationary Process

Of great importance is the case when the $y(t)$ process starts from $-\infty$, instead of $t = 0$, and $g(t,\lambda)$ is a function of the difference $t - \lambda$ only

$$y(t) = \int_{-\infty}^{t} g(t - \lambda)\, dz(\lambda) \qquad (8\text{-}45)$$

with the random function z having independent increments. Formally, this has been obtained from (8-30) by replacing $y(t)$ with $y(t) - y(t_0)$, taking the integral from t_0 to t, and letting $t_0 \to -\infty$. It follows from (8-31) that the cumulant generating function does not involve t:

$$C_y(\theta,t) = \int_0^{\infty} C_z(\theta g(s),ds) \qquad (8\text{-}46)$$

so that the random function $y(.)$ is stationary in the strict sense. This in turn implies that $R_y(t,t')$ depends only on the difference $\lambda = t' - t$, so that

$$E\{y(t + \lambda)\overline{y(t)}\} = R_y(\lambda)$$

with
$$R_y(-\lambda) = \overline{R_y(\lambda)} \qquad (8\text{-}47)$$

The continuous function $R_y(t)$ is positive-definite, and, according to Bochner's theorem, it can be expressed in the form

$$R_y(t) = \int_{-\infty}^{\infty} e^{ipt}\, dH(p) \qquad (8\text{-}48)$$

where $H(p)$ is the spectral function (monotone, nondecreasing, and bounded). The result (8-48) is known as the *Wiener-Khintchine theorem*. In the present case $H(p)$ possesses *spectral density* $h(p) = dH/dp$, which can be obtained by inversion

$$h(p) = \frac{1}{2\pi i} \int_{-\infty}^{\infty} e^{-ipt} R(t)\, dt \qquad (8\text{-}49)$$

From (8-33) it follows that

$$\text{cov}\,\{y(t+\lambda),y(t)\} = \int_0^\infty g(t+\lambda)\overline{g(t)}\,d\sigma^2(t) \tag{8-50}$$

The following special case occurs frequently. Suppose that $m(t) = ct$ and $\sigma^2(t) = \sigma^2 t$; then (8-32) yields

$$E\{y(t)\} = c \int_0^\infty g(s)\,ds \qquad \text{var}\,\{y(t)\} = \sigma^2 \int_0^\infty |g(s)|^2\,ds \tag{8-51}$$

These two formulas are known as the *Campbell theorem;* cf. (8-38).

Now let $m(t) = 0$, so that

$$R_y(\tau) = \sigma^2 \int_0^\infty g(t+\tau)\overline{g(t)}\,dt$$

with $R_z(t) = \sigma^2 t$. Substitution into (8-49) yields, by the convolution property, the spectral density of the form

$$h(p) = \frac{\sigma^2}{2\pi}\,\gamma(p)\,\overline{\gamma(p)} \tag{8-52}$$

where

$$\gamma(p) = \int_0^\infty e^{-ipt}g(t)\,dt$$

Suppose now that $z(t)$ represents a random input to the network with a transient response $g(t)$, $t \geq 0$, and the output $y(t)$ given by (8-45). Equation (8-52) indicates that the transfer function $\gamma(p)$ can be obtained from factorization of the spectral density $h(p)$ of the output, which in turn is obtained as the response to the input with constant spectral density. This is a special case of the Cramer-Loève representation theorem from harmonic analysis of stationary processes, which asserts that every mean-square-continuous process stationary in the wide sense $\{y(t),\ -\infty < t < \infty\}$ has the spectral representation

$$y(t) = \int_{-\infty}^\infty e^{itp}\,d\hat{y}(p) \tag{8-53}$$

where the $\hat{y}(p)$ process has orthogonal increments and

$$E\{|d\hat{y}(p)|^2\} = dH(p)$$

This provides the operational calculus for solving linear stochastic equations analogous to that for nonrandom equations. References can be found in "NM," chap. 6, where applications to prediction theory and filtering are discussed in sec. 6-7; for harmonic analysis of general mean-square-continuous random functions, see Loève.[69] Observe also the close analogy between (8-53) and the Stone representation theorem for unitary operators A as spectral integrals.

Exercise 8-8 Let

$$x(t) = \int_{-\infty}^{t} g(t - \tau) \, dz(\tau)$$

be a stationary solution of a linear stochastic equation $Lx(t) = dz/dt$, with Green's function $g(t)$ and random forcing function dz/dt, where $z(t)$ is the brownian motion (see below). Show that:

1. According to the mean-square criterion, the best linear prediction given whole past up to t, is

$$\hat{x}(t + h) = E\{x(t + h)|x(s), s \leq t\} = \int_{-\infty}^{t} g(t + h - \tau) \, dz(\tau)$$

and the prediction error is

$$\|x(t + h) - \hat{x}(t + h)\|^2 = \sigma^2 \int_{0}^{h} |g(\tau)|^2 \, d\tau$$

Remark: This is a complete and rigorous solution for those linear stochastic processes which arise from the solutions of linear stochastic equations. Discuss the markovian character; show that the best prediction is obtained as the solution of the corresponding homogeneous stochastic equation, but with conditions prevailing at t taken as initial conditions (see Exercise 8-13). Extend the formulas to the $z(t)$ process with orthogonal increments.

2. The prediction operator $K(t)$ defined by

$$\hat{x}(t + h) = \int_{0}^{\infty} x(t - \tau) \, dK(\tau)$$

satisfies the Wiener-Hopf equation

$$R(t + h) = \int_{0}^{\infty} R(t - \tau) \, dK(\tau)$$

where $R(t)$ is the covariance function of $x(t)$ process; the solution for

$$K^*(p) = \int_{0}^{\infty} e^{-ipt} \, dK(t)$$

is given by

$$K^*(p) = \frac{1}{\gamma(p)} \int_{0}^{\infty} e^{-ipt} \, dt \int_{-\infty}^{\infty} e^{is(t+h)} \gamma(s) \, ds$$

where $\gamma(p)$ is the Fourier transformation of $g(t)$. For prediction theory see Wiener,[88] Bartlett,[6] Yaglom,[92] and sec. 6–9 of "NM."

BROWNIAN MOTION

The mean-square properties of a solution of a differential equation have been given the most extensive study in the case when the forcing random function $z(.)$ is either a brownian-motion process or a Poisson process. Examples given below are taken from practical applications; see, for example, Moyal[74] and Yaglom.[92] The brownian-motion process will be discussed first, and the treatment of it will be continued from another point of view later on.

Wiener Process

Brownian motion has been the subject of intensive research by physicists and mathematicians, and its mathematical theory represents one of the major achievements in probability. In particular, the theory of the brownian motion has contributed significantly to such branches of stochastic processes as Markoff processes, diffusion theory, and modern potential theory, and it has many applications in physics, engineering, and communication theory.

In 1827, the botanist Robert Brown observed that particles suspended in fluids perform peculiarly irregular movements, which later were attributed to the impact of molecules of the liquid. This phenomenon is referred to as *brownian motion*, and the stochastic brownian-motion process is used to analyze this physical motion.

The process was first discussed mathematically by Bachelier in 1900. Since 1905 the Brownian motion has been treated statistically, on the basis of the pioneering work by Einstein and Smoluchowski. The first rigorous study of the brownian-motion process was made in 1923 and 1930 by Wiener, who explained the irregularities of the motion by showing that its trajectories are almost all nondifferentiable continuous functions. The brownian-motion process is also called the *Wiener process*, and the concepts of Wiener measure and the Wiener stochastic integral are associated with this process. On the mathematical side, further important contributions were made by Kolmogorov, Feller, Levy, Doob, Kac, Ito, and many others.

In this section only a brief account of the basic properties of brownian motion will be given. For details, further work, and newer, more refined theorems the reader should consult the references. The book by Kac[52] and the collection of papers edited by Wax[86] provide a good introduction (see also "NM," sec. 6-9); for more advanced accounts see the books by Doob,[28,29] Levy,[68] and Wiener.[88,89]

The brownian motion is a real stochastic process $\{z(t), 0 \leq t < \infty \}$ with independent gaussian increments and with

$$E\{z(t) - z(t')\} = 0$$
$$E\{|z(t) - z(t')|^2\} = \sigma^2|t - t'|$$

where σ is a positive constant. In applications, $z(t)$ represents the position of a particle in one-dimensional brownian motion at time t; the assumption of proportionality to t follows from Einstein's result, which related σ^2 to the diffusion constant. Evidently, the process also has orthogonal increments, because for nonoverlapping intervals Δt and $\Delta' t'$

$$E\{\Delta z(t) \, \Delta' z(t')\} = E\{\Delta z(t)\} E\{\Delta' z(t')\} = 0$$

It is further assumed that the process starts initially from zero, with probability one. It follows that the brownian motion is a Markoff process (and also a martingale). Hence, the transition probability density of $z(t)$, given $z(t')$, coincides with the gaussian density of the increment $z(t) - z(t')$

$$f(\lambda,t|\lambda',t') = f(\lambda - \lambda', t - t') = \frac{1}{\sigma\sqrt{2\pi|t - t'|}} \exp\left[-\frac{(\lambda - \lambda')^2}{2\sigma^2|t - t'|}\right]$$

where λ and λ' are values of $z(t)$ and $z(t')$, respectively.

The density f satisfies the familiar diffusion equation

$$\frac{\partial f}{\partial t} = \tfrac{1}{2}\sigma^2 \frac{\partial^2 f}{\partial \lambda^2}$$

with $f(\lambda,t|\lambda',t) = \vartheta(\lambda - \lambda')$. The markovian property of f implies the Chapman-Kolmogorov equation

$$f(\lambda - \mu, t - s) = \int_{-\infty}^{\infty} f(\alpha - \mu, u - s)f(\lambda - \alpha, t - u)\, d\alpha$$

with $s < u < t$, which in this case is simply the convolution property of the gaussian density.

Direct evaluation yields

$$R_z(t,t') = E\{z(t)z(t')\} = \sigma^2 \min(t,t')$$

and

$$R_z(t) = \sigma^2 t \qquad 0 \le t$$

By criteria from Sec. 8-3, the brownian-motion process is at each t continuous in mean square but is not mean-square-differentiable. It is also easily seen that with a mean-square norm

$$\left\| \frac{z(t + h) - z(t)}{h^\delta} \right\| = \sigma h^{\frac{1}{2}-\delta} \qquad h > 0, \delta \ge 0$$

so that the indicated quotient cannot converge in mean square for $\delta \ge \tfrac{1}{2}$.

Formally, the derivative $\dot{z}(t)$ process is stationary in the wide sense, with the covariance function

$$E\{\dot{z}(t)\dot{z}(t')\} = \sigma^2 \vartheta(t - t')$$

and constant spectral density

$$h(p) = \frac{\sigma^2}{2\pi}$$

(so-called *white noise*).

Langevin Equation

As the first example, consider the equation of the form

$$d\dot{x}(t) + \alpha\dot{x}(t)\, dt = dz(t) \qquad (8\text{-}54)$$

whose solution is obviously

$$\dot{x}(t) - \dot{x}_0 e^{-\alpha t} = \int_0^t e^{-\alpha(t-\tau)}\, dz(\tau)$$

$$x(t) - x_0 - \dot{x}_0\alpha^{-1}(1 - e^{-\alpha t}) = \frac{1}{\alpha}\int_0^t (1 - e^{-\alpha(t-\tau)})\, dz(\tau)$$

where x_0 and $\dot{x}_0$ are initial conditions at $t = 0$, and α is a positive constant. The above equation gives the response $x(t)$ of a linear oscillator to the brownian-motion input $z(t)$ when no restoring force is present; the equation itself is known as the *Langevin equation* for the velocity $\dot{x}(t)$. As already noted, the difference form in (8-54) is used because the mean-square derivative of $z(t)$ does not exist.

Suppose that the initial conditions are fixed to be

$$x_0 = \mu \qquad \dot{x}_0 = \dot{\mu}$$

Then, the cumulant generating function of $x(t)$ and $\dot{x}(t)$, conditional on μ and $\dot{\mu}$, is, from (8-44),

$$\begin{aligned}
C_{x,\dot{x}}(\theta_1,\theta_2,t|\mu,\dot{\mu}) &= \log E\left\{e^{i\theta_1 x(t)+i\theta_2\dot{x}(t)}|\mu,\dot{\mu}\right\} \\
&= i\theta_1[\mu + \dot{\mu}\alpha^{-1}(1 - e^{-\alpha t})] + i\theta_2\dot{\mu}e^{-\alpha t} \\
&\quad - \frac{\sigma^2}{2\alpha^2}\int_0^t [\theta_1 + (\alpha\theta_2 - \theta_1)e^{-\alpha(t-\tau)}]^2\, d\tau
\end{aligned}$$

And this is again the gaussian cumulant generating function; the moments, conditional on μ and $\dot{\mu}$, are read off to be

$$E\left\{\dot{x}(t)|\mu,\dot{\mu}\right\} = \dot{\mu}e^{-\alpha t}$$

$$E\left\{x(t)|\mu,\dot{\mu}\right\} = \mu + \dot{\mu}\alpha^{-1}(1 - e^{-\alpha t})$$

$$\text{var}\left\{\dot{x}(t)|\mu,\dot{\mu}\right\} = \frac{\sigma^2}{2\alpha}(1 - e^{-2\alpha t})$$

$$\text{var}\left\{x(t)|\mu,\dot{\mu}\right\} = \frac{\sigma^2}{2\alpha^3}(2\alpha t - 3 + 4e^{-\alpha t} - e^{-2\alpha t})$$

$$\text{cov}\left\{x(t),\dot{x}(t)|\mu,\dot{\mu}\right\} = \frac{\sigma^2}{2\alpha^2}(1 - e^{-\alpha t})^2$$

As explained earlier, the vector process $\{x(t),\dot{x}(t)\}$ is markovian; it is then characterized by the gaussian transition probability density $f(\lambda,\eta,t|\mu,\dot{\mu})$, with the above cumulant generating function, where λ and η are values of $x(t)$ and $\dot{x}(t)$. The equation for the density f can be obtained by the method explained above. In fact, after some algebra, it can be seen that the cumulant generating function $C_{x,\dot{x}}(\theta_1,\theta_2,t|\mu,\dot{\mu})$ satisfies the partial differential equation

$$\frac{\partial C}{\partial t} = (\theta_1 - \alpha\theta_2)\frac{\partial C}{\partial \theta_2} - \tfrac{1}{2}\sigma^2\theta_2{}^2C$$

From this, by inversion (of the double characteristic function), the equation for the density f is obtained in the form

$$\frac{\partial f}{\partial t} = -\eta\frac{\partial f}{\partial \lambda} + \alpha\frac{\partial}{\partial \eta}(\eta f) + \tfrac{1}{2}\sigma^2\frac{\partial^2}{\partial \eta^2}f$$

which is the Fokker-Planck equation for (two-dimensional) diffusion; cf. (8-42).

It is also of interest to find the cumulant generating function for velocity $\dot{x}(t)$ at two points, conditional on the initial value $\dot{\mu}$; the result is again gaussian $(t > s)$:

$$
\begin{aligned}
C_{\dot{x}}(\theta_1,\theta_2,t,s|\dot{\mu}) &= \log E\{e^{i\theta_1\dot{x}(t)+i\theta_2\dot{x}(s)}|\dot{\mu}\} \\
&= i\theta_1\dot{\mu}e^{-\alpha t} + i\theta_2\dot{\mu}_e{}^{-\alpha s} - \frac{1}{2}\frac{\sigma^2}{2\alpha} \\
&\quad [\theta_1{}^2(1 - e^{-2\alpha t}) + 2\theta_1\theta_2e^{-\alpha t}(e^{\alpha s} - e^{-\alpha s}) + \theta_2{}^2(1 - e^{-2\alpha s})]
\end{aligned}
$$

It follows that $E\{\dot{x}(t)|\dot{\mu}\}$ and var $\{\dot{\mu}(t)|\dot{\mu}\}$ agree with those given above, and the covariance is

$$\text{cov }\{\dot{x}(t),\dot{x}(s)|\dot{\mu}\} = \frac{\sigma^2}{2\alpha}(e^{-\alpha(t-s)} - e^{-\alpha(t+s)}) = \text{var }\{\dot{x}(t)\dot{\mu}\}e^{-\alpha(t-s)}$$

whereas the covariance function for the $x(t)$ process is $(t > s)$

$$E\{\dot{x}(t)\dot{x}(s)|\dot{\mu}\} = \frac{\sigma^2}{2\alpha}(1 - e^{-2\alpha s})e^{-\alpha(t-s)} + \dot{\mu}^2e^{-\alpha(t+s)}$$

The form of $\dot{x}(t)$ implies that the $\dot{x}(t)$ process is also markovian. For increasing t, $\dot{x}(t)$ tends to a stationary gaussian process (independent of initial conditions) with zero mean, variance $\sigma^2/2\alpha$, and covariance function

$$E\{\dot{x}(t + \tau)\dot{x}(t)\} \underset{t\to\infty}{\to} \frac{\sigma^2}{2\alpha}e^{-\alpha|\tau|}$$

which is obviously a covariance function of a non-mean-square-differentiable process. It follows from the Markoff property that if this limiting distribution is taken as the initial distribution of $\dot{x}_0$, then the absolute distribution of $\dot{x}(t)$ is independent of t and equals the limiting distribution.

There is another way to obtain the stationary process. Suppose that the basic assumptions on the $\dot{x}(t)$ process are changed by allowing it to start from $-\infty$ instead from $t = 0$. For the initial random variable $\dot{x}_0$ put

$$\dot{x}_0 = \int_{-\infty}^{0} e^{\alpha \tau} \, dz(\tau)$$

Hence, for $-\infty < t < \infty$,

$$\dot{x}(t) = \int_{-\infty}^{t} e^{-\alpha(t-\tau)} \, dz(\tau)$$

and obviously $E\{\dot{x}(t)\} = 0$. Direct evaluation yields for the covariance function

$$E\{\dot{x}(t)\dot{x}(s)\} = \frac{\sigma^2}{2\alpha} e^{-\alpha|t-s|}$$

and the $\dot{x}(t)$ process is stationary, with a continuous spectral density given by (8-49)

$$h(p) = \frac{1}{2\pi} \frac{\sigma^2}{\alpha^2 + p^2}$$

which corresponds to the response of a linear oscillator under impulse excitation (white-noise input).

Exercise 8-9   Verify the formulas in this subsection.

Exercise 8-10   Using the Ito equation (8-60) below, show that the density $f(\dot{\mu},t)$ for $\dot{x}(t)$ from the Langevin equation (8-54) satisfies the Fokker-Planck equation of the form

$$\frac{\partial f}{\partial t} = \alpha \frac{\partial \dot{\mu} f}{\partial \dot{\mu}} + \frac{1}{2}\sigma^2 \frac{\partial^2 f}{\partial \dot{\mu}^2}$$

Brownian Oscillator

As the second example, consider the brownian motion of a damped oscillator, with

$$d\dot{x}(t) + 2\alpha \dot{x}(t) \, dt + \omega_0^2 x(t) \, dt = dz(t) \tag{8-55}$$

In the case of damped oscillations ($\omega_0 > \alpha$) the solution is

$$x(t) = e^{-\alpha t}\left[x_0 \cos \omega t + \frac{\dot{x}_0 + \alpha x_0}{\omega} \sin \omega t \right]$$
$$+ \frac{1}{\omega} \int_0^t e^{-\alpha(t-\tau)} \sin \omega \, (t-\tau) \, dz(\tau)$$

where $\omega^2 = \omega_0^2 - \alpha^2$ and x_0 and $\dot{x}_0$ are initial random variables at $t = 0$. (The solution for the overdamped case is obtained simply by replacing sin and cos by sinh and cosh, respectively.) As before, $\{x(t),\dot{x}(t)\}$ form a vector markovian process with joint gaussian distribution. Its parameters will now be calculated in a manner slightly different from that in the previous example; cf. Moyal.[74]

The joint cumulant generating function of the marginal $x(t)$ process at two points $t > s$, conditional on $x_0 = \mu$, $\dot{x}_0 = \dot{\mu}$, is

$$
\begin{aligned}
C_x(\theta_1,t;\theta_2,s|\mu,\dot{\mu}) &= \log E \left\{ e^{i\theta_1 x(t) + i\theta_2 x(s)} \right\} \\
&= i\theta_1 e^{-\alpha t} \left[\mu \cos \omega t + \frac{\dot{\mu}_0 + \alpha\mu}{\omega} \sin \omega t \right] \\
&\quad + i\theta_2 e^{-\alpha s} \left[\mu \cos \omega s + \frac{\dot{\mu} + \alpha\mu}{\omega} \sin \omega s \right] \\
&\quad - \frac{\sigma^2}{2\omega^2} \left[\theta_1{}^2 \int_0^t e^{-2\alpha(t-\tau)} \sin^2 \omega(t-\tau)\, d\tau \right. \\
&\qquad + 2\theta_1\theta_2 \int_0^s e^{-\alpha(t+s-2\tau)} \sin \omega(t-\tau) \sin \omega(s-\tau)\, d\tau \\
&\qquad \left. + \theta_2{}^2 \int_0^s e^{-2\alpha(s-\tau)} \sin^2 \omega(s-\tau)\, d\tau \right]
\end{aligned}
$$

which again is of gaussian form. From this (or from direct computation) the conditional covariance assumes the form $(t \geq s)$

$$
\begin{aligned}
\text{cov} \left\{ x(t),x(s)|\mu,\dot{\mu} \right\} &= \frac{\sigma^2}{\omega^2} \int_0^s e^{-\alpha(t+s-2\tau)} \sin \omega(t-\tau) \sin \omega(s-\tau)\, d\tau \\
&= \frac{\sigma^2}{4\omega_0{}^2\alpha} \left\{ e^{-\alpha(t-s)} \left[\cos \omega(t-s) + \frac{\alpha}{\omega} \sin \omega(t-s) \right] \right. \\
&\quad + e^{-\alpha(t+s)} \left[\frac{\alpha^2}{\omega^2} \cos \omega(t+s) - \frac{\alpha}{\omega} \sin \omega(t+s) \right. \\
&\qquad \left. \left. - \frac{\omega_0{}^2}{\omega^2} \cos \omega(t-s) \right] \right\}
\end{aligned}
$$

Denote the above covariance by $\phi(t,s)$. Then the parameters for the joint gaussian distribution of $\{x(t),\dot{x}(t)\}$ are

$$
\begin{aligned}
\text{var} \left\{ x(t)|\mu,\dot{\mu} \right\} &= \phi(t,t) = \frac{\sigma^2}{4\omega_0{}^2\alpha} \left\{ 1 - e^{-2\alpha t} \left[\frac{\omega_0{}^2}{\omega^2} + \frac{\alpha}{\omega} \sin 2\omega t \right. \right. \\
&\qquad \left. \left. - \frac{\alpha^2}{\omega^2} \cos 2\omega t \right] \right\}
\end{aligned}
$$

$$
\begin{aligned}
\text{var} \left\{ \dot{x}(t)|\mu,\dot{\mu} \right\} &= \left(\frac{\partial^2 \phi}{\partial t\, \partial s} \right)_{s=t} = \frac{\sigma^2}{4\alpha} \left\{ 1 - e^{-2\alpha t} \left[\frac{\omega_0{}^2}{\omega^2} - \frac{\alpha}{\omega} \sin 2\omega t \right. \right. \\
&\qquad \left. \left. - \frac{\alpha^2}{\omega^2} \cos 2\omega t \right] \right\}
\end{aligned}
$$

$$
\text{cov} \left\{ x(t),\dot{x}(t)|\mu,\dot{\mu} \right\} = \left(\frac{\partial \phi}{\partial s} \right)_{s=t} = \frac{\sigma^2}{2\omega^2} e^{-2\alpha t} \sin^2 \omega t
$$

When $t \to \infty$, the $x(t)$ process tends to a stationary gaussian process (independent of initial conditions) with zero mean, variance $\sigma^2/(4\omega_0{}^2\alpha)$, and covariance function, obtained from $\phi(t,s)$,

$$E\{x(t+\tau)x(t)\} \to \frac{\sigma^2}{4\alpha\omega_0^2} e^{-\alpha|\tau|}\left(\cos\omega\tau + \frac{\alpha}{\omega}\sin\omega|\tau|\right) \equiv R(\tau)$$

Similarly, for the $\dot{x}(t)$ process, the above expression differentiated twice yields

$$E\{\dot{x}(t+\tau)\dot{x}(t)\} \to \frac{\sigma^2}{4\alpha} e^{-\alpha|\tau|}\left(\cos\omega\tau - \frac{\alpha}{\omega}\sin\omega|\tau|\right)$$

As before, it follows from criteria in Sec. 8-3 that the $x(t)$ process is mean-square-differentiable but the $\dot{x}(t)$ process is not.

Suppose now that the $x(t)$ process starts from $-\infty$, and let

$$x_0 = -\frac{1}{\omega}\int_{-\infty}^0 \sin\omega\tau\, dz(\tau) \qquad \dot{x}_0 + \alpha x_0 = \int_{-\infty}^0 \cos\omega\tau\, dz(\tau)$$

Consequently, for $-\infty < t < \infty$,

$$x(t) = \frac{1}{\omega}\int_{-\infty}^t e^{-\alpha(t-\tau)}\sin\omega(t-\tau)\, dz(\tau)$$

with $E\{x(t)\} = 0$. Direct evaluation yields for the covariance function $R(\tau)$ the expression given above.† The $x(t)$ process is stationary with a continuous spectral density, found from (8-49),

$$h(p) = \frac{\sigma^2}{8\pi\omega_0^2}\left[\frac{2+p/\omega}{\alpha^2+(p+\omega)^2} + \frac{2-p/\omega}{\alpha^2+(p-\omega)^2}\right]$$

$$= \frac{1}{2\pi}\frac{\sigma^2}{(p^2-\omega_0^2)^2+4\alpha^2 p^2}$$

with the transfer function $\gamma(p) = (-p^2 + 2i\alpha p + \omega_0^2)^{-1}$.

Exercise 8-11 Verify formulas in this subsection.

Exercise 8-12 Let $f(\mu,\dot{\mu},t|\mu_0,\dot{\mu}_0,0)$ be the joint transition density for the Markoff vector process $\{x(t),\dot{x}(t)\}$ obtained as solution of Eq. (8-55) for the brownian-motion oscillator. Using the Chapman-Kolmogorov equation for f, show that the Fokker-Planck equation for f is

$$\frac{\partial f}{\partial t} + \dot{\mu}\frac{\partial f}{\partial\mu} = \omega_0^2\mu\frac{\partial f}{\partial\dot{\mu}} + 2\alpha\frac{\partial\dot{\mu}f}{\partial\dot{\mu}} + \tfrac{1}{2}\sigma^2\frac{\partial^2 f}{\partial\dot{\mu}^2}$$

† Alternatively, multiplying (8-55) by $x(t-\tau)$, $\tau > 0$, and averaging yield, for $R(\tau)$, the equation

$$R'' + 2\alpha R' + w_0^2 R = 0$$

with the initial condition $R(0)$ and $R'(0) = 0$.

Exercise 8-13 Consider stationary solutions of the Langevin equation (8-54) or brownian-motion oscillator (8-55)

$$\dot{x}(t) = \int_{-\infty}^{t} e^{-\alpha(t-\tau)} \, dz(\tau)$$

$$x(t) = \frac{1}{\omega} \int_{-\infty}^{t} e^{-\alpha(t-\tau)} \sin \omega(t - \tau) \, dz(\tau)$$

Show that the best linear prediction at $t + h$ ($h > 0$), according to the mean-square criterion, is given by the solution at $t + h$ of the corresponding *homogeneous* stochastic equation

$$d\dot{x} + \alpha \dot{x} \, dt = 0 \qquad \text{or} \qquad d\dot{x} + 2\alpha \dot{x} \, dt + \omega_0^2 x \, dt = 0$$

with *random* initial conditions $\dot{x}(t)$ or $x(t)$ and $\dot{x}(t)$, respectively (cf. Sec. 8-2):

$$\hat{\dot{x}}(t + h) = e^{-\alpha h} \dot{x}(t)$$

$$\hat{x}(t + h) = e^{-\alpha h} \left(\cos \omega h + \frac{\alpha}{\omega} \sin \omega h \right) x(t) + e^{-\alpha h} \frac{\sin \omega h}{\omega} \dot{x}(t)$$

Discuss the form of the predictor operator $K(t)$ and the prediction error. Thus, to predict the velocity or the position, as the case may be, of a linear system with white-noise input, it suffices to assume that in future, i.e., after the instant t, the random forcing function will not act and to calculate the motion subject to random initial conditions prevailing at the instant t; see Yaglom[92] and Exercise 8-8.

Systems

For an extension of the ideas discussed in this subsection to the systems of differential equations, see, for example, Moyal,[72] Bartlett,[6] and Yaglom.[92]

Exercise 8-14 Discuss the generalization of the results in this subsection to the system of linear stochastic equations (see also Sec. 8-4).

Exercise 8-15 Consider a nonlinear second-order equation for the n-dimensional random vector $x(t)$

$$d\dot{x}(t) + \beta \dot{x}(t) \, dt + K[x(t)] \, dt = dz(t)$$

where K is a given n-dimensional vector with coordinates depending on $x(t)$ and $z(t)$ is the n-dimensional brownian motion (with the same σ^2). Assume that the vector $(x(t),\dot{x}(t))$ process is markovian, and let $f(\mu,\dot{\mu},t|\mu_0,\dot{\mu}_0,0)$ be its transition density, where $\mu = (\mu_1, \ldots, \mu_n)$ and $\dot{\mu} = (\dot{\mu}_1, \ldots, \dot{\mu}_n)$ are the values of $x(t)$ and $\dot{x}(t)$, respectively. Using the Chapman-Kolmogorov equation for f, show that the Fokker-Planck equation for density f takes the form

$$\frac{\partial f}{\partial t} + \sum_{i=1}^{n} \dot{\mu}_i \frac{\partial f}{\partial \mu_i} = \sum_{i=1}^{n} K_i \frac{\partial f}{\partial \dot{\mu}_i} + \beta \sum_{i=1}^{n} \frac{\partial \dot{\mu}_i f}{\partial \dot{\mu}_i} + \frac{1}{2}\sigma^2 \sum_{i=1}^{n} \frac{\partial^2 f}{\partial \dot{\mu}_i^2}$$

This is the generalization of the Liouville equation (8-11) to include brownian motion (diffusion terms); see also Moyal[74] and Chandrasekhar.[23]

Ornstein-Uhlenbeck Process

It is seen from the above examples that the solution $x(t)$ has essentially brownian-motion character after suitable rescaling. An essential modification of the Wiener process has been introduced by Uhlenbeck and Ornstein[85] in order to obtain a process that has velocities and thus is physically more plausible; see also Doob.[28]

The equation governing the Ornstein-Uhlenbeck process is obtained from the Langevin equation by a suitable change of the time scale

$$d\dot{x}(t) + \alpha \dot{x}(t) \, dt = e^{-\alpha t} \, dz(\epsilon^{2\alpha t}) \tag{8-56}$$

with solution for velocity

$$\dot{x}(t) = e^{-\alpha t} z(e^{2\alpha t}) \qquad -\infty < t < \infty$$

Here, $z(t)$ is brownian motion with parameter σ. It follows that the Ornstein-Uhlenbeck process $\{\dot{x}(t), -\infty < t < \infty\}$ is strictly stationary, markovian, and gaussian, with mean zero, variance σ^2, and covariance function

$$E\{\dot{x}(t)\dot{x}(t')\} = \sigma^2 e^{-\alpha|t-t'|}$$

so that the process is mean-square-continuous but not mean-square-differentiable. The conditional density $f(\lambda,t|\zeta,s)$ satisfies the (forward) diffusion equation

$$\frac{\partial f}{\partial t} = \frac{1}{2}\sigma^2 \alpha \frac{\partial^2 f}{\partial \lambda^2} + \alpha \frac{\partial}{\partial \lambda}(\lambda f)$$

with the boundary conditions $f = 0$ for $\lambda = \pm \infty$ and the initial condition $f \equiv \partial(\lambda - \zeta)$.

The distribution of the position coordinate $x(t)$ can be deduced from that of $\dot{x}(t)$ by use of the stochastic integral

$$x(t) - x(0) = \int_0^t \dot{x}(s) \, ds$$

It can be shown that $x(t) - x(0)$ has gaussian distribution with zero mean and variance

$$\text{var}\,\{x(t) - x(0)\} = \int_0^t \int_0^t E\{\dot{x}(s)\dot{x}(s')\} \, ds \, ds'$$
$$= \frac{2\sigma^2}{\alpha^2}(e^{-\alpha t} - 1 + \alpha t)$$

However, the $x(t)$ process is not markovian; its covariance function is

$$E\{x(t)x(t')\} = \int_0^t \int_0^{t'} E\{\dot{x}(s)\dot{x}(s')\} \, ds \, ds'$$
$$= \frac{2\sigma^2}{\alpha}\min(t,t') + \frac{\sigma^2}{\alpha^2}[e^{-\alpha \min(t,t')} + e^{-\alpha \max(t,t')} - e^{-\alpha|t-t'|} - 1]$$

It follows that the $x(t)$ process is mean-square-continuous and also mean-square-differentiable, as required. On the other hand, the vector process $\{x(t) - x(0), \dot{x}(t)\}$ is markovian. It has a gaussian joint density, with marginal means zero and with marginal variances, given above, and covariance (for $t > 0$)

$$E\{[x(t) - x(0)]\dot{x}(t)\} = \int_0^t E\{\dot{x}(t)\dot{x}(s)\}\,ds = \frac{\sigma^2}{\alpha}\,(1 - e^{-\alpha t})$$

The joint transition probability $f(\lambda,\eta,t|\mu,\dot{\mu})$ satisfies the (forward) diffusion equation

$$\frac{\partial f}{\partial t} = \frac{\sigma^2}{2}\,\alpha\,\frac{\partial^2 f}{\partial \lambda^2} + \alpha\,\frac{\partial}{\partial \lambda}\,(\lambda f) - \frac{\partial}{\partial \eta}\,(\lambda f)$$

and the corresponding backward equation.

Wiener Integrals

In connection with the Wiener process, mention must be made of Wiener measure and the corresponding integral, defined by brownian motion. Although a more fitting place would be below, where sample-function properties are discussed, a few remarks are given here for the sake of completeness. For details, the reader should consult the original publications by Wiener,[88,89] the last one giving an informal exposition. A very readable account is given by Kac[52] and also in a review article by Gelfand and Yaglom.[38]

The markovian character and the zero initial position imply that the joint density of the brownian-motion process $\{z(t),\ 0 \leq t < \infty\}$ is given by

$$f(\lambda_1,t_1)f(\lambda_2 - \lambda_1, t_2 - t_1)\ \cdots\ f(\lambda_n - \lambda_{n-1}, t_n - t_{n-1})$$

for $0 = t_0 < t_1 < \cdots < t_n$. Consequently, the n-dimensional distributions of the process, i.e., the probability that values of $x(t_i)$ lie between μ_i and λ_i for $i = 1, \ldots, n$, are given by the n-fold integrals of the form

$$\boldsymbol{P}\{\mu_i \leq x(t_i,\omega) < \lambda_i \qquad i = 1, \ldots, n\}$$

$$= \frac{1}{(2\pi)^{n/2}\sigma^n} \int_{\mu_1}^{\lambda_1} \cdots \int_{\mu_n}^{\lambda_n} \frac{1}{\sqrt{t_1}} \exp\left(-\frac{{\alpha_1}^2}{2\sigma^2 t_1}\right) \prod_{i=2}^{n} \frac{1}{\sqrt{t_i - t_{i-1}}}$$

$$\exp\left[-\frac{(\alpha_i - \alpha_{i-1})^2}{2\sigma^2(t_i - t_{i-1})}\right] d\alpha_1 \cdots d\alpha_n$$

This expression is the starting point of the Wiener approach.

Consider the space of all real-valued functions, and let Ω_0 be a subspace of all continuous functions $z(.,\omega)$ of t $(0 \leq t < \infty)$ normalized by the condition $z(0,\omega) = 0$ (see Sec. 8-6).   The n-dimensional cylinder sets in Ω_0 (*quasi intervals* in the terminology of Wiener) are assigned the probability measure defined by the above formula.   Since these n-dimensional distributions are consistent, this probability measure on cylinder sets can be extended (by the Daniel-Kolmogorov extension theorem) to the product σ field $\mathfrak{F}_0$ of the function space Ω_0 (sets in $\mathfrak{F}_0$ are determined by a countable number of coordinates, and contain only continuous functions).   This is the Wiener measure.   The above construction is that of Doob.[29]   In Wiener's original construction use is made of the explicit measure-preserving mapping of the space Ω_0 into the unit interval I, in which sets in Ω_0 with Wiener measure go into sets in I with Lebesgue measure.   Wiener's result (in 1922) was the first construction of a probability measure on a function space; the measure is actually concentrated on continuous nondifferentiable sample functions, vanishing at the origin.

The Wiener integral is now defined in the conventional manner as a Lebesgue integral with respect to Wiener measures.   Let V be a continuous bounded functional on Ω_0, and define the function $V\{z(t)\}$ of the process by its values $V\{z(t,\omega)\}$ at sample functions $z(t,\omega)$.   The process $\{V\{z(t)\}, \; 0 < t < \infty\}$ is called a functional of the brownian-motion process.   Let $\boldsymbol{P}$ be the Wiener measure constructed above.   The *Wiener integral* of the random variable $V\{z(t)\}$ with respect to $\boldsymbol{P}$ is written symbolically

$$E\{V\} = \int_{\Omega_0} V\{z(t)\}\boldsymbol{P}\{dz(t)\}$$

and the actual construction for the closed interval $[0,t]$ starts with a finite-dimensional approximation (with sample functions replaced by step functions) and then passage to the limit.   By Wiener's transformation, this integral can be expressed by the single integral with respect to Lebesgue measure over the unit interval.

The evaluation of Wiener's integrals always presents difficulties, except in extremely special isolated cases.   For example, when V depends only on the value of the sample function $z(t,\omega)$ at a finite number of points, the integral $E\{V\}$ reduces to an ordinary finite-dimensional integral.   In particular, when $V\{z(t,\omega)\} = \lambda_1 \cdots \lambda_n$, with $\lambda_i = z(t_i,\omega)$, $i = 1, \ldots, n$, the integral $E\{V\}$ yields the n-dimensional moments of brownian motion

$$E\{z(t_1) \cdots z(t_n)\} = \begin{cases} \displaystyle\sum_0 a(t_{i_1},t_{i_2}) \cdots a(t_{i_{n-1}},t_{i_n}) & n \text{ even} \\ & n \text{ odd} \end{cases}$$

where the summation extends over all possible partitions of the n indices into $n/2$ pairs, $(i_1,i_2), \ldots, (i_{n-1},i_n)$, and where

$$a(t_{i-1},t_i) = \sigma^2 \min (t_{i-1},t_i)$$

The case when
$$V\{z(t)\} = e^{i\theta y(t)}$$
with $y(t)$ a stochastic integral with respect to brownian motion, say that given by (8-30), is of great importance; in fact $E\{V\}$ becomes then the characteristic function, and the results found earlier can be obtained.

Wiener integrals have important applications in statistical mechanics and in quantum theory, and the literature devoted to their evaluation is extensive; see the references already mentioned, especially the paper by Gelfand and Yaglom and the book by Kac, where further references to the work of Feynman, Cameron, Montrol, and others can be found.

POISSON PROCESS

The Poisson process is a real stochastic process $\{z(t), 0 \leq t < \infty\}$ with independent integer-valued increments, with probability function

$$\boldsymbol{P}\{z(t,\omega) - z(s,\omega) = k\} = \frac{[c(t-s)]^k}{k!}\, e^{-c(t-s)} \qquad t > s, k = 0, 1, 2, \ldots$$

where $c \geq 0$ is a fixed parameter (called the *intensity* of the process). The Poisson process has stationary, but not orthogonal, increments, and

$$E\{z(t) - z(s)\} = c(t-s) \qquad t > s$$
$$E\{[z(t) - z(s) - c(t-s)]^2\} = c(t-s)$$

with the cumulant generating function given by (8-37), for $s = 0$. The differences $z(t) - z(0)$ form a Markoff process, and assuming $z(0) = 0$ with probability one, the transition probability is

$$\boldsymbol{P}\{z(t,\omega) = n + k | z(s,\omega)\} = n\} = \boldsymbol{P}\{z(t,\omega) - z(s,\omega) = k\}$$

Hence, the covariance function is

$$R_z(t,t') = E\{z(t)z(t')\} = c^2 tt' + c \min(t,t')$$
and $$R_z(t) = c^2 t^2 + ct \qquad 0 \leq t$$

By criteria from Sec. 8-3, the Poisson process is mean-square-continuous but not mean-square-differentiable.

Shot Effect

As the first example, consider the equation of the form

$$dx(t) + \alpha x(t)\, dt = dz(t) \qquad (8\text{-}57)$$

whose solution is

$$x(t) - x_0 e^{-\alpha t} = \int_0^t e^{-\alpha(t-\tau)}\, dz(\tau) \qquad t \geq 0$$

where x_0 is the initial random variable at $t = 0$ and α is a positive constant. The above equation occurs in the theory of *shot effect*, caused by poissonian impulsive input of electrons.

For simplicity, attention is now restricted to the random variable

$$y(t) = \int_0^t e^{-\alpha(t-\tau)}\, dz(\tau)$$

with $y(0) = 0$. The mean and the variance of the $y(t)$ process are easily found from (8-32)

$$E\{y(t)\} = \frac{c}{\alpha}(1 - e^{-\alpha t})$$

$$\text{var } \{y(t)\} = \frac{c}{2\alpha}(1 - e^{-2\alpha t})$$

and the covariance is, from (8-33),

$$\text{cov } \{y(t + \tau), y(t)\} = \frac{c}{2\alpha}(1 - e^{-2\alpha t})e^{-\alpha|\tau|}$$

which in the limit gives for the covariance function

$$R_y(\tau) = \frac{c}{2\alpha} e^{-\alpha|\tau|} + \frac{c^2}{\alpha^2}$$

which is the covariance function of a non-mean-square-differentiable process. The spectral density is found to be

$$h(p) = \frac{1}{2\pi}\left[\frac{c}{\alpha^2 + p^2} + \delta(p)\frac{c^2}{\alpha^2}\right]$$

Observe the analogy between these formulas and those in the previous subsection when one puts $\sigma^2 = c$.

The cumulant generating function for $y(t)$ is, according to (8-37),

$$C_y(\theta, t) = c \int_0^t (e^{i\theta e^{-\alpha s}} - 1)\, ds$$

$$= \frac{c}{\alpha}[C(\theta) - C(\theta e^{-\alpha t}) - iS(\theta) + iS(\theta e^{-\alpha t}) - \alpha t]$$

where
$$C(r) = -\int_r^\infty \frac{\cos r}{r}\, dr \qquad S(r) = \int_0^r \frac{\sin r}{r}\, dr$$

are the cosine and sine integrals, respectively.

The form of $x(t)$ again indicates that the $x(t)$ process is markovian, and its conditional probability of $x(t)$, given x_0, is the same as the distribution of $y(t)$ with the substitution $y(t) = x(t) - x_0 e^{-\alpha t}$. Hence the formula for $C_y(\theta, t)$ gives the exact solution for the conditional distribution of the $x(t)$ process with a Poisson input. For further analysis of the shot effect, especially for the passage of a Poisson process through a nonlinear network, see the paper by Rice.[78]

Fortet Integrals

In many applications one encounters stochastic integrals of the form

$$y(t) = \int_0^t n(t,\tau) \, dz(\tau) \qquad 0 \le t < \infty \qquad (8\text{-}58)$$

with $y(0) = 0$, where the $z(t)$ process is poissonian and $n(t,\tau)$ is a random variable with two parameters, having first and second moments, and assumed independent of $z(t)$. The existence of such mean-square integrals is covered by the theory discussed in Sec. 8-3. The cumulant generating function of $y(t)$ is computed in the manner indicated above, and the result is

$$C_y(\theta,t) = c \int_0^t [E(e^{i\theta n(t,\tau)}) - 1] \, d\tau$$

which reduces to (8-37) when $n(t,\tau)$ is degenerate. The mean and variance are easily found to be

$$E\{y(t)\} = c \int_0^t E\{n(t,\tau)\} \, d\tau$$

$$\text{var } \{y(t)\} = c \int_0^t E\{|n(t,\tau)|^2\} \, d\tau$$

which is (when $t \to \infty$) a generalized Campbell theorem.

In the special case when $n(t,s)$ assumes only values 1 and 0 with probability

$$\boldsymbol{P}\{n(t,\tau) = 1\} = 1 - F(t - \tau) \qquad t \ge \tau$$
$$\boldsymbol{P}\{n(t,\tau) = 0\} = F(t - \tau)$$

where $F(t)$, $t \ge 0$, is a given distribution function, then

$$C_y(\theta,t) = (e^{i\theta} - 1)c \int_0^t [1 - F(s)] \, ds$$

so that the $y(t)$ process is a nonhomogeneous Poisson process.

As another case, let

$$n(t,\tau) = a(\tau)g(t - \tau)$$

where $a(t)$ is a random variable with a given distribution function. This corresponds to the forcing function of the form $a(t) \, dz(t)$; that is, Poisson events produce secondary random effects. For stationary process one finds that

$$E\{y(t)\} = c\bar{a} \int_0^\infty g(\tau) \, d\tau \qquad \text{var } \{y(t)\} = c\bar{\bar{a}} \int_0^\infty g^2(\tau) \, d\tau$$

where $\bar{a}$ and $\bar{\bar{a}}$ are the first and second moment of the random variable $a(t)$, respectively.

Still another modification has been discussed by Fortet.[37] Suppose that $y(t)$ is restricted to lie within the range $[0,K]$. The *Fortet integral* can now be written as

$$y(t) = \int_0^t k(\tau)n(t,\tau)\,dz(\tau) \tag{8-59}$$

where
$$k(t) = \begin{cases} 1 & \text{if } 0 \le y(\tau) < K \\ 0 & \text{if } y(\tau) = K \end{cases}$$

The random variable $k(\tau)$ depends on $y(\tau)$, so that the above integral is in fact a nonlinear stochastic integral for $y(t)$ of the form

$$y(t) = \int_0^t V[y(\tau)]n(t,\tau)\,dz(\tau)$$

where V is a known function. The problem of finding the distribution function of $y(t)$ has not yet been solved, except in some particular cases; however, the expectation of $y(t)$ can be easily found, because the integrands are assumed to be independent.

When $n(t,\tau)$ assumes only values 0 and 1, with *arbitrary* distribution function F, so that $y(t)$ assumes integer values $0, 1, \ldots, K$, the famous theorem first proved by Pollaczek, and recently proved in abstract form by Sevastjanov, states that the stationary $y(t)$ process has Erlang distribution; see Syski[83] for references.

DIFFUSION PROCESS (ITO EQUATION)

The brownian-motion process was defined above and its mean-square properties were discussed. However, the sample-function behavior is of greater significance than the mean-square properties. The analysis of sample functions is of considerable difficulty and requires more advanced probabilistic argument. The following theorem summarizes the results: almost all sample functions of a separable brownian-motion process are continuous and nondifferentiable, and almost none has bounded variation. This theorem was proved by Wiener, Doob, and Levy, and its proof involves in an essential way the concept of separability of the process (cf. Sec. 8-6); for proof see Doob,[29] Levy,[68] and Paley and Wiener.[75] The implications of this theorem in physical applications partially explain the irregularities of the brownian motion. In fact, when $z(t)$ represents the position of the particle, the theorem asserts that velocities $\dot{z}(t)$ do not exist and that almost all trajectories are continuous but have infinite length for any finite time interval.

This subsection will deal with the important nonlinear stochastic equation whose forcing function is defined by brownian motion. The principal result asserts that the equation in question characterizes Markoff diffusion process (whose theory was earlier developed from an entirely different point of view). This result is due to Ito;[47] see also Doob[29] and Ito.[48]

The equation considered by Ito is written in the form

$$dx(t) = m[t,x(t)] \, dt + \sigma[t,x(t)] \, dz(t) \qquad a \leq t \leq b \qquad (8\text{-}60)$$

where the $z(t)$ process is the brownian-motion process with variance $\sigma^2 = 1$. The functions $m(.,.)$ and $\sigma^2(.,.)$ satisfy the following hypotheses:

1. $m(.,.)$ and $\sigma(.,.)$ are Baire functions of the pair (t,λ) for $a \leq t \leq b$ and $-\infty < \lambda < \infty$.

2.   There is a constant K' for which

$$|m(t,\lambda)| \leq K'(1 + \lambda^2)^{1/2}$$
$$0 \leq \sigma(t,\lambda) \leq K'(1 + \lambda^2)^{1/2}$$

3. $m(.,.)$ and $\sigma(.,.)$ satisfy a uniform Lipschitz condition in λ

$$|m(t,\lambda_2) - m(t,\lambda_1)| \leq K|\lambda_2 - \lambda_1|$$
$$|\sigma(t,\lambda_2) - \sigma(t,\lambda_1)| \leq K|\lambda_2 - \lambda_1|$$

where K is independent of t and λ.

The existence and uniqueness of a solution to Eq. (8-60) are established in the following theorem, due to Ito.

***Theorem* 8-11**   Under hypotheses 1 to 3 there exists a unique random function x with the following properties:

1.   Almost all sample functions ot the $x(t)$ process are continuous in $[a,b]$.

2. $\int_a^b E\{x^2(t)\} \, dt < \infty$.

3.   For each $t_0 \, \epsilon \, (a,b)$, $x(t_0) - x(a)$ is independent of the collection of differences $\{z(b) - z(s), s > t_0\}$.

4.   For each $t \, \epsilon \, [a,b]$, with probability one,

$$x(t) - x(a) = \int_a^t m[s,x(s)] \, ds + \int_a^t \sigma[s,x(s)] \, dz(s) \qquad (8\text{-}61)$$

Equation (8-60) is interpreted to mean the truth of Eq. (8-61), with probability one. Integrals in (8-61) can be defined in terms of integrals of individual sample functions of the $x(t)$ process. For any sample function $x(t,\omega)$ the first integral in (8-61) is an ordinary integral; the second assumes the form

$$y(t) = \int_0^t g(s,\omega) \, dz(s)$$

for a fixed ω (see below for a definition of this integral).

The proof of the above theorem cannot be given here, and a brief outline must suffice; for details see Doob.[29] Equation (8-61) is solved by successive approximation, starting with an arbitrary $x_0(t)$ process satisfying properties 1 to 3 and with an arbitrary $x(a) \in L_2$ independent of the differences $\Delta z(t)$.   Writing

$$x_n(t) = x(a) + \int_a^t m[s, x_{n-1}(s)] \, ds + \int_a^t \sigma[s, x_{n-1}(s)] \, dz(s)$$

one first proves that if the $x_{n-1}(t)$ process has properties 1 to 3, then the $x_n(t)$ process also has these properties.   Letting

$$\lim_{n \to \infty} x_n(t) = x(t) \qquad a \le t \le b$$

one next shows that this limit defines the $x(t)$ process with properties 1 to 3, which is a solution of (8-61).   This completes the existence proof, and the simple argument shows that $x(t)$ is uniquely determined by $x(a)$.

Any solution of (8-61) satisfies

$$x(t) - x(\tau) = \int_\tau^t m[s, x(s)] \, ds + \int_\tau^t \sigma[s, x(s)] \, dz(s)$$

and since the $z(t)$ process has independent increments, it follows that the $x(t)$ process is markovian with the transition probability

$$\boldsymbol{P}\{x(t, \omega) < \lambda | x(s, \omega) = \zeta\} = F(\lambda, t | \zeta, s)$$

satisfying the Chapman-Kolmogorov equation.   Furthermore, the following are shown.

Corollary 1:   The solutions of the stochastic equation (8-60) satisfy:
1.   Partial derivatives of $F(\lambda, t | \zeta, s)$

$$\frac{\partial F}{\partial \zeta} \qquad \text{and} \qquad \frac{\partial^2 F}{\partial \zeta^2}$$

exist and are continuous for arbitrary values of t, λ, ζ, and $s < t$.
2.   For any $\delta > 0$, the limits

$$\lim_{h \downarrow 0} E \left\{ \frac{x(t + h) - x(t)}{h} \,\bigg|\, x(t, \omega) = \zeta \right\} = m[t, \zeta]$$

$$\lim_{h \downarrow 0} E \left\{ \frac{[x(t + h) - x(t)]^2}{h} \,\bigg|\, x(t, \omega) = \zeta \right\} = \sigma^2[t, \zeta]$$

exist, and the convergence is uniform in ζ; these limits define functions $m(.,.)$ and $\sigma(.,.)$.
3.   For every $\epsilon > 0$,

$$\int_{|\lambda - \zeta| > \epsilon} d_\lambda F(\lambda, t | \zeta, s) = \boldsymbol{P}\{|x(t, \omega) - x(s, \omega)| > \epsilon | x(s, \omega) = \zeta\}$$
$$= o(t - s) \qquad t > s$$

It was shown earlier by Feller and by Kolmogorov that under conditions 1 to 3 of Corollary 1 the transition probability $F(\lambda, t | \zeta, s)$ satisfies the following parabolic partial differential equation:

$$\frac{\partial F}{\partial s} + m(s,\zeta)\frac{\partial F}{\partial \zeta} + \frac{\sigma^2(s,\zeta)}{2}\frac{\partial^2 F}{\partial \zeta^2} = 0 \tag{8-62}$$

known as the *backward equation* for a *diffusion* process, with the initial condition

$$F(\lambda,t|\zeta,t) = \begin{cases} 1 & \lambda > \zeta \\ 0 & \lambda < \zeta \end{cases}$$

Furthermore, when the conditional density

$$f(\lambda,t|\zeta,s) = \frac{\partial}{\partial \lambda} F(\lambda,t|\zeta,s)$$

exists, and the following condition holds:

1′.   Partial derivatives

$$\frac{\partial f}{\partial t} \qquad \frac{\partial}{\partial \lambda}\,(mf) \qquad \frac{\partial^2}{\partial \lambda^2}\,(\sigma^2 f)$$

exist and are continuous, then under conditions 1′, 2, and 3 the density f satisfies the following parabolic partial differential equation:

$$\frac{\partial f}{\partial t} + \frac{\partial}{\partial \lambda}\,[m(t,\lambda)f] = \frac{1}{2}\frac{\partial^2}{\partial \lambda^2}\,[\sigma^2(t,\lambda)f] \tag{8-63}$$

known as the *forward equation* for a *diffusion* process (also called the Fokker-Planck equation in physical applications).   The initial conditions are

$$f(\lambda,s|\zeta,s) = \vartheta(\lambda - \zeta)$$

Since the $x(t)$ process has the same sample-function properties as the diffusion process and its transition probabilities satisfy diffusion equations, it follows that the $x(t)$ process must be a diffusion process.   Ito's result can now be summarized in the following way:

Corollary 2:   The $x(t)$ process obtained as a solution of the stochastic differential equation (8-60), with brownian motion as the forcing function, is a Markoff diffusion process.

It has been also shown that, conversely, all $x(t)$ processes which can be written as solutions of Eq. (8-60) are Markoff diffusion processes.

Ito-Doob Integral

In connection with the Ito equation, it has been mentioned that stochastic integrals can be defined (in the mean-square sense) as generalized Stieltjes integrals, although the integrator function is not of bounded variation as a sample function; see Eq. (8-61).

This subsection is devoted to a brief description of the integral

$$\varphi = \int_{-\infty}^{\infty} \Phi(t,\omega) \, dz(t) \tag{8-64}$$

introduced by Ito[47] and generalized by Doob.[29]   For proofs, the reader should consult these references; see also Sec. 8-5.

Assumptions:

1.   The process $\{z(t), -\infty < t < \infty\}$ is a martingale, that is, for $s \leq t$

$$E\{z(t)|\mathfrak{F}_s\} = z(s) \qquad \text{almost everywhere} \tag{8-65}$$

There is a monotone nondecreasing function R such that for $s \leq t$

$$E\{|z(t) - z(s)|^2\} = E\{|z(t) - z(s)|^2|\mathfrak{F}_s\} = R(t) - R(s)$$
$$\text{almost everywhere} \tag{8-66}$$

2.   The (t,ω) function $\Phi(.,.)$ is measurable with respect to $dt \, d\boldsymbol{P}$ measure.   For each t, $\Phi(t,.)$ is $\mathfrak{F}_t$-measurable, and

$$\int_{-\infty}^{\infty} E\{|\Phi(t,\omega)|^2\} \, dR(t) < \infty \tag{8-67}$$

Remark: One has here the usual probability space $(\Omega,\mathfrak{F},\boldsymbol{P})$ and the Lebesgue measure on the t line; $\mathfrak{F}_t \subset \mathfrak{F}$ is the σ field induced by the random variables $z(s)$ for $s \leq t$ (cf. Sec. 8-6).   The expression $E\{z(t)|\mathfrak{F}_s\}$, defining the martingale property, is the conditional expectation of $z(t)$ given $\mathfrak{F}_s$.   It follows from assumption 1 that the $z(t)$ process has orthogonal increments; the brownian-motion process is the most important special case.

The integral is first defined for (t,ω) step functions

$$\Phi(t,\omega) = \begin{cases} 0 & t \leq t_1 \\ \Phi_j(\omega) & t_j \leq t < t_{j+1}, j \leq n - 1 \\ 0 & t_n \leq t \end{cases}$$

where $\Phi_j(\omega)$ is $\mathfrak{F}_{t_j}$-measurable, and $E\{|\Phi_j|^2\} < \infty$.   By definition

$$\varphi = \int_{-\infty}^{\infty} \Phi(t,\omega) \, dz(t) = \sum_j \Phi_j(\omega)[z(t_{j+1} - 0) - z(t_j - 0)]$$

where the random variable φ is determined uniquely, neglecting values on ω null sets.

Let integrands Φ and Ψ correspond to the integrals φ and ψ; then

$$E\{\varphi\} = E \int_{-\infty}^{\infty} \Phi(t,\omega) \, dz(t) = 0 \qquad E\{\varphi\bar{\psi}\} = \int_{-\infty}^{\infty} E\{\Phi(t,\omega)\overline{\Psi(t,\omega)}\} \, dR(t)$$
$$\tag{8-68}$$

Define distances in the space of Φ functions and in the space of random variables φ by

$$\|\Phi_1 - \Phi_2\|^2 = \int_{-\infty}^{\infty} E\{|\Phi_1(t,\omega) - \Phi_2(t,\omega)|^2\} \, dR(t) \qquad (8\text{-}69)$$

$$\|\varphi_1 - \varphi_2\|^2 = E\{|\varphi_1 - \varphi_2|^2\} \qquad (8\text{-}70)$$

It then follows from (8-68) that the correspondence between Φ and its integral φ is distance-preserving.

Let Φ be a limit (in the sense of Φ distance) of a sequence $\{\Phi_n\}$ of (t,ω) step functions. The integral φ of Φ is defined as the limit (in the sense of φ distance) of the corresponding integrals φ_n

$$\varphi = \underset{n \to \infty}{\text{l.i.m.}} \, \varphi_n \qquad (8\text{-}71)$$

For proof that the closure of the class of (t,ω) step functions does indeed include all functions satisfying assumption 2, see Doob.[29] Evidently φ satisfy relations (8-68).

Consider now integrals of the form

$$y(t) = \int_a^t \Phi(s,\omega) \, dz(s) \qquad t \geq a \qquad (8\text{-}72)$$

Doob proved the following.

Theorem 8-12

1. The stochastic integral (8-72) can be defined for each t in such a way that the $y(t)$ process is a separable martingale.

2. If the sample functions of the $z(t)$ process are almost all continuous, then those of a separable $y(t)$ process will almost all be continuous also.

3. The fixed points of discontinuity of the $y(t)$ process are points of discontinuity of R.

The Ito-Doob stochastic integral has several interesting properties. For example, let $\Phi(t,\omega) = z(t,\omega) - z(a,\omega), a \leq t \leq b$, and let $dR(t) = \sigma^2 \, dt$. Then, it can be shown that

$$\int_a^b [z(t) - z(a)] \, dz(t) = \tfrac{1}{2}[z(b) - z(a)]^2$$

$$- \tfrac{1}{2} \underset{\delta \to 0}{\text{l.i.m.}} \sum_{j=0}^{m-1} [z(t_{j+1}) - z(t_j)] \qquad (8\text{-}73)$$

where $\delta = \max(t_{j+1} - t_j)$ and $a = t_0 < \cdots < t_m = b$.

The mean-square limit on the right is shown to be equal to $\tfrac{1}{2}\sigma^2(b - a)$, for $z(t)$ being the brownian motion, and to $\tfrac{1}{2}\sigma^2[z(b) - z(a)]$, for

$$z(t) = p(t) - \sigma^2 t$$

where $p(t)$ is the Poisson process with the mean σ^2. Note that the formal integration would give only the first term on the right in (8-73).

8-4 RANDOM COEFFICIENTS

INTRODUCTION

Stochastic differential equations with random coefficients have received attention chiefly in engineering applications, where they describe the influence of fluctuation in system parameters on overall performance. Mathematically their theory is related to that described in the preceding section, but it also presents several new features, e.g., concepts of random differential operators or random Green's functions. For this reason interest in these equations has increased, and their study constitutes a branch of a new field of random operators. In fact, the term *random differential equation* usually means an equation with random coefficients (with a random or deterministic forcing function).

As an illustration consider a second-order nonlinear equation of the form

$$Lx \equiv \ddot{x} + a\dot{x} + bx^2 = 0$$

where the coefficients $a = a(t)$ and $b = b(t)$ are random variables with given joint distribution; here L is a random operator. Under reasonable conditions on coefficients, the equation can be integrated in the usual way, the solution being written in the form

$$x(t) = g(a,b,x^0,x_1^0,t)$$

Then, the random function $x = \{x(t), 0 \le t < \infty\}$ whose random variables are functionals of coefficients can be analyzed by methods previously described. However, except in some very simple special cases, this procedure is very difficult and even prohibitive. Therefore various methods developed in the literature (apart from the above-mentioned general studies) are applicable to particular cases (especially with gaussian random variables), but results are far from being complete. Hence this section will be restricted to a brief description of some selected results. Comments on the theory of random operators are postponed till Sec. 8-5 because of their difficulty; see, however, the review paper by Bharucha-Reid.[10]

In general, coefficients in random differential equations are taken from one or several stochastic processes, assumed independent of the $x(t)$ process. In particular, distribution of coefficients may be time-independent, e.g., when coefficients themselves are time-independent. Frequently, in addition, random forcing functions appear, and one must consider equations of mixed type, involving all three possibilities considered here. It also may happen that coefficients are probabilistically dependent on the random forcing function or indirectly on $x(t)$.

The basic problems concern the stochastic properties of solutions, uniqueness and existence of solutions, and the relationship between stochastic and sample-function equations. Of principal interest, however, is the question of how far random differential operators resemble the corresponding nonrandom operators. The following examples illustrate possibilities one may encounter.

Example 11: Consider

$$\dot{x} + ax = 0$$

whose solution is
$$x(t) = x^0 e^{-at} \qquad 0 \leq t < \infty$$

where x^0 and a are random variables, assumed independent. Suppose that the random variable a has uniform distribution over an interval $[0,A]$ and that $x^0 = \lambda$, with probability one. Then, by the method from Sec. 8-2, the density of $x(t)$ is found to be

$$f(\mu,t) = \begin{cases} \dfrac{1}{A\mu t} & \text{for } \lambda e^{-At} \leq \mu \leq \lambda \\ 0 & \text{otherwise} \end{cases}$$

Consequently,
$$E\{x(t)\} = \frac{\lambda}{At}\left[ei(\lambda t e^{-\lambda t}) - ei(\lambda t)\right]$$

where the exponential integral is defined by

$$ei(t) = \int_t^\infty \frac{e^{-s}}{s}\, ds \qquad t > 0$$

Example 12: Consider

$$\dot{x} + a(t)x = z(t)$$

whose solution with zero initial conditions is

$$x(t) = \int_0^t z(\tau)\, d\tau \exp\left[-\int_0^\tau a(\sigma)\, d\sigma\right] \qquad t \geq 0$$

Assume that the stochastic processes $\{a(t),\ t \geq 0\}$ and $\{z(t),\ t \geq 0\}$ are independent and mean-square-continuous. Then the distribution of the stochastic integral for $x(t)$ can be obtained, in principle, by the methods discussed in Sec. 8-3.

Example 13: Consider the Bessel equation

$$t^2\ddot{x} + t\dot{x} + (t^2 - n^2)x = 0 \qquad t \geq 0$$

Take the solution given by the Bessel function of order n

$$x(t) = J_n(t) = \sum_{r=0}^{\infty} (-1)^r \left(\frac{t}{2}\right)^{n+2r} \frac{1}{r!(n+r)!}$$

where $n \geq 0$ is the integer-valued random variable. Suppose that n has Poisson distribution with mean m. Then

$$E\{x(t)\} = e^{-m}J_0(\sqrt{t^2 - 2mt})$$

Observe, however, that formal averaging of the Bessel equation yields

$$t^2 \ddot{M} + t\dot{M} + (t^2 - m^2 - m)M = 0$$

whose solution is $M = J_k(t)$, with $k = (m + m^2)^{1/2}$, which is an incorrect value for $E\{x(t)\}$.

Example 14: Consider a system of n equations with n unknowns

$$a_{i1}x_1 + \cdots + a_{in}x_n = z_i \qquad i = 1, \ldots, n$$

where coefficients a_{ij} are random with given joint distribution. In matrix notation

$$\mathfrak{A}x = z$$

where $\mathfrak{A} = (a_{ij})$ is a random matrix.

Example 15: Consider

$$\ddot{x} + 2a\dot{x} + bx = 0$$

and suppose that coefficients a and b have the joint uniform distribution over the square $-1 \leq \alpha \leq 1$, $-1 \leq \beta \leq 1$, with density $\frac{1}{4}$. It is required to find the probability that every solution $x(t)$ tends to zero as $t \to \infty$ (stability of solutions). This event will occur when zeros of the random characteristic equation

$$p^2 + 2ap + b = 0$$

are either both real and both negative or both complex with negative real parts. These two disjoint events are described by random variables as follows. If $a^2 - b \geq 0$, then both roots are real, and they have the same sign if and only if $b > 0$, this sign being negative if and only if $a > 0$. If $a^2 - b < 0$, then the roots are complex conjugate, and they have negative real parts if and only if $a > 0$; in this case also $b > a^2 > 0$. Hence the required event will occur if and only if $a > 0$ and $b > 0$; the corresponding probability is

$$P\{\lim_{t \to \infty} x(t) = 0\} = P\{a > 0, b > 0\} = \frac{1}{4} \int_0^1 \int_0^1 d\alpha \, d\beta = \frac{1}{4}$$

Note also that the probability that zeros of the characteristic equation are complex is

$$P\{a^2 - b < 0\} = \frac{1}{4} \int_{-1}^{1} \int_{\alpha^2}^{1} d\alpha \, d\beta = \frac{1}{3}$$

whereas the probability that both zeros are equal is $P\{a^2 = b\} = 0$.

LINEAR EQUATION

The best-known case is that of a single linear equation, and even for it determination of the explicit distribution is prohibitive. The main source of difficulty is the fact that, despite the linearity, the dependence of a solution $x(t)$ on coefficients is nonlinear and no analytic formulas are available. Moreover, the average solution $E\{x(t)\}$ does not coincide with the solution of an equation for averages. In this subsection a brief summary will be given of some recent work, and the reader is referred to references for details.

First Approach

Samuels and Eringen[79] consider a linear equation of nth order with random coefficients $a_k(t)$ and random forcing function $z(t)$

$$L_a x \equiv a_n(t) x^{(n)}(t) + a_{n-1}(t) x^{(n-1)}(t) + \cdots + a_0(t) x(t) = P(t) \quad (8\text{-}74)$$

where the random variable $P(t)$ has the form $N(z)$, where N is a differential operator of the same type as L_a but operating on the given forcing function $z(t)$. It is assumed that

$$a_k(t) = E\{a_k(t)\} + \alpha_k(t) \qquad k = 0, 1, \ldots, n$$

Substitution yields

$$Lx = P(t) - L_\alpha x$$

where the operator L is not random whereas the operator L_α has random coefficients α_k.

A solution is assumed to be of the form

$$x(t) = u_0(t) + \epsilon u_1(t) + \epsilon^2 u_2(t) + \cdots$$

with $\quad \alpha_k(t) = \epsilon \tilde{\alpha}_k(t) \qquad E\{\tilde{\alpha}_k(t)\} = 0 \qquad k = 0, 1, \ldots, n$

Substitution and comparison of equal powers of ϵ lead to the system

$$Lu_0(t) = P(t) \qquad Lu_j(t) = -P_{\alpha, j-1}(t)$$
$$P_{\alpha, j-1}(t) \equiv L_\alpha u_{j-1}(t) \qquad j = 1, 2, \ldots$$

and the solution can be written in the form of stochastic integrals

$$u_0(t) = \int_0^t g(t,s)P(s)\,ds + \sum_{k=1}^n c_k w_k(t)$$

$$u_j(t) = \int_0^t g(t,s)P_{\alpha,j-1}(s)\,ds$$

where $g(t,s)$ is a one-sided Green's function associated with the operator L, w_k form a fundamental set of independent solutions of $Lx = 0$, and c_k are constants to be determined by the initial conditions at $t = 0$.

Stochastic properties of such integrals are difficult to investigate, and only special cases have been considered. They include (1) equations containing small randomly varying coefficients, (2) equations containing slowly varying coefficients, and (3) equations containing only one random coefficient. Even simple examples lead to analytical difficulties, but moments can sometimes be determined from appropriate equations.

Case 1: Terminating the solution after two terms, one gets

$$x(t) = u_0(t) + \epsilon u_1(t)$$

Suppose first that $P(t)$ and the initial conditions are not random, and let $\alpha_k(t)$ be independent gaussian. Then $P_{\alpha 1}(t) = L_\alpha u_0(t)$ is gaussian, as a linear combination of $\alpha_k(t)$, and $u_1(t)$ is also gaussian as the stochastic integral. Consequently, $x(t)$ is gaussian with

$$E\{x(t)\} = u_0(t)$$

and
$$E\{x(t)x(t')\} = u_0(t)u_0(t') + \epsilon^2 E\{u_1(t)u_1(t')\}$$

where the second moment should be computed as explained in Sec. 8-3. The n-dimensional gaussian distribution for $x(t_1), \ldots, x(t_n)$ can be written down when the covariance matrix is computed.

Next, let $P(t)$ be random, but suppose that the initial conditions are not random. Assuming that the random function $P(t)$ has a gaussian distribution, it follows that $u_0(t)$ will be gaussian. Further assumptions are needed, however, for $P_{\alpha,1}(t)$; some results can be obtained when $\alpha_k(t)$ are gaussian and independent of $u_0(t)$.

Case 2: It is assumed that all derivatives of a coefficient $a_k(t)$ are negligible with respect to the coefficient. Good approximation is obtained then by finding a one-sided Green's function for L_a on the assumption that the coefficients are constant and afterwards finding the distribution of the Green's function in terms of those coefficients.

The method will now be illustrated on a second-order equation

$$\ddot{x} + a(t)x = p(t) \tag{8-75}$$

where $a(t)$ is a random variable. The one-sided Green's function $g(t,s)$ is given by the solution of

$$\frac{d^2g}{dt^2} + a(t)g = \vartheta(t - s)$$

subject to $g = 0$ for all $t < s$. Regarding $a(t)$ as a constant, one finds

$$g(t,s) = \frac{1}{\sqrt{a}} \sin [\sqrt{a}\,(t - s)] \qquad t \geq s$$

Now suppose that $\alpha(t) = \sqrt{a(t)}$ has gaussian distribution with the mean m and the variance σ^2, so that $a(t)$ has chi-square distribution. The characteristic function of $g(t,s)$ is now

$$E\{e^{i\theta g(t,s)}\} = \frac{1}{\sqrt{2\pi}\,\sigma} \int_{-\infty}^{\infty} \exp \left[i\theta\,\frac{\sin \alpha(t - s)}{\alpha} - \frac{(\alpha - m)^2}{2\sigma^2} \right] d\alpha$$

from which the mean and the variance can be determined, but the results are too involved to be quoted here. The solution $x(t)$ is obtained in the usual way, but further assumptions are needed in order to find its distribution.

Case 3: The equation to be considered is

$$x^{(n)} + a_{n-1}(t)x^{(n-1)} + \cdots + a_1(t)x^{(1)} + [a_0(t) + \alpha_1(t)]x = \alpha_2(t) \quad (8\text{-}76)$$

where coefficients $a_k(t)$, $k = 0, 1, \ldots, n$, are nonrandom and $\alpha_1(t)$ and $\alpha_2(t)$ are random.

It is assumed that

$$x(t) = u_0(t) + \epsilon u_1(t) + \epsilon^2 u_2(t) \qquad -\infty < t < \infty$$

and that $\alpha_{1,2}(t)$ are both given by stochastic Stieltjes integrals with respect to brownian motion. Conditions are then imposed which allow one to express $x(t)$ as the sum of a series of first powers and a series of squares of independent gaussian random variables; the possibility of such an expansion depends on the existence of eigenvalues and eigenfunctions of a certain integral equation. From this, the characteristic function of $x(t)$ can be determined.

Other Approaches

The same equation (8-76) was considered by Caughey and Dienes,[22] with $\alpha_1(t)$ and $\alpha_2(t)$ being white-noise processes. The $x(t)$ process forms a continuous n-dimensional Markoff process, and its transition probability density $f(\mu,t|\lambda)$ satisfies an n-dimensional Fokker-Planck equation. The authors explicitly determined the coefficients of this equation and evaluated the moments, correlation function, and spectral density. In fact, they found that for $n > 1$,

$$\sum_{k=0}^{n} a_k \frac{d^k}{dt^k} E\{x(t)\} = 0$$

and this result can be checked by averaging equation (8-76), which then implies that $E\{\alpha_1(t)x(t)\} = 0$. However, Kozin and Bogdanoff[60] pointed out that the results presented by Caughey and Dienes are at variance with classical results; see the next subsection for comments.

Leibowitz[67] considered an equation

$$x^{(n)} + [a_{n-1} + \alpha_{n-1}(t)]x^{(n-1)} + \cdots + [a_0 + \alpha_0(t)]x = 0 \quad (8\text{-}77)$$

where a_i are constants and $\alpha_i(t)$ are gaussian noises with power $\sigma_i{}^2$, $i = 0, 1, \ldots, n - 1$, and found the equation for the mean value. His results, however, are also open to criticism; see the next subsection. Leibowitz also considered the system of equations

$$\dot{x}_i = \sum_{j=1}^{n} [a_{ij} + \alpha_{ij}(t)]x_j \qquad (8\text{-}78)$$

where a_{ij} are deterministic constants and $\alpha_{ij}(t)$ are mutually independent random variables (either shot noise or white noise). He obtained the Fokker-Planck equation for the joint density, but again the results are open to criticism.

Several authors investigated the analytic properties of a single equation. As an example of this type of study, Lyubarskii and Rabotnikov[70] examined the equation

$$x^{(2)} + a_1(t)x^{(1)} + [a_0(t) - \alpha(t)]x = 0 \qquad (8\text{-}79)$$

where $a_1(t)$ and $a_0(t)$ are real, piecewise-continuous, and periodic functions with the same period T and $\alpha(t)$ is a real random variable ($-\infty < t < \infty$) such that its correlation length L is much shorter than T and that $\alpha(t)$ does not exceed the value $\gamma/\sqrt{L}$ ($\gamma < 1$, constant), and found a necessary and sufficient condition for boundedness of the mean values $E\{x^2(t)\}$, $E\{\dot{x}(t)x(t)\}$ and $E\{\dot{x}^2(t)\}$ (see also Jirina[50,51]).

SINGLE RANDOM COEFFICIENT
Basic Equation

The following equation with random coefficients deserves special mention because it has caused some controversy:

$$\frac{dx(t)}{dt} = [a + \alpha(t)]x(t) \qquad (8\text{-}80)$$

where a is a nonrandom constant and $\alpha(t)$ is the white noise, which is formally the derivative of brownian motion

$$\alpha(t) \, dt = dz(t)$$

with $E\{\alpha(t)\} = 0$ and $E\{\alpha(t)\alpha(t')\} = \sigma^2 \vartheta(t - t')$.   More rigorously, Eq. (8-80) should be written in the differential form:

$$dx(t) = ax(t) \, dt + x(t) \, dz(t) \tag{8-81}$$

which is a special case of the Ito equation discussed in Sec. 8-3.   Evidently, the distinction between a linear equation (8-80) with random coefficient and a nonlinear equation (8-81) with random forcing function is superfluous in the present case.

Equation (8-80) has been the subject of controversy in the literature; since the problem concerns the validity of some of the methods described earlier in this chapter, brief comments are necessary.   The fact that $\alpha(t)$ does not exist requires that special care must be taken in evaluating the density and the mean of the $x(t)$ process.

It follows from Ito's theory that the density $f(\mu,t|\lambda)$ of the $x(t)$ process satisfies the Fokker-Planck equation

$$\frac{\partial f}{\partial t} = -a \frac{\partial \mu f}{\partial \mu} + \tfrac{1}{2}\sigma^2 \frac{\partial^2 \mu^2 f}{\partial \mu^2} \tag{8-82}$$

because, in the notation of Sec. 8-3, $m(t,\mu) = a\mu$, and $\sigma(t,\mu) = \sigma\mu$.   The mean

$$E\{x(t)|\lambda\} = \int_{-\infty}^{\infty} \mu f(\mu,t|\lambda) \, d\mu \qquad E\{x(0)|\lambda\} = \lambda$$

is obtained as follows.   Multiplying both sides of (8-82) by μ and integrating by parts, one finds

$$\frac{d}{dt} E\{x(t)|\lambda\} = atE\{x(t)|\lambda\}$$

with the solution

$$E\{x(t)|\lambda\} = \lambda e^{at} \qquad t \geq 0 \tag{8-83}$$

It is important to observe that the mean is not influenced by the brownian motion process and will be the same even if $z(t) = 0$ in (8-81).

The above correct results, based on Ito's theory, should be compared with the following results, based on methods described in Secs. 8-3 and 8-2.   The explicit solution of (8-80) is

$$x(t) = x(0)e^{y(t)} \qquad t \geq 0 \tag{8-84}$$

where $$y(t) = at + \int_0^t \alpha(\tau) \, d\tau = at + z(t)$$

Evidently $y(t)$ has gaussian density $w(\eta,t)$, with mean at and variance $\sigma^2 t$, which satisfies

$$\frac{\partial w}{\partial t} = -a\frac{\partial w}{\partial \eta} + \tfrac{1}{2}\sigma^2\frac{\partial^2 w}{\partial \eta^2} \tag{8-85}$$

Since $\mu = \lambda e^\eta$, it follows that the density of $x(t)$, assuming $x(0) = \lambda > 0$ with probability one, is

$$f(\mu,t) = w(\eta,t)\frac{d\eta}{d\mu} = w(\eta,t)\frac{1}{\mu} \qquad \mu > 0$$

Hence substitution yields the Fokker-Planck equation

$$\frac{\partial f}{\partial t} = -a\frac{\partial \mu f}{\partial \mu} + \tfrac{1}{2}\sigma^2\left[-\frac{\partial \mu f}{\partial \mu} + \frac{\partial^2 \mu^2 f}{\partial \mu^2}\right] \tag{8-86}$$

and the mean, found from (8-86) by integration by parts, is

$$E\{x(t)\} = \lambda e^{at+\frac{1}{2}\sigma^2 t} \tag{8-87}$$

which indicates dependence on $z(t)$.

Equations (8-82) and (8-86) differ by the additional term $\tfrac{1}{2}\sigma^2\,\partial(\mu f)/\partial\mu$. The discrepancy is explained by the fact that Ito's theory imposes the condition that the $x(t)$ process be independent of increments $\Delta z(t)$ of the Wiener process (which does not have the derivative); cf. Theorem 8-11. This implies that

$$E\{x(t)\,dz(t)\} = 0 \tag{8-88}$$

and the average (8-83) can be obtained by averaging Eq. (8-80). On the other hand, averaging of Eq. (8-80) implies, in view of (8-87), that

$$E\{x(t)\,dz(t)\} = \tfrac{1}{2}\sigma^2\,dt\,E\{x(t)\} \tag{8-89}$$

The following is the more precise explanation. Referring back to the Ito equation in the form

$$dx(t) = m[x(t)]\,dt + \sigma[x(t)]\,dz(t) \tag{8-90}$$

one has the following lemma, a special case of the lemma proved by Ito,[48] which yields the random analogue of the total derivative.

Lemma: Let $x(t)$ be a markovian solution, and let $\sigma^2[x(t)]$ and $|m[x(t)]|$ be integrable over a finite interval. Let $V(t,x)$ be any function continuously differentiable in t and continuously twice differentiable in x. Then, for $s < t$,

$$V[t,x(t)] - V[s,x(s)] = \int_s^t \frac{\partial}{\partial x}V[\tau,x(\tau)]\,dz(\tau)$$

$$+ \int_s^t \left\{\frac{\partial}{\partial t}V[\tau,x(\tau)] + \frac{\partial}{\partial x}V[\tau,x(\tau)]m[x(\tau)]\right.$$

$$\left. + \tfrac{1}{2}\sigma^2[x(\tau)]\frac{\partial^2}{\partial x^2}V[\tau,x(\tau)]\right\}d\tau \tag{8-91}$$

or

$$dV[t,x(t)] = \frac{\partial}{\partial x} V[t,x(t)]\, dz(t) + \left\{ \frac{\partial}{\partial t} V[t,x(t)] + \frac{\partial}{\partial x} V[t,x(t)]m[x(t)] \right.$$
$$\left. + \tfrac{1}{2}\sigma^2[x(t)] \frac{\partial^2}{\partial x^2} V[t,x(t)] \right\} dt \quad (8\text{-}92)$$

Apply the lemma to the equation $dz(t) = dz(t)$, so that

$$dV[t,z(t)] = \frac{\partial V}{\partial z} dz(t) + \left[\frac{\partial V}{\partial t} + \tfrac{1}{2}\sigma^2 \frac{\partial^2 V}{\partial z^2} \right] dt \quad (8\text{-}93)$$

Comparison with Eq. (8-81) implies that the solution will have the form $x(t) = V[t,z(t)]$, where

$$x(t) = x(0)e^{(a-\frac{1}{2}\sigma^2)t+z(t)} \qquad t \geq 0 \quad (8\text{-}94)$$

Thus the correct solution of Eq. (8-80) is given by (8-84) but with

$$y(t) = (a - \tfrac{1}{2}\sigma^2)t + z(t) \quad (8\text{-}95)$$

so that $y(t)$ has gaussian distribution with mean $(a - \tfrac{1}{2}\sigma^2)t$ and variance $\sigma^2 t$. If this value of $y(t)$ is used in (8-85), then the same transformation method leads to the correct Fokker-Planck equation (8-82).

Consequently, Eqs. (8-86), (8-87), and (8-89) are not correct; the error is caused by the nonexistence of the derivative $dz(t)/dt$.

Exercise 8-16 Show that the solution of the differential equation (8-86) for density f is

$$f(\mu,t|\lambda) = \begin{cases} \dfrac{1}{\sqrt{2\pi t}\,\sigma\mu} \exp\left\{ -\dfrac{[at - \log(\mu/\lambda)]^2}{2\sigma^2 t} \right\} & \text{for } \lambda > 0,\ \mu \geq 0 \\ 0 & \text{for } \lambda > 0,\ \mu < 0 \end{cases}$$

and verify directly that the mean is given by (8-87).

Exercise 8-17 Show that the solution of differential, equation (8-82) for density f is

$$f(\mu,t|\lambda) = \begin{cases} \dfrac{1}{\sqrt{2\pi t}\,\sigma\mu} \exp\left\{ -\dfrac{[at - \frac{1}{2}\sigma^2 t - \log(\mu/\lambda)]^2}{2\sigma^2 t} \right\} & \text{for } \lambda > 0,\ \mu \geq 0 \\ 0 & \text{for } \lambda > 0,\ \mu < 0 \end{cases}$$

and verify directly that the mean is given by (8-83).

For one differential equation (8-76), Caughey and Dienes[22] observed that (8-88) is true only for $n > 1$ but not for $n = 1$. Moreover, their Fokker-Planck equation reduces, for $n = 1$, to (8-86). Equation (8-86) was solved by Dienes in an earlier paper. Kozin and Bogdanoff[60] pointed out, on the basis of the equivalence of (8-80) and (8-81), that the correct Fokker-Planck equation is (8-82). They give the detailed derivation in an earlier paper;[15] see also Bogdanoff and Kozin[16] for further work and references to the literature. The relationship between ordinary and stochastic solutions of the Ito equation is discussed by Wong and Zakai.[90] The derivation of Eqs. (8-86) and (8-47) given above is due to Leibowitz,[67] who on this basis criticized the derivation of (8-82) (for $a = 0$) by Bogdanoff and Kozin.[15] Using (8-86), Leibowitz found that the nth moment $M_n(t) = E\{x^n(t)\}$ should satisfy the equation

$$\frac{d}{dt} M_n(t) = (a + \tfrac{1}{2}\sigma^2 n^2) M_n(t)$$

which gives (8-87) for $n = 1$. Leibowitz also considered Eq. (8-80) with $\alpha(t)$ being a shot noise (cf. Sec. 8-3), with

$$\int_0^t \alpha(s)\, ds = \beta z(t)$$

where $z(t)$ is from the Poisson process with mean ct and β is a constant magnitude of impulses. The equation for the density $f(\mu,t)$ of the $x(t)$ process was found to be

$$\frac{\partial f(\mu,t)}{\partial t} = -a\, \frac{\partial[\mu f(\mu,t)]}{\partial \mu} + cf(\mu e^{-\beta},t) - cf(\mu,t)$$

Moments found for the explicit relation for $x(t)$ are

$$E\{x^n(t)\} = E\{x^n(0)\}\, \exp\,[nat + ct(e^{n\beta} - 1)]$$

Leibowitz extended Eq. (8-86) to the case of system (8-78), and therefore his results are also open to criticism.

Systems

In order to consider a simple equation with one random coefficient, like Eq. (8-76), it is necessary to have a preliminary lemma on the system of equations. This is the extension of Ito's results (cf. Sec. 8-3) to the multidimensional case and formally should belong to Sec. 8-3.

Consider an n-dimensional Markoff process with vector

$$x = (x_1, \ldots, x_n)$$

where $x_i = x_i(t)$, $i = 1, \ldots, n$, and let $f(\mu,t|\lambda,s)$ be the n-dimensional density of vector x, where $\mu = (\mu_1, \ldots, \mu_n)$.

Suppose that the following coefficients exist

$$\rho_{k_1,\ldots,k_n}(\mu_1, \ldots, \mu_n, t) = \lim_{\Delta t \to 0} \frac{1}{\Delta t} \int_{-\infty}^{\infty} \cdots \int_{-\infty}^{\infty} (\zeta_1 - \mu_1)^{k_1}$$
$$\cdots (\zeta_n - \mu_n)^{k_n} f(\zeta, t + \Delta t|\mu,t)\, d\zeta_1 \cdots d\zeta_n \quad (8\text{-}96)$$

with $k_i = 0, 1, 2, \ldots$, $i = 1, \ldots, n$. It follows from the general theory of Markoff processes that the density $f = f(\mu,t|\lambda,s)$ satisfies the

Fokker-Planck equation

$$\frac{\partial f}{\partial t} = \sum_{k_1 + \cdots + k_n > 0} \frac{(-1)^{k_1 + \cdots + k_n}}{k_1! \cdots k_n!} \frac{\partial^{k_1 + \cdots + k_n}}{\partial \mu_1^{k_1} \cdots \partial \mu_n^{k_n}} (\rho f) \qquad (8\text{-}97)$$

For the derivation of this equation see, for example, Moyal[74] or Bartlett;[6] cf. also the similar discussion in Sec. 8-3.

The (conditional) moments

$$M(t) = \int_{-\infty}^{\infty} \cdots \int_{-\infty}^{\infty} \mu_1^{k_1} \cdots \mu_n^{k_n} f(\mu, t | \lambda, s) \, d\mu_1 \cdots d\mu_n \qquad (8\text{-}98)$$

can be obtained from

$$\frac{dM(t)}{dt} = \int_{-\infty}^{\infty} \int_{-\infty}^{\infty} \mu_1^{k_1} \cdots \mu_n^{k_n} \frac{\partial f}{\partial t} \, d\mu_1 \cdots d\mu_n \qquad (8\text{-}99)$$

where Eq. (8-97) should be substituted for $\partial f / \partial t$. This leads (after integration by parts) to a system of linear differential equations for $M(t)$.

The coefficients ρ can be evaluated from the derivative characteristic function

$$L(\theta_1, \ldots, \theta_n | \mu, t)$$

$$= \lim_{\Delta t \to 0} \frac{1}{\Delta t} \left[E \left(\exp \left\{ i \sum_{j=1}^{n} \theta_j [x_j(t + \Delta t) - x_j(t)] \right\} \Big| \mu, t \right) - 1 \right] \qquad (8\text{-}100)$$

where $E\{.|\mu, t\}$ denotes the conditional expectation. Differentiating L with respect to parameters θ_j and letting all $\theta = 0$ yields coefficients ρ; cf. Sec. 8-3.

Exercise 8-18 Using Eq. (8-100), obtain coefficients ρ given by (8-96); cf. Moyal.[74]

Consider now a system of stochastic equations

$$\dot{x}_j = K_j(x_1, \ldots, x_n, t) \qquad j = 1, \ldots, n \qquad (8\text{-}101)$$

and write it in the difference form

$$x_j(t + \Delta t) - x_j(t) = K_j(x_1, \ldots, x_n, t) \, \Delta t + o(\Delta t) \qquad (8\text{-}102)$$

Substitution of (8-102) into (8-100) yields:

$$L(\theta_1, \cdots, \theta_n | \mu, t) = \lim_{\Delta t \to 0} \frac{1}{\Delta t} \left[E \left\{ \exp \left(i \sum_{j=1}^{n} \theta_j K_j \, \Delta t \right) \Big| \mu, t \right\} - 1 \right] \qquad (8\text{-}103)$$

from which the coefficients ρ for system (8-101) can be evaluated.

The above analysis is due to Bogdanoff and Kozin.[15] For applications of a similar system to control problems see Kalman and Bucy.[53]

Special Case

It is now possible to consider the nth-order equation with a single random coefficient

$$x^{(n)} + a_{n-1}x^{(n-1)} + \cdots + a_1 x^{(1)} + [a_0 + \alpha(t)]x = 0 \quad (8\text{-}104)$$

where a_i $(i = 0, 1, \ldots, n)$ are nonrandom constants and $\alpha(t)$ is a white noise, as before

$$E\{\alpha(t)\,\Delta t\} = 0 \qquad E\{(\alpha(t)\,\Delta t)^2\} = \sigma^2\,\Delta t$$

This equation can be regarded as a generalization of the Ito equation to n dimensions. As remarked in Sec. 8-3, the vector process $\{x, x^{(1)}, \ldots, x^{(n-1)}\}$ is the n-dimensional Markoff process. Let $f(\mu, t|\lambda)$ be its density, with $\mu = (\mu_0, \mu_1, \ldots, \mu_{n-1})$.

Equation (8-104) is a special case of the system (8-101) and the theory developed above is applicable here. Define as in Sec. 8-2 (case 3)

$$\dot{x}_1 = x^{(1)}, \ldots, \dot{x}_{n-1} = x^{(n-1)}$$
$$\dot{x}_n = -[a_0 + \alpha(t)]x - a_1 x^{(1)} - \cdots - a_{n-1}x^{(n-1)} \quad (8\text{-}105)$$

Substituting in (8-103), one has

$$L(\theta_1, \ldots, \theta_n | \mu, t) = \lim_{\Delta t \to 0} \frac{1}{\Delta t} [E\{e^{iB\Delta t}|\mu, t\} - 1] \quad (8\text{-}106)$$

where $$B = \sum_{j=1}^{n-1} \theta_j x^{(j)} + \theta_n \left(-\sum_{j=0}^{n-1} a_j x^{(j)}\right) - \theta_n \alpha(t) x(t)$$

The coefficients $\rho_{k_1 \cdots k_n}(\mu_0, \mu_1, \ldots, \mu_{kn-1}, t)$ are obtained from (8-106) by differentiation

$$\rho_{k_1 \cdots k_n}(\mu_0, \mu_1, \ldots, \mu_{n-1}, t) = \frac{1}{i^{k_1 + \cdots + k_n}} \frac{\partial^{k_1 + \cdots + k_n} L}{\partial \theta_1^{k_1} \cdots \partial \theta_n^{k_n}} \bigg|_{\theta_1 = \cdots = \theta_n = 0}$$
$$(8\text{-}107)$$

For example, the first-order coefficients $(k_1 + \cdots + k_n = 1)$ are

$$\rho_{0 \cdots 1 \cdots 0}(\mu, t) = \lim_{\Delta t \to 0} \frac{1}{\Delta t} E\{\Delta t\, x^{(j)}|\mu, t\}$$

$$= \lim_{\Delta t \to 0} \int \cdots \int \zeta_j f(\zeta, t + \Delta t|\mu, t)\,d\zeta$$

$$= \mu_j \qquad \text{for } j = 1, \ldots, n-1$$

and for $j = n$

$$\rho_{0 \cdots 1}(\mu, t) = -\lim_{\Delta t \to 0} E\left\{\sum_{j=0}^{n-1} a_j x^{(j)}|\mu, t\right\} - \lim_{\Delta t \to 0} E\{\alpha(t)x|\mu, t\} = -\sum_{j=0}^{n-1} a_j \mu_j$$

All second-order coefficients $(k_1 + \cdots + k_n = 2)$ vanish, except the last one

$$\rho_{0\ldots2}(\mu,t) = \lim_{\Delta t \to 0} \frac{1}{\Delta t} E\{(\alpha(t)x)^2|\mu,t\} = \sigma^2\mu_0{}^2$$

Finally, the Fokker-Planck equation is given by (8-97), with $\mu = (\mu_0, \ldots, \mu_{n-1})$ and coefficients ρ obtained from (8-107).

Example 16: Consider

$$\ddot{x} + a_1\dot{x} + (a_0 + \alpha(t))x = 0 \tag{8-108}$$

Thus, from (8-107),

$$\rho_{10} = \mu_1 \qquad \rho_{01} = -a_0\mu_0 - a_1\mu_1$$
$$\rho_{20} = 0 \qquad \rho_{11} = 0 \qquad \rho_{02} = \sigma^2\mu_0{}^2$$

and all higher-order coefficients are zero. Equation (8-97) yields

$$\frac{\partial f}{\partial t} = -\mu_1 \frac{\partial f}{\partial \mu_0} + a_0\mu_0 \frac{\partial f}{\partial \mu_1} + a_1 \frac{\partial \mu_1 f}{\partial \mu_1} + \tfrac{1}{2}\mu_0{}^2 \frac{\partial^2 f}{\partial \mu_1{}^2} \tag{8-109}$$

Exercise 8-19 Referring to the second-order equation (8-108) consider the moments

$$M_{kj} = E\{x^k \dot{x}^j\}$$

Show that the first-order-moment equations are

$$\dot{M}_{10} = M_{01}$$
$$\dot{M}_{01} = -a_0 M_{10} - a_1 M_{01}$$

and the second-order-moment equations are

$$\dot{M}_{20} = 2M_{11}$$
$$\dot{M}_{11} = -a_0 M_{20} - a_1 M_{11} + M_{02}$$
$$\dot{M}_{02} = \sigma^2 M_{20} - 2a_0 M_{11} - 2a_1 M_{02}$$

Discuss the dependence of first-order moments on the $\alpha(t)$ process and the mean-square stability, using $E\{x^2\}$; cf. Bogdanoff and Kozin.[15]

STABILITY

Problems of stability in stochastic differential equations have been considered by only a few authors, and the results are far from being complete. The discussion here is restricted to brief remarks on the available results, and the reader is referred to the literature for details. Quite recently, Bucy[18,19] put the problem of stability on a rigorous basis by discovering a connection with martingale-convergence theorems.

By analogy with classical modes of convergence, the following modes of stability have been considered. Let $x(t)$ be a solution of a stochastic differential equation, and let O be the random-equilibrium point whose stability is to be treated. The solution $x(t)$ is said to be (asymptotically) stable in the pth mean, or almost everywhere, or in probability, if $x(t) \to 0$ in the pth means, or almost everywhere, or in probability, as $t \to \infty$. Similarly, let $x(t) = \{x_1(t), \ldots, x_n(t)\}$ be a solution of a system of stochastic equations, and let the origin O be the random equilibrium point. The system is said to be (asymptotically) stable in the pth mean if

$$E \left\{ \sum_{i=1}^{n} |x_i(t)|^p \right\} \to 0$$

or $t \to \infty$ for every choice of initial distribution for $x(0)$. In particular, for $p = 2$ the mean-square stability is obtained. The system is said to be (asymptotically) stable almost everywhere (or almost surely) if

$$\lim_{t \to \infty} \sum_{i=1}^{n} |x_i(t)| = 0$$

with probability one and is said to be (asymptotically) stable in probability if for every $\epsilon > 0$

$$\lim_{t \to \infty} P \left\{ \sum_{i=1}^{n} |x_i(t)| > \epsilon \right\} = 0$$

Almost-sure stability of linear systems with random coefficients was discussed by Kozin,[59] who considered a linear system

$$\dot{x} = [\mathfrak{A} + \mathfrak{B}(t)]x \qquad (8\text{-}110)$$

where $x = (x_1, \ldots, x_n)$ is an n-dimensional vector of random variables $x_i = x_i(t)$, $\mathfrak{A}$ is a constant matrix, $\mathfrak{A} = (a_{ij})$, $i, j = 1, \ldots, n$, and $\mathfrak{B}(t)$ is an $n \times n$ random matrix whose elements are stochastic processes

$$b_{ij} = \{b_{ij}(t), 0 \le t < \infty\}$$

It is assumed that (1) the random functions b_{ij} are measurable and continuous on $[0, \infty]$ with probability one and (2) that the random functions b_{ij} are strictly stationary and ergodic. The norms of the vector x and of a random matrix $\mathfrak{M}$ are defined respectively by

$$\|x\| = \sum_{i=1}^{n} |x_i|$$

$$\|\mathfrak{M}\| = \sum_i \sum_j |m_{ij}|$$

and both are random variables.

Let $x = g(x^0, t)$ be the solution vector of the system (8-110); condition (1) guarantees the existence, uniqueness, and continuity of the solution with probability one for any initial x^0.

Definition: The trivial solution $x_1 = x_2 = \cdots = x_n = 0$ is said to be almost surely asymptotically stable in the large if for all solutions of (8-110)

$$\lim_{t \to \infty} \|g(x^0,t)\| = 0 \tag{8-111}$$

with probability one.   This is analogous to almost-sure Lyapunov asymptotic stability.

***Theorem* 8-13**   If the solution of the system

$$\dot{x} = \mathfrak{A}x$$

is asymptotically stable in the large, and if $E\{\|\mathfrak{B}(t)\}\|\}$ exists, where $\mathfrak{B}(t)$ satisfies conditions (1) and (2), then there exists a constant $c > 0$, depending upon the matrix $\mathfrak{A}$, such that

$$E\{\|\mathfrak{B}(t)\|\} < c$$

implies that the solution of the system (8-110) is almost surely asymptotically stable in the large.

Proof (outline):   Rewrite (8-110) as

$$x(t) = y(t) + \int_0^t \mathfrak{U}(t - \tau)\mathfrak{B}(\tau)x(\tau) \, d\tau$$

where $\mathfrak{U}(t)$ is the fundamental matrix satisfying

$$\dot{\mathfrak{U}}(t) = \mathfrak{A}\mathfrak{U}(t) \qquad \mathfrak{U}(0) = I$$

and the vector $y(t)$ is the solution of $\dot{y} = \mathfrak{A}y$, with $y(0) = x^0$.   Hence

$$\|x(t)\| \leq \|y(t)\| + \int_0^t \|\mathfrak{U}(t - \tau)\| \, \|\mathfrak{B}(\tau)\| \, \|x(\tau)\| \, d\tau$$

For a constant matrix, there exists a number $\beta > 0$ such that

$$\|y(t)\| \leq \beta\|x^0\|e^{-\alpha t}$$

where $-\alpha$ is an upper bound of the real parts of the characteristic values of the matrix $\mathfrak{A}$ for an asymptotically stable system.   From this, and from properties of $\mathfrak{U}(t)$, it follows that

$$\|x(t)\| \leq \beta\|x^0\|e^{-\alpha t} + \beta \int_0^t e^{-\alpha(t-\tau)}\|\mathfrak{B}(\tau)\| \, \|x(\tau)\| \, d\tau$$

At this point, the so-called *Bronwall-Bellman lemma* is used, which asserts that the above inequality implies that

$$\|x(t)\| \leq \beta\|x^0\| \exp\left[-\alpha t + \beta \int_0^t \|\mathfrak{B}(\tau)\| \, d\tau\right]$$

with probability one.   Condition (2) allows one to use the ergodic theorem

$$\lim_{t \to \infty} \frac{1}{t} \int_0^t \|\mathfrak{B}(\tau)\| \, d\tau = E\{\|\mathfrak{B}(0)\|\}$$

with probability one.   Assuming that $E\{\|\mathfrak{B}(0)\| < \alpha/\beta$, one has, for sufficiently large t,

$$\|x(t)\| \leq \beta \|x^0\| e^{-\alpha t + \alpha}$$

Hence
$$\lim_{t \to \infty} \|x(t)\| = 0$$

with probability one, independent of x^0.   Q.E.D.

Kozin applied this result to two important classes of ergodic processes, viz., the linear process (defined by convolution integrals with respect to the process with stationary independent increments) and the gaussian process with continuous covariance and continuous spectrum.

The mean-square stability $E\{x^2(t)\} \to 0$ has been considered by several authors, mostly in connection with the stabilizing effect of gaussian noise in an equation with a single random coefficient, like (8-80) or (8-104).   Since equations for the second moment are linear with constant coefficients, the stability can be determined by the Routh-Hurwitz procedure (cf. "NM," page 196).   Contradictory results found in the literature reflect the controversy concerning the dependence of moments of $x(t)$ on the moments of the noise $\alpha(t)$, as discussed earlier.

Exercise 8-20   Consider a single nth-order homogeneous differential equation with random coefficients (not involving time) with given joint distribution.   Using the Routh-Hurwitz criterion for all roots of the characteristic equation to have negative real parts, attempt to evaluate the probability that $x(t) \to 0$ when $t \to \infty$ (cf. Example 15).

Following earlier work by Samuels, Samuels and Eringen[79] considered the mean-square stability of a linear differential equation with a single gaussian noise coefficient and mean-square-bounded forcing function and noted that the mean-square stability is determined by the character of roots of the characteristic equation for the Fourier transform of the Green's function.   Leibowitz[67] examined stability in the pth mean and in the probability of Eq. (8-80) and conjectured that the system (8-78) with gaussian noise coefficients will be stable in probability if the deterministic system of the same form will be stable in the usual sense. Boganoff and Kozin[15] criticized earlier results on mean-square stability. For additional references on stability see, for example, papers by Bertram and Sarachik,[9] Kats and Krasovskii,[57] Bunke,[21] and Kalman and Bucy.[53]

Bucy[18,19] considered almost-sure asymptotic stability of the system of stochastic difference equations of the Ito type and put forward the idea that the stochastic Lyapunov functions are nonnegative super-martingales; Kushner[64] used this approach for the continuous case.

ALGEBRAIC EQUATIONS

Probabilistic analysis of algebraic equations, which has been considered by several authors, has many interesting features. Although the problem to be considered here does not refer explicitly to differential equations, there are obvious direct connections.

The problem can be stated as follows: What is the average number of *real* roots of an algebraic polynomial

$$p(t) = a_0 + a_1 t + \cdots + a_{n-1}(t^{n-1}) \qquad (8\text{-}112)$$

with *real* coefficients?

According to a fundamental theorem of algebra, every polynomial of nth degree has exactly n complex (real or imaginary) roots; on the other hand, a number of real roots in an open interval for a given algebraic equation can be determined by Sturm's rule of signs. However, in the above problem one considers not the individual polynomial but all polynomials of a specified type. It may also be useful to recall that any polynomial equation with real coefficients of degree ≤ 4 is soluble by radicals, but the Galois theorem asserts the impossibility of solution of polynomial equations of degree ≥ 5. Hence, for example, for a homogeneous system of linear differential equations with constant coefficients the solution is given explicitly by spectral expansion [cf. "NM," eq. (4-24)], but it is impossible to obtain explicit expressions for roots of the characteristic equation in terms of the coefficients; this fact illustrates the difficulties in the study of stochastic equations with random coefficients (cf. also Example 15).

To make the problem precise, assume that the coefficients a_i in Eq. (8-112) are *independent random variables* with the same gaussian distribution, with mean zero and variance one.

Let $N(a) = N(a_0, \ldots, a_{n-1})$ be a random variable representing the number of real roots. The required average $M_n = E\{N(a)\}$ is given by

$$M_n = \frac{1}{(2\pi)^{n/2}} \int_{-\infty}^{\infty} \cdots \int_{-\infty}^{\infty} N(\alpha) e^{-r^2/2} \, d\alpha_0 \cdots d\alpha_{n-1} \qquad (8\text{-}113)$$

where α_i, α are values of a_i, a, respectively, and $r^2 = \sum_{t=0}^{n-1} \alpha_k^2$.

It is first shown that M is twice the average number of real roots in the interval $(-1, 1)$ and that this number is given by

$$N(a; -1, 1) = \frac{1}{2\pi} \int_{-\infty}^{\infty} d\zeta \int_{-1}^{1} \cos\left[\zeta p(t)\right] |p'(t)| \, dt$$

Substituting, one can obtain, after complicated manipulation, an expression for M_n in the form

$$M_n = \frac{4}{\pi} \int_0^1 \frac{(1 - h_n^2)^{1/2}}{1 - t^2} \, dt$$

where

$$h_n = n^{n-1} \frac{1 - t^2}{1 - t^{2n}}$$

One then obtains an asymptotic formula

$$M \approx \frac{2}{\pi} \log n \tag{8-114}$$

which is of great interest. In fact, determination of a number of real roots of a given polynomial of high degree is almost impossible, but this simple formula nevertheless provides some useful information about the set of all real polynomials of that degree.

The above solution (8-114) is due to Kac,[52] who showed that roots tend to concentrate very strongly around ± 1; in fact, the density of roots is

$$n(t) = \frac{(1 - h_n^2)^{1/2}}{\pi(1 - t^2)} \qquad -\infty < t < \infty$$

with peaks at $t = \pm 1$. Kac also proved that formula (8-114) is still valid when coefficients are assumed to be independent and uniformly distributed in $(-1,1)$. Erdös and Offord[33] proved the validity of (8-114) even when the coefficients are allowed only the values $+1$ and -1 and all sequences $\pm 1, \ldots, \pm 1$ are assigned an equal probability 2^{-n}. This insensitivity of the asymptotic form (8-114) expresses the tendency of polynomials with real coefficients to have relatively few real roots. For further results see Hammersley,[44] who examined the distribution of real roots under general assumptions on coefficients.

It may be added that as early as 1782 Waring used a probabilistic argument for ascertaining the number of imaginary roots in an equation (of second and third degree); cf. page 618 of Ref. 84.

The general problem of zero crossings of a stochastic process is of great importance both for theory and applications, but it presents serious difficulties. For a very important study see Rice;[78] cf. also Doob's[29] submartingale crossing inequality and its many applications.

8-5 GENERAL PROPERTIES

INTRODUCTION

It has already been indicated that stochastic differential equations are, in fact, differential equations for abstract-valued functions. This point of view allows one to prove existence and uniqueness of a solution by the methods of classical theory suitably adapted for abstract-valued functions.

From the probabilistic point of view, an important question is: To what extent can properties of random functions (solutions of equations) be inferred from the properties of their sample functions?

This section is devoted to a brief discussion of equations for functions taking values in Banach spaces, but only those aspects relevant to stochastic equations will be mentioned. Even so, the full discussion involves topics from the theory of stochastic processes that are too specialized to be given here. It may be recalled, however, that the most important special case of Hilbert space has been discussed at length in Sec. 8-3.

It is convenient to begin with the existence and uniqueness theorem for solutions of stochastic differential equations. No attempt is made to achieve generality, and the theorem given in the next subsection is an adaptation of the Picard theorem to restricted situations. First, however, a few preliminary concepts must be introduced.

Let $(\Omega, \mathfrak{F}, P)$ be a given probability space, and let $\mathfrak{IC}$ be a space of random variables on Ω. The restriction is imposed that $\mathfrak{IC}$ is a Banach space with a norm related to the probability structure. For example, $\mathfrak{IC}$ may be a space of all essentially bounded random variables or L_2 space; cf. Edwards and Moyal,[31] who considered the weakly complete B space $\mathfrak{IC}$. It is not always evident how any nonartificial relation between norm convergence and probabilistic convergence can be established. On the other hand, it is well known that a metric can be introduced into $\mathfrak{IC}$ in such a way that $\mathfrak{IC}$ becomes merely an F space with convergence in the metric equivalent to convergence in probability.

Let T be a real line or its subset, and let x be a vector-valued function defined on T to $\mathfrak{IC}$, with values $x(t) \; \epsilon \; \mathfrak{IC}$ at a point $t \; \epsilon \; T$; thus

$$x = \{x(t), \; t \; \epsilon \; T\}$$

is a random function, and its values $x(t)$ are random variables on Ω.

The function x is said to be *strongly continuous* at t if $x(s)$ converges to $x(t)$ in the norm of $\mathfrak{IC}$

$$\lim_{s \to t} \|x(s) - x(t)\| = 0$$

The function x is said to be *strongly differentiable* at t if there exists an $\dot{x}(t) \; \epsilon \; \mathfrak{IC}$ such that

$$\lim_{h \to 0} \left\| \frac{x(t + h) - x(t)}{h} - \dot{x}(t) \right\| = 0$$

The random variable $\dot{x}(t)$ is called a strong derivative of x at t. The function x is strongly continuous (or differentiable) in a subset of T if it is strongly continuous (or differentiable) at each point of the subset.

Let $\mathcal{K}^*$ be a space of all linear continuous functionals on $\mathcal{K}$. The function x is said to be *weakly continuous* at t if $x(s)$ converges weakly to $x(t)$; that is,

$$\lim_{s \to t} |\zeta^*[x(s) - x(t)]| = 0$$

for each functional $\zeta^* \epsilon \mathcal{K}^*$. The function x is said to be *weakly differentiable* at t if there exists an $\dot{x}(t) \epsilon \mathcal{K}$ such that $[x(t + h) - x(t)]/h$ tends weakly to $\dot{x}(t)$; the random variable $\dot{x}(t)$ is called a *weak derivative* at t. Note that x weakly differentiable at t implies that $\zeta^*[x(.)]$ is a differentiable real-valued function at t for all $\zeta^* \epsilon \mathcal{K}^*$ (the converse is not true in general).

The concept of a *stochastic integral*, i.e., an integral of the vector-valued function x, is essential for further considerations. For Banach-space-valued functions, the Bochner integral (also called a *strong integral*) of a strongly measurable function x, which extends the Lebesgue integral, is most frequently used. For present purposes, however, the Riemann-Stieltjes integral will suffice; see Hille and Phillips[46] for details of construction of these integrals.

The stochastic integral of a random function x, defined on an interval $[a,b]$, to be denoted by

$$y = \int_a^b x(t)\, dt$$

is defined as the limit in strong topology of $\mathcal{K}$ of Riemann-Stieltjes sums

$$y_n = \sum_{i=1}^{n} x(t_i')\, \Delta t_i$$

Evidently the strong integral $y \epsilon \mathcal{K}$. It is known that when x is strongly continuous on $[a,b]$, the above integral exists relative to the strong topology of $\mathcal{K}$, and the indefinite integral

$$y(t) = \int_a^t x(s)\, ds \qquad a \le t \le b$$

is strongly differentiable, its derivative being $x(t)$. The definition of a stochastic integral can be extended to integrals of the form

$$y(t) = \int_a^t x(s)\, dz(s) \qquad a \le t \le b$$

where $x(s)$ and $z(s)$ are elements of $\mathcal{K}$, but in addition $\mathcal{K}$ must be a Banach algebra; i.e., the product $x(s)z(s)$ must also be in $\mathcal{K}$, and $\|x(s)z(s)\| \le \|x(s)\|\, \|z(s)\|$.

It is very important to distinguish equations whose solutions are random functions from equations for sample functions. Thus one must distinguish between stochastic integrals and the integrals of sample functions.

Rigorous justification of the integrals of sample functions requires several concepts from the theory of stochastic processes, especially the concept of a measurable process. Basic definitions and a brief summary of results will now be given, but it is impossible to enter into details; the interested reader should consult the references, especially the treatises by Doob[29] and by Loève.[69]

A stochastic process $x = \{x(t), t \in T\}$, also called a *random function*, has been defined as a function on T with values in the space $\mathcal{K}$ of all random variables on a probability space $(\Omega, \mathfrak{F}, P)$. A random function x is said to be *continuous* in probability, or almost everywhere, or in the pth mean at point t according as $x(t')$ converges to $x(t)$ in probability, or almost everywhere, or in the pth mean, respectively, when $t' \to t$. Two random functions x and $\bar{x}$ are called *equivalent* if $x(t) = \bar{x}(t)$ almost everywhere for every $t \in T$; that is,

$$P\{x(t,\omega) = \bar{x}(t,\omega)\} = 1 \qquad t \in T$$

A random function x is said to be sample-continuous, or sample-measurable, or sample-integrable at a point $\omega \in \Omega$ if the corresponding property is true for sample functions, i.e., for functions $x(.,\omega)$ obtained by fixing ω and letting t vary. Sample properties hold almost everywhere if they hold apart from a set of probability zero.

In general, analytical properties of x relative to sample functions are stronger than the corresponding properties relative to the space $\mathcal{K}$. For example, almost-everywhere sample continuity of x implies almost-everywhere continuity of x, but the converse need not be true; in fact, almost-everywhere continuity of x excludes fixed discontinuity points, but almost-everywhere sample continuity excludes fixed and moving discontinuity points (depending on ω) outside a null set.

A stochastic process x can also be regarded as a real-valued function $x(.,.)$ of two variables, with values $x(t,\omega)$. A stochastic process x is said to be *measurable* if x is a *real-valued* function on $T \times \Omega$, measurable with respect to the product σ field $\overline{\mathfrak{B} \times \mathfrak{F}}$, where $\mathfrak{B}$ is a Borel field on T and the bar denotes completion of $\mathfrak{B} \times \mathfrak{F}$; the measure is the usual product measure of Lebesgue measure and P.

The modern theory of stochastic processes involves the basic concept of a separability of the process, introduced by Doob in order to overcome difficulties in assigning probability measure to a wider class of sets than $\mathfrak{F}$-measurable sets. The stochastic process $x = \{x(t), t \in T\}$ is said to be *separable* relative to closed sets if there is a sequence $S = (s_j)$ of parameter values and a set $N \subset \Omega$ of probability zero such that for any closed set C and any open interval I on real line, sets

$$\{x(t,\omega) \ \epsilon \ C, \ t \ \epsilon \ IT\} \subset \{x(s_j,\omega) \ \epsilon \ C, \ s_j \ \epsilon \ IS\}$$

differ by at most a subset of N. In effect, the set on the left, which is determined by a noncountable number of coordinates (and thus is not in $\mathfrak{F}$), is assigned the same probability as the set on the right, which is determined by a countable number of coordinates (and thus belongs to $\mathfrak{F}$).

Separability leads to the important consequences listed below; the theorems are due to Doob (for proofs see references).

Theorem 8-14   For any stochastic process x there exists a stochastic process $\bar{x}$ defined on the same space Ω, separable relative to the class of closed sets, and equivalent to x.

Theorem 8-15   If the separable process x is continuous in probability, then any countable dense set $S = (s_j)$ separates x.

Theorem 8-16   If the process x is separable and almost everywhere continuous, then it is measurable.

Theorem 8-17   If the process x is continuous in probability, then there exists an equivalent process $\bar{x}$ which is separable and measurable.

Several analytic conditions for almost-everywhere sample continuity are known, but they are too involved to be stated here. However, the following sufficient condition, due to Kolmogorov, has many applications.

Theorem 8-18   A separable random function x on $T = [0,a]$ is almost everywhere sample-continuous when for some p

$$E\{|x(t + h) - x(t)|^{1+p}\} \leq c|h|^{1+\alpha}$$

for all t, $t + h$ in T, and c, α, $p > 0$.

The restriction to separable processes is really necessary in order to obtain fruitful theory. However, if the original process is not separable, it can always be replaced by an equivalent separable process, according to Theorem 8-14.

EXISTENCE THEOREM

A stochastic differential equation to be considered here is the differential equation of the first order involving the unknown function x and its derivative $\dot{x}$ in the form of the following relation written for values of these functions at $t \ \epsilon \ T$:

$$\frac{dx(t)}{dt} = K[t,x(t)] \tag{8-115}$$

where K is a given operator on $T \times \mathfrak{X}$ to $\mathfrak{X}$ (thus both sides of this equation are random variables in the space $\mathfrak{X}$). As in the case of real functions, an equation of higher order is equivalent to the system of first-order equations.

The above equation embraces all three types of equations considered here. When K does not involve other random variables besides $x(t)$, then an equation with random initial conditions is obtained. If K depends on other random variables besides $x(t)$, two other types are obtained. For example, $K = -a(t)x(t) + z(t)$, where $a(t)$ is an ordinary function and $z(t) \in \mathfrak{K}$, yields a linear equation with random forcing function $z(t)$. Reversing the roles and letting $z(t)$ be an ordinary function and $a(t) \in \mathfrak{K}$, one obtains a nonhomogeneous equation with random coefficient $a(t)$.

The following is the basic existence (and uniqueness) theorem for stochastic differential equations.

***Theorem* 8-19**   Let the function $K(t,\zeta)$ on $T \times \mathfrak{K}$ with values in $\mathfrak{K}$ be defined and continuous in each variable separately for $|t - t_0| \leq a$, $\|\zeta - \zeta_0\| \leq b$, and satisfy $\|K(t,\zeta)\| \leq k$, and the Lipschitz condition with respect to ζ

$$\|K(t,\zeta_1) - K(t,\zeta_2)\| \leq M\|\zeta_1 - \zeta_2\|$$

for t, ζ, ζ_1, ζ_2 in the indicated regions. Here a, b, k, and M are fixed positive numbers with

$$ak \leq b \qquad Ma < 1$$

Then there exists a unique strongly continuously differentiable function x of t such that

$$\frac{dx(t)}{dt} = K[t,x(t)]$$

in $|t - t_0| \leq a$ and $x(t_0) = \zeta_0$.

For proof, paralleling that of classical case, see Hille and Phillips[46] or Bourbaki.[17] Alternatively, proof can be reduced to the application of the Banach fixed-point theorem for complete metric spaces (for fixed-point theorems see Sec. 1.4). In fact, the differential equation is equivalent to the integral equation

$$x(t) = x(t_0) + \int_{t_0}^{t} K[\tau,x(\tau)]\, d\tau$$

and the transformation A, defined by

$$Ax(t) = \zeta_0 + \int_{t_0}^{t} K[\tau,x(\tau)]\, d\tau \qquad |t - t_0| \leq a$$

on the subspace (determined by the hypotheses of the theorem) of a space of continuous functions x, possesses, by the Banach theorem, a fixed point x such that $Ax = x$. Details of the proof are omitted here.

Let $x(t)$ be the solution of (8-115) satisfying the initial conditions $x^0 = x(t_0) = \zeta_0$. Write

$$x(t) = g(x^0, t_0, t) \qquad (8\text{-}116)$$

It can be shown that the mapping $g:(x^0, t_0, t) \rightarrow g(x^0, t_0, t)$ is uniformly continuous. Furthermore, in the neighborhood of x^0, the equation $x^0 = h[t_0, t, x(t)]$ has the unique solution $x(t) = g(x^0, t_0, t)$. For proofs of these statements see Bourbaki.[17]

SAMPLE FUNCTIONS

The Banach space $\mathcal{K}$ of random variables is itself a function space of equivalence classes of real-valued functions defined on probability space Ω and measurable with respect to $\mathfrak{F}$. (Recall that two random variables are said to be equivalent if they differ on a set of probability zero). For the purpose of the present argument denote by $\varphi(t) = \varphi(t, .)$ any such function on Ω, and reserve the symbol $x(t) = x(t, .)$ for functions in $\mathcal{K}$ which are solutions of stochastic differential equations. Suppose further that parameter t ranges through the interval $I = [a, b]$.

One could expect the differential equation (8-115) for $x(t)$ to be related somehow to the partial differential equation for the function $\varphi(., .)$ $I \times \Omega$:

$$\frac{\partial \varphi(t, \omega)}{\partial t} = K[t, \varphi(t, .)](\omega) \qquad (8\text{-}117)$$

with the initial condition $\varphi(t_0, \omega) = \varphi_0(\omega)$, $\omega \in \Omega$, t and t_0 in I.

If $\mathcal{K}$ were the space of individual functions, then for fixed t it would be enough to set $x(t, .) = \varphi(t, .)$, and thus $x(t)$ would have a unique representation. The fact that x is strongly continuously differentiable implies that $\partial \varphi(t, \omega)/\partial t$ exists uniformly in ω and satisfies (8-117).

Since $\mathcal{K}$ is the space of equivalence classes, the connection between (8-115) and (8-117) is less obvious, because the solution $x(t)$ of the differential equation does not indicate how to choose a representative of the equivalence class (for t fixed); such a choice is needed, however, to give meaning to Eq. (8-117).

Hille and Phillips (page 68 of Ref. 46) distinguish a class of spaces $\mathcal{K}$ (which they call spaces of type L) for which a precise connection can be established. These are Banach spaces of equivalence classes of real-valued functions, possessing the following two properties:

1. If x is a strongly continuous vector-valued function defined on the interval $I = [a, b]$, then there exists a function $\varphi(., .)$ measurable on the product set $I \times \Omega$ such that

$$x(t) = \varphi(t, .) \qquad \text{for each } t \in I$$

2.   Let x and $\varphi(.,.)$ have the properties stated in 1.   Then

$$\left[\int_a^b x(t) \, dt \right] (\omega) = \int_a^b \varphi(t,\omega) \, dt$$

where the integral on the left is the abstract Riemann integral and the integral on the right is the ordinary Lebesgue integral for real-valued functions.

Hille and Phillips showed that spaces $L_p(\Omega,\mathfrak{F},\boldsymbol{P})$ for $1 \le p \le \infty$ are of type L and proved the general theorem, a special case of which is given below:

Theorem 8-20   Let the space $\mathfrak{K}$ satisfy properties 1 and 2.   If x (on I to $\mathfrak{K}$) is continuously differentiable, then there exists a real-valued function $\varphi(.,.)$ measurable on $I \times \Omega$ and such that $\varphi(.,\omega)$ is absolutely continuous for each $\omega \,\epsilon\, \Omega$ and

$$\varphi(t,.) = x(t) \qquad \text{for each } t \,\epsilon\, I$$

Furthermore, $\partial\varphi(t,\omega)/\partial t$ exists almost everywhere in $I \times \Omega$ and

$$\frac{\partial}{\partial t} \varphi(t,.) = \frac{d}{dt} x(t)$$

for almost all $t \,\epsilon\, I$.

Proof (outline):   To $dx(t)/dt$ there corresponds by property 1 a function $\psi_0(t,.)$.   Define

$$\psi(t,\omega) = \int_a^t \psi_0(\tau,\omega) \, d\tau$$

It is shown that $\psi(t,\omega)$ is absolutely continuous in t for all ω and is measurable on $I \times \Omega$ and that

$$\frac{\partial\psi(t,\omega)}{\partial t} = \psi_0(t,\omega) \qquad \text{almost everywhere on } I \times \Omega$$

so that $\partial\psi(t,.)/\partial t = dx(t)/dt$ for almost all $t \,\epsilon\, I$.   By property 2

$$\psi(t,.) = \int_a^t dx(\tau) = x(t) - x(a)$$

Taking any realization of $x(a)$, one has for the desired realization of $x(t)$

$$\varphi(t,\omega) = \psi(t,\omega) + x(a,\omega)$$

For details, see Hille and Phillips.   Q.E.D.

This theorem implies that the stochastic equation (8-115) for $x(t)$ can be obtained from the corresponding equation for sample functions $x(.,\omega)$:

$$\frac{dx(t,\omega)}{dt} = K[t,x(t,\omega)] \tag{8-118}$$

for almost ω, and consequently its integral can be also obtained from the integral of sample functions. This procedure is commonly used in practical applications, especially when explicit forms of sample functions are known.

It is necessary therefore to reexamine the probabilistic interpretation of conditions 1 and 2. In fact, condition 2 requires that stochastic integrals be defined in terms of sample integrals, whereas condition 1 demands choice of a norm such that continuity in this norm implies measurability of the random function x. These requirements are difficult to meet, but under rather mild conditions less restrictive situations are treated in the theory of stochastic processes, as shown in the following theorem, which is concerned with the implications of measurability on sample integrals.

Theorem 8-21   Let x be a measurable process. Then almost all sample functions are Lebesgue-measurable functions of t; and if, for a Lebesgue-measurable subset A of T,

$$\int_A E\{|x(t,\omega)|\}\, dt < \infty$$

then almost all sample functions are Lebesgue-integrable over A, and

$$E \int_A x(t,\omega)\, dt = \int_A E\{x(t,\omega)\}\, dt$$

Proof: Since $x(.,.)$ is (t,ω)-measurable, the section $x(.,\omega)$ at ω is measurable in t for almost all ω. By the Fubini theorem, the iterated integral in reversed order is also finite

$$E \int_A |x(t,\omega)|\, dt < \infty$$

Hence, the inner integral is finite for almost all ω

$$\int_A |x(t,\omega)|\, dt < \infty$$

This means that almost all sample functions are Lebesgue-integrable over A. Since the iterated integrals are absolutely convergent, their order can be interchanged, and the last assertion follows. Q.E.D.

This theorem justifies integrals of sample functions. It equates the integral of the t function $E\{x(.,\omega)\}$ with the expectation of a random variable

$$y = \int_A x(t,.)\, dt$$

defining the stochastic integral by its sample values. Doob (page 64 of Ref. 29) has also shown that the integral of a Lebesgue-measurable and -integrable sample function $x(.,\omega)$ over a finite interval I can be approximated by Riemann sums in such a way that the set of those points $t \in I$ for which approximation is not good has negligible measure.

The following argument has also been tried. Suppose that a norm on the function space $\mathfrak{IC}$ of random variables has been selected in such a way that the strong continuity implies continuity in probability of the random function x on T; for example the mean-square continuity satisfies this condition. The following analogue of Theorem 8-20 holds.

Theorem 8-22 Let x be a strongly continuously differentiable random function on T satisfying the differential equation

$$\dot{x}(t) = K[t, x(t)] \tag{8-119}$$

and let $E\{|x(t)|\}$ be Lebesgue-integrable over any Lebesgue subset of T. Then there exists an equivalent measurable random function $\tilde{x}$ such that almost all sample functions $\tilde{x}(.,\omega)$ are almost everywhere differentiable and satisfy the corresponding equation

$$\frac{\partial \tilde{x}(t,\omega)}{\partial t} = K[t, x(t,.)](\omega) \tag{8-120}$$

for almost all (t, ω).

Proof (Outline): Strong differentiability implies strong continuity and hence continuity in probability; by Theorem 8-17, there exists a separable and measurable random function $\tilde{x}$, equivalent to x. Thus, there exists a real-valued function $\tilde{x}(.,.)$ measurable on $T \times \Omega$. Make $\tilde{x}(.,.)$ correspond to x in the sense that $x(t) = \tilde{x}(t,.)$ almost everywhere for each t; this provides a modified form of condition 1. Next, this condition 1 implies condition 2 by Theorem 8-21, because sample functions $x(.,\omega)$ and $\tilde{x}(.,\omega)$ are equal for almost all t, with probability one, and the corresponding sample integrals

$$\int_A x(t,\omega)\, dt = \int_A \tilde{x}(t,\omega)\, dt \qquad A \subset T$$

define a unique random variable $\int_A x\, dt$.

Let $\tilde{\dot{x}}_0(.,.)$ correspond to the strong derivative $\dot{x}(t)$, and define $\tilde{x}_1(.,.)$ by

$$\tilde{x}_1(t,\omega) = \int_0^t \tilde{\dot{x}}_0(\tau,\omega)\, d\tau$$

As in the proof of Theorem 8-20, it can be shown that $\tilde{x}(.,.)$ corresponding to $x(t)$ has the form

$$\tilde{x}(t,\omega) = \tilde{x}_1(t,\omega) + \tilde{x}(0,\omega)$$

so that $\partial x(t,\omega)/\partial t$ exists almost everywhere on $T \times \Omega$ and

$$\frac{\partial}{\partial t}\, \tilde{x}(t,.) = \dot{x}(t)$$

for almost all $t \in T$.

For almost all ω, the ω sections (sample functions) are Lebesgue-measurable and differentiable for almost all t, and Eq. (8-120) is valid for sample functions.   Q.E.D.

CONCLUDING REMARKS

To conclude this discussion, brief comments on other work on stochastic differential equations are in order.   However, the present account does not claim to be complete.   Very useful survey papers in this field are the old one by Moyal[74] and the recent one by Kampé de Ferriet,[54] where an extensive list of references is given.   The books edited by Bellman[7] and by Bogdanoff and Kozin[16] present extensive surveys of random equations and their applications.   See also Bharucha-Reid[11] and Adomian.[2]

The earliest study of stochastic differential equations is due to Bernstein,[8] whose idea of considering increments $\Delta x(t)$ was later used effectively by Ito.[47]   The earliest study of stochastic integrals is due to Slutsky[80] who considered mean-square integrals.   Interest in this type of problem arose in connection with the curious properties of brownian motion.   Important contributions to stochastic equations and stochastic integrals were obtained by Levy,[68] Bochner,[12] Cramer,[25] Wiener,[88] Doob,[29] and Loève,[69] as already mentioned.   Recently, McShane[71] systematically developed stochastic integrals for general processes with independent increments.   Most attention has been paid to diffusion processes, and here the work of Ito and Doob on stochastic integrals and equations is of primary importance.[49]   The recent book by Dynkin[30] examines in detail the interrelation between Markoff processes and stochastic equations, defining a stochastic integral in terms of functionals of a Markoff process.   Gelfand[39] discusses the brownian-motion process as a generalized process defined in terms of theory of distributions.

Equations for the density, e.g., the Liouville or Fokker-Planck equation, discussed in this chapter are special cases of backward and forward equations for general Markoff processes.   Let $(T_t, t \geq 0)$ be a semigroup of Markoff endomorphisms on the Banach space G of all bounded Borel functions on a real line; let $(T_t^*, t \geq 0)$ be the semigroup of dual Markoff endomorphisms on the Banach space Φ of all bounded signed measures on a Borel field on a real line.   Then the backward and forward equations are respectively given by

$$\frac{d}{dt} T_t g = D T_t g \qquad g \in G$$

$$\frac{d}{dt} T_t^* \mu = D T_t^* \mu \qquad \mu \in \Phi$$

where D is the infinitesimal generator (in strong or weak operator topology).   Equations for transition probabilities $P(\zeta;t,B)$ are obtained by letting g or μ be the characteristic function $I(\zeta,B)$, where ζ is a point on real line and B is a Borel set.   For a full account, see Loève.[69]

The interpretation of stochastic equations as equations for abstract-valued functions made it possible to apply the methods of functional analysis.   This new approach is responsible for significant results of great generality.   Stochastic integrals can now be taken in a Bochner or Pettis sense.   Edwards and Moyal[31] considered existence theorems for stochastic linear equations (with one- or two-point boundary conditions) for a random variable belonging to weakly complete Banach spaces; the existence theorem in Sec. 8-3 is the specialization of their results to Hilbert space of random variables.

Equations for Banach-space-valued functions were discussed by Hille and Phillips[46] and by Bourbaki,[17] who gave existence and uniqueness proofs (cf. Sec. 8-5).   Aziz and Diaz[3,4] proved uniqueness of solutions of $\dot{x} = K(t,x)$, with $x(t)$ in a linear normed space, by application of the (extended) mean-value theorem of the differential calculus for vector-valued functions.

Much attention has been devoted to equations of the type

$$\dot{x}(t) = Dx(t) \tag{8-121}$$

where D is an infinitesimal generator (with domain $\mathcal{H}_d \subset \mathcal{H}$ and range $\mathcal{H}$) of a semigroup $(T_t, t \geq 0)$:

$$T_{t+s} = T_t T_s$$

Here $\mathcal{H}$ is a function space of random variables $x(t)$, and T_t is an endomorphism on $\mathcal{H}$.   The operator D is taken in one of three usual operator topologies (weak, strong, or uniform).   The above equation was investigated at length by Hille and Phillips[46] in connection with semigroup theory.   In particular, for T_t being a contraction, $\|T_t\| \leq 1$, and $\mathcal{H}_d$ being the set of strong differentiability, the unique solution of Eq. (8-121), $x(t) \in \mathcal{H}_d$, which is bounded and reduces to $x(0) \in \mathcal{H}_d$ for $t = 0$ and has a strongly continuous strong derivative, is given by

$$x(t) = T_t x(0) \qquad t \geq 0$$

A similar result holds for the weak topology, and $T_t x(0)$ is shown to be measurable in (t,ω).   For the uniform-operator topology the generator D is bounded, and the solution can be expressed in the form

$$x(t) = x(O)e^{Dt}$$

cf. Sec. 8-2.   For strong topology, D is unbounded, and the solution can be expressed as a (strong) limit of exponential solutions; cf. Hille and Phillips[46] for details.   More generally, for the conditions of existence and uniqueness of solution of the abstract Cauchy problem see Hille and Phillips (page 617 of Ref. 46).

Kato[56] considered some existence and uniqueness theorems for solutions of linear differential equations in Banach spaces:

$$\frac{dx(t)}{dt} = D(t)x(t) + z(t) \tag{8-122}$$

for $a \leq t \leq b$, in connection with the Cauchy problem and represented a solution by

$$x(t) = T(t,a)x(a) + \int_a^t T(t,s)\, z(s)\, ds$$

where $T(t,s)$ is a semigroup operator; for D time-independent,

$$T(t - s) = e^{(t-s)D}$$

Agmon and Nirenberg[3] studied the behavior of solutions of equations

$$\frac{dx(t)}{dt} - iDx(t) = iz(t) \tag{8-123}$$

where $x(t) \,\epsilon\, \mathcal{H}_0 \subset \mathcal{H}$ and D is a closed operator with domain in $\mathcal{H}_0$ and range in $\mathcal{H}$. When D represents a partial differential operator, one obtains a partial differential equation

$$\frac{\partial x(t,\omega)}{\partial t} - iDx(t,\omega) = iz(t,\omega)$$

and the authors applied their theory to the questions of uniqueness in the Cauchy problem.

Krasnosel'skii et al.[61] also considered nonlinear equations. Mlak[72,73] investigated the limitations of solutions of such equations; see also Lakshmikantham[65] for a study of the boundedness of solutions of nonlinear equations.

In order to extend the Gibbs-Liouville theory, i.e., a method used in equations with random initial conditions, to more general stochastic integrals, Kampé de Feriet[54] considered the same equation

$$\dot{x}(t) = Dx(t) \tag{8-124}$$

where D is an infinitesimal generator of a semigroup $(T_t, t \geq 0)$ on the Banach space $\mathcal{H}$, but in addition introduced a probability measure on $\mathcal{H}$ itself. This leads to difficult problems of construction of measure on Banach spaces; the best-known case is the Wiener measure, described in Sec. 8-3.

Random Operators

The alternative approach to stochastic equations makes use of a concept of a random operator introduced in an attempt to extend operator theory to operators involving random elements. The systematic theory of such operators was built up by Špaček and his school in a series of papers; cf. Refs. 81 and 7. The detailed exposition can be found in the survey paper by Bharucha-Reid.[10] Here, only brief remarks can be given.

Let R be a real line (or, more generally, a Banach space), and let Ω be a probability space. The operator A, defined on $\Omega \times R$ with range in R, is called a *random operator* if, for each $\lambda \epsilon R$, $A(\omega)\lambda$ is a random variable with values $A(\omega)\lambda$ in R, for $\omega \epsilon \Omega$.

It is then proved that if $x(t)$ is a random variable with values in R, and if A is a random operator on $\Omega \times R$ to R, then the mapping $z(t)$ of Ω into R defined for every $\omega \epsilon \Omega$ by

$$z(t,\omega) = A(\omega)x(t,\omega) \tag{8-125}$$

is a random variable with values in R; in symbols,

$$z(t) = Ax(t) \tag{8-126}$$

Every random variable with values in R for which (8-125) holds almost everywhere is said to be a *random solution* of the operator equation (8-126). Špaček and Hanš obtained stochastic analogues of contraction operators and fixed-point theorems and also investigated the resolvent and spectrum of random operators.

For the case when A is a random differential operator, Eq. (8-126) yields the most general form of a stochastic differential equation (with random coefficients and random forcing function). In this connection, the concept of a random Green's function is of importance; see the paper by Adomian[1] devoted to this topic.

It should be added that there is no essential difference (apart from the shift of emphasis) between the random-operator approach and the approach through equations for abstract-valued functions. As already mentioned, the operator K may involve other random variables from $\mathfrak{X}$, and the equation $\dot{x}(t) = K[x(t),t]$ is clearly of the form $Ax(t) = z(t)$, with $x(t)$ considered as a vector random variable if necessary. However, the use of random operators permits one to handle random elements in stochastic equations in a more direct fashion.

8-6 APPENDIX: PROBABILITY THEORY

To facilitate the reading of this chapter, a brief summary of basic concepts from probability theory and stochastic processes is given here.

Those readers who require a detailed exposition should consult the various available texts, e.g., Feller,[34] Fisz,[35] and Parzen[76] or, on a more advanced level, Loève[69] and Doob[29]; see also "NM," sec. 6-1.

Probability space is a triple $(\Omega, \mathfrak{F}, \boldsymbol{P})$, where Ω is an abstract space of points ω, $\mathfrak{F}$ is a σ field of sets ("events") in Ω, and $\boldsymbol{P}$ is a probability measure on $\mathfrak{F}$, i.e., a nonnegative, countably additive set function with $\boldsymbol{P}(\Omega) = 1$. A set of probability zero is called a *null* set; any property which holds on Ω except at a null set is said to hold *almost everywhere*, or with probability one. The probability measure is *complete* if every subset of a null set belongs to $\mathfrak{F}$.

A *random variable* is an $\mathfrak{F}$-measurable almost-everywhere-finite real-valued function x, defined on Ω. Two random variables are called *equivalent* if they differ on a null set. A function F on real line, defined by the relation

$$F(\lambda) = \boldsymbol{P}\{x(\omega) < \lambda\}$$

is called a *distribution function* of the random variable x. F is non-decreasing, left-continuous (with at most a denumerable number of discontinuities), and such that $F(-\infty) = 0$, $F(+\infty) = 1$.

For any Borel set B on real line

$$\boldsymbol{P}\{x(\omega) \,\epsilon\, B\} = \int_B dF(\lambda)$$

If $F(\lambda)$ is absolutely continuous, then

$$F(\lambda) = \int_{-\infty}^{\lambda} f(t)\, dt$$

where the nonnegative function $f(t)$ is called a *density*.

The *expectation* of a random variable x is defined by the Lebesgue-Stieltjes integral

$$E\{x\} = \int_{\Omega} x(\omega)\, d\boldsymbol{P}(\omega) = \int_{-\infty}^{\infty} \lambda\, dF(\lambda)$$

The random variable x is said to be integrable if $E\{x\} < \infty$. The moments and absolute moments of order p of the random variable x are defined as $E\{x^p\}$ and $E\{|x|^p\}$, respectively.

The *characteristic function* of the random variable x is defined as

$$L(\theta) = E\{e^{i\theta x}\} = \int_{-\infty}^{\infty} e^{i\theta\lambda}\, dF(\lambda)$$

The inversion formula is

$$F(\lambda + h) - F(\lambda - h) = \lim_{T \to \infty} \frac{1}{\pi} \int_{-T}^{T} \frac{\sin \theta h}{\theta} e^{-i\theta\lambda} L(\theta)\, d\theta$$

where $\lambda \pm h$ are continuity points of F and $h > 0$. $\log L(\theta)$ is called the *cumulant generating function*.

The above concepts are easily extended to finitely dimensional random vectors $x = (x_1, \ldots, x_n)$. The function F defined by

$$F(\lambda_1, \ldots, \lambda_n) = \boldsymbol{P}\{x_1(\omega) < \lambda_j, j = 1, \ldots, n\}$$

is called the *joint* (or n-dimensional) distribution function. For any Borel set B in the n-dimensional product of real lines

$$\boldsymbol{P}\{x(\omega) \, \epsilon \, B\} = \int \cdot \cdot \cdot \int_B d_{\lambda_1} \cdots_{\lambda_n} F(\lambda_1, \ldots, \lambda_n)$$

The joint characteristic function is defined as

$$L(\theta_1, \ldots, \theta_n) = E\left\{e^i \sum_{k=1}^{n} \theta_k x_k\right\}$$

The random variables $x_1, \ldots, x_n$ are (mutually) *independent* if

$$F(\lambda_1, \ldots, \lambda_n) = F_1(\lambda_1) \cdots F_n(\lambda_n)$$

The *mean* and the *variance* of a random variable x are

$$m_1 = E\{x\} \qquad \text{var } x = E(x - m_1)^2$$

and the *covariance* of two random variables x and y with means m_1 and m_2 is

$$\text{cov } (x,y) = E\{(x - m_1)(y - m_2)\} = E\{xy\} - m_1 m_2$$

If x and y are independent, they are uncorrelated; i.e., their covariance vanishes.

Typical distributions are as follows:
1.   Poisson:

$$\boldsymbol{P}\{x(\omega) = j\} = \frac{c^j}{j!} e^{-c} \qquad j = 0, 1, 2, \ldots$$
$$E\{x\} = \text{var } x = c > 0$$
$$\log L(\theta) = c(e^{i\theta} - 1)$$

2.   Gaussian (or normal):

$$f(\lambda) = \frac{1}{\sqrt{2\pi}\,\sigma} \exp\left[-\frac{(\lambda - m)^2}{2\sigma^2}\right] \qquad -\infty < \lambda < \infty$$
$$E\{x\} = m \qquad \text{var}(x) = \sigma^2$$
$$\log L(\theta) = im\theta - \tfrac{1}{2}\sigma^2\theta^2$$

3.   Joint gaussian:

$$f(\lambda,\mu) = \frac{1}{2\pi\sigma_1\sigma_2 \sqrt{1 - \rho^2}} \exp\left[-\frac{1}{2(1 - \rho^2)} M(\lambda,\mu)\right]$$

where

$$M(\lambda,\mu) = \left[\frac{(\lambda - m_1)^2}{\sigma_1^2} - \frac{2\rho(\lambda - m_1)(\mu - m_2)}{\sigma_1\sigma_2} + \frac{(\mu - m_2)^2}{\sigma_2^2}\right]$$

and

$$E\{x\} = m_1 \qquad E\{y\} = m_2 \qquad \text{var } x = \sigma_1^2 \qquad \text{var } y = \sigma_2^2$$

and the correlation coefficient ρ is defined by $\sigma_1\sigma_2\rho = E\{xy\}$, and the cumulant generating function is

$$\log L(\theta_1,\theta_2) = i\theta_1 m_1 + i\theta_2 m_2 - \tfrac{1}{2}(\sigma_1{}^2\theta_1{}^2 + 2\rho\sigma_1\sigma_2\theta_1\theta_2 + \sigma_2{}^2\theta_2{}^2)$$

The *conditional expectation* $E\{x|y\}$ of an integrable random variable x, given a random variable y, is defined as the Radom-Nikodym derivative

$$\int_B E\{x|\mathfrak{W}\}\, d\boldsymbol{P} = \int_B x\, d\boldsymbol{P}$$

$B \in \mathfrak{W} \subset \mathfrak{F}$, where $\mathfrak{W}$ is the σ field induced by the random variable y; $E\{x|y\} = E\{x|\mathfrak{W}\}$ is an $\mathfrak{W}$-measurable function on Ω.

Conditional covariance:

$$E[(x - E(x|y))^2|y] = E\{x^2|y\} - E^2\{x|y\}$$

In the case when densities exist, the conditional distribution function $F(\lambda|\mu)$ of x, given that y assumes value μ, is

$$\boldsymbol{P}\{x(\omega) < \lambda | y(\omega) = \mu\} = \int_{-\infty}^{\lambda} f(\alpha|\mu)\, d\alpha$$

where
$$f(\lambda|\mu)f_2(\mu) = f(\lambda,\mu)$$

so that
$$F_1(\lambda) = \int_{-\infty}^{\infty} F(\lambda|\mu)f_2(\mu)\, d\mu$$

For example, in case 3 above the conditional density is again gaussian and (for $m_1 = m_2 = 0$)

$$f(\lambda|\mu) = \frac{1}{\sqrt{2\pi}\, s_1} \exp\left[-\frac{1}{2}\left(\frac{\lambda}{s_1} - \rho\,\frac{\mu}{s_2}\right)^2 \right]$$

where
$$s_1 = \sigma_1\sqrt{1 - \rho^2} \qquad s_2 = \sigma_2\sqrt{1 - \rho^2}$$

In the text, use has been made of a concept of convergence for a sequence of random variable to a random variable; three principal types of convergence are defined below.

The sequence $(x_1,x_2, \ldots)$ of random variables converges in probability (or stochastically) to a random variable x if, for every $\epsilon > 0$,

$$\lim_{n\to\infty} \boldsymbol{P}\{|x_n(\omega) - x(\omega)| \geq \epsilon\} = 0$$

in symbols, $x_n \xrightarrow{P} x$. A sequence converges mutually in probability if $x_n - x_m \xrightarrow{P} 0$, when $n, m \to \infty$; a sequence of random variables converges in probability to a random variable if and only if it converges mutually in probability.

The sequence $(x_1,x_2, \ldots)$ of random variables converges almost everywhere (or with probability one) to a random variable x if

$\lim\limits_{n \to \infty} x_n(\omega) = x(\omega)$ for almost all ω; that is, the set of divergence is null; in symbols, $x_n \xrightarrow{\text{a.e.}} x$.   There is such convergence if and only if, for every $\epsilon > 0$,

$$\lim_{n \to \infty} \boldsymbol{P}\{\sup_{m \geq n} |x_n(\omega) - x(\omega)| \geq \epsilon\} = 0$$

A sequence converges mutually almost everywhere if $x_n - x_m \xrightarrow{\text{a.e.}} 0$, when $n, m \to \infty$; a sequence of random variables converges almost everywhere to a random variable if and only if it converges mutually almost everywhere.

The sequence $(x_1, x_2, \ldots)$ of random variables converges in the rth mean to a random variable x if x_n and x have finite absolute rth moments, and

$$\lim_{n \to \infty} E\{|x_n - x|^r\} = 0$$

in symbols, $x_n \xrightarrow{r} x$.   A sequence of random variables converges in the mean to a random variable if and only if it converges mutually in the mean.

The convergence in probability is the weakest of these three types and is implied either by convergence almost everywhere or by convergence in the mean; converse implications do not hold in general.   The limit function x is unique almost everywhere; i.e., x is determined up to an equivalence.

Let $(\Omega, \mathfrak{F}, \boldsymbol{P})$ be a probability space, and let T be an index set (real line or its subset).   The family of random variables $x(t)$ defined on Ω and indexed by T

$$x = \{x(t), t \, \epsilon \, T\}$$

is called a *stochastic process* (or a random function); x is a function on T with values $x(t)$ in the range space $\mathfrak{K}$ of all random variables on the probability space.   The values at $\omega \, \epsilon \, \Omega$ of the process x, obtained by fixing ω and letting t vary, form a *sample function*, $x(.,\omega)$, or a realization of the process.   Hence, the stochastic process x can also be regarded as a measurable function from the space $(\Omega, \mathfrak{F})$ to the *sample space* $(R^r, \mathfrak{B}^r)$ where R^r is the product space of range spaces R of every $x(t)$ and $\mathfrak{B}^r$ is the product Borel field (generated by finite-dimensional cylinder sets with Borel bases), with $\mathfrak{B}$ being a Borel field in R.

Let $\mathfrak{F}(x) = x^{-1}(\mathfrak{B}^r) \subset \mathfrak{F}$ be the smallest σ field which contains sets of the form $\{\omega : x(t, \omega) \, \epsilon \, B\}$ where $t \, \epsilon \, T$ and B is a linear Borel set; $\mathfrak{F}(x)$ is the smallest σ field with respect to which all random variables $x(t)$ of the family x are measurable. $\mathfrak{F}(x)$ contains events determined by, at most, a countable number of random variables from the family x.

The n dimensional distribution functions for any finite collection of n random variables of the family x are defined by

$$F(\lambda,t_1; \; \ldots \; ;\lambda_n,t_n) = \boldsymbol{P}\{x(t_j,\omega) < \lambda_j, j = 1, \; \ldots \; , n\}$$

Conversely, by the Daniell-Kolmogorov extension theorem, any collection of such functions F satisfying consistency conditions determines uniquely some probability measure $\boldsymbol{P}$ on the events in $\mathfrak{F}(x)$. However, sets determined by a noncountable number of random variables, for example, $\{\omega : \limsup_{t \in T} |x(t,\omega)| > \lambda\}$, will in general not belong to $\mathfrak{F}$, and thus will have no probabilities defined. The concept of *separability* (stated in Sec. 8-5) imposes restrictions on sets determined by a noncountable number of random variables and allows one to determine their probabilities.

Typical stochastic processes include the following:

1. Markoff process:

$$\boldsymbol{P}\{x(t_n,\omega) \leq \lambda | x(t_1), \; \ldots \; ,x(t_{n-1})\}$$
$$= \boldsymbol{P}\{x(t_n,\omega) \leq \lambda | x(t_{n-1})\} \qquad \text{almost everywhere}$$

for $t_1 < \cdots < t_n$, for any integer $n \geq 1$.

2. Martingale:

$$E\{x(t_n) | x(t_1), \; \ldots \; ,x(t_{n-1})\} = x(t_{n-1}) \qquad \text{almost everywhere}$$

whenever $n \geq 1$ and $t_1 < \cdots < t_n$, and $E\{|x(t)|\} < \infty$ for each t.

3. Stationary process:

$$\boldsymbol{P}\{x(t_i,\omega) \leq \lambda_i, i = 1, \; \ldots \; , n\} = \boldsymbol{P}\{x(t_i + h, \omega) < \lambda_i, i = 1, \; \ldots \; , n\}$$

for any finite set $(t_1, \; \ldots \; ,t_n)$ and any h.

REFERENCES

(Starred items have not been discussed in the text.)

1. Adomian, G.: Stochastic Green Functions, in Ref. 7, pp. 1–39.
2. ———: "Stochastic Operators and Physical Problems," Academic Press Inc., New York, in press.
3. Agmon, S. and L. Nirenberg: Properties of Solutions of Ordinary Differential Equations in Banach Space, *Comm. Pure Appl. Math.*, vol. 16, pp. 121–239, 1963.
4. Azis, A. K., and J. B. Diaz: On a Mean Value Theorem . . . , *Contrib. Differential Equations*, vol. 1, no. 2, pp. 251–269, 1963.
5. ——— and ———: On a Mean Value Theorem . . . , *Contrib. Differential Equations*, vol. 7, pp. 271–273, 1963.
6. Bartlett, M. S.: "Stochastic Processes," Cambridge University Press, London, 1962.

7. Bellman, R. (ed.): "Stochastic Processes in Mathematical Physics and Engineering," Proceedings of Symposia in Applied Mathematics, vol. 16, American Mathematical Society, Providence, R.I., 1964.

8. Bernstein, S.: Equations différentielles stochastiques, *Actualités Sci. Indust.*, vol. 738, pp. 5–31, 1938.

9. Bertram, J. E., and P. E. Sarachik: On the Stability of Circuits with Randomly Varying Parameters, *IRE Trans.*, PGIT-5, Special Suppl. 5, pp. 260–270, 1959.

10. Bharucha-Reid, A. T.: On the Theory of Random Equations in Ref. 7, pp. 40–69.

11. ———: "Random Equations: An Introduction to the Probabilistic Analysis of Operator Equations," Academic Press Inc., New York, 1966.

12. Bochner, S.: Stochastic Processes, *Ann. of Math.*, vol. 48, pp. 1014–1061, 1942.

13. ———: "Harmonic Analysis and the Theory of Probability," University of California Press, Berkeley, Calif., 1955.

14. ———: Stationarity, Boundedness, Almost Periodicity of Random-valued Functions, in "Proceedings of the Third Berkeley Symposium on Mathematical Statistics and Probability," vol. 2, "Probability Theory," edited by J. Neyman, pp. 7–27, University of California Press, Berkeley, Calif., 1956.

15. Bogdanoff, J. L., and F. Kozin: Moments of the Output of Linear Random Systems, *J. Acoust. Soc. Amer.*, vol. 34, pp. 1063–1066, 1960.

16. ——— and ——— (eds.): "Proceedings of the First Symposium on Engineering Applications of Random Function Theory and Probability," John Wiley & Sons, Inc., New York, 1963.

17. Bourbaki, N.: Fonctions d'une variable réelle, chap. 4, in "Equations différentielles," Herman & Cie, Paris, 1951.

18. Bucy, R. S.: Stability and Positive Supermartingales, *Notices Amer. Math. Soc.*, abs. 64T-233, vol. 11, p. 381, 1964.

19. ———: Stability and Positive Supermartingales, *J. Differential Equations*, vol. 1, pp. 151–155, April, 1965.

20. ——— and R. E. Kalman: "Optimal Stochastic Control Theory," in press.

21. Bunke, H.: Stabilität bei stochastischen Differentialgleichungssystemen, *Z. Angew. Math. Mech.*, vol. 43, pp. 63–70, 1963.

22. Caughey, T. K., and J. K. Dienes: The Behavior of Linear Systems with Random Parametric Excitation, *J. Math. and Phys.*, vol. 41, pp. 300–318, 1962.

23. Chandrasekhar, S.: Stochastic Problems in Physics and Astronomy, *Rev. Modern Phys.*, vol. 15, no. 1 (reprinted in Ref. 86).

24. *Courrege, P.: Intégrales stochastiques associées à une martingale de carré intégrable, *C. R. Acad. Sci. Paris*, vol. 256, pp. 867–870, 1963.

25. Cramer, H.: On the Theory of Stationary Random Processes, *Ann. of Math.*, vol. 41, pp. 215–230, 1940.

26. *Deutsch, R.: Non-linear Transformations of Random Processes, Prentice-Hall, Inc., Englewood Cliffs, N.J., 1962.

27. *Dolph, C. L.: Positive Real Resolvents and Linear Passive Hilbert Systems, *Ann. Acad. Sci. Fenn. Ser. A. I*, no. 336/9, p. 33, 1963.

28. Doob, J. L.: The Brownian Motion and Stochastic Equations, *Ann. of Math.*, vol. 43, pp. 351–369, 1942 (reprinted in Ref. 86).

29. ———: "Stochastic Processes," John Wiley & Sons, Inc., New York, 1953.

30. Dynkin, E. B.: "Markov Processes," vols. 1, 2, Academic Press Inc., New York, 1965.

31. Edwards, D. A., and J. E. Moyal: Stochastic Differential Equations, *Proc. Cambridge Philos. Soc.*, vol. 51, pp. 663–677, 1955.

32. Eisenhart, L. P.: "Continuous Groups of Transformations," Dover Publications, Inc., New York, 1961.

33. Erdös, P., and A. C. Offord: On the Number of Real Roots of a Random Algebraic Equation, *Proc. London Math. Soc.*, ser. 3, vol. 6, pp. 139–160, 1956.

34. Feller, W.: An Introduction to Probability Theory and Its Applications," 2d ed., John Wiley & Sons, Inc., New York, 1961.

35. Fisz, M.: "Probability Theory and Mathematical Statistics," John Wiley & Sons, Inc., New York, 1963.

36. *Freidlin, M. I.: Ito's Stochastic Equations and Degenerate Elliptic Equations, *Jzv. Akad. Nauk. SSSR Ser. Mat.*, vol. 26, pp. 653–676, 1962.

37. Fortet, R. M.: Random Distributions with an Application to Telephone Engineering, in "Proceedings of the Third Berkeley Symposium on Mathematical Statistics and Probability," vol. 2, "Probability Theory," edited by J. Neyman, pp. 81–88, University of California Press, Berkeley, Calif., 1956.

38. Gelfand, I. M., and A. M. Yaglom: Integration in Functional Spaces and Its Applications in Quantum Physics, *J. Math. Phys.*, vol. 1, pp. 48–69, 1960.

39. ——— and N. Y. Vilenkin: "Generalized Functions," vol. 4, "Applications of Harmonic Analysis," Academic Press Inc., New York, 1964.

40. Gibbs, J. W.: "Elementary Principles in Statistical Mechanics," Dover Publications, Inc., New York, 1960.

41. Goursat, E.: "Differential Equations," Dover Publications, Inc., New York, 1959.

42. Halmos, P.: "Lectures on Ergodic Theory," Mathematic Society of Japan, Tokyo, 1956.

43. ———: "Introduction to Hilbert Space and the Theory of Spectral Multiplicity," Chelsea Publishing Company, New York, 1951.

44. Hammersley, J. M.: The Zeros of a Random Polynomial, "Proceedings of the Third Berkeley Symposium on Mathematical Statistics and Probability," vol. 2, "Probability Theory," edited by J. Neyman, pp. 89–111, University of California Press, Berkeley, Calif., 1956.

45. Hans, O.: Random Operator Equations, pp. 185–202, "Proceedings of the Fourth Berkeley Symposium on Mathematical Statistics and Probability," vol. 2, University of California Press, Berkeley, Calif., 1961.

46. Hille, E., and R. S. Phillips: "Functional Analysis and Semi-groups," University of California Press, Berkeley, Calif., Colloquium Publications, American Mathematical Society, Providence, R.I., 1957.

47. Ito, K.: On Stochastic Differential Equations, *Mem. Amer. Math. Soc.*, vol. 4, 1961.

48. ———: "Lectures on Stochastic Processes," Tata Institute of Fundamental Research, Bombay, 1961.

49. ——— and H. P. McKean: "Diffusion Processes and Their Sample Paths," Springer-Verlag OHG, Berlin, 1965.

50. Jirina, M.: Ordinary Differential and Difference Equations with Random Coefficients and Random Right-hand Side, *Czechoslovak Math. J.*, vol. 12, pp. 457–474, 1962.

51. ———: Harmonisable Solutions of Ordinary Differential Equations with Random Coefficients and Random Right-hand Side (in Russian, English summary), *Czechoslovak Math. J.*, vol. 13, pp. 360–371, 1963.

52. Kac, M.: "Probability and Related Topics in Physical Sciences," Interscience Publishers, Inc., New York, 1959.

53. Kalman, R. E., and R. S. Bucy: New Results in Linear Filtering and Prediction Theory, *Trans. ASME Ser. D J. Basic Engrg.*, 1961.

54. Kampé de Feriet, J.: Random Integrals of Differential Equations, in "Lectures on Modern Mathematics," edited by T. L. Saaty, John Wiley & Sons, Inc., New York, 1965.

55. *Kampen, N. G. van: Thermal Fluctuations in Non-linear Systems, *J. Math. Phys.*, vol. 4, pp. 190–194, 1963.

56. Kato, T.: On Linear Differential Equations in Banach Spaces, *Comm. Pure Appl. Math.*, vol. 9, pp. 479–486, 1956.

57. Kats, I. I., and N. N. Krasovskij: On the Stability of Systems with Random Parameters, *Appl. Math. Mech. (PPM)*, vol. 24, pp. 809–823, 1960.

58. Khinchin, A. I.: "Statistical Mechanics," Dover Publications, Inc., New York, 1949.

59. Kozin, F.: On Almost Sure Stability of Linear Systems with Random Coefficients, *J. Math. and Phys.*, vol. 42, pp. 59–67, 1963.

60. ―――― and J. L. Bogdanoff: A Comment on "The Behavior of Linear Systems with Random Parametric Excitation," *J. Math. and Phys.*, vol. 42, pp. 336–337, 1963.

61. Krasnosel'skii, M. A., S. G. Krein, and P. E. Sobolevskii: On Differential Equations with Unbounded Operations in Banach Spaces, *Dokl. Akad. Nauk SSSR*, vol. 111, pp. 19–22, 1956.

62. *Kubo, R.: Stochastic Liouville Equations, *J. Math. Phys.*, vol. 4, pp. 174–183, 1963.

63. Kurth, R.: "Axiomatics of Classical Statistical Mechanics," Pergamon Press, New York, 1960.

64. Kushner, H. J.: On the Stability of Stochastic Dynamical Systems, *Proc. Nat. Acad. Sci. U.S.A.*, vol. 53, pp. 8–12, 1965.

65. Lakshmikantham, V.: On the Boundedness of Solutions of Non-linear Differential Equations, *Proc. Amer. Math. Soc.*, vol. 8, pp. 1044–1048, 1957.

66. *Laning, J. H., Jr., and R. H. Battin: Random Processes in Automatic Control, McGraw-Hill Book Company, New York, 1956.

67. Leibowitz, M. A.: Statistical Behavior of Linear Systems with Randomly Varying Parameters, *J. Math. Phys.*, vol. 4, pp. 852–858, 1963.

68. Levy, P.: "Processus stochastiques et mouvement Brownien," Gauthier-Villars, Paris, 1948.

69. Loève, M.: "Probability Theory," 3d ed., D. Van Nostrand Company, Inc., Princeton, N.J., 1963.

70. Lyubarskii, G. Y., and Y. L. Rabotnikov: On the Theory of Differential Equations with Random Coefficients, *Theory Probability Appl.*, vol. 8, pp. 290–298, 1963.

71. McShane, E. J.: Stochastic Integrals and Non-linear Processes, *J. Math. Mech.*, vol. 11, pp. 235–283, 1962.

72. Mlak, W.: Limitations and Dependence on Parameter of Solutions of Nonstationary Differential Operator Equations, *Ann. Polon. Math.*, vol. 6, pp. 305–322, 1959.

73. ―――― : Differential Inequalities with Unbounded Operators in Banach Spaces, *Ann. Polon. Math.*, vol. 9, pp. 101–111, 1960.

74. Moyal, J. E.: Stochastic Processes and Statistical Physics, *J. Roy. Statist. Soc. Ser. (B)*, vol. 11, pp. 150–210, 1949.

75. Paley, R. E. A. C., and N. Wiener: "Fourier Transforms in the Complex Domain," Colloquium Publications, American Mathematical Society, Providence, R.I., 1954.

76. Parzen, E.: "Stochastic Processes," Holden-Day, Inc., San Francisco, Calif., 1962.

77. Pontrjagin, L.: "Topological Groups," Princeton University Press, Princeton, N.J., 1946.

78. Rice, S. O.: Mathematical Analysis of Random Noise, *Bell System Tech. J.*, vol. 23, pp. 282–332, 1944; vol. 24, pp. 46–136, 1945 (reprinted in Ref. 86).

79. Samuels, J. C., and A. C. Eringen: On Stochastic Linear Systems, *J. Math. and Phys.*, vol. 38, pp. 83–103, 1959.

80. Slutsky, E.: Sur les fonctions éventuelles continués, intégrables et dérivables dans le sens stochastique, *C. R. Acad. Sci. Paris*, vol. 187, pp. 878–880, 1928.

81. Špaček, A.: Zufällige Gleichungen, *Czechoslovak Math. J.*, vol. 5, no. 80, pp. 462–466, 1955.

82. Srinivasan, S. K.: On a Class of Stochastic Differential Equations, *Z. Angew. Math. Mech.*, vol. 43, pp. 259–265, 1963.

83. Syski, R.: "Introduction to Congestion Theory in Telephone Systems," Oliver and Boyd Ltd., Edinburgh, 1960.

84. Todhunter, I.: "A History of the Theory of Probability," Chelsea Publishing Company, New York, 1949.

85. Uhlenbeck, G. E., and L. S. Ornstein: On the Theory of the Brownian Motion, *Phys. Rev.*, vol. 36, no. 3, pp. 823–841, 1930 (reprinted in Ref. 86).

86. Wax, N.: "Noise and Stochastic Processes," Dover Publications, Inc., New York, 1954.

87. Whittaker, E. T.: "Analytical Dynamics," 4th ed., Cambridge University Press, London, 1937.

88. Wiener, N.: "Stationary Time Series," John Wiley & Sons, Inc., New York, 1949.

89. ———: "Non-linear Problems in Random Theory," John Wiley & Sons, Inc., New York, 1958.

90. Wong, E., and M. Zakai, On the Relation between Ordinary and Stochastic Differential Equations, *Internat. J. Engrg. Sci.*, vol. 3, pp. 213–230, 1965.

91. *Woodword, D. A.: Continuous Transformations and Stochastic Differential Equations, *Trans. Amer. Math. Soc.*, vol. 115, pp. 76–82, 1965.

92. Yaglom, A. M.: "Theory of Stationary Random Functions," Prentice-Hall, Inc., Englewood Cliffs, N.J., 1962.

INDEX

A CATALOGUE OF
SELECTED DOVER BOOKS
IN ALL FIELDS OF INTEREST

A CATALOGUE OF SELECTED DOVER
BOOKS IN ALL FIELDS OF INTEREST

RACKHAM'S COLOR ILLUSTRATIONS FOR WAGNER'S RING. Rackham's finest mature work—all 64 full-color watercolors in a faithful and lush interpretation of the *Ring*. Full-sized plates on coated stock of the paintings used by opera companies for authentic staging of Wagner. Captions aid in following complete Ring cycle. Introduction. 64 illustrations plus vignettes. 72pp. 8⅝ x 11¼.   23779-6 Pa. $6.00

CONTEMPORARY POLISH POSTERS IN FULL COLOR, edited by Joseph Czestochowski. 46 full-color examples of brilliant school of Polish graphic design, selected from world's first museum (near Warsaw) dedicated to poster art. Posters on circuses, films, plays, concerts all show cosmopolitan influences, free imagination. Introduction. 48pp. 9⅜ x 12¼.
23780-X Pa. $6.00

GRAPHIC WORKS OF EDVARD MUNCH, Edvard Munch. 90 haunting, evocative prints by first major Expressionist artist and one of the greatest graphic artists of his time: *The Scream, Anxiety, Death Chamber, The Kiss, Madonna,* etc. Introduction by Alfred Werner. 90pp. 9 x 12.
23765-6 Pa. $5.00

THE GOLDEN AGE OF THE POSTER, Hayward and Blanche Cirker. 70 extraordinary posters in full colors, from Maitres de l'Affiche, Mucha, Lautrec, Bradley, Cheret, Beardsley, many others. Total of 78pp. 9⅜ x 12¼.   22753-7 Pa. $5.95

THE NOTEBOOKS OF LEONARDO DA VINCI, edited by J. P. Richter. Extracts from manuscripts reveal great genius; on painting, sculpture, anatomy, sciences, geography, etc. Both Italian and English. 186 ms. pages reproduced, plus 500 additional drawings, including studies for *Last Supper,* Sforza monument, etc. 860pp. 7⅞ x 10¾. (Available in U.S. only)
22572-0, 22573-9 Pa., Two-vol. set $15.90

THE CODEX NUTTALL, as first edited by Zelia Nuttall. Only inexpensive edition, in full color, of a pre-Columbian Mexican (Mixtec) book. 88 color plates show kings, gods, heroes, temples, sacrifices. New explanatory, historical introduction by Arthur G. Miller. 96pp. 11⅜ x 8½. (Available in U.S. only)   23168-2 Pa. $7.95

UNE SEMAINE DE BONTÉ, A SURREALISTIC NOVEL IN COLLAGE, Max Ernst. Masterpiece created out of 19th-century periodical illustrations, explores worlds of terror and surprise. Some consider this Ernst's greatest work. 208pp. 8⅛ x 11.   23252-2 Pa. $5.00

DRAWINGS OF WILLIAM BLAKE, William Blake. 92 plates from Book of Job, *Divine Comedy, Paradise Lost,* visionary heads, mythological figures, Laocoon, etc. Selection, introduction, commentary by Sir Geoffrey Keynes. 178pp. 8⅛ x 11. 22303-5 Pa. $4.00

ENGRAVINGS OF HOGARTH, William Hogarth. 101 of Hogarth's greatest works: *Rake's Progress, Harlot's Progress, Illustrations for Hudibras, Before and After, Beer Street and Gin Lane,* many more. Full commentary. 256pp. 11 x 13¾. 22479-1 Pa. $12.95

DAUMIER: 120 GREAT LITHOGRAPHS, Honore Daumier. Wide-ranging collection of lithographs by the greatest caricaturist of the 19th century. Concentrates on eternally popular series on lawyers, on married life, on liberated women, etc. Selection, introduction, and notes on plates by Charles F. Ramus. Total of 158pp. 9⅜ x 12¼. 23512-2 Pa. $5.50

DRAWINGS OF MUCHA, Alphonse Maria Mucha. Work reveals drafts-man of highest caliber: studies for famous posters and paintings, render-ings for book illustrations and ads, etc. 70 works, 9 in color; including 6 items not drawings. Introduction. List of illustrations. 72pp. 9⅜ x 12¼. (Available in U.S. only) 23672-2 Pa. $4.00

GIOVANNI BATTISTA PIRANESI: DRAWINGS IN THE PIERPONT MORGAN LIBRARY, Giovanni Battista Piranesi. For first time ever all of Morgan Library's collection, world's largest. 167 illustrations of rare Piranesi drawings—archeological, architectural, decorative and visionary. Essay, detailed list of drawings, chronology, captions. Edited by Felice Stampfle. 144pp. 9⅜ x 12¼. 23714-1 Pa. $7.50

NEW YORK ETCHINGS (1905-1949), John Sloan. All of important American artist's N.Y. life etchings. 67 works include some of his best art; also lively historical record—Greenwich Village, tenement scenes. Edited by Sloan's widow. Introduction and captions. 79pp. 8⅜ x 11¼. 23651-X Pa. $4.00

CHINESE PAINTING AND CALLIGRAPHY: A PICTORIAL SURVEY, Wan-go Weng. 69 fine examples from John M. Crawford's matchless private collection: landscapes, birds, flowers, human figures, etc., plus calligraphy. Every basic form included: hanging scrolls, handscrolls, album leaves, fans, etc. 109 illustrations. Introduction. Captions. 192pp. 8⅞ x 11¾. 23707-9 Pa. $7.95

DRAWINGS OF REMBRANDT, edited by Seymour Slive. Updated Lipp-mann, Hofstede de Groot edition, with definitive scholarly apparatus. All portraits, biblical sketches, landscapes, nudes, Oriental figures, classical studies, together with selection of work by followers. 550 illustrations. Total of 630pp. 9⅛ x 12¼. 21485-0, 21486-9 Pa., Two-vol. set $15.00

THE DISASTERS OF WAR, Francisco Goya. 83 etchings record horrors of Napoleonic wars in Spain and war in general. Reprint of 1st edition, plus 3 additional plates. Introduction by Philip Hofer. 97pp. 9⅜ x 8¼. 21872-4 Pa. $3.75

THE EARLY WORK OF AUBREY BEARDSLEY, Aubrey Beardsley. 157 plates, 2 in color: *Manon Lescaut, Madame Bovary, Morte Darthur, Salome,* other. Introduction by H. Marillier. 182pp. 8⅛ x 11. 21816-3 Pa. $4.50

THE LATER WORK OF AUBREY BEARDSLEY, Aubrey Beardsley. Exotic masterpieces of full maturity: *Venus and Tannhauser, Lysistrata, Rape of the Lock, Volpone,* Savoy material, etc. 174 plates, 2 in color. 186pp. 8⅛ x 11. 21817-1 Pa. $4.50

THOMAS NAST'S CHRISTMAS DRAWINGS, Thomas Nast. Almost all Christmas drawings by creator of image of Santa Claus as we know it, and one of America's foremost illustrators and political cartoonists. 66 illustrations. 3 illustrations in color on covers. 96pp. 8⅜ x 11¼. 23660-9 Pa. $3.50

THE DORÉ ILLUSTRATIONS FOR DANTE'S DIVINE COMEDY, Gustave Doré. All 135 plates from Inferno, Purgatory, Paradise; fantastic tortures, infernal landscapes, celestial wonders. Each plate with appropriate (translated) verses. 141pp. 9 x 12. 23231-X Pa. $4.50

DORÉ'S ILLUSTRATIONS FOR RABELAIS, Gustave Doré. 252 striking illustrations of *Gargantua and Pantagruel* books by foremost 19th-century illustrator. Including 60 plates, 192 delightful smaller illustrations. 153pp. 9 x 12. 23656-0 Pa. $5.00

LONDON: A PILGRIMAGE, Gustave Doré, Blanchard Jerrold. Squalor, riches, misery, beauty of mid-Victorian metropolis; 55 wonderful plates, 125 other illustrations, full social, cultural text by Jerrold. 191pp. of text. 9⅜ x 12¼. 22306-X Pa. $7.00

THE RIME OF THE ANCIENT MARINER, Gustave Doré, S. T. Coleridge. Dore's finest work, 34 plates capture moods, subtleties of poem. Full text. Introduction by Millicent Rose. 77pp. 9¼ x 12. 22305-1 Pa. $3.50

THE DORE BIBLE ILLUSTRATIONS, Gustave Doré. All wonderful, detailed plates: Adam and Eve, Flood, Babylon, Life of Jesus, etc. Brief King James text with each plate. Introduction by Millicent Rose. 241 plates. 241pp. 9 x 12. 23004-X Pa. $6.00

THE COMPLETE ENGRAVINGS, ETCHINGS AND DRYPOINTS OF ALBRECHT DURER. "Knight, Death and Devil"; "Melencolia," and more—all Dürer's known works in all three media, including 6 works formerly attributed to him. 120 plates. 235pp. 8⅜ x 11¼. 22851-7 Pa. $6.50

MAXIMILIAN'S TRIUMPHAL ARCH, Albrecht Dürer and others. Incredible monument of woodcut art: 8 foot high elaborate arch—heraldic figures, humans, battle scenes, fantastic elements—that you can assemble yourself. Printed on one side, layout for assembly. 143pp. 11 x 16. 21451-6 Pa. $5.00

THE COMPLETE WOODCUTS OF ALBRECHT DURER, edited by Dr. W. Kurth. 346 in all: "Old Testament," "St. Jerome," "Passion," "Life of Virgin," Apocalypse," many others. Introduction by Campbell Dodgson. 285pp. 8½ x 12¼. 21097-9 Pa. $7.50

DRAWINGS OF ALBRECHT DURER, edited by Heinrich Wolfflin. 81 plates show development from youth to full style. Many favorites; many new. Introduction by Alfred Werner. 96pp. 8⅛ x 11. 22352-3 Pa. $5.00

THE HUMAN FIGURE, Albrecht Dürer. Experiments in various techniques—stereometric, progressive proportional, and others. Also life studies that rank among finest ever done. Complete reprinting of *Dresden Sketchbook.* 170 plates. 355pp. 8⅜ x 11¼. 21042-1 Pa. $7.95

OF THE JUST SHAPING OF LETTERS, Albrecht Dürer. Renaissance artist explains design of Roman majuscules by geometry, also Gothic lower and capitals. Grolier Club edition. 43pp. 7⅞ x 10¾ 21306-4 Pa. $3.00

TEN BOOKS ON ARCHITECTURE, Vitruvius. The most important book ever written on architecture. Early Roman aesthetics, technology, classical orders, site selection, all other aspects. Stands behind everything since. Morgan translation. 331pp. 5⅜ x 8½. 20645-9 Pa. $4.50

THE FOUR BOOKS OF ARCHITECTURE, Andrea Palladio. 16th-century classic responsible for Palladian movement and style. Covers classical architectural remains, Renaissance revivals, classical orders, etc. 1738 Ware English edition. Introduction by A. Placzek. 216 plates. 110pp. of text. 9½ x 12¾. 21308-0 Pa. $10.00

HORIZONS, Norman Bel Geddes. Great industrialist stage designer, "father of streamlining," on application of aesthetics to transportation, amusement, architecture, etc. 1932 prophetic account; function, theory, specific projects. 222 illustrations. 312pp. 7⅞ x 10¾. 23514-9 Pa. $6.95

FRANK LLOYD WRIGHT'S FALLINGWATER, Donald Hoffmann. Full, illustrated story of conception and building of Wright's masterwork at Bear Run, Pa. 100 photographs of site, construction, and details of completed structure. 112pp. 9¼ x 10. 23671-4 Pa. $5.50

THE ELEMENTS OF DRAWING, John Ruskin. Timeless classic by great Viltorian; starts with basic ideas, works through more difficult. Many practical exercises. 48 illustrations. Introduction by Lawrence Campbell. 228pp. 5⅜ x 8½. 22730-8 Pa. $3.75

GIST OF ART, John Sloan. Greatest modern American teacher, Art Students League, offers innumerable hints, instructions, guided comments to help you in painting. Not a formal course. 46 illustrations. Introduction by Helen Sloan. 200pp. 5⅜ x 8½. 23435-5 Pa. $4.00

THE ANATOMY OF THE HORSE, George Stubbs. Often considered the great masterpiece of animal anatomy. Full reproduction of 1766 edition, plus prospectus; original text and modernized text. 36 plates. Introduction by Eleanor Garvey. 121pp. 11 x 14¾.   23402-9 Pa. $6.00

BRIDGMAN'S LIFE DRAWING, George B. Bridgman. More than 500 illustrative drawings and text teach you to abstract the body into its major masses, use light and shade, proportion; as well as specific areas of anatomy, of which Bridgman is master. 192pp. 6½ x 9¼. (Available in U.S. only)
22710-3 Pa. $3.50

ART NOUVEAU DESIGNS IN COLOR, Alphonse Mucha, Maurice Verneuil, Georges Auriol. Full-color reproduction of *Combinaisons ornementales* (c. 1900) by Art Nouveau masters. Floral, animal, geometric, interlacings, swashes—borders, frames, spots—all incredibly beautiful. 60 plates, hundreds of designs. 9⅜ x 8-1/16.   22885-1 Pa. $4.00

FULL-COLOR FLORAL DESIGNS IN THE ART NOUVEAU STYLE, E. A. Seguy. 166 motifs, on 40 plates, from *Les fleurs et leurs applications decoratives* (1902): borders, circular designs, repeats, allovers, "spots." All in authentic Art Nouveau colors. 48pp. 9⅜ x 12¼.
23439-8 Pa. $5.00

A DIDEROT PICTORIAL ENCYCLOPEDIA OF TRADES AND IN-DUSTRY, edited by Charles C. Gillispie. 485 most interesting plates from the great French Encyclopedia of the 18th century show hundreds of working figures, artifacts, process, land and cityscapes; glassmaking, papermaking, metal extraction, construction, weaving, making furniture, clothing, wigs, dozens of other activities. Plates fully explained. 920pp. 9 x 12.
22284-5, 22285-3 Clothbd., Two-vol. set $40.00

HANDBOOK OF EARLY ADVERTISING ART, Clarence P. Hornung. Largest collection of copyright-free early and antique advertising art ever compiled. Over 6,000 illustrations, from Franklin's time to the 1890's for special effects, novelty. Valuable source, almost inexhaustible.
Pictorial Volume. Agriculture, the zodiac, animals, autos, birds, Christmas, fire engines, flowers, trees, musical instruments, ships, games and sports, much more. Arranged by subject matter and use. 237 plates. 288pp. 9 x 12.
20122-8 Clothbd. $14.50

Typographical Volume. Roman and Gothic faces ranging from 10 point to 300 point, "Barnum," German and Old English faces, script, logotypes, scrolls and flourishes, 1115 ornamental initials, 67 complete alphabets, more. 310 plates. 320pp. 9 x 12.   20123-6 Clothbd. $15.00

CALLIGRAPHY (CALLIGRAPHIA LATINA), J. G. Schwandner. High point of 18th-century ornamental calligraphy. Very ornate initials, scrolls, borders, cherubs, birds, lettered examples. 172pp. 9 x 13.
20475-8 Pa. $7.00

ART FORMS IN NATURE, Ernst Haeckel. Multitude of strangely beautiful natural forms: Radiolaria, Foraminifera, jellyfishes, fungi, turtles, bats, etc. All 100 plates of the 19th-century evolutionist's *Kunstformen der Natur* (1904). 100pp. 9⅜ x 12¼.　　　　　　　22987-4 Pa. $5.00

CHILDREN: A PICTORIAL ARCHIVE FROM NINETEENTH-CENTURY SOURCES, edited by Carol Belanger Grafton. 242 rare, copyright-free wood engravings for artists and designers. Widest such selection available. All illustrations in line. 119pp. 8⅜ x 11¼.
23694-3 Pa. $3.50

WOMEN: A PICTORIAL ARCHIVE FROM NINETEENTH-CENTURY SOURCES, edited by Jim Harter. 391 copyright-free wood engravings for artists and designers selected from rare periodicals. Most extensive such collection available. All illustrations in line. 128pp. 9 x 12.
23703-6 Pa. $4.50

ARABIC ART IN COLOR, Prisse d'Avennes. From the greatest ornamentalists of all time—50 plates in color, rarely seen outside the Near East, rich in suggestion and stimulus. Includes 4 plates on covers. 46pp. 9⅜ x 12¼.　　　　　　　23658-7 Pa. $6.00

AUTHENTIC ALGERIAN CARPET DESIGNS AND MOTIFS, edited by June Beveridge. Algerian carpets are world famous. Dozens of geometrical motifs are charted on grids, color-coded, for weavers, needleworkers, craftsmen, designers. 53 illustrations plus 4 in color. 48pp. 8¼ x 11. (Available in U.S. only)　　　　　　　23650-1 Pa. $1.75

DICTIONARY OF AMERICAN PORTRAITS, edited by Hayward and Blanche Cirker. 4000 important Americans, earliest times to 1905, mostly in clear line. Politicians, writers, soldiers, scientists, inventors, industrialists, Indians, Blacks, women, outlaws, etc. Identificatory information. 756pp. 9¼ x 12¾.　　　　　　　21823-6 Clothbd. $40.00

HOW THE OTHER HALF LIVES, Jacob A. Riis. Journalistic record of filth, degradation, upward drive in New York immigrant slums, shops, around 1900. New edition includes 100 original Riis photos, monuments of early photography. 233pp. 10 x 7⅞.　　　　　　　22012-5 Pa. $7.00

NEW YORK IN THE THIRTIES, Berenice Abbott. Noted photographer's fascinating study of city shows new buildings that have become famous and old sights that have disappeared forever. Insightful commentary. 97 photographs. 97pp. 11⅜ x 10.　　　　　　　22967-X Pa. $5.00

MEN AT WORK, Lewis W. Hine. Famous photographic studies of construction workers, railroad men, factory workers and coal miners. New supplement of 18 photos on Empire State building construction. New introduction by Jonathan L. Doherty. Total of 69 photos. 63pp. 8 x 10¾.
23475-4 Pa. $3.00

THE DEPRESSION YEARS AS PHOTOGRAPHED BY ARTHUR ROTH-STEIN, Arthur Rothstein. First collection devoted entirely to the work of outstanding 1930s photographer: famous dust storm photo, ragged children, unemployed, etc. 120 photographs. Captions. 119pp. 9¼ x 10¾.
23590-4 Pa. $5.00

CAMERA WORK: A PICTORIAL GUIDE, Alfred Stieglitz. All 559 illus-trations and plates from the most important periodical in the history of art photography, Camera Work (1903-17). Presented four to a page, re-duced in size but still clear, in strict chronological order, with complete captions. Three indexes. Glossary. Bibliography. 176pp. 8⅜ x 11¼.
23591-2 Pa. $6.95

ALVIN LANGDON COBURN, PHOTOGRAPHER, Alvin L. Coburn. Re-vealing autobiography by one of greatest photographers of 20th century gives insider's version of Photo-Secession, plus comments on his own work. 77 photographs by Coburn. Edited by Helmut and Alison Gernsheim. 160pp. 8⅛ x 11.
23685-4 Pa. $6.00

NEW YORK IN THE FORTIES, Andreas Feininger. 162 brilliant photo-graphs by the well-known photographer, formerly with Life magazine, show commuters, shoppers, Times Square at night, Harlem nightclub, Lower East Side, etc. Introduction and full captions by John von Hartz. 181pp. 9¼ x 10¾.
23585-8 Pa. $6.00

GREAT NEWS PHOTOS AND THE STORIES BEHIND THEM, John Faber. Dramatic volume of 140 great news photos, 1855 through 1976, and revealing stories behind them, with both historical and technical in-formation. Hindenburg disaster, shooting of Oswald, nomination of Jimmy Carter, etc. 160pp. 8¼ x 11.
23667-6 Pa. $5.00

THE ART OF THE CINEMATOGRAPHER, Leonard Maltin. Survey of American cinematography history and anecdotal interviews with 5 masters—Arthur Miller, Hal Mohr, Hal Rosson, Lucien Ballard, and Conrad Hall. Very large selection of behind-the-scenes production photos. 105 photo-graphs. Filmographies. Index. Originally Behind the Camera. 144pp. 8¼ x 11.
23686-2 Pa. $5.00

DESIGNS FOR THE THREE-CORNERED HAT (LE TRICORNE), Pablo Picasso. 32 fabulously rare drawings—including 31 color illustrations of costumes and accessories—for 1919 production of famous ballet. Edited by Parmenia Migel, who has written new introduction. 48pp. 9⅜ x 12¼. (Available in U.S. only)
23709-5 Pa. $5.00

NOTES OF A FILM DIRECTOR, Sergei Eisenstein. Greatest Russian filmmaker explains montage, making of Alexander Nevsky, aesthetics; com-ments on self, associates, great rivals (Chaplin), similar material. 78 illus-trations. 240pp. 5⅜ x 8½.
22392-2 Pa. $4.50

HOLLYWOOD GLAMOUR PORTRAITS, edited by John Kobal. 145 photos capture the stars from 1926-49, the high point in portrait photography. Gable, Harlow, Bogart, Bacall, Hedy Lamarr, Marlene Dietrich, Robert Montgomery, Marlon Brando, Veronica Lake; 94 stars in all. Full background on photographers, technical aspects, much more. Total of 160pp. 8⅜ x 11¼. 23352-9 Pa. **$6.00**

THE NEW YORK STAGE: FAMOUS PRODUCTIONS IN PHOTO-GRAPHS, edited by Stanley Appelbaum. 148 photographs from Museum of City of New York show 142 plays, 1883-1939. *Peter Pan, The Front Page, Dead End, Our Town,* O'Neill, hundreds of actors and actresses, etc. Full indexes. 154pp. 9½ x 10. 23241-7 Pa. **$6.00**

DIALOGUES CONCERNING TWO NEW SCIENCES, Galileo Galilei. Encompassing 30 years of experiment and thought, these dialogues deal with geometric demonstrations of fracture of solid bodies, cohesion, leverage, speed of light and sound, pendulums, falling bodies, accelerated motion, etc. 300pp. 5⅜ x 8½. 60099-8 Pa. $4.00

THE GREAT OPERA STARS IN HISTORIC PHOTOGRAPHS, edited by James Camner. 343 portraits from the 1850s to the 1940s: Tamburini, Mario, Caliapin, Jeritza, Melchior, Melba, Patti, Pinza, Schipa, Caruso, Farrar, Steber, Gobbi, and many more—270 performers in all. Index. 199pp. 8⅜ x 11¼. 23575-0 Pa. $6.50

J. S. BACH, Albert Schweitzer. Great full-length study of Bach, life, background to music, music, by foremost modern scholar. Ernest Newman translation. 650 musical examples. Total of 928pp. 5⅜ x 8½. (Available in U.S. only) 21631-4, 21632-2 Pa., Two-vol. set $11.00

COMPLETE PIANO SONATAS, Ludwig van Beethoven. All sonatas in the fine Schenker edition, with fingering, analytical material. One of best modern editions. Total of 615pp. 9 x 12. (Available in U.S. only) 23134-8, 23135-6 Pa., Two-vol. set $15.00

KEYBOARD MUSIC, J. S. Bach. Bach-Gesellschaft edition. For harpsichord, piano, other keyboard instruments. English Suites, French Suites, Six Partitas, Goldberg Variations, Two-Part Inventions, Three-Part Sinfonias. 312pp. 8⅛ x 11. (Available in U.S. only) 22360-4 Pa. **$6.95**

FOUR SYMPHONIES IN FULL SCORE, Franz Schubert. Schubert's four most popular symphonies: No. 4 in C Minor ("Tragic"); No. 5 in B-flat Major; No. 8 in B Minor ("Unfinished"); No. 9 in C Major ("Great"). Breitkopf & Hartel edition. Study score. 261pp. 9⅜ x 12¼. 23681-1 Pa. $6.50

THE AUTHENTIC GILBERT & SULLIVAN SONGBOOK, W. S. Gilbert, A. S. Sullivan. Largest selection available; 92 songs, uncut, original keys, in piano rendering approved by Sullivan. Favorites and lesser-known fine numbers. Edited with plot synopses by James Spero. 3 illustrations. 399pp. 9 x 12. 23482-7 Pa. **$9.95**

PRINCIPLES OF ORCHESTRATION, Nikolay Rimsky-Korsakov. Great classical orchestrator provides fundamentals of tonal resonance, progression of parts, voice and orchestra, tutti effects, much else in major document. 330pp. of musical excerpts. 489pp. 6½ x 9¼.   21266-1 Pa. **$7.50**

TRISTAN UND ISOLDE, Richard Wagner. Full orchestral score with complete instrumentation. Do not confuse with piano reduction. Commentary by Felix Mottl, great Wagnerian conductor and scholar. Study score. 655pp. 8⅛ x 11.   22915-7 Pa. $13.95

REQUIEM IN FULL SCORE, Giuseppe Verdi. Immensely popular with choral groups and music lovers. Republication of edition published by C. F. Peters, Leipzig, n. d. German frontmaker in English translation. Glossary. Text in Latin. Study score. 204pp. 9⅜ x 12¼.
23682-X Pa. $6.00

COMPLETE CHAMBER MUSIC FOR STRINGS, Felix Mendelssohn. All of Mendelssohn's chamber music: Octet, 2 Quintets, 6 Quartets, and Four Pieces for String Quartet. (Nothing with piano is included). Complete works edition (1874-7). Study score. 283 pp. 9⅜ x 12¼.
23679-X Pa. **$7.50**

POPULAR SONGS OF NINETEENTH-CENTURY AMERICA, edited by Richard Jackson. 64 most important songs: "Old Oaken Bucket," "Arkansas Traveler," "Yellow Rose of Texas," etc. Authentic original sheet music, full introduction and commentaries. 290pp. 9 x 12.   23270-0 Pa. **$7.95**

COLLECTED PIANO WORKS, Scott Joplin. Edited by Vera Brodsky Lawrence. Practically all of Joplin's piano works—rags, two-steps, marches, waltzes, etc., 51 works in all. Extensive introduction by Rudi Blesh. Total of 345pp. 9 x 12.   23106-2 Pa. $14.95

BASIC PRINCIPLES OF CLASSICAL BALLET, Agrippina Vaganova. Great Russian theoretician, teacher explains methods for teaching classical ballet; incorporates best from French, Italian, Russian schools. 118 illustrations. 175pp. 5⅜ x 8½.   22036-2 Pa. $2.50

CHINESE CHARACTERS, L. Wieger. Rich analysis of 2300 characters according to traditional systems into primitives. Historical-semantic analysis to phonetics (Classical Mandarin) and radicals. 820pp. 6⅛ x 9¼.
21321-8 Pa. $10.00

EGYPTIAN LANGUAGE: EASY LESSONS IN EGYPTIAN HIERO-GLYPHICS, E. A. Wallis Budge. Foremost Egyptologist offers Egyptian grammar, explanation of hieroglyphics, many reading texts, dictionary of symbols. 246pp. 5 x 7½. (Available in U.S. only)
21394-3 Clothbd. $7.50

AN ETYMOLOGICAL DICTIONARY OF MODERN ENGLISH, Ernest Weekley. Richest, fullest work, by foremost British lexicographer. Detailed word histories. Inexhaustible. Do not confuse this with *Concise Etymological Dictionary*, which is abridged. Total of 856pp. 6½ x 9¼.
21873-2, 21874-0 Pa., Two-vol. set $12.00

A MAYA GRAMMAR, Alfred M. Tozzer. Practical, useful English-language grammar by the Harvard anthropologist who was one of the three greatest American scholars in the area of Maya culture. Phonetics, grammatical processes, syntax, more. 301pp. 5⅜ x 8½. 23465-7 Pa. $4.00

THE JOURNAL OF HENRY D. THOREAU, edited by Bradford Torrey, F. H. Allen. Complete reprinting of 14 volumes, 1837-61, over two million words; the sourcebooks for *Walden*, etc. Definitive. All original sketches, plus 75 photographs. Introduction by Walter Harding. Total of 1804pp. 8½ x 12¼. 20312-3, 20313-1 Clothbd., Two-vol. set $50.00

CLASSIC GHOST STORIES, Charles Dickens and others. 18 wonderful stories you've wanted to reread: "The Monkey's Paw," "The House and the Brain," "The Upper Berth," "The Signalman," "Dracula's Guest," "The Tapestried Chamber," etc. Dickens, Scott, Mary Shelley, Stoker, etc. 330pp. 5⅜ x 8½. 20735-8 Pa. $4.50

SEVEN SCIENCE FICTION NOVELS, H. G. Wells. Full novels. *First Men in the Moon, Island of Dr. Moreau, War of the Worlds, Food of the Gods, Invisible Man, Time Machine, In the Days of the Comet.* A basic science-fiction library. 1015pp. 5⅜ x 8½. (Available in U.S. only) 20264-X Clothbd. $8.95

ARMADALE, Wilkie Collins. Third great mystery novel by the author of *The Woman in White* and *The Moonstone*. Ingeniously plotted narrative shows an exceptional command of character, incident and mood. Original magazine version with 40 illustrations. 597pp. 5⅜ x 8½. 23429-0 Pa. $6.00

MASTERS OF MYSTERY, H. Douglas Thomson. The first book in English (1931) devoted to history and aesthetics of detective story. Poe, Doyle, LeFanu, Dickens, many others, up to 1930. New introduction and notes by E. F. Bleiler. 288pp. 5⅜ x 8½. (Available in U.S. only) 23606-4 Pa. $4.00

FLATLAND, E. A. Abbott. Science-fiction classic explores life of 2-D being in 3-D world. Read also as introduction to thought about hyperspace. Introduction by Banesh Hoffmann. 16 illustrations. 103pp. 5⅜ x 8½. 20001-9 Pa. $2.00

THREE SUPERNATURAL NOVELS OF THE VICTORIAN PERIOD, edited, with an introduction, by E. F. Bleiler. Reprinted complete and unabridged, three great classics of the supernatural: *The Haunted Hotel* by Wilkie Collins, *The Haunted House at Latchford* by Mrs. J. H. Riddell, and *The Lost Stradivarius* by J. Meade Falkner. 325pp. 5⅜ x 8½. 22571-2 Pa. $4.00

AYESHA: THE RETURN OF "SHE," H. Rider Haggard. Virtuoso sequel featuring the great mythic creation, Ayesha, in an adventure that is fully as good as the first book, *She*. Original magazine version, with 47 original illustrations by Maurice Greiffenhagen. 189pp. 6½ x 9¼. 23649-8 Pa. $3.50

UNCLE SILAS, J. Sheridan LeFanu. Victorian Gothic mystery novel, considered by many best of period, even better than Collins or Dickens. Wonderful psychological terror. Introduction by Frederick Shroyer. 436pp. 5⅜ x 8½. 21715-9 Pa. $6.00

JURGEN, James Branch Cabell. The great erotic fantasy of the 1920's that delighted thousands, shocked thousands more. Full final text, Lane edition with 13 plates by Frank Pape. 346pp. 5⅜ x 8½.
23507-6 Pa. $4.50

THE CLAVERINGS, Anthony Trollope. Major novel, chronicling aspects of British Victorian society, personalities. Reprint of Cornhill serialization, 16 plates by M. Edwards; first reprint of full text. Introduction by Norman Donaldson. 412pp. 5⅜ x 8½. 23464-9 Pa. $5.00

KEPT IN THE DARK, Anthony Trollope. Unusual short novel about Victorian morality and abnormal psychology by the great English author. Probably the first American publication. Frontispiece by Sir John Millais. 92pp. 6½ x 9¼. 23609-9 Pa. $2.50

RALPH THE HEIR, Anthony Trollope. Forgotten tale of illegitimacy, inheritance. Master novel of Trollope's later years. Victorian country estates, clubs, Parliament, fox hunting, world of fully realized characters. Reprint of 1871 edition. 12 illustrations by F. A. Faser. 434pp. of text. 5⅜ x 8½. 23642-0 Pa. $5.00

YEKL and THE IMPORTED BRIDEGROOM AND OTHER STORIES OF THE NEW YORK GHETTO, Abraham Cahan. Film *Hester Street* based on *Yekl* (1896). Novel, other stories among first about Jewish immigrants of N.Y.'s East Side. Highly praised by W. D. Howells—Cahan "a new star of realism." New introduction by Bernard G. Richards. 240pp. 5⅜ x 8½. 22427-9 Pa. $3.50

THE HIGH PLACE, James Branch Cabell. Great fantasy writer's enchanting comedy of disenchantment set in 18th-century France. Considered by some critics to be even better than his famous *Jurgen*. 10 illustrations and numerous vignettes by noted fantasy artist Frank C. Pape. 320pp. 5⅜ x 8½. 23670-6 Pa. $4.00

ALICE'S ADVENTURES UNDER GROUND, Lewis Carroll. Facsimile of ms. Carroll gave Alice Liddell in 1864. Different in many ways from final Alice. Handlettered, illustrated by Carroll. Introduction by Martin Gardner. 128pp. 5⅜ x 8½. 21482-6 Pa. $2.00

FAVORITE ANDREW LANG FAIRY TALE BOOKS IN MANY COLORS, Andrew Lang. The four Lang favorites in a boxed set—the complete *Red, Green, Yellow* and *Blue* Fairy Books. 164 stories; 439 illustrations by Lancelot Speed, Henry Ford and G. P. Jacomb Hood. Total of about 1500pp. 5⅜ x 8½. 23407-X Boxed set, Pa. $14.95

HOUSEHOLD STORIES BY THE BROTHERS GRIMM. All the great Grimm stories: "Rumpelstiltskin," "Snow White," "Hansel and Gretel," etc., with 114 illustrations by Walter Crane. 269pp. 5⅜ x 8½.
21080-4 Pa. $3.50

SLEEPING BEAUTY, illustrated by Arthur Rackham. Perhaps the fullest, most delightful version ever, told by C. S. Evans. Rackham's best work. 49 illustrations. 110pp. 7⅞ x 10¾. 22756-1 Pa. $2.50

AMERICAN FAIRY TALES, L. Frank Baum. Young cowboy lassoes Father Time; dummy in Mr. Floman's department store window comes to life; and 10 other fairy tales. 41 illustrations by N. P. Hall, Harry Kennedy, Ike Morgan, and Ralph Gardner. 209pp. 5⅜ x 8½. 23643-9 Pa. $3.00

THE WONDERFUL WIZARD OF OZ, L. Frank Baum. Facsimile in full color of America's finest children's classic. Introduction by Martin Gardner. 143 illustrations by W. W. Denslow. 267pp. 5⅜ x 8½.
20691-2 Pa. $3.50

THE TALE OF PETER RABBIT, Beatrix Potter. The inimitable Peter's terrifying adventure in Mr. McGregor's garden, with all 27 wonderful, full-color Potter illustrations. 55pp. 4¼ x 5½. (Available in U.S. only)
22827-4 Pa. $1.25

THE STORY OF KING ARTHUR AND HIS KNIGHTS, Howard Pyle. Finest children's version of life of King Arthur. 48 illustrations by Pyle. 131pp. 6⅛ x 9¼. 21445-1 Pa. $4.95

CARUSO'S CARICATURES, Enrico Caruso. Great tenor's remarkable caricatures of self, fellow musicians, composers, others. Toscanini, Puccini, Farrar, etc. Impish, cutting, insightful. 473 illustrations. Preface by M. Sisca. 217pp. 8⅜ x 11¼. 23528-9 Pa. $6.95

PERSONAL NARRATIVE OF A PILGRIMAGE TO ALMADINAH AND MECCAH, Richard Burton. Great travel classic by remarkably colorful personality. Burton, disguised as a Moroccan, visited sacred shrines of Islam, narrowly escaping death. Wonderful observations of Islamic life, customs, personalities. 47 illustrations. Total of 959pp. 5⅜ x 8½.
21217-3, 21218-1 Pa., Two-vol. set $12.00

INCIDENTS OF TRAVEL IN YUCATAN, John L. Stephens. Classic (1843) exploration of jungles of Yucatan, looking for evidences of Maya civilization. Travel adventures, Mexican and Indian culture, etc. Total of 669pp. 5⅜ x 8½. 20926-1, 20927-X Pa., Two-vol. set $7.90

AMERICAN LITERARY AUTOGRAPHS FROM WASHINGTON IRVING TO HENRY JAMES, Herbert Cahoon, et al. Letters, poems, manuscripts of Hawthorne, Thoreau, Twain, Alcott, Whitman, 67 other prominent American authors. Reproductions, full transcripts and commentary. Plus checklist of all American Literary Autographs in The Pierpont Morgan Library. Printed on exceptionally high-quality paper. 136 illustrations. 212pp. 9⅛ x 12¼. 23548-3 Pa. $12.50

AN AUTOBIOGRAPHY, Margaret Sanger. Exciting personal account of hard-fought battle for woman's right to birth control, against prejudice, church, law. Foremost feminist document. 504pp. 5⅜ x 8½.
20470-7 Pa. $5.50

MY BONDAGE AND MY FREEDOM, Frederick Douglass. Born as a slave, Douglass became outspoken force in antislavery movement. The best of Douglass's autobiographies. Graphic description of slave life. Introduction by P. Foner. 464pp. 5⅜ x 8½.
22457-0 Pa. $5.50

LIVING MY LIFE, Emma Goldman. Candid, no holds barred account by foremost American anarchist: her own life, anarchist movement, famous contemporaries, ideas and their impact. Struggles and confrontations in America, plus deportation to U.S.S.R. Shocking inside account of persecution of anarchists under Lenin. 13 plates. Total of 944pp. 5⅜ x 8½.
22543-7, 22544-5 Pa., Two-vol. set $12.00

LETTERS AND NOTES ON THE MANNERS, CUSTOMS AND CONDITIONS OF THE NORTH AMERICAN INDIANS, George Catlin. Classic account of life among Plains Indians: ceremonies, hunt, warfare, etc. Dover edition reproduces for first time all original paintings. 312 plates. 572pp. of text. 6⅛ x 9¼.
22118-0, 22119-9 Pa.. Two-vol. set $12.00

THE MAYA AND THEIR NEIGHBORS, edited by Clarence L. Hay, others. Synoptic view of Maya civilization in broadest sense, together with Northern, Southern neighbors. Integrates much background, valuable detail not elsewhere. Prepared by greatest scholars: Kroeber, Morley, Thompson, Spinden, Vaillant, many others. Sometimes called Tozzer Memorial Volume. 60 illustrations, linguistic map. 634pp. 5⅜ x 8½.
23510-6 Pa. $7.50

HANDBOOK OF THE INDIANS OF CALIFORNIA, A. L. Kroeber. Foremost American anthropologist offers complete ethnographic study of each group. Monumental classic. 459 illustrations, maps. 995pp. 5⅜ x 8½.
23368-5 Pa. $13.00

SHAKTI AND SHAKTA, Arthur Avalon. First book to give clear, cohesive analysis of Shakta doctrine, Shakta ritual and Kundalini Shakti (yoga). Important work by one of world's foremost students of Shaktic and Tantric thought. 732pp. 5⅜ x 8½. (Available in U.S. only)
23645-5 Pa. $7.95

AN INTRODUCTION TO THE STUDY OF THE MAYA HIEROGLYPHS, Syvanus Griswold Morley. Classic study by one of the truly great figures in hieroglyph research. Still the best introduction for the student for reading Maya hieroglyphs. New introduction by J. Eric S. Thompson. 117 illustrations. 284pp. 5⅜ x 8½.
23108-9 Pa. $4.00

A STUDY OF MAYA ART, Herbert J. Spinden. Landmark classic interprets Maya symbolism, estimates styles, covers ceramics, architecture, murals, stone carvings as artforms. Still a basic book in area. New introduction by J. Eric Thompson. Over 750 illustrations. 341pp. 8⅜ x 11¼.
21235-1 Pa. $6.95

GEOMETRY, RELATIVITY AND THE FOURTH DIMENSION, Rudolf Rucker. Exposition of fourth dimension, means of visualization, concepts of relativity as Flatland characters continue adventures. Popular, easily followed yet accurate, profound. 141 illustrations. 133pp. 5⅜ x 8½.
23400-2 Pa. $2.75

THE ORIGIN OF LIFE, A. I. Oparin. Modern classic in biochemistry, the first rigorous examination of possible evolution of life from nitrocarbon compounds. Non-technical, easily followed. Total of 295pp. 5⅜ x 8½.
60213-3 Pa. $4.00

PLANETS, STARS AND GALAXIES, A. E. Fanning. Comprehensive introductory survey: the sun, solar system, stars, galaxies, universe, cosmology; quasars, radio stars, etc. 24pp. of photographs. 189pp. 5⅜ x 8½. (Available in U.S. only)
21680-2 Pa. $3.75

THE THIRTEEN BOOKS OF EUCLID'S ELEMENTS, translated with introduction and commentary by Sir Thomas L. Heath. Definitive edition. Textual and linguistic notes, mathematical analysis, 2500 years of critical commentary. Do not confuse with abridged school editions. Total of 1414pp. 5⅜ x 8½.
60088-2, 60089-0, 60090-4 Pa., Three-vol. set $18.50

Prices subject to change without notice.

Available at your book dealer or write for free catalogue to Dept. GI, Dover Publications, Inc., 180 Varick St., N.Y., N.Y. 10014. Dover publishes more than 175 books each year on science, elementary and advanced mathematics, biology, music, art, literary history, social sciences and other areas.